State of the Apes

Infrastructure Development and Ape Conservation

Infrastructure development in Africa and Asia is expanding at breakneck speed, largely in biodiversity-rich developing nations. The trend reflects governments' efforts to promote economic growth in response to increasing populations, rising consumption rates and persistent inequalities. Large-scale infrastructure development is regularly touted as a way to meet the growing demand for energy, transport and food—and as a key to poverty alleviation. In practice, however, road networks, hydropower dams and "development corridors" tend to have adverse effects on local populations, natural habitats and biodiversity. Such projects typically weaken the capacity of ecosystems to maintain ecological functions on which wildlife and human communities depend, particularly in the face of climate change.

This volume—*State of the Apes: Infrastructure Development and Ape Conservation*—presents original research and analysis, topical case studies and emerging tools and methods to inform debate, practice and policy with the aim of preventing and mitigating the harmful impacts of infrastructure projects on biodiversity. Using apes as a proxy for wildlife and ecosystems themselves, it identifies opportunities for reconciling economic and social development with environmental stewardship.

This title is available as an open access eBook via Cambridge Core and at www.stateoftheapes.com.

State of the Apes

Series editors

Helga Rainer	Arcus Foundation
Alison White	
Annette Lanjouw	Arcus Foundation

The world's primates are among the most endangered of all tropical species. All great ape species – gorilla, chimpanzee, bonobo and orangutan – are classified as either Endangered or Critically Endangered. Furthermore, nearly all gibbon species are threatened with extinction. Whilst linkages between ape conservation and economic development, ethics and wider environmental processes have been acknowledged, more needs to be done to integrate biodiversity conservation within broader economic, social and environmental communities if those connections are to be fully realized and addressed.

Intended for a broad range of policymakers, industry experts and decision-makers, academics, researchers and NGOs, the *State of the Apes* series will look at the threats to these animals and their habitats within the broader context of economic and community development. Each publication presents a different theme, providing an overview of how these factors interrelate and affect the current and future status of apes, with robust statistics, welfare indicators, official and various other reports providing an objective and rigorous analysis of relevant issues.

Other Titles in this Series

Arcus Foundation. 2015. *State of the Apes: Industrial Agriculture and Ape Conservation*. Cambridge: Cambridge University Press.

Arcus Foundation. 2014. *State of the Apes: Extractive Industries and Ape Conservation*. Cambridge: Cambridge University Press.

Other Language Editions

Bahasa Indonesia

Arcus Foundation. 2018. *Negara Kera: Pembangunan Infrastruktur dan Konservasi Kera*.

Arcus Foundation. 2015. *Negara Kera: Pertanian Industri dan Konservasi Kera*.

Arcus Foundation. 2014. *Negara Kera: Industri Ekstraktif dan Konservasi Kera*.

French

Arcus Foundation. 2018. *La planète des grands singes: Le développement des infrastructures et la conservation des grands singes*.

Arcus Foundation. 2015. *La planète des grands singes: L'agriculture industrielle et la conservation des grands singes*.

Arcus Foundation. 2014. *La planète des grands singes: Les industries extractives et la conservation des grands singes*.

State of the Apes
Infrastructure Development and
Ape Conservation

CAMBRIDGE
UNIVERSITY PRESS

University Printing House, Cambridge CB2 8BS, United Kingdom

Cambridge University Press is part of the University of Cambridge.

It furthers the University's mission by disseminating knowledge in the pursuit of education, learning and research at the highest international levels of excellence.

www.cambridge.org

Information on this title:
www.cambridge.org/9781108423212
DOI: 10.1017/9781108436427

© Cambridge University Press 2018

This work is in copyright. It is subject to statutory exceptions and to the provisions of relevant licensing agreements; with the exception of the Creative Commons version, the link for which is provided below, no reproduction of any part of this work may take place without the written permission of Cambridge University Press.

An online version of this work is published at http://dx.doi.org/10.1017/9781108436427 under a Creative Commons Open Access license CC-BY-NC-ND 4.0 which permits re-use, distribution and reproduction in any medium for non-commercial purposes providing appropriate credit to the original work is given. You may not distribute derivative works without permission. To view a copy of this license, visit https://creativecommons.org/licenses/by-nc-nd/4.0.

All versions of this work may contain content reproduced under license from third parties. Permission to reproduce this third-party content must be obtained from these third-parties directly. When citing this work, please include a reference to the DOI 10.1017/9781108436427.

First published 2018

Printed in the United Kingdom by Clays, St Ives plc, Elcograf S.p.A.

A catalogue record for this publication is available from the British Library

ISBN 978-1-108-42321-2 Hardback
ISBN 978-1-108-43641-0 Paperback

Cambridge University Press has no responsibility for the persistence or accuracy of URLs for external or third-party internet websites referred to in this publication, and does not guarantee that any content on such websites is, or will remain, accurate or appropriate.

Credits

Editors
Helga Rainer, Alison White and Annette Lanjouw

Production Coordinator
Alison White

Editorial Consultant and Copy-editor
Tania Inowlocki

Designer
Rick Jones, StudioExile

Cartographer
Jillian Luff, MAP*grafix*

Fact-checker
Rebecca Hibbin

Proofreader
Sarah Binns

Referencing
Eva Fairnell

Indexer
Caroline Jones, Osprey Indexing

Cover photographs:

Background: © Jabruson
Bonobo: © Takeshi Furuichi
Gibbon: © IPPL
Gorilla: © Annette Lanjouw
Orangutan: © Jurek Wajdowicz, EWS
Chimpanzee: © Nilanjan Bhattacharya/Dreamstime.com

Foreword

Our world is on the verge of unprecedented economic and environmental change. While access to technology and opportunity are dramatically improving many parts of the planet, we also see climate change and widening inequality putting that progress in jeopardy. Even though new infrastructure investments—in roads, dams, pipelines and railways—promise economic prosperity for poorer countries, the risks such projects pose threaten to outweigh the benefits.

Look no further than what is happening to the world's population of apes across Africa and southeast Asia. The fragmentation and exploitation of tropical forests is a direct threat to apes—compromising their habitat, making food scarce, and bringing dangers like poaching and disease. As a result, their numbers are decreasing everywhere. Today, many ape species are on the precipice of extinction.

As this latest volume of *State of the Apes* argues, our well-being is inextricably linked to the well-being of our environment—and the well-being of all species that call our planet home.

That is in part because this same forest destruction is equally devastating to forest-dependent *people*. Industrial-scale infrastructure development has a significant, detrimental impact on local communities that have a long-term relationship with the forest. These rural and indigenous communities frequently don't see the economic benefits brought by new roads and power plants; instead they see their land taken away, without any just compensation or respect for their voices or rights.

Moreover, these developments have even wider ramifications for our environment. When we eliminate forests and fail to lift up the communities that protect them, we dramatically increase the amount of CO_2 in our atmosphere. When we abuse our lands, we severely weaken our position in the fight against climate change.

In other words, when apes are displaced —when their forests are degraded and their lives are devalued—humans are too. When we ignore the larger consequences of these massive infrastructure projects, especially in the name of inequitable and unjust gain, the whole world suffers.

Our planet and our communities are in urgent need of a more sustainable, equitable kind of economic development—one that empowers everyone, while protecting the planet's life and resources.

At the Ford Foundation, we understand that all these issues are deeply intertwined, and to address them comprehensively, the solutions must be similarly interconnected. The question is: how do we strike a balance that enables development while also allowing apes and other species, local communities, the environment and the economy to thrive?

This book is aimed at helping us answer that difficult question. Through reasoned, peer-reviewed science and practical examples, *State of the Apes* shows that though there may always be trade-offs, smart policy comes from considering what will create long-term benefits for all. It offers real solutions for how we plan, organize and educate to produce socially inclusive and green infrastructure. And it reminds us that both long-term environmental sustainability and long-term economic progress are the result of *equitable* and just solutions—not unsustainable or corrupt investments.

Most importantly, this volume demonstrates how equitable development is not only a possibility, but a necessity.

We know we can't stop the world from developing, but we can make sure inevitable and necessary developments in infrastructure also contribute to the larger march of progress for all people, and protect the environment. It's up to all of us to ensure that these projects are executed thoughtfully, responsibly and sustainably—that they are not destructive, but truly constructive.

At this critical moment—when governments, businesses and civil society organizations across the world are coming up against the dual threats of climate change and economic inequality—it has never been more essential that we keep our shared future in mind.

The *State of the Apes* series makes clear that charting a path forward is not about the state, or fate, of any one species, but about the fair and sustainable solutions that our world so desperately needs.

Darren Walker
President
Ford Foundation

Contents

The Arcus Foundation ix
Notes to Readers ix
Acknowledgments x
Apes Overview xii

Section 1
Infrastructure Development and Ape Conservation

Introduction 1

1. Towards More Sustainable Infrastructure: Challenges and Opportunities in Ape Range States of Africa and Asia 11

Introduction 11
Infrastructure: A Game Changer 13
Drivers of Infrastructure Expansion 18
Emerging Threats to Ape Habitats 22
Social and Political Concerns 27
A Dire Need for Better Infrastructure Planning 28
Priorities for Change 31

2. Impacts of Infrastructure on Apes, Indigenous Peoples and Other Local Communities 41

Introduction 41
Ecological Impacts of Infrastructure on Apes 43
Steps Forward 54
Social Impacts of Infrastructure 60
Overall Conclusion 78

3. Deforestation Along Roads: Monitoring Threats to Ape Habitat 81

Introduction 81
Proposed New Approaches to Road Monitoring 83
Case Study Approach 86
Recommendations for Road Infrastructure in Ape Habitat 86
The Potential of Remote Sensing Tools to Detect and Monitor Changes in Ape Habitat 104

4. Apes, Protected Areas and Infrastructure in Africa 107

Introduction 107
African Ape Ranges and Protected Areas 109
Threats to Protected Areas from Infrastructure 111
Protected Area Downgrading, Downsizing and Degazettement (PADDD) in Africa 116
The Mitigation Hierarchy: Reconciling Infrastructure and Ape Conservation 119
Future Threats and Prospects 128

5. Roads, Apes and Biodiversity Conservation: Case Studies from the Democratic Republic of Congo, Myanmar and Nigeria 137

Introduction 137
Overall Conclusion 164

6. Renewable Energy and the Conservation of Apes and Ape Habitat 167

Introduction 167
Global Hydropower: Drivers and Trends 171
Impacts of Hydropower 172
Hydropower and Apes 174
Conclusion 195

Section 2
The Status and Welfare of Great Apes and Gibbons

Introduction 198

7. Mapping Change in Ape Habitats: Forest Status, Loss, Protection and Future Risk 201

Introduction 201
A Summary of the State of the Apes through the Lenses of Forest Cover and Protection, 2000–14 207
Forest Dynamics and Loss from 2000 to 2014 207
Annual Forest Loss Trends in Ape Habitat 216
Regular Monitoring of Forest Change 220
Conclusion 222

8. The Status of Captive Apes 225

Introduction 225
I. Beyond Capacity: Sanctuaries and the Status of Captive Apes in Shrinking Natural Habitats 227
Conclusion 254
II. The Status of Captive Apes: A Statistical Update 255
Conclusion 262

Annexes 264

Acronyms and Abbreviations 279

Glossary 282

References 291

Index 344

The Arcus Foundation

The Arcus Foundation is a private grant-making foundation that advances social justice and conservation goals. The Foundation works globally and has offices in New York City, USA and Cambridge, UK. For more information visit:

- arcusfoundation.org.

Or connect with Arcus at:

- twitter.com/ArcusGreatApes; and
- facebook.com/ArcusGreatApes.

Great Apes Program

The long-term survival of humans and the great apes is dependent on how we respect and care for other animals and our shared natural resources. The Arcus Foundation seeks to increase respect for and recognition of the rights and value of the great apes and gibbons, and to strengthen protection from threats to their habitats. The Arcus Great Apes Program supports conservation and policy advocacy efforts that promote the survival of great apes and gibbons in the wild and in sanctuaries that offer high-quality care, safety and freedom from invasive research and exploitation.

Contact details

New York office:

44 West 28th Street, 17th Floor
New York, New York 10001, United States

+1 212 488 3000 / phone
+1 212 488 3010 / fax

Cambridge office (Great Apes program):

CB1 Business Centre
Leda House, 20 Station Road
Cambridge CB1 2JD, United Kingdom

+44 (0)1223 653040 / phone
+44.1223.451100 / fax

Notes to Readers

Acronyms and abbreviations

A list of acronyms and abbreviations can be found at the back of the book, starting on p. 279.

Annexes

All annexes can be found at the back of the book, starting on p. 264, except for the Abundance Annex, which is available from the *State of the Apes* website:

- www.stateoftheapes.com.

Glossary

There is a glossary of scientific terms and keywords at the back of the book, starting on p. 282.

Chapter cross-referencing

Chapter cross-references appear throughout the book, either as direct references in the body text or in brackets.

Ape Range Maps

The ape range maps throughout this edition show the extent of occurrence (EOO) of each species. An EOO includes all known populations of a species contained within the shortest possible continuous imaginary boundary. It is important to note that some areas within these boundaries are unsuitable and unoccupied.

The Arcus Foundation commissioned the ape distribution maps in the Apes Overview, Figures AO1 and AO2, to provide the most accurate and up-to-date illustration of range data. These maps were created by the Max Planck Institute for Evolutionary Anthropology, who manage the A.P.E.S. portal and database. This volume also features maps created by contributors who used ape range data from other sources. As a consequence, the maps may not all align exactly.

Photographs

We aim to include photographs that are relevant to each theme and illustrate the content of each chapter. If you have photographs that you are willing to share with the Arcus Foundation, for use in this series, or for multiple purposes, please contact the Production Coordinator (awhite@arcusfoundation.org) or the Cambridge office.

Acknowledgments

The aim of the *State of the Apes* series is to facilitate critical engagement on conservation, industry and government practice and to expand support for great apes and gibbons. We are grateful to everyone who played a part, including meeting participants, our authors, contributors and reviewers, and those involved in the production of the book.

The support of Jon Stryker and the Arcus Foundation Board of Directors is essential to the production of this publication. We thank them for their ongoing support.

A key element outside of the thematic content is the overview of the status of apes, both in their natural habitats and in captivity. We extend our thanks to the captive-ape organizations that provided detailed information and to all the great ape and gibbon scientists who contribute their valuable data to build the A.P.E.S. database. Such collaborative efforts are key to effective and efficient conservation action.

Authors, contributors, reviewers and those who provided essential data and support are acknowledged at the end of each chapter. We could not have produced this book without them, and we thank them for their enduring patience and participation, in sickness and in health. Many of the photographs included were generously shared by their creators, who are credited alongside each one. Respect and thanks to those who read and reviewed the whole publication—no small task: Mihai Coroi, Cindy Rizzo and Tommaso Savini.

Particular thanks go to the following individuals, organizations and agencies: Marc Ancrenaz, Iain Bray, Stanley Brunn, Genevieve Campbell, the Center for International Forestry Research (CIFOR), Susan Cheyne, Bruce Davidson, Eric Dinerstein, the Ford Foundation, Forest Peoples Programme (FPP), Getty Images, Hao Chunxu, Matthew Hatchwell, Randy Hayes, Tatyana Humle, Jack Hurd, the Jane Goodall Institute (JGI), Lin Ji, Anup Joshi, Justin Kenrick, Josh Klemm, Bill Laurance, Cath Lawson, Liz Macfie, the Max Planck Institute for Evolutionary Anthropology, Linda May, Adriana Gonçalves Moreira, Mott MacDonald, Steve Peedell, Adam Phillipson, Refuge for Wildlife, RESOLVE, Martha Robbins, Ian Singleton, Tenekwetche Sop, Gideon Suharyanto, the Sumatran Orangutan Conservation Programme (SOCP), Bob Tansey, The Biodiversity Consultancy (TBC), The Nature Conservancy (TNC), Anne Trainor, University of Minnesota, Steve Volkers, Darren Walker, Wildlife Conservation Society (WCS), Laura Wilkinson, Jake Willis and the World Wide Fund for Nature (WWF).

Our thanks also to Katrina Halliday and the team at Cambridge University Press for their support and commitment to this series.

We are committed to ensuring that these books are available to as many stakeholders as possible, not least by providing them in English, French and Bahasa Indonesia. Our thanks go to our cartographer, copy editor, graphic designers, indexer, proofreaders, reference editor and translators: Sarah Binns, Eva Fairnell, Tania Inowlocki, Caroline Jones, Rick Jones, Hyacinthe Kemp, Jillian Luff, Anton Nurcahyo, Islaminur Pempasa, Hélène Piantone, Erica Taube, Beth Varley and Rumanti Wasturini. All language editions are provided on the *State of the Apes* website, and our thanks go to the Arcus Communications team for managing this site, especially Stephanie Myers, Sebastian Naidoo and Bryan Simmons.

Many others contributed in a number of ways that cannot be attributed to specific content, by providing introductions, anonymous comments or strategic advice, and by helping with essential administrative tasks. Marie Stevenson deserves particular thanks for her logistical support. We also thank everyone who provided much-appreciated moral support.

Helga Rainer, Alison White and Annette Lanjouw
Editors

Apes Overview

APES INDEX

Bonobo (*Pan paniscus*)

Distribution and Numbers in the Wild

The bonobo is only present in the Democratic Republic of Congo (DRC), biogeographically separated from chimpanzees and gorillas by the Congo River (see Figure AO1). The population size is unknown, as only 30% of the species' historic range has been surveyed; however, estimates from the four geographically distinct bonobo strongholds suggest a minimum population of 15,000–20,000 individuals, with numbers decreasing (Fruth et al., 2016).

The bonobo is included in the Convention on International Trade in Endangered Species of Wild Fauna and Flora (CITES) Appendix I and is categorized as endangered on the International Union for Conservation of Nature (IUCN) Red List (Fruth et al., 2016; see Box AO1). The causes of population decline include poaching; habitat loss and degradation; disease; and people's lack of awareness that hunting and eating bonobos is unlawful. Poaching, which is mainly carried out as part of the commercial wild meat trade and for some medicinal purposes, has been exacerbated by the ongoing effects of armed conflict, such as military-sanctioned hunting and the accessibility of modern weaponry and ammunition (Fruth et al., 2016).

Physiology

Male adult bonobos reach a height of 73–83 cm and weigh 37–61 kg, while females are slightly smaller, weighing 27–38 kg. Bonobos are moderately sexually dimorphic and similar in size and appearance to chimpanzees, although with a smaller head and lither appearance. The maximum life span in the wild is 50 years (Hohmann et al., 2006; Robson and Wood, 2008).

The bonobo diet is mainly frugivorous (more than 50% fruit), supplemented with leaves, stems, shoots, pith, seeds, bark, flowers, honey and fungi, including truffles. Only a very small part of their diet consists of animal matter—such as insects, small reptiles, birds and medium-sized mammals, including other primates.

Social Organization

Bonobos live in fission–fusion communities of 10–120 individuals, consisting of multiple males and females. When foraging, they split into smaller mixed-sex subgroups, or parties, averaging 5–23 individuals.

Male bonobos cooperate with and tolerate one another; however, lasting bonds between adult males are rare, in contrast to the bonds between adult females, which are strong and potentially last for years. A distinguishing feature of female bonobos is that they are co-dominant with males and form alliances against certain males within the community. Among bonobos, the bonds between mother and son are the strongest, prove highly important for the social status of the son and last into adulthood.

Together with chimpanzees, bonobos are the closest living relatives to humans, sharing 98.8% of our DNA (Smithsonian Institution, n.d.; Varki and Altheide, 2005).

Chimpanzee (*Pan troglodytes*)

Distribution and Numbers in the Wild

Chimpanzees are widely distributed across equatorial Africa, with discontinuous populations from southern Senegal to western Uganda and Tanzania (Humle et al., 2016b; see Figure AO1).

Chimpanzees are listed in CITES Appendix I, and all four subspecies are categorized as endangered on the IUCN Red List. There are approximately 140,000 central chimpanzees (*Pan troglodytes troglodytes*); 18,000–65,000 western chimpanzees (*Pan t. verus*); 181,000–256,000 eastern chimpanzees (*Pan t. schweinfurthii*); and probably fewer than 6,000–9,000 Nigeria–Cameroon chimpanzees (*Pan t. ellioti*). Populations are believed to be declining, but the rate has not yet been quantified (Humle et al., 2016b).

Decreases in chimpanzee numbers are mainly attributed to increased poaching for the commercial wild meat trade, habitat loss and degradation, and disease (particularly Ebola) (Humle et al., 2016b).

Physiology

Male chimpanzees are 77–96 cm tall and weigh 28–70 kg, while females measure 70–91 cm and weigh 20–50 kg. They share many facial expressions with humans, although forehead musculature is less pronounced and they have more flexible lips. Chimpanzees live for up to 50 years in the wild.

Chimpanzees are mainly frugivorous and opportunistic feeders. Some communities include 200 species of food items in a diet of fruit supplemented by herbaceous vegetation and animal prey, such as ants and termites, but also small mammals, including other primates. Chimpanzees are the most carnivorous of all the apes.

Social Organization

Chimpanzees show fission–fusion, multi-male–multi-female grouping patterns. A large community includes all individuals who regularly associate with one another; such communities comprise an average of 35 individuals, with the largest-known group counting 150, although this size is rare. The community separates into smaller, temporary subgroups, or parties. The parties can be highly fluid, with members moving in and out quickly or a few individuals staying together for a few days before rejoining the community.

Typically, home ranges are defended by highly territorial males, who may attack or even kill neighboring chimpanzees. Male chimpanzees are dominant over female chimpanzees and are generally the more social sex, sharing food and grooming each other more frequently. Chimpanzees are noted for their sophisticated forms of cooperation, such as in hunting and territorial defense; the level of cooperation involved in social hunting activities varies across communities, however.

Gorilla (*Gorilla* species (spp.))

Distribution and Numbers in the Wild

The western gorilla (*Gorilla gorilla*) is distributed throughout western equatorial Africa and has two subspecies: the western lowland gorilla (*Gorilla g. gorilla*) and the Cross River gorilla (*Gorilla g. diehli*). The eastern gorilla (*Gorilla beringei*) is found in the DRC and across the border in Uganda and Rwanda. There are two subspecies of the eastern gorilla: the mountain gorilla (*Gorilla b. beringei*) and Grauer's gorilla (*Gorilla b. graueri*), also referred to as the eastern lowland gorilla (see Figure AO1).

All gorillas are listed as critically endangered on the IUCN Red List. Population estimates for the western gorilla range from 150,000 to 250,000, while as few as 250–300 Cross River gorillas remain in the wild (Bergl, 2006; Oates *et al.*, 2007; Sop *et al.*, 2015; Williamson *et al.*, 2013). The most recent population estimate for Grauer's gorilla is 3,800, which indicates a 77% loss since 1994 (Plumptre, Robbins and Williamson, 2016c). Mountain gorillas are estimated to number at least 880 individuals (Gray *et al.*, 2013; Roy *et al.*, 2014). The main threats to both species are poaching for the commercial wild meat trade, habitat destruction and degradation, and disease (for the western gorilla, the Ebola virus in particular). The eastern gorilla is also threatened by civil unrest (Maisels, Bergl and Williamson, 2016a; Plumptre *et al.*, 2016c).

Physiology

The adult male of the eastern gorilla is slightly larger (159–196 cm, 120–209 kg) than the western gorilla (138–180 cm, 145–191 kg). Both species are highly sexually dimorphic, with females being about half the size of males. Their life span ranges from 30 to 40 years in the wild. Mature males are known as "silverbacks" due to the development of a gray saddle with maturity.

The gorillas' diet consists predominantly of ripe fruit and terrestrial, herbaceous vegetation. More herbaceous vegetation is ingested while fruit is scarce, in line with seasonality and fruit availability, and protein gain comes from leaves and bark of trees; gorillas do not eat meat but occasionally consume ants and termites. Mountain gorillas have less fruit in their environment than lowland gorillas, so they feed mainly on leaves, pith, stems, bark and, occasionally, ants.

Social Organization

Western gorillas live in stable groups with multiple females and one adult male (silverback); in contrast, eastern gorillas are polygynous and can be polygynandrous, with groups that comprise one or more silverbacks, multiple females, their offspring and immature relatives. The average group consists of ten individuals, but eastern gorillas can live in groups of up to 65 individuals, whereas the maximum group size for the western gorilla is 22. Gorillas are not territorial and home ranges overlap extensively. Chest beats and vocalizations typically are used when neighboring silverbacks come into contact, but intergroup encounters may escalate into physical fights. Mutual avoidance is normally the adopted strategy for groups that live in the same areas.

Orangutan (*Pongo* spp.)

Distribution and Numbers in the Wild

The orangutan range is now limited to the forests of Sumatra and Borneo, but these great apes were once present throughout much of southern Asia (Wich *et al.*, 2008, 2012a; see Figure AO2). Survey data indicate that in 2015 fewer than 15,000[1] Sumatran orangutans (*Pongo abelii*) and just under 105,000[2] Bornean orangutans (*Pongo pygmaeus* spp.) remained in the wild (Ancrenaz *et al.*, 2016; Wich *et al.*, 2016). As a result of continuing habitat loss and hunting, both the Sumatran orangutan and the Bornean orangutan are classified as critically endangered (Ancrenaz *et al.*, 2016; Singleton *et al.*, 2016). Both species are listed in Appendix I of CITES.

In November 2017, a new species of orangutan was described in three forest fragments in Sumatra's Central, North and South Tapanuli districts, which are part of the Batang Toru Ecosystem (Nater *et al.*, 2017).[3] The Tapanuli orangutan (*Pongo tapanuliensis*) has a total distribution of about 1,100 km² (110,000 ha) and a population size of fewer than 800 individuals (Wich *et al.*, 2016).[4]

The main threats to all orangutan species are habitat loss and fragmentation, and killings due to human–ape conflict, hunting and the international pet trade (Ancrenaz *et al.*, 2016; Gaveau *et al.*, 2014; Singleton *et al.*, 2016; Wich *et al.*, 2008).*

Physiology

Adult males can reach a height of 94–99 cm and weigh 60–85 kg (flanged) or 30–65 kg (unflanged). Females are 64–84 cm tall and weigh 30–45 kg, meaning that orangutans are highly sexually dimorphic. In the wild, males on Sumatra have a life expectancy of 58 years and females 53 years. No accurate data exist for the Bornean orangutan.

Fully mature males develop a short beard and protruding cheek pads, termed "flanges." Some male orangutans experience "developmental arrest," maintaining a female-like size and appearance for many years past sexual maturity; they are known as "unflanged" males. Orangutans are the only great ape to exhibit male bimaturism.

The orangutan diet mainly consists of fruit, but they also eat leaves, shoots, seeds, bark, pith, flowers, eggs, soil and invertebrates (termites (*Isoptera*) and ants (*Formicidae*)). Carnivorous behavior has also been observed, but at a low frequency (preying on species such as slow lorises (*Nycticebus*)).

Social Organization

The mother–offspring unit is the only permanent social unit among orangutans, yet social groupings between independent individuals do occur, although their frequency varies across populations (Wich, de Vries and Ancrenaz, 2009a). While females are usually relatively tolerant of each other, flanged males are intolerant of other flanged and unflanged males (Wich *et al.*, 2009a). Orangutans on Sumatra are generally more social than those on Borneo and live in overlapping home ranges, with flanged males emitting "long calls" to alert others to their location (Delgado and Van Schaik, 2000; Wich *et al.*, 2009a). Orangutans are characterized by an extremely slow life history, with the longest interbirth interval (6–9 years) of any primate species (Wich *et al.*, 2004, 2009a).

Gibbons (*Hoolock* spp.; *Hylobates* spp.; *Nomascus* spp.; *Symphalangus* spp.)

All four genera of gibbon generally share ecological and behavioral attributes, such as social monogamy in territorial groups; vocalization through elaborate song (including complex duets); frugivory and brachiation (moving through the canopy using only the arms). Gibbons primarily consume fruit but have a varied diet including insects, flowers, leaves and seeds. Female gibbons have a single offspring every 2.5–3 years (S. Cheyne, personal communication, 2017). Gibbons are diurnal and sing at sunrise and sunset; they dedicate a significant part of the day to finding fruit trees within their territories.

Hoolock genus

Distribution and Numbers in the Wild

Three species comprise the *Hoolock* genus: the western hoolock (*Hoolock hoolock*), the eastern hoolock (*Hoolock leuconedys*) and the newly discovered Gaoligong or Skywalker hoolock (*Hoolock tianxing*) (Fan *et al.*, 2017). The Mishmi Hills hoolock (*Hoolock h. mishmiensis*), the most recently discovered subspecies of western hoolock, was officially named in 2013 (Choudhury, 2013).

The western hoolock's distribution spans Bangladesh, India and Myanmar. The eastern hoolock lives in China, India and Myanmar (see Figure AO2). To date, the Gaoligong hoolock has only been seen in eastern Myanmar and southwestern China (Fan *et al.*, 2017).

With an estimated population of 2,500 individuals, the western hoolock is listed as endangered on the IUCN Red List. The eastern hoolock has a much larger population, numbering 293,200–370,000, and is listed as vulnerable on the IUCN Red List. Both species are listed in CITES Appendix I, with the main threats identified as habitat loss and fragmentation, and hunting for food, pets and medicinal purposes. The Gaoligong hoolock is likely to be categorized as endangered but has not yet been formally listed on the IUCN Red List (Fan et al., 2017).

Physiology

An individual hoolock can have a head and body length of 45–81 cm and weigh 6–9 kg, with males slightly heavier than females. Like most gibbons, the *Hoolock* genus is sexually dichromatic, with the pelage (coat) of females and males differing in terms of patterning and color. Pelage also differs across species: unlike the western hoolock, the eastern one features a complete separation between the white brow markings and a white preputial tuft.

The diet of the western hoolock is primarily frugivorous, supplemented with vegetative matter such as leaves, shoots, seeds, moss and flowers. While little is known about the diet of the eastern hoolock, it most likely resembles that of the western hoolock.

Social Organization

Hoolocks live in family groups of 2–6 individuals, consisting of a mated adult pair and their offspring. They are presumably territorial, although no specific data exist. Hoolock pairs vocalize a "double solo" rather than the more common "duet" of various gibbons.

Hylobates genus

Distribution and Numbers in the Wild

Nine species are currently included in the *Hylobates* genus, although there remains some dispute about whether Abbott's gray gibbon (*Hylobates abbottii*), the Bornean gray gibbon (*Hylobates funereus*) and Müller's gibbon (*Hylobates muelleri*) represent full species (see Table AO1).

This genus of gibbon occurs discontinuously in tropical and subtropical forests from southwestern China, through Indochina, Thailand and the Malay Peninsula to the islands of Sumatra, Borneo and Java (Wilson and Reeder, 2005; see Figure AO2). The overall estimated minimum population for the *Hylobates* genus is about 360,000–400,000, with the least abundant species being the moloch gibbon (*Hylobates moloch*) and most abundant being, collectively, the "gray gibbons" (Abbott's, the Bornean and Müller's gibbons), although no accurate population numbers are available for Abbott's gray gibbon.

All *Hylobates* species are listed as endangered on the IUCN Red List and are in CITES Appendix I. Three hybrid zones occur naturally and continue to coexist with the unhybridized species in the wild. The main collective threats facing the genus are deforestation, hunting and the illegal pet trade (S. Cheyne, personal communication, 2017; Mittermeier et al., 2013).

Physiology

Average height for both sexes of all species is approximately 46 cm and their weight ranges between 5 kg and 7 kg. With the exception of the pileated gibbon (*Hylobates pileatus*), species in the genus are not sexually dichromatic, although the lar gibbon (*Hylobates lar*) has two color phases, which are not related to sex or age.

Gibbons are mainly frugivorous. Figs are an especially important part of their diet and are supplemented by leaves, buds, flowers, shoots, vines and insects, while small animals and bird eggs form the protein input.

Social Organization

Hylobates gibbons are largely socially monogamous, forming family units of two adults and their offspring; however, polyandrous and polygynous units have been observed, especially in hybrid zones. Territorial disputes are predominantly led by males, who become aggressive towards other males, whereas females tend to lead daily movements and ward off other females.

Nomascus genus

Distribution and Numbers in the Wild

Seven species make up the *Nomascus* genus (see Table AO1).

The *Nomascus* genus, which is somewhat less widely distributed than the *Hylobates* genus, is present in Cambodia, Laos, Vietnam and southern China (including Hainan Island; see Figure AO2). Population estimates exist for some taxa: there are approximately 1,500 western black crested gibbons (*Nomascus concolor*), 130 Cao Vit gibbons (*Nomascus nasutus*) and 23 Hainan gibbons (*Nomascus hainanus*). Population estimates for the white-cheeked gibbons (*N. leucogenys* and *N. siki*) are not available except for some sites, yet overall numbers are known to be severely depleted. The yellow-cheeked gibbons (*N. annamensis* and *N. gabriellae*) have the largest populations among the *Nomascus* gibbons.

All species are listed in CITES Appendix I; in the IUCN Red List, four are categorized as critically endangered (*N. concolor, nasutus, hainanus* and *leucogenys*) and two as endangered (*N. siki* and *N. gabriellae*), while one—the northern yellow-cheeked crested gibbon (*Nomascus annamensis*)—is yet to be assessed (IUCN, 2017). Major threats to these populations include hunting for food, pets and for medicinal purposes, as well as habitat loss and fragmentation.

Physiology

Average head and body length across all species of this genus, for both sexes, is approximately 47 cm; individuals weigh around 7 kg. All *Nomascus* species have sexually dimorphic pelage; adult males are predominantly black while females are a buffy yellow. Their diet is much the same as that of the *Hylobates* genus: mainly frugivorous, supplemented with leaves and flowers.

Social Organization

Gibbons of the *Nomascus* genus are mainly socially monogamous; however, most species have also been observed in polyandrous and polygynous groups. More northerly species appear to engage in polygyny to a greater degree than southern taxa. Copulations outside monogamous pairs have been recorded, although infrequently.

Symphalangus genus

Distribution and Numbers in the Wild

Siamang (*Symphalangus syndactylus*) are found in several forest blocks across Indonesia, Malaysia and Thailand (see Figure AO2); the species faces severe threats to its habitat across its range. No accurate estimates exist for the total population size. The species is listed in CITES Appendix I and is classified as endangered on the IUCN Red List.

Physiology

The siamang's head and body length is 75–90 cm, and adult males weigh 10.5–12.7 kg, while adult females weigh 9.1–11.5 kg. The siamang is minimally sexually dimorphic, and the pelage is the same across sexes: black. The species has a large inflatable throat sac.

Siamang rely heavily on figs and somewhat less on leaves—a diet that allows them to be sympatric with *Hylobates* gibbons in some locations, since the latter focus more on fleshy fruits. The siamang diet also includes flowers and insects.

Social Organization

Males and females call territorially, using their large throat sacs, and males will give chase to neighboring males. One group's calls will inhibit other groups nearby, and they will consequently take turns to vocalize. The groups are usually based on monogamous pairings, although polyandrous groups have been observed. Males may also adopt the role of caregiver for infants.

Notes:

All information is drawn from the *Handbook of the Mammals of the World, Volume 3: Primates* (Mittermeier, Rylands and Wilson, 2013), unless otherwise cited.

* For the Bornean orangutan, additional threats include forest fires and people's lack of awareness that they are protected by law. For the Sumatran orangutan, the current most important threat is a land use plan issued by the government of Aceh in 2013. The plan does not recognize that the Leuser Ecosystem is a National Strategic Area, a legal status that prohibits cultivation, development and other activities that would degrade the ecosystem's environmental functions (Singleton *et al*., 2016).

Photo Credits:

Bonobo: © Takeshi Furuichi, Wamba Committee for Bonobo Research

Chimpanzee: © Arcus Foundation and Jabruson, 2014. All rights reserved. www.jabruson.photoshelter.com

Gorilla: © Annette Lanjouw

Orangutan: © Perry van Duijnhoven 2013

Gibbons: *Hoolock*: © Dr. Axel Gebauer/naturepl.com; *Hylobates*: © International Primate Protection League (IPPL); *Nomascus*: © IPPL; *Symphalangus*: © Pete Oxford/naturepl.com

Ape Socioecology

This section presents an overview of the socioecology of the seven non-human apes: bonobos, chimpanzees, gibbons (including siamangs), eastern gorillas, western gorillas, and Bornean and Sumatran orangutans.[5]

Gorillas, whose range extends across ten central African countries, are the largest living primate species and the most terrestrial of all the apes. Chimpanzees are the most wide-ranging ape species in Africa, occurring in 21 countries (Humle *et al.*, 2016b). Orangutans are found in Asia—in both Indonesia and Malaysia—and are the only ape species to have two distinct male types. Gibbons are the most numerous of the apes, with 20 species across Asia (see Table AO2).

Great Ape Socioecology

Social organization differs considerably across the three great ape genera.

Chimpanzees and bonobos form dynamic communities, fissioning into smaller parties or coming together (fusioning) according to food availability and the presence of reproductively active females (Wrangham, 1986). Chimpanzee communities average 35 members, with a known maximum of 150 members (Mitani, 2009). Bonobo communities are usually composed of between 30 and 80 individuals (Fruth, Williamson and Richardson, 2013).

Gorillas live in cohesive social groups. The median group size in eastern gorillas is ten, with groups consisting of one or more silverback males with several females and their offspring, although group sizes can be much larger. Western lowland gorillas differ from eastern gorillas, with groups frequently comprising more than 20 individuals and about 40% of groups having a multi-male structure. Their large body size and largely vegetation-based diet enable them to live in

---

BOX AO1
IUCN Red List Categories and Criteria, and CITES Appendices

The IUCN Species Survival Commission assesses the conservation status of each species and subspecies using IUCN Red List categories and criteria. As all great apes and gibbons are categorized as vulnerable, endangered or critically endangered, this box presents details on a selection of the criteria for these three categories (see Table AO1). Full details of the IUCN Red List categories and criteria (in English, French and Spanish) can be viewed and downloaded at:

http://www.iucnredlist.org/technical-documents/categories-and-criteria.

Detailed guidelines on their use are available at:

http://www.iucnredlist.org/documents/RedListGuidelines.pdf.

Appendices I, II and III to the Convention on International Trade in Endangered Species of Wild Fauna and Flora (CITES) are lists of species afforded different levels or types of protection from overexploitation.

All non-human apes are in Appendix I, which comprises species that are the most endangered among CITES-listed animals and plants. As they are threatened with extinction, CITES prohibits international trade in specimens of these species, except when the purpose of the import is not commercial, for instance for scientific research. In these exceptional cases, trade may take place, provided it is authorized by both an import permit and an export permit (or re-export certificate). Article VII of the Convention provides for a number of exemptions to this general prohibition. For more information, see:

https://www.cites.org/eng/disc/text.php#VII.

Table AO1
Criteria for Categorization as Vulnerable, Endangered and Critically Endangered

IUCN Red List category	Risk of extinction in the wild	Number of mature individuals in the wild	Rate of population decline over the past 10 years or 3 generations
Vulnerable	High	<10,000	>50%
Endangered	Very high	<2,500	>50%
Critically endangered	Extremely high	<250	>80%

Table AO2
Great Apes and Gibbons

GREAT APES		
Pan genus		
Bonobo (a.k.a. pygmy chimpanzee)	*Pan paniscus*	■ Democratic Republic of Congo (DRC)
Central chimpanzee	*Pan troglodytes troglodytes*	■ Angola ■ Cameroon ■ Central African Republic ■ DRC ■ Equatorial Guinea ■ Gabon ■ Republic of Congo
Eastern chimpanzee	*Pan troglodytes schweinfurthii*	■ Burundi ■ Central African Republic ■ DRC ■ Rwanda ■ Sudan ■ Tanzania ■ Uganda
Nigeria–Cameroon chimpanzee	*Pan troglodytes ellioti*	■ Cameroon ■ Nigeria
Western chimpanzee	*Pan troglodytes verus*	■ Ghana ■ Guinea ■ Guinea-Bissau ■ Ivory Coast ■ Liberia ■ Mali ■ Senegal ■ Sierra Leone[6]
Gorilla genus		
Cross River gorilla	*Gorilla gorilla diehli*	■ Cameroon ■ Nigeria
Grauer's gorilla (a.k.a. eastern lowland gorilla)	*Gorilla beringei graueri*	■ DRC
Mountain gorilla	*Gorilla beringei beringei*	■ DRC ■ Rwanda ■ Uganda
Western lowland gorilla	*Gorilla gorilla gorilla*	■ Angola ■ Cameroon ■ Central African Republic ■ Equatorial Guinea ■ Gabon ■ Republic of Congo
Pongo genus		
Northeast Bornean orangutan	*Pongo pygmaeus morio*	■ Indonesia ■ Malaysia
Northwest Bornean orangutan	*Pongo pygmaeus pygmaeus*	■ Indonesia ■ Malaysia
Southwest Bornean orangutan	*Pongo pygmaeus wurmbii*	■ Indonesia
Sumatran orangutan	*Pongo abelii*	■ Indonesia
Tapanuli orangutan[7]	*Pongo tapanuliensis*	■ Indonesia
GIBBONS (excluding subspecies)		
Hoolock genus		
Eastern hoolock	*Hoolock leuconedys*	■ China ■ India ■ Myanmar
Gaoligong hoolock (a.k.a. Skywalker hoolock)	*Hoolock tianxing*	■ China ■ Myanmar
Western hoolock	*Hoolock hoolock*	■ Bangladesh ■ India ■ Myanmar

Table AO2
Continued

Hylobates genus		
Abbott's gray gibbon	*Hylobates abbotti*	■ Indonesia ■ Malaysia
Agile gibbon (a.k.a. dark-handed gibbon)	*Hylobates agilis*	■ Indonesia ■ Malaysia
Bornean gray gibbon (a.k.a. northern gray gibbon)	*Hylobates funereus*	■ Indonesia ■ Malaysia ■ Brunei
Bornean white-bearded gibbon (a.k.a. Bornean agile gibbon)	*Hylobates albibarbis*	■ Indonesia
Kloss's gibbon (a.k.a. Mentawai gibbon)	*Hylobates klossii*	■ Indonesia
Lar gibbon (a.k.a. white-handed gibbon)	*Hylobates lar*	■ China ■ Indonesia ■ Lao People's Democratic Republic (PDR) ■ Malaysia ■ Myanmar ■ Thailand
Moloch gibbon (a.k.a. Javan gibbon, silvery gibbon)	*Hylobates moloch*	■ Indonesia
Müller's gibbon (a.k.a. Müller's gray gibbon, southern gray gibbon)	*Hylobates muelleri*	■ Indonesia
Pileated gibbon (a.k.a. capped gibbon, crowned gibbon)	*Hylobates pileatus*	■ Cambodia ■ Lao PDR ■ Thailand
Nomascus genus		
Cao Vit gibbon (a.k.a. eastern black crested gibbon)	*Nomascus nasutus*	■ China ■ Viet Nam
Hainan gibbon (a.k.a. Hainan black crested gibbon, Hainan black gibbon, Hainan crested gibbon)	*Nomascus hainanus*	■ China (Hainan Island)
Northern white-cheeked crested gibbon (a.k.a. northern white-cheeked gibbon, white-cheeked gibbon)	*Nomascus leucogenys*	■ Lao PDR ■ Viet Nam
Northern yellow-cheeked crested gibbon (a.k.a. northern buffed-cheeked gibbon)	*Nomascus annamensis*	■ Cambodia ■ Lao PDR ■ Viet Nam
Southern white-cheeked crested gibbon (a.k.a. southern white-cheeked gibbon)	*Nomascus siki*	■ Lao PDR ■ Viet Nam
Southern yellow-cheeked crested gibbon (a.k.a. red-cheeked gibbon, buff-cheeked gibbon, buffy-cheeked gibbon)	*Nomascus gabriellae*	■ Cambodia ■ Lao PDR ■ Viet Nam
Western black crested gibbon (a.k.a. black crested gibbon, black gibbon, concolor gibbon, Indochinese gibbon)	*Nomascus concolor*	■ China ■ Lao PeDR ■ Viet Nam
Symphalangus genus		
Siamang	*Symphalangus syndactylus*	■ Indonesia ■ Malaysia ■ Thailand

Sources: Susan Cheyne, personal communication, 2017; Elizabeth Macfie, personal communication, 2017; Mittermeier *et al*. (2013); Serge Wich, personal communication, 2017

environments with little fruit and to maintain stable groups, due to a lack of competition over highly nutritious foods.

Orangutans have loosely defined communities. Flanged males, characterized by fatty cheek pads and large size, lead a semi-solitary existence (Emery Thompson, Zhou and Knott, 2012). Compared to flanged individuals, the smaller, unflanged adult males are more tolerant of other orangutans and,

Figure AO1

Ape Distribution in Africa[8]

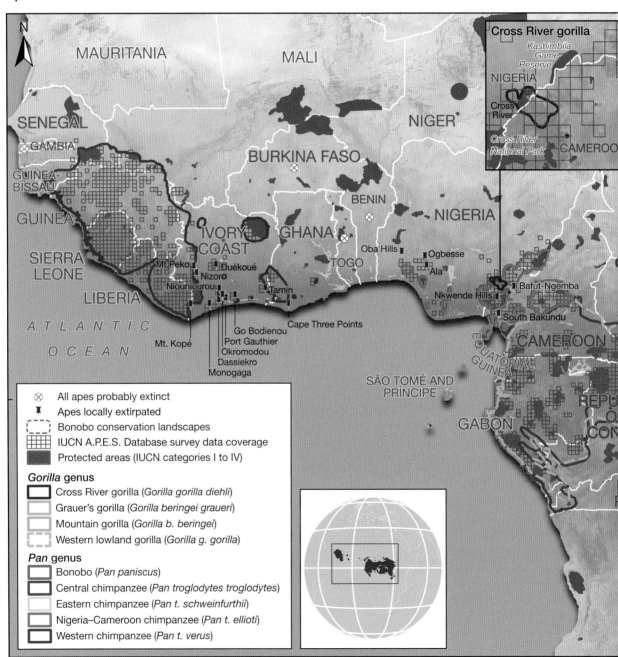

like some adult females, they can travel together for a few hours and up to several days. Sumatran orangutans occasionally congregate when food is abundant (Wich *et al.*, 2006).

Habitat

Most great apes live in closed, moist, mixed tropical forest, occupying a range of various forest types, including lowland, swamp, seasonally inundated, gallery, coastal, submontane, montane and secondary regrowth. Some bonobo populations and eastern and western chimpanzees also live in savannah–woodland mosaic landscapes. The largest populations of great apes are found below 500 m elevation, in the lowland forests of Asia and Africa (Morrogh-Bernard *et al.*, 2003; Stokes *et al.*, 2010). Bonobos have a restricted, discontinuous range in the DRC, south of the Congo River (Fruth *et al.*, 2016). Eastern chimpanzees and eastern gorillas can range above 2,000 m altitude, and orangutans well above 1,000 m in both Sumatra and Borneo (Payne, 1988; Wich *et al.*, 2016).

Most chimpanzees and bonobos inhabit evergreen forests, but some populations exist in deciduous woodland and drier savannah–woodland-dominated habitats interspersed with gallery forest. Although many populations inhabit protected areas, a great number of chimpanzee communities, especially on the western and eastern coasts of Africa, live outside of protected areas, including the majority of individuals in countries such as Guinea, Liberia and Sierra Leone (Brncic, Amarasekaran and McKenna, 2010; Kormos *et al.*, 2003; Tweh *et al.*, 2014). In Indonesian Borneo, more than half of the remaining orangutan populations are currently found outside of protected areas and large numbers of Sumatran orangutans also occur outside of protected areas (Wich *et al.*, 2011, 2012b).

Daily Behaviour Patterns

Great apes are adapted to a plant diet, but all taxa consume insects, and some kill and eat small mammals. Succulent fruits are the main source of nutrition for bonobos, chimpanzees and orangutans, except at

Figure AO2

Ape Distribution in Asia[9]

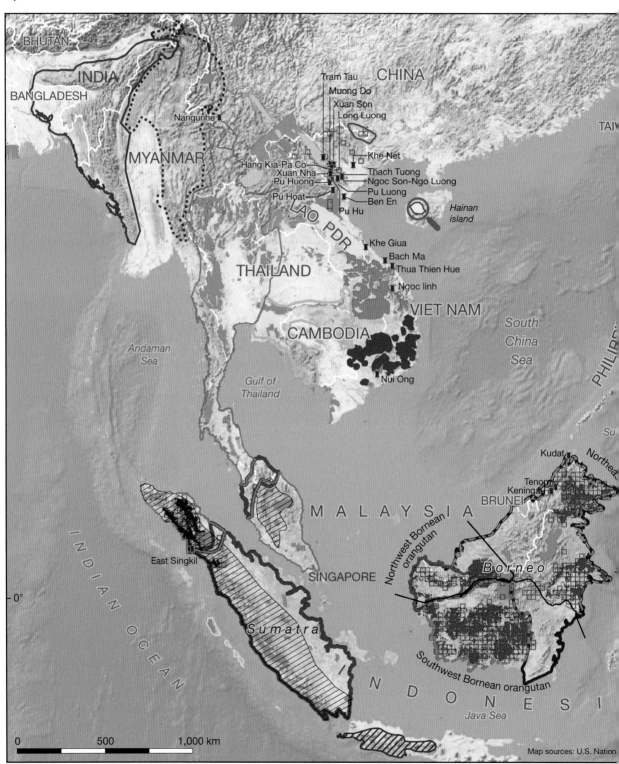

- Apes locally extirpated
- IUCN A.P.E.S. Database survey data coverage
- Protected areas (IUCN categories I to IV)

Hoolock gibbons
- Eastern hoolock (*Hoolock leuconedys*)
- Western hoolock (*Hoolock hoolock*)

Hylobates gibbons
- Abbott's gray gibbon (*Hylobates abbotti*)
- Agile gibbon (*Hylobates agilis*)
- Bornean gray gibbon (*Hylobates funereus*)
- Bornean white-bearded gibbon (*Hylobates albibarbis*)
- Kloss's gibbon (*Hylobates klossii*)
- Lar gibbon (*Hylobates lar*)
- Moloch gibbon (*Hylobates moloch*)
- Müller's gibbon (*Hylobates muelleri*)
- Pileated gibbon (*Hylobates pileatus*)

Nomascus gibbons
- Cao Vit gibbon (*Nomascus nasutus*)
- Hainan gibbon (*Nomascus hainanus*)
- Northern white-cheeked crested gibbon (*Nomascus leucogenys*)
- Northern yellow-cheeked crested gibbon (*Nomascus annamensis*)
- Southern white-cheeked crested gibbon (*Nomascus siki*)
- Southern yellow-cheeked crested gibbon (*Nomascus gabriellae*)
- Western black-crested gibbon (*Nomascus concolor*)

Symphalangus genus
- Siamang (*Symphalangus syndactylus*)

Pongo genus
- Bornean orangutan subspecies boundaries
- Bornean orangutan (*Pongo pygmaeus*)
- Sumatran orangutan (*Pongo abelii*)

altitudes where few fleshy fruits are available (Wright *et al.*, 2015). Gorillas rely heavily on herbaceous vegetation but consume a significant amount of fruit in nearly all locations (Robbins, 2011). During certain periods, African apes concentrate on terrestrial herbs or woody vegetation, such as bark. Similarly, in Asia, orangutans consume more bark and young leaves when fruits are scarce. Sumatran orangutans are more frugivorous than their Bornean relatives (Russon *et al.*, 2009).

The area used habitually by an individual, group or community of a species is referred to as a home range. Establishment of a home range helps secure access to resources within it (Delgado, 2010). Foraging in complex forest environments requires spatial memory and mental mapping. The great apes' daily searches for food are generally restricted to a particular location, an area of forest that a group or individual knows well. Chimpanzees are capable of memorizing the individual locations of thousands of trees over many years (Normand and Boesch, 2009); the other great ape species are likely to possess similar mental capacities.

Most great apes not only feed in trees, but also rest, socialize and sleep in them (although gorillas are largely terrestrial). Being large-brained, highly intelligent mammals, they need long periods of sleep. With the exception of gorillas, who nest primarily on the ground, great apes tend to spend the night in nests that they build high up in the trees, 10–30 m above ground (Morgan *et al.*, 2006). African apes are semi-terrestrial and often rest on the ground during the daytime, but orangutans are almost exclusively arboreal.

More or less restricted to the canopy, orangutans typically do not travel great distances. Bornean flanged adult males and adult females move 200 m each day, unflanged adult males usually double that distance. Sumatran orangutans move farther, but still less than 1 km each day on average (Singleton et al., 2009). The semi-terrestrial African apes range considerably longer distances and the most frugivorous roam several kilometers each day: bonobos and western lowland gorillas average 2 km, but sometimes 5–6 km, while chimpanzees travel 2–3 km, with occasional excursions up to 8 km. Savannah-dwelling chimpanzees generally range farther daily than their forest-dwelling counterparts.

Reproduction

Male apes reach sexual maturity between the ages of 8 and 18 years: chimpanzees attain adulthood at 8–15 years, bonobos at 10, eastern gorillas around 12–15 and western gorillas at 18. Orangutan males mature between the ages of 8 and 16 years, but they may not develop flanges for another 20 years (Wich et al., 2004). Female great apes become reproductively active between the ages of 6 and 12 years: gorillas at 6–7 years, chimpanzees at 7–8, bonobos at 9–12 and orangutans at 10–11. They tend to give birth to their first offspring between the ages of 8 and 16: gorillas at 10 (with a range of 8–14 years), chimpanzees at 13.5 years (with a mean of 9.5–15.4 years at different sites), bonobos at 13–15 years and orangutans at 15–16 years.

Gestation periods in gorillas and orangutans are about the same as for humans; they are slightly shorter in chimpanzees and bonobos, at 7.5–8.0 months. Apes usually give birth to one infant at a time, although twin births do occur (Goossens et al., 2011). Births are not seasonal; however, conception requires females to be in good health. Chimpanzees and bonobos are more likely to ovulate when fruit is abundant, so in some populations there are seasonal peaks in the numbers of conceiving females, with contingent peaks in the birth rate during particular months (Anderson, Nordheim and Boesch, 2006a; Emery Thompson and Wrangham, 2008). Bornean orangutans living in highly seasonal dipterocarp forests are most likely to conceive during mast fruiting events, when fatty seeds are plentiful (Knott, 2005). Sumatran orangutans do not face such severe constraints (Marshall et al., 2009; Wich et al., 2006). Meanwhile, gorillas show no seasonality in their reproduction, as they are less dependent on seasonal foods.

All great apes have slow reproductive rates, due to the mother's high investment in a single offspring and the infant's slow development and maturation. Infants sleep with their mother until they are weaned (4–5 years in African apes; 5–6 years in Bornean orangutans; 7 years in Sumatran orangutans) or a subsequent sibling is born. Weaning marks the end of infancy for African apes, but orangutan infants remain dependent on their mothers until they reach 7–9 years of age (van Noordwijk et al., 2009). Females cannot become pregnant while an infant is nursing because suckling inhibits the reproductive cycle (Stewart, 1988; van Noordwijk et al., 2013). Consequently, births are widely spaced, occurring on average every 4–7 years in African apes, every 6–8 years in Bornean orangutans and every 9 years in Sumatran orangutans. Interbirth intervals can be shortened by the death or killing of unweaned offspring. Infanticide, typically by an unrelated adult male, has been observed in gorillas (Harcourt and Greenberg, 2001; Watts, 1989). Infanticide has not been observed in orangutans or bonobos, but if a female chimpanzee with an infant transfers to a different group, her offspring is likely to be killed by a male in her new group, resulting in early resumption of her reproductive cycle (Wilson and Wrangham, 2003).

Long-term research on mountain gorillas and chimpanzees has allowed female lifetime reproductive success to be evaluated. The mean birth rate is 0.2–0.3 births per adult female per year, or one birth per adult female every 3.3–5.0 years. Mountain gorilla females produce an average of 3.6 offspring during their lifetimes (Robbins *et al.*, 2011). Although female chimpanzees occasionally have twins, they usually give birth to a single surviving offspring approximately every 5 to 6 years. However, in their lifetime, female chimpanzees typically produce only 1 to 4 offspring who will survive to reproductive age (Thompson, 2013).

Key points to be noted are (1) that documenting the biology of long-lived species takes decades of research due to their slow rates of reproduction, and (2) that great ape populations that have declined in numbers are likely to take several generations to recover (generation time in the great apes is 20–25 years) (IUCN, 2014d). These factors make great apes far more vulnerable than smaller, faster-breeding species. Orangutans have the slowest life history of any mammal, with later age at first reproduction, longer interbirth intervals and longer generation times than African apes (Wich *et al.*, 2009a, 2009b); as a result, they are the most susceptible to loss.

Gorillas

Gorillas live in a broad range of habitats across Africa. As a result of their dietary patterns, they are restricted to moist forest habitats (at altitudes ranging from sea level to more than 3,000 m) and are not found in forest–savannah mosaics or gallery forests inhabited by chimpanzees and bonobos.

Across their range, gorillas rely more heavily than any other ape species on herbaceous vegetation, such as the leaves, stems and pith of understory vegetation, as well as leaves from shrubs and trees (Doran-Sheehy *et al.*, 2009; Ganas *et al.*, 2004; Masi, Cipolletta and Robbins, 2009; Wright *et al.*, 2015; Yamagiwa and Basabose, 2009). Early research suggested that gorillas ate very little fruit, a finding that can be attributed to the fact that initial studies of their dietary patterns were conducted in the Virunga Volcanoes, the only habitat in which gorillas eat almost no fruit as it is virtually unavailable (Watts, 1984). These conclusions were adjusted once detailed studies were conducted on gorillas living in lower-altitude habitats (Doran-Sheehy *et al.*, 2009; Masi *et al.*, 2015; Rogers *et al.*, 2004; Wright *et al.*, 2015; Yamagiwa *et al.*, 2003).

Gorillas incorporate a notable amount of fruit into their diets when it is available, but they are less frugivorous than bonobos and chimpanzees, preferring vegetative matter even at times of high fruit availability (Head *et al.*, 2011; Morgan and Sanz, 2006; Yamagiwa and Basabose, 2009). They rely heavily on terrestrial herbaceous vegetation, which often increases in availability in disturbed landscapes, such as abandoned farmland or plantations, selectively logged land, and areas that border on human settlements.

Mountain gorillas are primarily terrestrial. Although western gorillas are more arboreal, they still primarily travel on the ground and not through the tree canopy. The distance traveled per day by gorillas declines with increasing availability of understory vegetation, varying between approximately 500 m and 3 km per day (Robbins, 2011).

Eastern gorillas range over areas of 6–34 km^2 (600–3,400 ha) (Robbins, 2011; Williamson and Butynski, 2013a); western gorilla home ranges average 10–20 km^2 (1,000–2,000 ha), and potentially up to 50 km^2 (5,000 ha) (Head *et al.*, 2013). Gorillas are not territorial but have overlapping home ranges that they do not actively defend. However, there is evidence that they have non-overlapping, exclusive core areas (the zone used the most by a group), suggesting that groups do partition their habitat (Seiler *et al.*, 2017).

As the density of gorillas increases, the degree of home range overlap can increase dramatically, as can the frequency of intergroup encounters (Caillaud *et al.*, 2014);

results can include increased fighting, injuries and mortality. Encounters between groups can occur without visual contact, as silverbacks may exchange vocalizations and chestbeats until one or both groups move away. Some encounters between groups involve more than auditory contact, however, and can escalate to include aggressive displays or fights (Bradley et al., 2004; Robbins and Sawyer, 2007). Physical aggression is rare, but if contests escalate, fighting between silverbacks can be intense. Some gorillas have died from infections of injuries sustained during such interactions (Williamson, 2014).

Chimpanzees and Bonobos

Chimpanzees eat mainly fruit, although they have an omnivorous diet, which may include plant pith, bark, flowers, leaves and seeds, as well as fungi, honey, insects and mammal species, depending on the habitat and the community; some groups may consume as many as 200 plant species (Humle, 2011). Terrestrial and arboreal, chimpanzees live in multi-male–multi-female, fission–fusion communities. A single community will change size by fissioning into smaller parties according to resource availability and activity (food and access to reproductive females). Parties thus tend to be smaller during periods of fruit scarcity. Adult female chimpanzees frequently spend time alone with their offspring or in a party with other females.

Chimpanzees have home ranges of 7–41 km² (700–4,100 ha) in forest habitats and more than 65 km² (6,500 ha) in savannah (Emery Thompson and Wrangham, 2013; Pruetz and Bertolani, 2009). Male chimpanzees are highly territorial and patrol the boundaries of their ranges. Parties of males may attack members of neighboring communities and some populations are known for their aggression (Williams et al., 2008).

After a fight, victors may seize females or territory from the vanquished.

Bonobo communities share home ranges of 20–60 km² (2,000–6,000 ha) (Fruth et al., 2013). Bonobos exhibit neither territorial defense nor cooperative patrolling; encounters between members of different communities involve excitement rather than conflict (Hohmann et al., 1999).

Chimpanzees and bonobos both live in multi-male and multi-female groups and are semi-terrestrial. The size of their home ranges varies in line with their group size, the quality of the habitat and food availability, which may change from season to season. Bonobos are not territorial, whereas chimpanzees are generally highly intolerant of neighboring groups; intergroup encounters can result in lethal aggressive attacks among males in particular. The frequency of such encounters can be exacerbated by shifts in home ranges linked to habitat loss, changes in habitat quality and disruptions in their environment (e.g. roads, logging) (Watts et al., 2006; Wilson et al., 2014b).

Bonobos are generally frugivorous but are more dependent on terrestrial herbaceous vegetation, including aquatic plants, than chimpanzees (Fruth et al., 2016).

Wherever gorillas and chimpanzees are sympatric, dietary divisions between the species limit direct competition for food (Head et al., 2011). If the area of available habitat is restricted, such mechanisms for limiting competition are compromised (Morgan and Sanz, 2006).

Orangutans

Male orangutans are the dispersing sex: upon reaching sexual maturity, they leave the area where they were born to establish their own range. A male orangutan's range encompasses several (smaller) female ranges. High-status flanged males are able to monopolize both food and females to a degree and may thus temporarily reside in a relatively small area—typically 4–8 km² (400–800 ha) for Bornean males (Mittermeier *et al.*, 2013). Orangutan home range overlap is usually extensive, but flanged orangutans establish personal space by emitting long calls. As long as distance is maintained, physical conflicts are rare; however, close encounters between adult males trigger aggressive displays that sometimes lead to fights. If an orangutan inflicts serious injury on his opponent, infection of the wounds can result in death (Knott, 1998).

Although they are primarily fruit-eaters, orangutans are able to adapt their diet to what is available in the forest. In Borneo, they feed on more than 500 plant species (Russon *et al.*, 2009). The resilience of orangutans and their ability to cope, albeit temporarily, with drastic habitat changes is further illustrated by recent records of species presence in acacia plantations in East Kalimantan (Meijaard *et al.*, 2010a); a mosaic of mixed agriculture in Sumatra (Campbell-Smith *et al.*, 2011a); oil palm plantations in Borneo (Ancrenaz *et al.*, 2015b); and in forests exploited for timber (Ancrenaz *et al.*, 2010; Wich *et al.*, 2016).

It must be noted, however, that orangutan presence in these human-altered landscapes does not imply the species' long-term survival. Orangutan survival is still dependent on a landscape mosaic with adequate forest patches for food, shelter and other needs. Today, half of the wild orangutan populations in Indonesian Borneo are surviving outside of protected forests, in areas that are prone to human development and transformation (Wich *et al.*, 2012b).

Orangutans are the largest arboreal mammals in the world, but recent studies have shown that they also walk on the ground for considerable distances in all types of natural and man-made habitats (Ancrenaz *et al.*, 2014; Loken, Boer and Kasyanto, 2015; Loken, Spehar and Rayadin, 2013). Consequently, orangutans are able to cross open artificial infrastructures to a certain extent. In Sabah, Malaysian Borneo, for example, orangutans have been seen crossing sealed and dust roads when traffic is not too heavy. Greater terrestriality in orangutans will increase the risk of contracting diseases that the animals are not usually exposed to when they live in the tree canopy; however, there is a dearth of information about such new risks. When territories of resident individuals are destroyed, it is difficult for them to establish a new territory if other animals already reside in nearby areas. Indeed, resident animals who have lost their territory, and cannot easily establish a new range, slowly die off. However, adult unflanged males do not have a territory and can thus move away from a disturbed area and return after the source of nuisance has been eliminated (Ancrenaz *et al.*, 2010).

Gibbon Socioecology

Gibbons are the most diverse and widespread group of apes. Currently, 20 species of gibbon in 4 genera are recognized:

- 9 *Hylobates* species;
- 7 *Nomascus* species;
- 3 *Hoolock* species; and
- the single *Symphalangus* species (Fan *et al.*, 2017; IUCN, 2017).

Gibbons inhabit a wide range of habitats, predominantly lowland, submontane and montane broadleaf evergreen and semi-evergreen forests, as well as dipterocarp-dominated and mixed-deciduous (that is, non-evergreen) forests. Some members of the *Nomascus* genus also occur in limestone karst forests and some populations of *Hylobates* live in peat-swamp forest (Cheyne, 2010). Gibbons occur from sea level up to around 1,500–2,000 m above sea level, although their distribution is taxon- and location-specific; *Nomascus concolor*, for example, has been recorded at up to 2,900 m above sea level in China (Fan, Jiang and Tian, 2009).

All gibbons are heavily impacted by the extent and quality of forest as they are arboreal. Only rarely do they move bipedally and terrestrially across forest gaps or to access isolated fruiting trees in more degraded and fragmented habitats (Bartlett, 2007).

Gibbons are reliant on forest ecosystems for food. Gibbon diets are characterized by high levels of fruit intake, dominated by figs and supplemented with young leaves, mature leaves and flowers (Bartlett, 2007; Cheyne, 2008; Elder, 2009). Siamangs are more folivorous than other gibbons (Palombit, 1997). Reliance on other protein sources, such as insects, birds' eggs and small vertebrates, is probably underrepresented in the literature. The composition of the diet changes with the seasons and habitat type, with flowers and young leaves dominating during the dry season in peat-swamp forests and figs dominating in dipterocarp forests (Cheyne, 2010; Fan and Jiang, 2008; Lappan, 2009; Marshall and Leighton, 2006). Since gibbons are important seed dispersers, their frugivorous nature is significant in maintaining forest diversity (McConkey, 2000, 2005; McConkey and Chivers, 2007).

Gibbons are highly territorial and live in semi-permanent family groups defending a core area to the exclusion of other gibbons. Their territories average 0.42 km² (42 ha) (Bartlett, 2007); however, there is considerable variation and some indication that the more northerly *Nomascus* taxa maintain larger territories, possibly related to lower resource abundance at some times of year in these more seasonal forests.

Gibbons have been typified as forming socially monogamous family groups. Some studies, however, reveal that they are not necessarily sexually monogamous (Palombit, 1994). Notable exceptions include extra-pair copulations (mating outside of the pair bond), individuals leaving the home territory to take up residence with neighboring individuals and male care of infants (Lappan, 2008; Palombit, 1994; Reichard, 1995). Research also indicates that the more northerly Cao Vit, Hainan and western black crested gibbons commonly form polygynous groups with more than one breeding female (Fan and Jiang, 2010; Fan *et al.*, 2010; Zhou *et al.*, 2008). There is no conclusive argument regarding these variable social and mating structures; they may be natural or a by-product of small population sizes, compression scenarios or suboptimal habitats.

Both males and females disperse from their natal groups and establish their own territories (Leighton, 1987); females have their first offspring at around 9 years of age. Data from captivity suggest that gibbons become sexually mature as early as 5.5 years of age (Geissmann, 1991). Interbirth intervals are in the range of 2–4 years, with 7 months gestation (Bartlett, 2001; Geissmann, 1991). Captive individuals have lived upwards of 40 years; gibbon longevity in the wild is unknown but thought to be considerably shorter. Due to gibbons' relatively late age of maturation and long interbirth intervals, their reproductive lifetime may be only 10–20 years (Palombit, 1992). Population replacement in gibbons is therefore relatively slow.

Group demography changes only in the event of a death of one of the adults, as there is no regular immigration or emigration into these social groups. Gibbons in habitat fragments are isolated from other groups and dispersal is compromised, which may cause long-term issues regarding the sustainability of these populations. There is insufficient information about dispersal distances for subadult gibbons to determine maximum distances over which they could disperse (perhaps with the assistance of canopy bridges). Gibbons have not been observed to crop-raid—either from plantations or small-scale farms—but this lack of information does not mean gibbons will not exploit disturbed areas if necessary.

Acknowledgments

Principal authors: Annette Lanjouw, Helga Rainer and Alison White

Socioecology section: Marc Ancrenaz, Susan M. Cheyne, Tatyana Humle, Benjamin M. Rawson, Martha M. Robbins and Elizabeth A. Williamson

Reviewers: Elizabeth J. Macfie and Serge Wich

Endnotes

1 This estimate for Sumatran orangutans is higher than that of about 6,500 individuals cited in the previous volume of *State of the Apes* because it considers three new factors: "a) orangutans were found in greater numbers at higher altitudes than previously supposed (i.e., up to 1,500 m asl [above sea level] not just to 1,000 m asl), b) they were found to be more widely distributed in selectively-logged forests than previously assumed, and c) orangutans were found in some previously unsurveyed forest patches. The new estimate does not, therefore, reflect a real increase in Sumatran orangutan numbers. On the contrary, it reflects only much improved survey techniques and coverage, and hence more accurate data. It is extremely important to note, therefore, that overall numbers continue to decline dramatically" (Singleton *et al.*, 2016).

2 This estimate for Bornean orangutans is higher than the figure cited in the previous volume of *State of the Apes*, which suggests that about 54,000 individuals inhabited 82,000 km² (8.2 million ha) of forest (Wich *et al.*, 2008). Modeling and the latest field data available for Borneo were used to revise the map of the current distribution of Bornean orangutans; the range now covers an estimated 155,000 km² (15.5 million ha), or 21% of Borneo's landmass (Gaveau *et al.*, 2014; Wich *et al.*, 2012b). As Ancrenaz *et al.* (2016) explain: "If the mean average orangutan density recorded in 2004 (0.67 individuals/km²) is applied to the updated geographic range, then the total population estimate would be 104,700 individuals. This represents a decline from an estimated 288,500 individuals in 1973 and is projected to decline further to 47,000 individuals by 2025. [...] many populations will be reduced or become extinct in the next 50 years (Abram *et al.*, 2015)."

3 The Tapanuli orangutan was described and distinguished from the Sumatran orangutan as this volume of *State of the Apes* was being finalized for publication. As a consequence, this new species is only mentioned in the Apes Index, Table AO2 of this Overview and Case Study 6.4, and not in the rest of the volume.

4 The distribution and population estimates for the Tapanuli orangutan are based on earlier surveys in the area where the species occurs. Since these individuals were still being identified as Sumatran orangutans at the time of the surveys, the cited reference does not mention the Tapanuli orangutan.

5 For more detailed information, refer to Emery Thompson and Wrangham (2013), Reinartz, Ingmanson and Vervaecke (2013), Robbins (2011), Wich *et al.* (2009b), Williamson and Butynski (2013a, 2013b), and Williamson *et al.* (2013).

6 Some of these countries were erroneously omitted in the previous volume of *State of the Apes*. Benin, Burkina Faso, Gambia and Togo have been removed from the list as *Pan troglodytes verus* are extinct/possibly extinct in these countries.

7 See Endnote 3.

8 The Arcus Foundation commissioned the ape distribution maps (Figures AO1 and AO2) for this publication, so as to provide the most accurate and up-to-date illustration of range data. This volume also features maps created by contributors who used ape range data from different sources. As a consequence, the maps may not all align exactly.

9 See Endnote 8.

Photo: Large dams affect the political, social and environmental landscape of a region. Bakun Hydroelectric Dam, Malaysia © MOHD RASFAN/AFP/Getty Images

State of the Apes Infrastructure Development and Ape Conservation

INTRODUCTION

Section 1: Infrastructure Development and Ape Conservation

This, the third in the *State of the Apes* series, focuses on the impact of infrastructure—such as roads, railways and hydroelectric power plants—on ape conservation and welfare. While infrastructure for transport, energy and other purposes may be designed to improve peoples' lives, it often has negative consequences for local communities and biodiversity. The first two volumes of *State of the Apes* briefly considered the impact of infrastructure on apes and their habitat in relation to extractive industries and industrial agriculture; this volume explores that relationship more explicitly, featuring in-depth analysis of large-scale infrastructure projects.

The *State of the Apes* Series

Commissioned by the Arcus Foundation, the *State of the Apes* series strives to raise awareness of the impacts of human activities on all great ape and gibbon populations. Apes are vulnerable to a range of threats that are primarily driven by humans, including hunting associated with the trade in wild meat, body parts and live animals; deforestation and degradation of habitat; and the transmission of disease. Interactions between humans and apes continue to increase as development and human population growth drive further incursions into spaces that are inhabited by apes. By using apes as an example, this publication series also aims to underscore the importance of wider species conservation.

State of the Apes covers all non-human ape species, namely bonobos, chimpanzees, gibbons, gorillas and orangutans, as well as their habitats. Ape ranges are found throughout the tropical belt of Africa and South and Southeast Asia. Robust statistics on the status and welfare of apes are derived from the Ape Populations, Environments and Surveys (A.P.E.S.) Portal (Max Planck Institute, n.d.-a). Abundance estimates of the different ape taxa are presented in the Abundance Annex, available on the *State of the Apes* website at www.stateoftheapes.com. The annex is updated with each new volume in the series, to allow for comparisons over time. Details on the socioecology and geographic range of each species are provided in the Apes Overview.

Each volume in the *State of the Apes* series is divided into two sections. Section 1 focuses on the thematic topic of interrogation, which in this case is infrastructure development (see Box I.1). The immediate objectives are to provide accurate information on the current situation, present various perspectives and, where applicable, highlight best practice. In the longer term, the key findings and messages are intended to stimulate debate, multi-stakeholder collaboration and changes to policies and practice that can facilitate the reconciliation of economic development and the conservation of biodiversity. Section 2 presents more general details on the status and welfare of apes, in their natural habitat and in captivity.

> " Infrastructure encompasses bridges, geothermal power plants, hydropower dams, power lines and distribution networks, ports and industrial installations, such as mines and pipelines, railways, roads and tunnels. "

Infrastructure Development and Ape Conservation

Both Africa and Asia are facing a number of development challenges, with expanding populations and increasing urbanization; rising demand for water, energy, food and other commodities on a local, regional and global scale; predicted hydrological variability due to climate change; and persistent poverty and inequality.

Dams appear to offer a tempting range of benefits to meet development needs—they can reduce floods, store water for irrigation, provide energy for burgeoning populations and contribute to regional integration. The social, environmental and economic costs and benefits of dams are not distributed evenly, however, and oftentimes they are not viable investments due to excessive cost and time overruns (International Rivers, n.d.-b). Large dams also affect the political, social and environmental landscape of a region.

Similarly, the development of road networks is promoted as contributing to economic and social development by providing

---

BOX I.1

Definition of Infrastructure

State of the Apes focuses on physical infrastructure and defines the term to refer to large, diverse structures that are built to enable the provision of services for households, industry and other entities (e.g., government buildings, state hospitals and schools) and that are closely aligned with economic development. For the purposes of this publication, *infrastructure* refers to fixed assets that can form part of a large network. The term encompasses bridges, geothermal power plants, hydropower dams, power lines and distribution networks, ports and industrial installations, such as mines and pipelines, railways, roads and tunnels.

access to markets and resources, without taking account of the environmental and social costs. At least 25 million kilometers of additional roads are anticipated worldwide by 2050—90% of them in developing nations, including many regions with exceptional biodiversity and vital ecosystem services (Global Road Map, n.d.). As much of the planned infrastructure is to be built in developing countries, ape habitat across the tropical belt of Africa and Asia will certainly be affected.

Before presenting chapter-by-chapter highlights of Section 1, this introduction explores the factors that influence the rate and extent of infrastructure development. Summaries of Chapters 7–8 appear in the introduction to Section 2 (see p. 198).

This volume describes various efforts to mitigate the effects of infrastructure, such as roads and hydropower, across specific sectors, including activism, planning, ecology, legislation and advocacy. To understand and be able to address the adverse impacts of infrastructure development, it is important to know where these investments are likely to occur and how rapidly they are developing. The following sections explore the role of incentives, capacity, institutions, corruption and finance in shaping infrastructure.

Incentives and Capacity

Most reports about infrastructure investment depend on government budgets, policy documents, official pronouncements and company press releases for specific figures. These sources have often proved unreliable, however, as many planned projects never materialize, while others have large cost overruns. Moreover, both proponents and critics of infrastructure investments can benefit from exaggerating the rate at which investors support development projects. In some cases, the rate actually exceeds expectations.

What makes potential investors *want* to invest and what makes them *able* to do so? To arrive at meaningful answers, it is helpful to split the determinants of infrastructure investment into two broad categories: incentives and capacity. Factors that increase incentives and capacity to invest while also reducing disincentives can be expected to accelerate investment—and vice versa.

Incentives can be economic, political or both. Economic motives include generating export revenues, opening land for agriculture, accessing raw materials and moving goods between locations. Common political rationales are establishing government presence, populating border regions, building geopolitical alliances and capturing votes. Key disincentives include high construction costs and political or local opposition. Even when elites want infrastructure, they will not get it unless they have the capacity to produce and maintain it. That requires political support, funds, technical and managerial capacity, and the ability to overcome regulatory and administrative hurdles. Generating new sources of tax revenue and implementing fiscal decentralization provide both incentives and capacity for infrastructure investment (Kis-Katos and Suharnoko Sjahrir, 2014).

Institutions, Instability and Corruption

Political instability, inadequate planning, limited administrative capacity, a lack of trained staff and bureaucratic delays typically reduce a government's ability to provide infrastructure, while also undermining private interest in partnerships (Berg *et al.*, 2012; Galinato and Galinato, 2013; Gillanders, 2013; Kikawasi, 2012; Percoco, 2014). These factors cause delays, disruptions and poor maintenance, impeding effective investment (see Case Study 5.3). While opportunities for bribes may motivate officials to promote projects, corruption raises the costs and slows progress (Collier, Kirchberger and Soderbom, 2015).

In principle, strong judicial and regulatory systems, which ensure that projects meet environmental and social standards, can deter harmful infrastructure investments in forest regions. That has certainly happened on some occasions. On balance, however, instability, institutional limitations and corruption are probably greater constraints on infrastructure investments than well-functioning regulatory systems (Collier *et al.*, 2015; Galinato and Galinato, 2013).

Political Support and Opposition

All the dominant economic paradigms see infrastructure investments as inherently positive. That applies as much to the more developmental state visions as to the more neoliberal, free-market views. This consensus gives these investments legitimacy and makes it easier to promote them. Nonetheless, in some regions, indigenous peoples and rural communities adamantly oppose such investments, particularly when they are linked to large-scale mining and energy projects or plantations. National and international environmental groups often support such opposition. Through demonstration, litigation, advocacy and other strategies, they have blocked or delayed many projects (see Case Study 6.2).

Changes in Investment Sources

Most funds for infrastructure come from the governments of developing countries, multilateral development banks (MDBs), bilateral aid agencies, emerging-market development banks and private companies. Each type of agency or lender has different goals, strengths and weaknesses and operates in distinct environments. For decades, national governments in developing countries usually had to secure some funding from MDBs and/or bilateral development agencies if they wished to undertake large infrastructure investment projects. Their weak tax base limited their ability to finance large projects on their own. Conversely, the MDBs were interested in making large loans and had few resource constraints.

After the World Bank and other MDBs adopted environmental and social safeguards in the 1980s, the environmental impacts of large infrastructure projects came under greater scrutiny. It became harder for national governments to borrow for projects that were likely to harm the environment (Currey, 2013). The MDBs were concerned about their reputation and pressure from non-governmental organizations (NGOs).

Over the last decade or so, however, various trends have made it easier for national governments to obtain funds for controversial projects. Emerging-market development banks—such as the Asian Infrastructure Investment Bank, Brazilian Development Bank, China Development Bank, Development Bank of Southern Africa and New Development Bank—have partially replaced the traditional MDBs. These new banks put a premium on geopolitical considerations, such as gaining political allies, securing access to markets and raw materials, and supporting national companies. They tend to be less concerned about environmental considerations and less susceptible to pressure from NGOs (Kahler *et al.*, 2016). There has also been an upswing in private funding, as market-friendly ideologies and low international interest rates have led governments to work with private banks and construction companies. Meanwhile, to remain competitive, some believe that the World Bank has weakened its own safeguard policies (see Box 1.4 and Box 5.1).

> " In principle, strong judicial and regulatory systems, which ensure that projects meet environmental and social standards, can deter harmful infrastructure investments in forest regions. "

These evolving dynamics can greatly affect levels of investment, as can domestic instability. In Brazil, for example, corruption scandals and the national political and economic crisis recently forced the Brazilian Development Bank to curtail its activities beyond the country's borders (Molina *et al.*, 2015).

Chapter Highlights: Infrastructure Development and Ape Conservation

The first six chapters of this volume of *State of the Apes* interrogate the interface between ape conservation and large-scale infrastructure development. **Chapter 1** presents an overview of proposed infrastructure projects in the ape habitats of Asia and Africa. It explores the role of major economies such as China and multilateral financial institutions in the expansion of infrastructure in the tropical belt and considers the potential impacts of specific planned infrastructure projects. **Chapter 2** assesses the impacts of infrastructure development on apes and people, highlighting issues ranging from displacement and loss of ancestral land, and habitat destruction and forest degradation, to disruptions in access to food, clean water and shelter, to road kills, increased poaching and the introduction of disease. **Chapter 3** discusses the findings of a historical analysis of road construction in three ape sites and describes how these infrastructure projects have affected ape forest habitat over time. The chapter proposes approaches that can serve to minimize environmental damage, as well as tools that allow for effective forest monitoring. **Chapter 4** explores the robustness of one of the most commonly used conservation strategies—the establishment of protected areas—in the face of large-scale infrastructure development. Results suggest that as roads spread across sub-Saharan Africa, they will cut through one-third of all existing protected areas. The chapter encourages a more considered approach to land use and infrastructure planning, as well as the application of the "mitigation hierarchy" to reduce threats to critical habitats. **Chapter 5** presents three case studies on proposed road developments in the ape ranges of Cross River State, Nigeria; the Dawei region connecting Thailand and Myanmar; and the northern region of the Democratic Republic of Congo (DRC). In documenting the role of conservation organizations in these cases, the chapter identifies a variety of approaches and common challenges. **Chapter 6** considers the engagement of social and environmental actors in relation to energy development. It presents case studies involving dam construction projects in Cameroon and Sarawak, Malaysian Borneo, a geothermal project in Indonesia's Leuser ecosystem, as well as a planning approach developed by a conservation organization to mitigate the impacts of hydropower development.

Section 2 provides updates on *in situ* ape conservation in Africa and Asia (Chapter 7) and the welfare of apes in captivity (Chapter 8). The highlights for these two chapters are included in the Introduction to Section 2 (see p. 198).

Chapter 1: Challenges and Opportunities in Sustainable Infrastructure Development

This chapter considers the current unprecedented rate of global infrastructure expansion and the factors that typically prevent resulting benefits from being distributed equitably. It explores the role of multilateral financial institutions and major economies, such as China, in backing proposed

> ❝ Over the last decade or so, various trends have made it easier for national governments to obtain funds for controversial infrastructure projects through emerging-market development banks and private funding. ❞

> This volume demonstrates the value of anticipating development, planning early, forming partnerships, establishing robust monitoring and relying on empirical evidence to reconcile conservation objectives with those of infrastructure development.

infrastructure projects in the ape range states of Africa and Asia. Specifically, it examines the extent to which ape habitat is likely to be degraded by "development corridors" such as the LAPSSET (Lamu Port, South Sudan, Ethiopia Transport) corridor, which is to slice through the Congo Basin; the Central African Iron Ore Corridor, which will cross the Republic of Congo, Cameroon and Gabon, comprising road, rail and hydropower components; and the Simandou iron ore project in southeastern Guinea. The chapter identifies promising alternatives to such destructive development projects, highlighting the advantages of "leapfrogging" traditional grid-based energy infrastructure in favor of decentralized renewables, as well as the benefits of carrying out strategic land use planning to protect ape habitat and biodiversity more broadly.

Chapter 2: Impacts of Infrastructure on Apes and People

This chapter assesses the environmental and social impacts of infrastructure development, highlighting issues ranging from displacement, loss of land and habitat, and forest degradation to disruptions in access to food, clean water and shelter. Among the most serious environmental impacts is the increased access that infrastructure-related roads and settlements provide to critical habitat. Such access tends to exacerbate illegal hunting, habitat loss and fragmentation, degradation of ecological integrity, the frequency of disease outbreaks, and wildlife mortality and injury rates. Projections show that, by 2030, fewer than 10% of ape ranges in Africa and only about 1% of those in Asia will remain untouched by infrastructure development and the associated habitat disturbance. If this trajectory is to be avoided, greater incorporation of species ecology into infrastructure planning is required. Significant knowledge gaps remain, however.

In assessing the social impacts of infrastructure development, the chapter considers road and rail projects in southern Cameroon, as well as the Chad–Cameroon pipeline. When undertaken in customary land, such infrastructure development can have a negative impact on the livelihoods, cultural practices and norms of indigenous peoples, who traditionally manage and utilize forest resources sustainably. Conservation efforts designed to mitigate and offset adverse effects of infrastructure development on biodiversity can also have negative impacts on indigenous peoples.

Chapter 3: Effects of Road Projects in Ape Landscapes

This chapter presents analysis of changes in forest cover around roads that were substantially upgraded between 2000 and 2014 in ape forest habitat in Northern Sumatra, Indonesia, and western Tanzania, as well as in Peru's tropical forest, which is home to primates but not apes. In these case studies, satellite imagery and associated spatial data analysis tools are used to reveal changes in canopy cover. The studies demonstrate that geospatial data can serve to inform road siting and the design of measures to minimize the impact of infrastructure on wildlife habitat.

The findings show that forested areas near roads are highly vulnerable to deforestation. In particular, roads facilitate the development of uncontrolled settlements, which tends to be accompanied by a rise in poaching and farming; they also enable illegal access to protected areas, such as the Leuser Ecosystem. The chapter argues that an integrated approach to infrastructure planning is required if critical habitats are to be preserved. In instances where roads cannot be rerouted to avoid protected areas, such planning can ensure that road design incorporates measures to mitigate negative impacts on natural areas. The chapter illustrates the value of satellite imagery and platforms such as Global Forest Watch to both forest monitoring and sustainable road development.

Chapter 4: Apes, Protected Areas and Infrastructure in Africa

In Africa, many development corridors are being planned or are already under construction in areas of high environmental value, including critical ape habitat. This chapter shows that as road networks and related infrastructure spread across equatorial Africa, they are likely to cut through more than one-third of all existing protected areas in sub-Saharan Africa. Bwindi Impenetrable National Park, a stronghold of the mountain gorilla, is among the areas at risk. Across the continent, protected areas that are considered to be obstacles to large-scale infrastructure development are particularly vulnerable to reduction or degazzettement.

The chapter encourages a more considered approach to land use and infrastructure planning. It argues for an expanded application of the "mitigation hierarchy" to reduce threats to critical habitats while also calling for viable financial strategies to help developing nations meet pressing economic and food-production needs. The chapter presents the DRC's Virunga National Park as an example of a successful approach to balancing economic and environmental priorities; as part of its program of socioeconomic development, the park provides energy and tourism revenue to local communities and businesses.

Chapter 5: Case Studies of Large-Scale Road Development

This chapter explores how advance planning that is evidence-based and inclusive can help to minimize the negative impacts of road development on biodiversity. To that end, it presents three case studies on proposed road development in ape ranges in Africa and Asia: the Cross River superhighway of Cross River State, Nigeria; the Dawei road between Thailand and Myanmar; and the Pro-Routes project in the DRC. By interrogating how conservation actors are engaging with different road projects that present major threats to great ape and gibbon habitat, the analysis reveals a range of approaches and common challenges.

The case studies demonstrate that sustainable infrastructure development requires the active participation of a range of stakeholders. Specifically, the chapter highlights the importance of advocacy by local and international conservation NGOs in Nigeria, civil society engagement with industry and government actors in Myanmar, and the inclusion of conservation actors early on in the planning and implementation of mitigation measures in the DRC. All case studies underscore the importance of integrating ecosystem and wildlife considerations in the planning and design of roads. Unless political actors and decision-makers prioritize environmental concerns, however, conservationists will remain reliant on standards and safeguards that may be weakly enforced or poorly applied, if at all.

Chapter 6: Case Studies of Renewable Energy Projects

Hydropower is by far the largest source of renewable energy, and projections suggest that its global capacity may double by 2040.

This expansion is likely to entail the construction of thousands of new large dams and tens of thousands of small dams. The plans are proceeding despite the availability of more sustainable, more cost-effective energy alternatives, and notwithstanding evidence that the oft-touted economic benefits of dams rarely materialize for the vulnerable sectors of society. The rapid growth of hydropower is certain to have substantial environmental and social ramifications, ranging from the disruption of hydrological connectivity and the destruction of upstream terrestrial habitats to the emission of high levels of greenhouse gases. The chapter indicates that hydropower development is likely to impact apes in Asia more significantly than in Africa, with gibbons identified as particularly vulnerable.

Two of the chapter's case studies explore the environmental and social impacts of dam development in ape ranges in Cameroon and in Sarawak, Malaysian Borneo. The first considers the challenges of implementing best practices designed to protect apes once a project shifts from the planning to the construction phase; the second explores how community activism and collaboration between communities and scientists can block the construction of destructive dams. Given that hydropower is not the only form of renewable energy associated with adverse impacts, the chapter also features a case study on the implications of a proposed geothermal plant in Sumatra's Leuser Ecosystem. The chapter also presents a system-scale hydropower planning and design framework—"Hydropower by Design"—which was developed to mitigate the impacts of hydropower development.

Conclusion

On a spectrum of government capacity to undertake infrastructure development, the conservation of ape habitat appears to be most likely at the opposite extremes. At one end of the spectrum, weak, unstable and corrupt governments are unable to fund, construct or maintain projects that would threaten forests, thus inadvertently preserving habitat. At the other end, in stable countries with transparent governments and effective regulatory systems, opposition forces and civil society can put a brake on harmful projects. The greatest risks to wildlife and their habitat lie between these extremes, in countries where institutions are weak, rulers and officials are corrupt, and conservation actors are silenced or treated with indifference.

Many ape range states sit in the middle of this spectrum. This volume of *State of the Apes* seeks to avert situations in which the natural world is particularly vulnerable to infrastructure development, by providing accurate information on the current situation, identifying challenges and possible solutions, and leveraging the iconic status of apes to contribute to the overall conservation of tropical forest ecosystems. It demonstrates the value of anticipating development, planning early, forming partnerships, establishing robust monitoring and relying on empirical evidence to reconcile conservation objectives with those of infrastructure development in the ape ranges of Africa and Asia—and in wildlife habitat elsewhere.

Acknowledgments

Principal authors: David Kaimowitz[1] and Helga Rainer[2]

Endnotes

1 Ford Foundation (www.fordfoundation.org)
2 Arcus Foundation (www.arcusfoundation.org)

SECTION 1

Photo: By 2050, an additional 25 *million kilometers* of paved roads are expected to crisscross the earth—enough to encircle the planet more than 600 times. © Waldo Swiegers/Bloomberg via Getty Images

CHAPTER 1

Towards More Sustainable Infrastructure: Challenges and Opportunities in Ape Range States of Africa and Asia

Introduction

We are living in one of the most dramatic eras of infrastructure expansion in human history. By 2050, an additional *25 million kilometers* of paved roads are expected to crisscross the earth—enough to encircle the planet more than 600 times. In addition to the growth in road networks, work on other infrastructure projects—such as railroads, hydroelectric dams, power lines, gas lines and industrial mines—is expected to increase sharply over the next few decades (Laurance and Balmford, 2013; Laurance and Peres, 2006).

Roads and other infrastructure have strong and intimate links with economic growth, frontier expansion, globalization, land colonization, agriculture and economic and social integration (Hettige, 2006; Weinhold and Reis, 2008; Weng et al., 2013). Unfortunately, such projects can also have severe impacts on many ecosystems and species (Adeney, Christensen and Pimm, 2009; Blake et al., 2007; Fearnside and Graça, 2006; Forman and Alexander, 1998; Laurance, Goosem and Laurance, 2009; Laurance et al., 2001; see Chapter 2). Roads that penetrate into wilderness areas, for example, often have profound and proliferating environmental effects—such as promoting habitat loss and fragmentation, poaching, illegal mining and wildfires (Adeney et al., 2009; Laurance et al., 2001, 2009; see Chapter 3). Even relatively narrow (10–100-m wide) clearings associated with forest roads can hinder or completely halt the movements of some ecologically specialized fauna, such as forest-interior or strictly arboreal species that require a continuous canopy (Laurance, Stouffer and Laurance, 2004; Laurance et al., 2009).

The remarkable pace of infrastructure expansion in developing nations—and its very real potential to provoke profound environmental harm—underscores an urgent need for better planning and management of new infrastructure projects to allow for the mitigatation of their adverse effects (Laurance and Balmford, 2013). This chapter identifies key issues revolving around the proliferation of large-scale infrastructure, focusing in particular on their potential effects on critical ape habitats in equatorial Africa and Asia.

Key Findings

- The contemporary pace of infrastructure expansion is unprecedented. A majority of the projects are planned or underway in biodiversity-rich developing nations, including all ape range states in the African and Asian tropics.

- Roads and other infrastructure often open up remote areas to a range of human pressures, such as deforestation, poaching, illegal mining and land speculation.

- Rising demands for natural resources and energy, as well as the rapid growth of multinational transportation networks, are providing a key impetus for building new infrastructure.

- The explosive pace of infrastructure development is partly the result of ambitious schemes to promote economic growth via increased access to land and natural resources, and partly an indirect symptom of more fundamental drivers, such as rising population growth, increased per capita consumption, economic disparity and the heavy national-level focus on extractive industries.

- Via its ambitious international policies, China is having a dramatic impact on infrastructure expansion in developing nations. This expansion is designed to gain access to natural resources.

- Environmental assessment and mitigation efforts for many infrastructure projects are inadequate, often seriously so.

- Alarmingly, major multilateral lenders are loosening some environmental and social safeguards. In target nations, large influxes of foreign capital for infrastructure projects and extractive industries often provoke a variety of negative economic and social consequences, unless managed carefully.

- Innovative solutions, such as an increased emphasis on "green" energy sources and natural capital, could lessen the negative impacts of some infrastructure.

- In view of the rapid pace of infrastructure expansion, two urgent priorities emerge: the need for (1) strategic regional planning, and (2) efforts to prevent infrastructure from expanding into remaining wilderness and protected areas.

Infrastructure: A Game Changer

Global Infrastructure

The contemporary scale of global infrastructure expansion is unprecedented. From 2010 to 2050, the total length of paved roads worldwide is expected to increase by more than 60% (Dulac, 2013). In Asia, scores of hydroelectric dams and associated energy and transportation projects are planned for the Mekong River and its tributaries (Grumbine, Dore and Xu, 2012). Meanwhile, several mega-dams are planned for Africa's Congo Basin (Laurance *et al.*, 2015a). In fact, Africa is currently experiencing unprecedented foreign investment for mineral exploitation, with China alone investing more than US$100 billion annually (Edwards *et al.*, 2014). Such investments are a key economic impetus for 35 planned or ongoing "development corridors" that would exceed 53,000 km in length and crisscross sub-Saharan Africa, opening up vast areas for economic exploitation (Laurance *et al.*, 2015b; Weng *et al.*, 2013; see Figure 1.1).

Environmental Impacts

The rapid proliferation of infrastructure is having substantial and often irreversible impacts on many ecosystems and species (Adeney *et al.*, 2009; Blake *et al.*, 2007; Clements *et al.*, 2014; Fearnside and Graça, 2006; Laurance *et al.*, 2001, 2009). In the Brazilian Amazon, the construction of new roads, hydroelectric dams, power lines and gas lines is projected to cause major increases in the rates of forest loss, fragmentation and degradation (Laurance *et al.*, 2001). In the Congo Basin, more than 50,000 km of logging and other roads have been built since 2000, greatly increasing access to forests for poachers and hunters armed with modern rifles and cable snares (Kleinschroth *et al.*, 2015; Laporte *et al.*, 2007).

The threats to wildlife from humans entering into their habitats are indisputable. From 2002 to 2011 alone, nearly two-thirds of Africa's forest elephants were slaughtered (Maisels *et al.*, 2013). Ape populations are particularly vulnerable to hunting because they are highly desirable as wild meat in some areas, are diurnal and conspicuous, have delayed maturation and slow rates of reproduction, and have restricted geographic distributions (Chapman, Lawes and Eeley, 2006; Cowlishaw and Dunbar, 2000; Robinson, Redford and Bennett, 1999; Struhsaker, 1999; see Chapter 2).

Photo: A researcher examines the skull of a western lowland gorilla found in Nouabalé-Ndoki National Park, Republic of Congo, November 2016. The cause of the ape's death is unknown, although poachers are increasingly detected inside the park near upgraded roads that skirt its boundaries.
© William Laurance

FIGURE 1.1

Status of Major Development Corridors in Sub-Saharan Africa, 2015

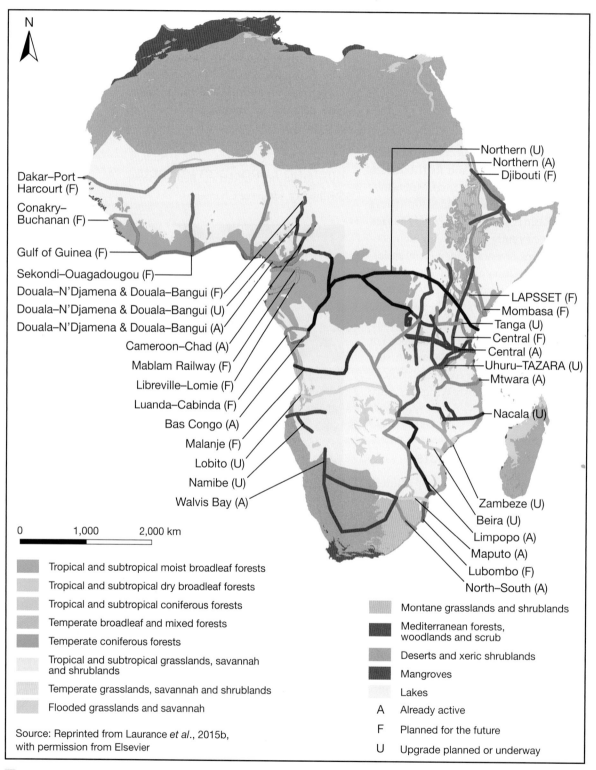

Infrastructure projects related to natural resource exploitation, such as mining, fossil fuel and hydroelectric projects, have direct environmental impacts and also provide a key economic impetus for road building (Edwards *et al.*, 2014; Laurance *et al.*, 2009; WWF, 2006; see Box 1.1). Consequently, such projects and roads cannot be planned or studied independently of one another. In the Amazon–Andes region, for instance, proposals currently envision more than 330 hydroelectric dams (with a total capacity of more than 1 megawatt); these projects would require extensive road networks for both the dams and associated power line construction (Fearnside, 2016b; Laurance *et al.*, 2015a). In the southeastern Amazon, new dams planned for the Tapajós River alone are projected to increase deforestation by nearly 10,000 km² (1 million hectares (ha)), predominantly by increasing access to remote forests for colonists and land speculators (Barreto *et al.*, 2014). Scores of new dams planned for Southeast Asia might have comparably serious impacts on great ape and gibbon habitats (Grumbine *et al.*, 2012).

Inferring Long-Term Impacts of Infrastructure

In the wet and humid tropical forests that serve as ape habitat, rivers are a conspicuous feature. Used as natural "highways" for

---

BOX 1.1

Infrastructure for Extractive Industries

Escalating Demand

Starting in 2003, sharply rising prices for oil, gas and minerals—spurred in particular by growing demand from China and other developing Asian nations—made it economically feasible to exploit ever more remote regions of the world. Such conditions can create a powerful economic impetus for building new roads, railroads and waterways, especially to ferry low-value, high-volume commodities such as iron ore, copper and coal over long distances to ports, refineries and smelters. Conflicts with nature conservation can easily arise because many natural resources are located in remote regions with high conservation value—including, in some cases, critical habitats for apes (Nellemann and Newton, 2002).

Since 2014, declines in commodity prices have slowed the expansion of new mining ventures, but this is probably just a temporary respite.[1] As demand and prices are likely to rise again in the future, the current economic slowdown may be seen as a "window of opportunity" in which to implement direly needed environmental and social safeguards wherever possible (Hobbs and Kumah, 2015).

Development Corridors

The construction of large-scale infrastructure, such as roads, railroads, power lines and gas lines, is increasingly being planned and concentrated along so-called "development corridors" (Hobbs and Butkovic, 2016). Political support for such corridors revolves around their potential to catalyze economic growth and trade, unlock private sector and development finance, encourage regional integration, improve logistical efficiency and increase frontier security (AgDevCo, 2013; Weng *et al.*, 2013). Development corridors can also be the legacy of investment in extractives long after closure of the initial resource extraction project.

In Africa, the 35 planned and initiated development corridors are sure to be transformational (Laurance *et al.*, 2015b; WWF, 2015b). In East Africa, for example, the Lamu Port, South Sudan, Ethiopia Transport (LAPSSET) corridor is to comprise port facilities, airports, cities, tourist resorts, highways, railways, pipelines and fossil fuel, hydropower and water-reticulation schemes. In 2013, projected costs for this venture were estimated at more than US$29 billion (Warigi, 2015).

In Asia, the massive "Belt and Road" initiative, launched in 2013, is a prominent feature of China's current Five-Year Plan (2016–20). This scheme aims to reinvent ancient silk trade routes between China and Europe and to expand Beijing's political, economic and cultural influence. It also extends to Africa, via a "21st-Century Maritime Silk Road." Fueled by massive investments from both China (US$40 billion) and the Asian Infrastructure Investment Bank (AIIB), this landmark venture will involve more than 70 nations. To date, the AIIB has been authorized to disburse US$100 billion to promote new global infrastructure (Honjiang, 2016).

Similarly, the Initiative for the Integration of Regional Infrastructure of South America is etching new highways and other transportation and energy infrastructure across South America (Killeen, 2007; Laurance *et al.*, 2001). Many of the initiatives' projects are penetrating into remote regions of the Amazon, Andes and beyond, where they are likely to provoke sharp increases in rates of forest loss, fragmentation, hunting and illegal gold mining. In the Brazilian Amazon, for instance, 95% of all deforestation occurs within 5.5 km of a legal or illegal road (Barber *et al.*, 2014).

millennia, rivers facilitate human movement, settlement, trade and hunting. They also form long-term biogeographic barriers for apes and other species, promoting genetic isolation and the evolution of distinctive new species or subspecies (Gascon *et al.*, 2000; Harcourt and Wood, 2012).

Rivers can thus be considered ecological analogs to roads—but ones that have existed for many millennia. Rivers might provide long-term insight into road impacts, just as land-bridge islands have been used to provide long-term perspectives on rates of population extinction in fragmented habitats, since

Photo: Illegal dwellings along a river in the interior of the Leuser Ecosystem in northern Sumatra, Indonesia—critical habitat for the Sumatran orangutan (*Pongo abelii*) and two gibbon species, 2016.
© Suprayudi

they were once linked to mainland areas—during past ice ages, when sea levels were lower—but have been isolated for millennia since (MacArthur and Wilson, 1967; Wilcox, 1978). Although rivers differ from roads in several respects, they might yield insight that is otherwise very difficult to infer (see Box 1.2).

BOX 1.2

Can Rivers Teach Us About Infrastructure?

As human activities penetrate ever deeper into ape habitats, maintaining ecological connectivity within intact forest blocks—particularly across linear infrastructures such as roads, railways, pipelines and power lines—is vital to prevent the fragmentation of larger wildlife populations into many smaller, isolated ones. Rivers have been used as human transportation corridors for millennia and can also halt or hinder animal movements; in that sense, they may share certain characteristics with roads.

Given the explosive rate of infrastructure expansion, linear infrastructure will increasingly allow for human access to remote areas, facilitating hunting and wildlife trafficking, and hindering animal movements (Blake *et al.*, 2008; Laurance *et al.*, 2004, 2008, 2009; Van Der Hoeven, De Boer and Prins, 2010; Vanthomme *et al.*, 2013, 2015). Navigable rivers play comparable roles as natural arteries for human movements. In the rainforests of Central Africa, for example, many human settlements are located along navigable rivers or their estuaries, including major cities such as Bangui, Brazzaville, Douala, Libreville, Kinshasa and Kisangani. In addition to being corridors, however, rivers can also hinder human movement, as crossing them requires bridges, rafts or boats.

In biogeographic terms, larger rivers have more profound impacts on the distribution of wildlife than do smaller rivers. This "river-width effect" was first noted in the 19th century in Amazonian monkeys and has since been studied in detail (Ayres and Clutton-Brock, 1992; Wallace, 1849). Ape distributions have been strongly influenced by river barriers. While the Oubangui River marks the eastern limit of the western gorilla (*Gorilla gorilla*), other rivers divide up genetically distinct subpopulations of this species (Anthony *et al.*, 2007; Fünfstück *et al.*, 2014; Mitchell *et al.*, 2015; Williamson and Butynski, 2013b). Similarly, the Congo River has separated bonobos (*Pan paniscus*) from other African ape populations for around two million years (Prüfer *et al.*, 2012; Reinartz, Ingmanson and Vervaecke, 2013).

In terms of their effects on wildlife, rivers and roads appear functionally similar in many respects. Wildlife responses to rivers are species-specific; while gorillas are unwilling to cross deep rivers, elephants will readily swim across them. Regardless of such distinctions, however, bonobos, chimpanzees, elephants and a number of other wildlife species all show consistent trends in terms of declines in population density near roads and rivers used by poachers (Blake *et al.*, 2007; Hickey *et al.*, 2013; Laurance *et al.*, 2008; Maisels *et al.*, 2013; Stokes *et al.*, 2010; WCS, 2015c). In a positive sense, the barrier effects of roads and rivers can slow the spread of infectious diseases such as Ebola in apes (Cameron *et al.*, 2016; Walsh, Biek and Real, 2005). Such barriers may be linked to the inability of apes or a disease-reservoir species to traverse rivers or roads efficiently (Cameron *et al.*, 2016).

Rivers can offer important analogs for roads, particularly as avenues that are readily used by poachers. For non-swimming species, rivers are likely to be stronger barriers than roads of comparable width, whereas the two may be roughly similar for swimming species. Wildlife managers could potentially learn much from studying river systems and how they have affected the distributions of apes and other fauna over large time periods.

Photo: Increasingly, China is linking infrastructure investments with policies that promote overseas trade, economic and political influence, and the acquisition of large stocks of minerals, fossil fuels, timber and other natural resources. Kaleta, Guinea © Waldo Swiegers/Bloomberg via Getty Images

Drivers of Infrastructure Expansion

Rapid Economic Growth in Asia

Contemporary investment in infrastructure is unprecedented in terms of its scale and pace. Since just after 2000, rapid economic growth in Asia—and especially in China (see Box 1.3)—has been a major driver of new infrastructure projects both within and outside the continent. In recent decades, China's gross domestic product has grown at an average rate of 10% annually, from just over US$200 billion in 1980 to US$8.6 trillion in 2013 (*The Guardian*, n.d.).

China is now the world's second-largest economy, having contributed one-quarter of all global economic growth over the period 2011–15 (NBS of China, n.d.). Increasingly, China is linking infrastructure investments from its corporations and multilateral lenders with policies that promote overseas trade, economic and political influence, and the acquisition of large stocks of minerals, fossil fuels, timber and other natural resources.

Multilateral Financial Institutions

China is far from the only driver of infrastructure expansion around the globe. During their 2014 global summit, the heads of state of the G20 nations—comprising the world's largest economies—committed to invest US$60–US$70 trillion in new infrastructure

---

BOX 1.3

China's Growth and Global Infrastructure

Economic Expansion

China's remarkable economic growth, together with its ambitious development and international outreach policies, has had major impacts on global infrastructure expansion. The nation's growth rate began to accelerate in 1978 with the government's landmark "reform and opening up" policy, which planted the seeds of private enterprise. Growth was further promoted in the 1980s and 1990s by rapid internal infrastructure expansion, and in the following decade by international expansion under the country's "going global" policy. The latter was prompted in part by China's huge trade surpluses and accumulation of foreign reserves, which it decided to use to invest abroad and to obtain overseas assets (GEI, 2013).

China's push to expand and improve its internal infrastructure began when the government, realizing that weak infrastructure was hindering its socioeconomic development, began investing heavily in its energy, telecommunications and transportation sectors. The slogan "building roads is the first step to becoming rich" became popular across China's villages and cities. The length of the country's roads nearly doubled from 1987 (0.89 million km) to 2000 (1.68 million km), giving China the second-highest national road mileage in the world (Liu, 2003; NBS of China, n.d.). Chinese hydropower, bridge, rail and telecommunications industries underwent similarly rapid expansion and upgrades (Liu, 2003).

China's "going global" strategy subsequently liberalized investment policies and provided financial incentives to encourage Chinese overseas investments and contracts. As a result, China's direct international investment multiplied rapidly, from US$2.7 billion in 2002 to US$118 billion in 2015 (MoC, 2016b). During this period, the country became the second-largest foreign investor worldwide, after the United States (MoC, 2014, 2016a).

The national government of Xi Jinpeng is continuing to promote the Chinese model of infrastructure development as the first step to development internationally. Starting in 2013, Xi announced three major initiatives: (1) domestic supply-side reform, (2) an acceleration of strategic adjustment of China's economic structure, and (3) the "Belt and Road" initiative, named after the Chinese term for "one belt, one road." The government also established two major financial institutions to support these initiatives, the Silk Road Fund and the Asian Infrastructure Investment Bank (Knowledge@Wharton, 2017).

On the strength of such ambitious efforts, the Chinese role in developing international infrastructure has expanded rapidly. In 2014, for example, Chinese "build–operate–transfer projects"—in which the private sector *builds* an infrastructure project, *operates* it and eventually *transfers its* ownership to the host government—generated 70% of Cambodia's hydropower electricity (GEI, 2016). In 2015, Chinese companies signed US$210 billion in new foreign-project contracts; transportation, electrical engineering and telecommunications are the top three sectors, accounting for 60% of the contracted value for the year (MoC, 2016c).

Addressing Social and Environmental Concerns

Many Chinese firms invest in Southeast Asia and Africa, regions rich in biodiversity but weak in environmental governance.

These investments have generated widespread environmental and social concerns (Edwards et al., 2014; Grumbine et al., 2012; Laurance et al., 2015b). A case in point is the Myitsone Dam, a US$3.6 billion project in Myanmar that was halted because local communities believed the project would destroy natural landscapes and their livelihoods (Chan, 2016). In response to this fiasco, the Chinese government developed guidelines on environmental and social responsibility, including:

- *A Guide for Chinese Enterprises on Sustainable Silviculture Overseas* (2007). This manual was developed by the Chinese Ministry of Commerce and the State Forestry Administration (MoC, 2007).
- *Green Credit Guidelines* (2012). Published by the China Banking Regulatory Commission, this document stipulates that operational practices of financial institutions must be consistent with international good practice standards, including environmental protection, land, and health and safety laws and regulations. Financial institutions are also required to establish green credit strategies and policies, abide by local laws requiring disclosure of significant environmental and social impact risks, and accept market and stakeholder supervision (GEI, 2015).
- *Guidelines for Environmental Protection in Foreign Investment and Cooperation* (2013). Published by the Ministries of Commerce and of Environmental Protection, these guidelines require companies that invest overseas to comply with the relevant local laws and regulations. The guidance relates specifically to environmental impact assessments, pollutant discharge standards, emergency management and other accepted environmental obligations. Companies are also encouraged to implement practices such as "clean production, circular economy and green procurement" (GEI, 2015, p. 18).
- *Measures for Overseas Investment Management* (2014). Published by the Ministry of Commerce, this guidance stipulates that foreign-funded enterprises must abide by local laws, respect local customs, and perform social responsibility and effect measures for environmental and labor protection and corporate-culture development (GEI, 2015).

Challenges and Limitations

While these guidelines demonstrate the Chinese government's commitment to promoting sustainable foreign investment, the policies remain weak at the implementation level, with poor policy promotion and a lack of compliance by Chinese industries (GEI, 2015). Environmental organizations and researchers have begun to address these problems by conducting policy field studies and training Chinese companies and local communities to strengthen their capacity for effective policy action.

Another challenge is the inoperability of some of China's current policies. Policy effectiveness depends on the framework and implementation of environmental safeguarding policies in host countries, as well as information disclosure, transparency and public participation. To achieve these goals, Chinese and host-country governments, civil society organizations, Chinese financiers and local communities must work together more effectively (GEI, 2015).

Photo: The presence of apes, such as Sumatran orangutans, should trigger extra environmental safeguards for multinational lenders. © Perry van Duijnhoven, 2013

by the year 2030 (Alexander, 2014). This would not only be the single largest financial transaction in human history, but would more than double the current value of global infrastructure (Laurance *et al.*, 2015a).

Large infrastructure investments are often disbursed via multilateral lenders. These lenders are playing a major role in infrastructure projects in ape range states in Africa and the Asia–Pacific region (ICA, 2014; Ray, 2015).

Meanwhile, the landscape of infrastructure investment is changing. Large infrastructure investments were traditionally disbursed via multilateral lenders such as the African, Asian and Inter-American Development Banks, the European Investment Bank and the World Bank Group. While these lenders continue to play a major role in infrastructure projects, including in ape range states in Africa and the Asia–Pacific region, their strongholds are being challenged (ICA, 2014; Ray, 2015). The Asian Infrastructure Investment Bank (AIIB), which opened for business in 2016, the Chinese Import–Export Bank and the expanding Brazilian Development Bank are all poised to become major international lenders.

As a consequence, the nature of infrastructure funding is undergoing worrying changes. After drawing criticism for years, the big traditional lenders had elaborated and implemented a number of environmental and social safeguards. Since the emerging banks generally consider environmental and social constraints a lower priority, however, they represent a formidable challenge to the traditional lenders (Laurance *et al.*, 2015a; Wade, 2011; Withanage *et al.*, 2006). In 2015 the World Bank decided to "streamline" its environmental and social safeguards in order to remain competitive with the emerging lenders, especially the AIIB (see Box 1.4).

BOX 1.4

Multilateral Lenders and Ape Conservation

Safeguards

To improve the sustainability outcomes of their investments, multilateral lenders such as the World Bank and regional development banks have developed environmental and social safeguards that identify standards and procedures for project screening. These frameworks determine the level of assessment and mitigation or management the lenders and their clients should apply.[2] High-risk projects or initiatives are subject to environmental and social impact assessments or strategic environmental assessments.

Critical Habitats

Environmental and social safeguards specify habitat value classifications that are determined through assessments of the critical nature of biodiversity and ecosystems. "Critical habitat"[3] is the most sensitive criterion and demands the most stringent avoidance or mitigation measures (EIB, 2013; IFC, 2012a, 2012c). Habitat that is important for apes would typically be classed as critical habitat because of the imperiled status of ape species and their keystone role in supporting ecosystem functioning. Ecological processes that support ape populations are also considered critical habitat by many multilateral lenders "where feasible."

In some project applications, the presence of apes represents a fatal flaw—one that can cause a bank to decline investment or withdraw. Alternatively, the bank could require demonstration that the project will produce no adverse effects (AfDB, 2013); no reduction in the ape population (ADB, 2012); a positive conservation outcome (EIB, 2013); or a net-gain outcome (IFC, 2012a, 2012c; World Bank, 2017). Such outcomes demand a comprehensive assessment of the direct, indirect and cumulative impacts of the project and rigorous application of impact reduction measures (see the discussion of the mitigation hierarchy in Chapter 4, p. 119). For ape landscapes, such assessments call for an appreciation of the complex

to reduce landscape-scale impacts and better inform project design or location (ADB, 2008).

Limitations and Risks

Multilateral lenders acknowledge major deficiencies in data and capacity. While a precautionary approach supported by long-term monitoring is considered ideal, it is not always applied. Time pressures coupled with sparse data can result in inadequate baselines, which, in turn, constrain management responses (see Box 1.6). Stakeholder engagement and expert input are highly valued by many lenders, but may be inadequate. The conservation community and species specialists have a vital role to play in ensuring that assessments of critical habitats and environmental impacts are based on sound ecological principles and the best available information. It is vital that civil society helps lenders to uphold their environmental and social impact requirements and holds them to account should they fail to do so.

The rapid rise of the Asian Infrastructure Investment Bank (AIIB) as a more streamlined and borrower-friendly lender, and the release of its environmental and social framework—soon followed by the announcement of simplified safeguards from the World Bank—have generated concerns about a potential "race to the bottom" in environmental and social protections (AIIB, 2016; CEE Bankwatch Network, 2015; Humphrey et al., 2015; World Bank, 2016c, 2017). Some consider the World Bank's anticipated transition from a rules-based compliance system towards one of "unprecedented flexibility that favors using a borrower's own laws and policies" in lieu of the Bank's traditional safeguards as especially worrying (BIC, 2016). However, others believe that the World Bank's new Environmental and Social Standard (ESS) 6[4], and the International Finance Corporation's (IFC) Performance Standard (PS) 6, continue to represent best practice in the protection of biodiversity and habitats (TBC, n.d.).

There is deep concern about the impacts of the weakening of some lenders' environmental safeguards. In ape range states, this relaxed approach is of particular concern when combined with borrowers' limited commitment and capacity, as well as weak national regulatory frameworks, and enforcement, which tend to be unable to prevent or mitigate the complex social and environmental impacts of high-risk infrastructure projects (BIC, 2016). Under these circumstances, approving a mega-infrastructure project seems analogous to pressing a car's accelerator to the floor while unbuckling the seatbelt.

socio-ecology of affected apes, their role in maintaining ecosystem integrity and the potential of habitats to support viable populations in the future; in practice, however, these factors are often addressed poorly (see Box 1.6 and the Apes Overview, p. xii).

The timing and duration of a lender's engagement, as well as its commitment and capacity to uphold environmental and social safeguards, can strongly affect its influence on a project. In some cases, lenders take more of a lead by requiring cumulative and strategic environmental assessments

The World Bank's shift in approach reflects deep internal conflicts within all multilateral lenders, as they seek to reconcile their primary business as profit-driven financial institutions with the fundamentals of long-term sustainability. Lenders have the ability to improve their environmental and social frameworks by developing detailed guidance notes, proper tools and well-resourced support for the critical implementation process (BIC, 2016). A great deal is going to depend on how their environmental and social frameworks are implemented in the future.

Photo: Forest clearing for a Chinese-operated road construction camp in the northern Republic of Congo. © William Laurance

Emerging Threats to Ape Habitats

Impacts on African Ape Habitats

In broad terms, there are many reasons to be concerned about Africa's environment. Indeed, nearly one-third of Africa's protected areas could face degradation if the entire suite of proposed and ongoing development corridors proceeds (Sloan, Bertzky and Laurance, 2016). The specific threats posed to apes by infrastructure projects and the further developments they catalyze are less certain, but one modeling study suggests that fewer than one-tenth of African ape habitats would remain free from infrastructure impacts by 2030 (Nellemann and Newton, 2002).

As currently being constructed, the Lamu Port, South Sudan, Ethiopia Transport (LAPSSET) project in East Africa will not directly threaten ape range states, but it will affect Kenya's imperiled Tana River Primate Reserve, which harbors the highly endangered Tana River red colobus (*Procolobus rufomitratus*) and Tana River crested mangabey (*Cercocebus galeritus*) (Kabukuru, 2016; see Figure 1.1). But LAPSSET is nothing if not ambitious. The long-term plan is to provide a "great equatorial land bridge" that would traverse Africa, linking Kenya on the east coast with Cameroon on the west coast (LAPSSET, 2017). If realized, this great bridge would slice through the Congo Basin and have substantial impacts on a number of ape range states.

Several other development corridors aim to access the mineral-rich region of the eastern Democratic Republic of Congo, Rwanda and Uganda, as well as the goldfields of western Tanzania (see Figure 1.1). The results could increase human pressures on bonobos (*Pan paniscus*), eastern chimpanzees (*Pan troglodytes schweinfurthii*), Grauer's gorillas

(*Gorilla beringei graueri*) and mountain gorillas (*Gorilla beringei beringei*).

In Africa, corridors that penetrate into equatorial forests are the greatest concern for ape conservation (see Box 1.5). Chief among these is the Central African Iron Ore Corridor. The backbone of this project is the M'Balam railway, which will stretch for more than 500 km and traverse the equatorial rainforests of Cameroon, Gabon and the Republic of Congo. The corridor will also include a new highway linking Brazzaville in the Republic of Congo with Yaoundé in Cameroon. Key components of this project include the Chollet Hydropower Dam near the Dja Biosphere Reserve, the Mekin Dam inside the Dja Reserve and the Memve'ele Dam near the Campo Ma'an Reserve, all of which are located in southern Cameroon (Halleson, 2016).

The greater Congo Basin harbors the second-largest expanse of rainforest on earth. It includes the vast (146,000 km², or 14.6 million ha) Tri-National Dja–Odzala–Minkébé (TRIDOM) landscape, which is jointly managed under an agreement by Cameroon, Gabon and the Republic of Congo. TRIDOM contains a complex of

---

BOX 1.5

Africa's Integrated Resource Corridors

Development corridors are not new concepts in Africa. In fact, corridors such as the Maputo Development Corridor, the Walvis Bay Development Corridor and TRIDOM have been promoted to varying degrees in different regions for many years. The potential for such multinational infrastructure projects to support sustainable development has been widely discussed and debated (ASI, 2015).

Many organizations tout development corridors as transformative vehicles through which to ensure equitable distribution of benefits from sector-specific operations. Corridor proponents include: the New Partnership for Africa's Development; the mining policy framework developed for the United Nations by the Intergovernmental Forum on Mining, Minerals, Metals and Sustainable Development; and, more recently, the Africa Mining Vision (AU, 2009; IGF, 2013; NEPAD, n.d.). Development corridors are also on the agendas of regional entities such as the African Development Bank, the Asian Development Bank, and the East African and Southern African Development Communities (AfDB, OECD and UNDP, 2015).

Opportunities

Ideally, development corridors should be able to leverage large extractive industry investment in infrastructure, goods and services to bring about sustainable, inclusive economic development and diversification for a specific geographic area. Potential opportunities include:

- Increasing prospects for governments and the private sector to work together.
- Developing supply chains that encircle the extractive industry, such as a major mine at the heart of a corridor. The direct procurement of local supplies can have a multiplier effect on the local economy, increasing local demand and employment. The use of local resources can also stimulate industrialization and domestic value-adding, which can promote transformational economic growth.
- Bringing together stakeholders from the government, private and community sectors, aligning their incentives and improving coordination. Such synergies can provide opportunities to embed robust environmental standards and practices into the project.
- Benefiting landlocked countries and their neighbors, enabling both to gain from resources in the landlocked country and their export via coastal states.
- Spreading benefits away from the anchor project to provide opportunities, such as shared-cost infrastructure, for isolated towns and villages. Such infrastructure is vital for remote communities, which can find themselves cut off from economic opportunities and political processes or dominated by local patronage systems that inhibit development.
- Allowing affected communities to have a seat at the negotiating table. Large-scale extractives and infrastructure projects can generate high expectations around jobs and the role of companies to provide services that should be the mandate of the state. Inclusion can improve understanding and help to manage the expectations of local communities.
- Allowing planners to concentrate linear infrastructure (such as roads, railroads, pipelines and power lines) along shared corridors, thereby reducing the overall impact by leaving other areas intact (ASI, 2015).

Challenges

While the potential benefits of Africa's development corridors may be considerable, they are far from fully realized. Key challenges include:

seven protected areas and harbors critically endangered western lowland gorillas (*Gorilla gorilla gorilla*) and chimpanzees (*Pan troglodytes*) (Ngano, 2010). The corridor is adding to stresses that the estimated 40,000 gorillas and chimpanzees in the region already face from industrial logging, agro-industrial concessions and poaching. A combination of threats—including ongoing forest loss and fragmentation, the increasing isolation of protected areas, expanding human settlements and now large-scale infrastructure projects—suggest that the TRIDOM region may be facing imminent demise as a contiguous forest landscape (Halleson, 2016).

In the imperiled forests of West Africa, a global biodiversity hotspot, a major concern is the massive Simandou iron ore project. Rights to explore the Simandou deposit were first granted in 1997 and following a number of issues and disputes, mining rights have been held by the Aluminum Corporation of China Limited (Chinalco), Beny Steinmetz Group Resources (BSGR), Rio Tinto Corporation and Vale. The largest integrated mining and infrastructure project in Africa, it is situated at the southern end of a biologically

---

- Poor planning and inadequate community engagement often plague corridor projects. Most active and planned corridors are currently unlikely to achieve sustainable development outcomes, particularly in relation to local economic benefits and environmental and social impacts.

- Government agencies are often ill equipped, ill informed and unable to apply an integrated approach to planning. They fail to consider the cumulative impacts of numerous ad hoc developments or synergies that could be created among them. They do not or evidently cannot take advantage of resource efficiencies that would result from economies of scale.

- Cross-national corridors are bedeviled by a lack of coordination when key agencies work in relative isolation. Limited dialog among government agencies, donors, civil society, the private sector and communities results in conflicts and inefficiencies.

- Corridors are often planned without adequate assessments of their potential social and environmental impacts, such as:
    - demographic shifts and the subsequent demand for additional services and infrastructure;
    - resilience considerations in relation to climate change;
    - protection of areas with high conservation value; and
    - effects on water supplies.

  This suite of factors can ultimately undermine the value of a corridor, particularly for the poor and vulnerable.

- Where they are carried out, assessments are usually restricted to site-specific environmental and social impact assessments of individual projects and therefore miss opportunities for key strategic decision-making through the integration of environmental and social considerations (ASI, 2015, p. 12).

A Success Story?

Despite such challenges, some corridors appear promising. The Maputo Development Corridor in southern Mozambique is often highlighted as a positive example (AfDB et al., 2015). Providing a 500 km-long link between Maputo and the landlocked provinces of Gauteng, Limpopo and Mpumalanga in South Africa, it will provide Swaziland with an alternative to the port of Durban, South Africa, for international trade. The corridor's anchor is the Mozal aluminum smelter, on the outskirts of Maputo (Byiers and Vanheukelom, 2014).

Arguably, the reported success of the Maputo Corridor is attributable in part to an alignment of national and cross-border interests. "From the perspective of the Mozambican government, the MDC was as an important signal to the outside world of stability and viability of carrying out major foreign investments" (Byiers and Vanheukelom, 2014, p. 18). Challenges remain, however. Operational inefficiency—including deficient rail infrastructure and capacity, high prices and unequal trade flows within the corridor (given that the volume of goods South Africa exports to Mozambique is 120 times greater than the volume it imports from its trade partner)—highlights the importance of effective planning and political will at all levels (Bowland and Otto, 2012).

As illustrated by the Maputo Development Corridor, five factors stand out as being most critical to the goals of development corridors to achieve sustainable economic progress and reduce poverty:

1. government support up to and including the highest level;
2. private sector involvement from the outset;
3. community engagement and capacity building throughout the project;
4. access to geospatial data; and
5. good governance.

Photo: Plans for large-scale highway expansion across Borneo could degrade some of the last virgin and unhunted forests on the island, such as these in eastern Sabah, Malaysia.
© William Laurance

critical region—the Simandou Mountains in southeastern Guinea. Transportation infrastructure needed to link the mine to the coast for shipping ore overseas would span about 700 km and would bisect and fragment the habitat of the western chimpanzee (*Pan troglodytes verus*). Although it is not yet at the production phase, the Simandou project demonstrates how large-scale infrastructure associated with industrial mines can have considerably greater environmental impacts than mines themselves.

Impacts on Asian Ape Habitats

Mapping the impacts of large-scale infrastructure on great ape and gibbon range states in Asia, as well as the array of ancillary developments such projects can catalyze, is a daunting challenge. If all of the proposed projects proceed, then the overall impacts are sure to be substantial.

China's scheme to construct an Asian "Belt and Road"—including a "21st-Century Maritime Silk Road" that is to traverse Asia, Europe and Africa—is certain to be world-changing (see Box 1.1). This spate of projects would have an impact on the habitats of orangutans, in parts of Borneo and Sumatra, and gibbons, whose ranges extend from the islands of Southeast Asia northward to Indochina, southern China and northeastern South Asia. Projects such as the planned high-speed railway linking southern China (Kunming) to Singapore would cut across Thailand and Peninsular Malaysia, affecting important ecosystems for gibbons, including parts of Malaysia's critical Central Forest Spine (Wu, 2016).

Ambitious plans for infrastructure expansion are also afoot in insular Southeast Asia. Indonesia's large-scale development is being structured around a "six-corridor" scheme that would traverse large swaths of Sumatra, Java, Indonesian Borneo (Kalimantan), Sulawesi, the island chain from Bali to West Timor and Indonesian Papua (Indonesia Investments, 2011). The forests of Malaysian Borneo will be further reduced and fragmented by the "Pan-Borneo Highway" plan, which is expanding highway networks across much of Sarawak and Sabah (Property Hunter, 2016).

Expanding infrastructure could affect Asian apes and other wildlife in a diversity of ways, such as by promoting extractive industries. Mining concessions already overlap with 15% of the current distribution of Borneo's orangutans (*Pongo pygmaeus*) and 9% of Sumatra's (*P. abelii*) (Lanjouw, 2014, p. 155; Meijaard and Wich, 2014, pp. 18–19). Case studies illustrating the impacts of infrastructure projects on Asian ape habitats are provided in chapters 3, 5 and 6.

Social and Political Concerns

Inequitable Social and Economic Benefits

Large-scale foreign investment is driving much of the ongoing expansion in infrastructure and extractive industries in developing nations (see Boxes 1.3–1.5). A common assumption is that these types of investment typically yield broad societal benefits for developing nations; in practice, such benefits rarely materialize, for five main reasons.

First, influxes of foreign capital, such as the major investments in infrastructure and extractive industries in African nations, typically elevate the value of the nation's currency relative to other currencies (Ebrahimzadeh, 2003). By increasing costs for foreign consumers, higher currency values reduce the competitiveness of agricultural and manufacturing exports, tourism, higher education and some other economic sectors. The economy then becomes less diversified and more reliant on a few extractive industries or large projects, and therefore more vulnerable to shocks from commodity price fluctuations or boom-and-bust cycles when key natural resources are depleted (Venables, 2016).

Second, the benefits of foreign capital are rarely distributed equitably. A few individuals, such as those in politically powerful positions, can benefit dramatically, whereas many others see little benefit (Edwards *et al.*, 2014; Venables, 2016). Even nations with strong governance, taxation and resource rent-capture mechanisms, such as Australia, have had much difficulty in distributing benefits from large foreign investments equitably. As a result, many people and sectors of the economy have struggled. Developing nations with weaker institutions and governance can be greatly challenged and even destabilized under such conditions (Venables, 2016). The catchphrases "blood diamonds" and "blood gold" vividly illustrate this concept.

Third, inflation typically increases in the developing nation because demand for goods and services rises. Wealthy elites are troubled little by such inflation but those struggling to meet their daily rent and food costs can suffer greatly. As a result, economic and social disparity can increase, rather than decline (Auty, 2002).

Fourth, corruption is a serious problem in many developing nations, including virtually all ape range states (Laurance, 2004). Even projects that are socially and environmentally ill advised may be approved by decision-makers who stand to reap large personal rewards from bribery or other illicit benefits. Decision-makers may also borrow from international lenders to advance projects for personal or political gain, knowing that future governments and taxpayers will have to bear the burden of servicing and repaying the loan. Documented examples of such corruption-driven environmental mismanagement are far too numerous to detail here (Collier, Kirchberger and Söderbom, 2015; Shearman, Bryan and Laurance, 2012; Smith *et al.*, 2003).

Finally, environmental damage resulting from large-scale developments is typically an economic externality borne by the entire population and domestic economy. Even in the most advanced nations, mechanisms to compensate the public for deforestation, water and air pollution, and mining damage are often far from adequate (Daily and Ellison, 2012). In turn, the absence of effective compensation measures creates perverse incentives in favor of polluting industries, as they do not bear the full costs of their activities (Myers, 1998).

Risks to Project Proponents and Investors

Risks from large-scale infrastructure and extractive projects are not confined only to the target nations. Multilateral lenders, corporations and investors are also exposed to considerable financial and reputational risks when projects go awry. For example, the reputation of Asia Pulp and Paper, an Indonesia-based corporation that caused massive forest loss in Borneo and Sumatra, became so toxic that it lost a considerable share of the market and suffered widespread international condemnation. Along with a number of major oil palm and wood pulp corporations operating in Southeast Asia, Asia Pulp and Paper has since made a "no deforestation" pledge to limit public criticism and avoid threatened boycotts (Arcus Foundation, 2015, p. 159; Laurance, 2014).

Large infrastructure and extractive projects also face other risks. These can arise from political instability, project cost overruns, labor disputes, liability for environmental disasters and an almost infinite variety of "unknown unknowns" that can bedevil major projects (Garcia *et al.*, 2016; Laurance, 2008). The failure of a large project can lead to "stranded assets," whereby major investments are lost or offset by unanticipated costs that outweigh the project's benefits. In Aceh, Indonesia, for example, deforestation associated with road expansion has increased downstream flooding that is estimated to cost landowners US$15 million per year (Cochard, 2017). Similarly, oil palm and wood pulp plantations on tropical peatlands are likely to incur long-term ecological restoration costs that could exceed the value of the plantations (Bonn *et al.*, 2016).

Advocates of major infrastructure projects often downplay the risks to investors

> " Advocates of major infrastructure projects often downplay the risks to investors and host nations while overstating their potential to yield large profits and societal benefits. "

and host nations while overstating their potential to yield large profits and societal benefits. The University of Oxford economist Bent Flyvberg describes how deceptions and an incessant "optimism bias" by proponents create a dynamic in which megaprojects continually proceed despite being "over budget, over time, over and over again" (Ansar *et al.*, 2014; Flyvberg, 2009).

A Dire Need for Better Infrastructure Planning

Optimizing Infrastructure Costs and Benefits

Not all infrastructure is inherently "bad" for the environment. In appropriate contexts, new infrastructure can yield sizeable social and economic benefits with only limited environmental costs. For instance, road improvements in already settled areas can facilitate increases in agricultural production and improve rural livelihoods, as they give farmers better access to urban markets, fertilizers and new agricultural technologies (Laurance and Balmford, 2013; Laurance *et al.*, 2014a; Weinhold and Reis, 2008). Such roads can also provide rural residents with better access to health care, schools and employment opportunities, while encouraging private investment (Laurance *et al.*, 2014a).

In developing regions, those areas with improved roads might actually function like "magnets," attracting settlers away from vulnerable forests and frontiers (Laurance and Balmford, 2013; Rudel *et al.*, 2009). In this way, improving transportation in suitable areas could help to concentrate and improve agricultural production, raising farm yields while potentially promoting land "sparing" for nature conservation (Hettige, 2006; Laurance and Balmford, 2013; Laurance *et al.*, 2014a; Phalan *et al.*, 2011; Weinhold and Reis, 2008).

However, efforts to plan roads strategically to optimize their benefits and limit their costs face practical challenges. First, environmental impact assessments (EIAs) often place the burden of proof on road opponents, who rarely have sufficient information on rare species, biological resources and ecosystem services to determine the actual environmental costs of roads (Gullett, 1998; Laurance, 2007; Wood, 2003). Second, many road assessments are limited in scope, focusing only on the direct effects of road building while ignoring their critical indirect effects, such as the promotion of deforestation, fires, poaching and land speculation (Laurance et al., 2014a, 2015a). Finally, until recently, there was no strategic system for zoning roads regionally, and thus road projects had to be assessed with little information on their broader context. As the pace of contemporary road expansion intensified, road planners and evaluators thus carried a growing burden (Laurance and Balmford, 2013).

For these reasons, a *strategic* scheme for prioritizing road building was recently devised (Laurance et al., 2014a). This approach has two components:

- an environmental values layer that estimates the natural importance of ecosystems, and
- a road benefits layer that estimates the potential for increased agricultural production, in part via new or improved roads.

The environmental values layer integrates data sets on species richness and endemism, threatened species, key habitats for wildlife, wilderness attributes, ecosystem representativeness and important ecosystem services.

The road benefits layer focuses on the role of new or improved roads for enhancing agricultural production—which is a crucial priority due to four main reasons:

- First, agriculture is by far the dominant form of human land use globally (Foley et al., 2005).
- Second, global food demand is expected to increase by 60%–100% from 2005 to 2050 (Alexandratos and Bruinsma, 2012; Tilman et al., 2001).
- Third, vast areas of land, especially in developing nations, have already been settled but support relatively unproductive agriculture (Mueller et al., 2012).
- Fourth, the amount of additional farmland needed to meet global food demand by 2050 is projected to reach up to 1 billion hectares—an area the size of Canada—unless production on under-yielding agricultural lands can be improved (Tilman et al., 2001).

In this context, strategic road improvements are a key prerequisite for achieving the needed increase in agricultural production (Laurance and Balmford, 2013; Laurance et al., 2014a; Weng et al., 2013). With concerted improvements in transportation, farming technologies and crop varieties, global food demand this century could be met with a far smaller amount of new farmland than if a "business-as-usual" approach were employed (Alexandratos and Bruinsma, 2012).

Combining the environmental values and road benefits layers allows areas to be grouped into three categories:

- areas where roads or road upgrades could have large benefits;
- areas where road building should be avoided; and
- "conflict areas," where the potential costs and benefits of roads are both sizeable.

An example of this analysis at the global scale demonstrates its potential for strategic road zoning, although planning of roads in

> Many road assessments focus only on the direct effects of road building while ignoring their critical indirect effects, such as the promotion of deforestation, fires, poaching and land speculation.

real-world contexts would be undertaken at a smaller scale, be it regional, national or local (Laurance et al., 2014a; see Figure 1.2).

Promoting Green Energy

Developing tropical nations such as those that sustain great apes and gibbons often have considerable potential to develop solar, wind and other smaller-scale energy sources. Sustainable energy sources could help to meet their growing energy demands, reducing the need for expensive, large-scale energy infrastructure such as hydropower or gas- or coal-fired electricity plants, which also require extensive road and power line networks. Decentralized solar and wind technologies could be particularly suitable for remote villages and settlements (McCarthy, 2017).

Thanks to its proximity to the equator, the tropical Asia–Pacific region has high solar-energy intensity, indicating a large potential for solar energy expansion. In 2010, the Asia Solar Energy Initiative of the Asian Development Bank announced plans to install 3,000 megawatts of solar capacity in the region, reflecting robust confidence and employment potential in this sector (ADB, 2011; McCarthy, 2017). By 2015, total wind power capacity had reached 175,000 megawatts in Asia, showcasing faster growth than in any other region except the Middle East (Global Wind Report, 2015). In addition, geothermal energy is being proposed or developed in a number of locales, although several proposed plants would be in remote regions, such as forested areas of Sumatra that are prime habitats for Sumatran orangutans (see Case Study 6.4). Since such installations require road networks for plant and power line construction, they are far less desirable than decentralized solar and wind energy in areas of high conservation significance.

Equatorial Africa also has strong potential for solar, wind, geothermal and biomass

FIGURE 1.2

A Global Map for Prioritizing Road Building

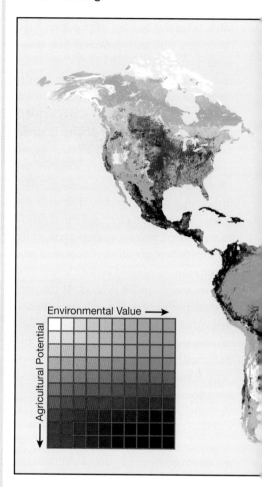

power (ESI Africa, 2016; IRENA, 2015). As Africa's energy demand is expected to double or even triple between 2015 and 2030, renewable energy advocates are urging African nations to "leapfrog" large-scale energy infrastructure in favor of solar, wind, geothermal and biomass energy sources (IRENA, 2015). At present, however, such technologies have limitations in terms of energy storage and meeting base-load demand, and it is likely that hydropower, coal-fired energy and other large-scale projects will also expand rapidly. Nonetheless, there is much potential for growth in solar, wind, biomass and other small-scale energy technologies, especially in

Notes: Green areas have high conservation values. In red areas, transportation improvements have a high potential to improve agriculture. Dark areas are "conflict zones," where environmental and agricultural values are both high.

Source: Laurance *et al*. (2014a, p. 231)

the rural areas of Central and West Africa, which harbor vital ape habitats (IRENA, 2015).

Priorities for Change

This final section highlights six urgent priorities for improving infrastructure finance, planning and environmental sustainability.

1. Avoiding new infrastructure construction in and near critical habitat. From a nature conservation perspective, infrastructure is going many places it should not. Infrastructure expansion is promoting large increases in the human footprint worldwide, intensifying human pressures on protected areas and driving rapid declines in the extent of remaining wilderness, especially in the tropics (Laurance *et al.*, 2012; Venter *et al.*, 2016; Watson *et al.*, 2016).

A key priority is "avoiding the first cut" into remaining wilderness areas by keeping them road-free wherever possible. This goal recognizes that deforestation is highly contagious spatially, in that forest loss tends to expand along new roads and then spread farther afield as the initial road spawns secondary and tertiary roads (Boakes *et al.*, 2010). Once the first road goes in, forest loss

Photo: Oil palms spread to the horizon in central Sumatra, Indonesia. © William Laurance

will typically increase exponentially unless robust safeguards are in place to halt it. Such safeguards require long-term expenditures for forest monitoring and protection.

The environmental impacts of new roads and other infrastructure are often amplified in developing nations where land use zoning and the rule of law are limited, especially in the remote frontier regions that are crucial for wildlife. In the Brazilian Amazon, for instance, there are nearly three kilometers of illegal roads for every kilometer of legal road (Barber *et al.*, 2014). Such roads can facilitate a range of illegal activities, including timber theft, poaching, illicit drug production and illegal gold mining, all of which can defraud governments of needed revenues while provoking serious environmental harm (Asner *et al.*, 2013; McSweeny *et al.*, 2014).

2. Addressing the drivers of unsustainable infrastructure expansion. Unsustainable infrastructure expansion reflects deeper challenges. We desire sustainability and environmental quality—yet the average per capita consumption of the human population, which could exceed 11 billion people this century, continues to rise (UN Population Division, 2015). Ultimately, life on earth is a zero-sum game: when humanity consumes land, water and other natural resources, the planet's health is typically degraded to a similar extent.

While infrastructure expansion is among the most important impacts of humankind on nature, it is a proximate rather than an ultimate driver—a symptom of a broader malady revolving around a rapidly growing human population and extractive economies, including in the developing nations that harbor apes. Failing to confront the broader drivers of unsustainable behavior is nonsensical and dangerous.

3. Requiring strategic environmental and social impact assessments. Too many impact assessments are rubber-stamping exercises.

All too frequently, environmental and social assessments for large infrastructure projects rely on inadequate data on ecosystems and biodiversity. They often fail to examine indirect, secondary or cumulative impacts of a project, and they do not assess the "bigger picture" because the project is evaluated in isolation from other human influences that affect the same ecosystem. Indeed, most major infrastructure corridors develop incrementally on a project-by-project basis, with little regional-scale planning (Laurance *et al.*, 2014a, 2015a). Many such assessments fail to anticipate potential cumulative and secondary impacts of projects; they may also be subordinated to the priorities of different government agencies with inconsistent or even opposing interests.

Experts in financial institutions that fund large projects argue that civil society and expert knowledge can play a vital role in the EIA process (see Boxes 1.3 and 1.4 and Case Study 5.1). Yet many EIAs are conducted too late in the project approval process to allow for fundamental changes or to lead to the cancellation of a project, even if they reflect sound expert knowledge. Furthermore, EIAs are often not made widely available to interested parties outside of the project area (Laurance *et al.*, 2015a). When combined with limited time frames for public comment, such measures increase the likelihood that a proposed project is effectively a *fait accompli*—with modest "tweaking" of the project and limited mitigation the only alternatives. The weakening of environmental and social safeguards by major multilateral lenders will only exacerbate this problem (see Box 1.4).

Some EIAs are essentially boilerplate documents that are written in dense bureaucratic language and lack key information. In a striking example, an EIA that was carried out for a large housing estate in Panama claimed that 12 bird species were present in the project area. Two experienced bird-

Photo: Today's infrastructure projects must not become tomorrow's environmental disasters. Nam Ou Cascade Hydropower Project, Lao PDR.
© In Pictures Ltd/Corbis via Getty Images

watchers surveyed the same area for two hours and documented 121 bird species, including several rare and threatened species (Laurance, 2007). EIAs for some major equatorial African and Amazonian highway projects have been similarly inadequate (Fearnside, 2006; Laurance, Mahmoud and Kleinschroth, 2017b; see Case Study 5.1). Not all EIAs are as weakly implemented as these, but only a minority are truly robust (Laurance, 2007; Laurance et al., 2015a).

One way to address the broader suite of impacts that are often missed in localized EIAs is to carry out strategic environmental assessments at an appropriate landscape scale (see Box 1.4). Box 1.6 provides a checklist of best practice in impact assessments to enable developers to minimize adverse impacts and to avert a *net* loss of biodiversity, given that infrastructure development in ape ranges, by its very nature, degrades landscapes and habitats. As illustrated above and throughout this publication, these best practice actions are seldom fully or even partially implemented; and sometimes, EIAs are rather used as tools to greenwash destructive projects. Effective implementation of EIA best practice can contribute to the conservation of biodiversity, including apes and ape habitat, while also ensuring that financing is effectively allocated to preventive action, rather than costly mitigation expenses.

4. Carrying out strategic land use planning for agriculture. Many observers call for an increase in the productivity of agriculture in developing nations in order to "spare" land for nature (Laurance et al., 2014a; Mueller et al., 2012; Phalan et al., 2011). Yet more productive agriculture is also more profitable, and highly profitable agriculture is likely to spread widely unless constrained in some manner. An apt example is the dramatic expansion of oil palm across the humid tropics, where the crop is promoting forest destruction both directly and

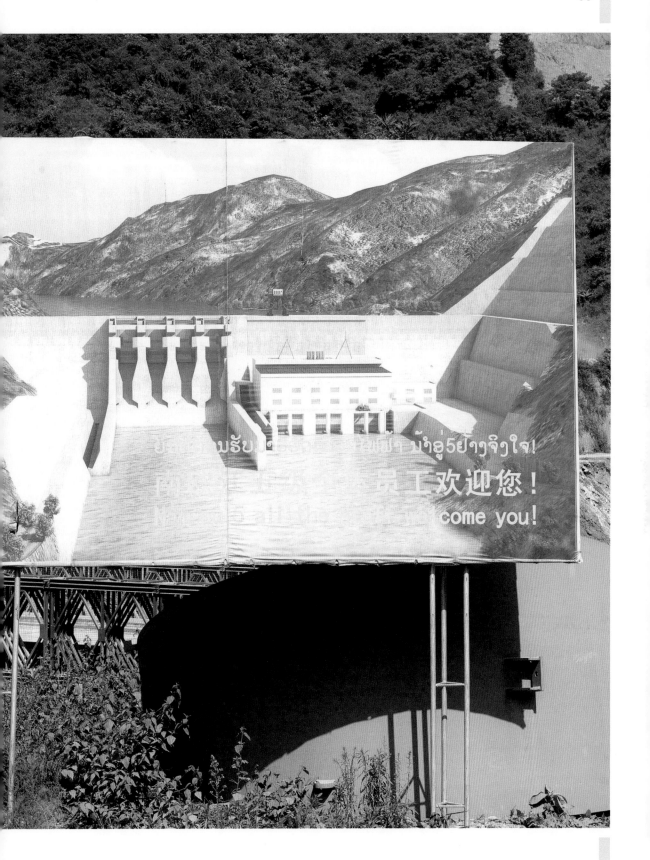

Chapter 1 Challenges and Opportunities

BOX 1.6
Best Practice in Impact Assessment: A Checklist for Developers

An infrastructure project may have significant adverse impacts on biodiversity and local communities throughout its lifetime—from its planning phase through to the construction and operation periods, and, if it ceases to operate, during its decommissioning. Impact assessments can serve to identify, evaluate and mitigate such negative effects. More often than not, carrying out such assessments is a statutory requirement or a condition for disbursements from financial lenders.

The following measures can assist developers as they seek to achieve the objective of causing no net loss of biodiversity:

- **Building and accessing expertise**. Although some developers have in-house expertise to undertake impact assessments, few, if any, have specialists to cover all relevant areas and most will be obliged to seek external support and advice, often through private-sector consultancies that specialize in ecological and related services. If a project is likely to have a significant impact on sensitive habitats and species, such as by causing the loss or fragmentation of areas that support ape populations, building early relationships and trust with experts is crucial. A developer organization that contracts external consultants needs dedicated internal support staff to provide a bridge to outside agencies and other departments. Such project managers can help to provide clear justifications for actions, as external stakeholders may not always understand or support the need for detailed studies or mitigation, often on financial or timescale grounds. Project managers also ensure continuity when contracted work is staggered or consultants are only engaged for limited periods.

- **Planning for impact assessments**. How much time is required to carry out an impact assessment is often dependent on the capacity of the developer organization, applicable legal requirements regarding the provision of independent, impartial advice, and technical needs associated with each stage of a project, from the planning through to the implementation phase. It is important to consider project-related impacts as early as possible to ensure favorable outcomes for biodiversity. Prompt action will reduce a developer's risk of incurring costly delays and constraints at later stages, such as construction stoppage if legally protected habitats or species are identified once a project is under way. Assessing the situation early also allows biodiversity specialists to implement the mitigation hierarchy to its full potential, by ensuring that the project design entails measures to avoid and minimize adverse impacts. These types of measures can prevent the need for expensive alternative mitigations, including changes to ongoing construction, such as the rerouting of roads, and complex, often less effective offset schemes.

- **Assessing baselines**. Initial baseline scoping studies are useful tools for identifying which key species may be affected by an infrastructure project. By covering both the immediate development zone as well as the surrounding area, they can reveal which parts of a landscape may be harmed during the various project stages. Baselines are always required with respect to ape populations; additional assessments are typically needed to fill any knowledge gaps regarding ape numbers, habitat use or distribution. Consultation with local conservation NGOs, academic institutions and state agencies can help to establish what type of data is available. Field surveys are usually necessary to assess the state of species in project areas if they have not been studied in detail.

- **Collecting data**. In the planning stages of impact assessments, the importance of gathering relevant baseline data that is robust and measurable, and allowing sufficient time for this collection and analysis, is essential. To capture seasonal variations in species behavior, surveyors require at least one calendar year to collect and analyze relevant data. If less time is allocated to the task or if inappropriate survey methods are employed, it will not be possible to determine the project's impact on target species with any degree of accuracy, with the result that all future stages of the impact assessment will be compromised. The chance to apply appropriate mitigation measures may therefore be missed, or measures may be applied on a speculative basis, which could lead to unpredicted detrimental impacts or costly—and potentially unnecessary—actions.

- **Collaborating**. Undertaking field surveys can provide a good opportunity for ecologists and sustainability and corporate social responsibility teams from private sector developers to collaborate with environmental consultancies, academic institutions, NGOs and state organizations (such as national park authorities). Collectively, these stakeholders can more readily establish, at an early stage, the likely impacts of a project, as well as appropriate mitigation measures. Private-sector environmental consultants usually have extensive experience drawing up ecological content for impact assessments and meeting financial lender requirements; academic institutions and NGOs can provide science-led research expertise; and state agencies generally contribute invaluable local knowledge and insight into what is achievable within regional and national legal frameworks. At the same time, the data collected can contribute to the ongoing study of habitats, biodiversity and the socioecology of particular species.

- **Mitigating effects**. Once baseline studies are complete and the impacts of an infrastructure project have been considered, developers and other stakeholders can begin to mitigate any subsequent effects—and to monitor the effectiveness of the mitigation measures. Ideally, such measures meet two requirements: they are tailored to address specific impacts, and their outcomes are measurable. If permanent habitat loss is a likely consequence

Photo: There is a pressing need to limit the expansion of new infrastructure into remaining wilderness, protected areas and biodiversity hotspots. Western lowland gorillas, Dzanga, Central African Republic.
© David Greer, WWF

of an infrastructure project, habitat amelioration within the remaining range of affected ape communities may be able to preserve populations at pre-construction levels. In some cases, however, predicted or observed residual effects require offsite mitigation measures within the wider landscape. In these cases, measures can be applied following established protocols, such as the Business and Biodiversity Offsets Programme (BBOP, 2009–2012). For information on the mitigation hierarchy, a set of guidelines established in the IFC's Performance Standard 6, see Chapter 4, page 119.

- **Applying additional measures.** In addition to direct mitigation, supplementary measures may be employed, such as awareness raising and community engagement—to reduce hunting pressure, for example. These strategies can be effective in contributing to the overall objective of achieving no net loss; however, it is not appropriate to use to use them as primary forms of mitigation or as replacements for key mitigation measures, such as habitat reinstatement and creation.

- **Producing biodiversity action plans (BAPs).** The process of implementing the above-mentioned steps and measures is commonly described in a BAP, a document that many lenders require. Under the IFC's PS6, for instance, a BAP is required if critical habitat may be affected by infrastructure development (IFC, 2012c). The standard covers habitat that supports endangered and critically endangered species, meaning that a BAP is required if a project threatens any great ape habitat and most gibbon habitats. Designed to help achieve the aims and objectives of a mitigation and monitoring program, a BAP serves as a single working reference of a given project, pulling together all related studies and reports. The document sets out clear guidance on how each action is to be carried out, by whom and in what time frame. Unlike other associated documents, such as the environmental statement, the BAP is a "living" report that is updated as actions are completed, and modified as new data come to light or if mitigation measures are not as effective as anticipated.

In practice, the environmental considerations and measures presented here are often overlooked or sidestepped, with potentially detrimental repurcussions for developers' finances as well as affected fauna and flora. By making a conscious effort to integrate these considerations into their planning, however, infrastructure developers can play an active role in seeking to avoid both going over budget and a net loss of biodiversity. It is as important for developers to factor social considerations into their activities to prevent harm to—and, ideally, to ensure benefits for—indigenous populations and local communities that may be affected by an infrastructure project (see Chapter 2). In so doing, they can seek to harness local support for a project and any related conservation actions and initiatives.

Chapter 1 Challenges and Opportunities

indirectly—by displacing other land uses, such as rice production, which then leads to further forest loss.

Only when coupled with strategic land use planning and backed by the rule of law will productive and profitable agriculture actually promote the "sparing" of land for nature. The most effective way to constrain the expansion of agriculture into environmentally sensitive areas is arguably by halting the spread of roads and other infrastructure into those areas.

5. Encouraging China to require compliance with its established development guidelines. Of all nations, China is currently the most ambitious and aggressive in terms of advancing large-scale infrastructure projects, often in concert with schemes to exploit and access natural resources in developing nations. Such projects are funded by Chinese public–private partnerships, corporations and lenders. Compared to projects that are underwritten by industrialized nations in the Organisation for Economic Co-operation and Development, Chinese-funded initiatives are significantly more likely to create "pollution havens" (areas where pollution or environmental damage are concentrated) in developing nations (Dean, Lovely and Wang, 2009). In this way, China exports its environmental degradation and pollution to poorer countries.

Having acknowledged these problems, China has devised a series of "green" guidelines and operating principles for Chinese ventures operating internationally (see Box 1.3). Nevertheless, the Chinese government has failed to accept any responsibility for the lack of enforcement of its stated principles. Instead, the recurring problems are being blamed on intransigence by its corporations, a lack of general transparency and weaknesses in the governing frameworks of the host countries (see Box 1.3). Beijing could take a firmer hand in promoting environmental sustainability, notably by requiring that Chinese firms and ventures operating overseas increase compliance with China's development guidelines.

6. Taking advantage of the current window of opportunity. For those striving to promote better infrastructure, the current global economic slowdown offers a limited window of opportunity (Hobbs and Kumah, 2015). The stakes are high: today's infrastructure projects must not become tomorrow's environmental disasters. Advocates of sustainable infrastructure will find it effective to address a broad constituency of environmental, economic, civil society and political stakeholders—emphasizing, for instance, the enormous value of biodiversity, ecosystem services, natural capital and climate regulation, as well as the primacy of sustainability for human welfare (Meijaard *et al.*, 2013). They can also build on the infrastructure sector's aim to avoid financial and reputational risks.

Moreover, researchers and land use planners must respond to a growing demand from businesses and private investors for guidance in determining the best locations for new infrastructure (Green *et al.*, 2015; Laurance *et al.*, 2015b; Natural Capital Coalition, 2016; see Box 4.5). There is a pressing need, in particular, to limit the rapid expansion of new infrastructure into remaining wilderness, protected areas and biodiversity hotspots. As noted above, "avoid the first cut" into wild places should become a clarion call for biodiversity and sustainability advocates.

It is difficult to overstate the urgency of the task at hand. We have rapidly shrinking opportunities to help steer infrastructure expansion in directions that meet human needs while promoting greater sustainability for critical ape habitats. It is time for decisive action—for the protection of great apes and nature in general.

> " We have rapidly shrinking opportunities to help steer infrastructure expansion in directions that meet human needs while promoting greater sustainability for critical ape habitats. "

Acknowledgments

Principal author: William F. Laurance[5]

Contributors: Adam Smith International, Iain Bray, Neil David Burgess, Fauna and Flora International (FFI), Global Environmental Institute (GEI), Matthew Hatchwell, Jon Hobbs, Pippa Howard, Nicky Jenner, Lin Ji, Fiona Maisels, Emily McKenzie, Tom Mills, Mott MacDonald, United Nations Environment Programme World Conservation Monitoring Centre (UNEP-WCMC), Wildlife Conservation Society (WCS), World Wide Fund for Nature (WWF), WWF International and Rong Zhu

Text Box 1.1: Jon Hobbs
Text Box 1.2: Matthew Hatchwell and Fiona Maisels
Text Box 1.3: Rong Zhu and Lin Ji
Text Box 1.4: Pippa Howard and Nicky Jenner
Text Box 1.5: Tom Mills
Text Box 1.6: Iain Bray

Author acknowledgments: Mason Campbell and Mohammed Alamgir provided useful comments on the manuscript.

Reviewers: Stanley D. Brunn, Miriam Goosem, Matthew Hatchwell and Wijnand de Wit

Endnotes

1. As predicted, since this content was provided in 2017, commodity prices have generally recovered, resulting in increasing demand for infrastructure development (J. Hobbs, personal communication, 2018).

2. This generalized description is derived from a review of multilateral lender safeguard documents and author interviews with lender environmental staff, conducted in late 2016.

3. "Critical habitats are areas with high biodiversity value, including (i) habitat of significant importance to Critically Endangered and/or Endangered species; (ii) habitat of significant importance to endemic and/or restricted-range species; (iii) habitat supporting globally significant concentrations of migratory species and/or congregatory species; (iv) highly threatened and/or unique ecosystems; and/or (v) areas associated with key evolutionary processes" (IFC, 2012c, p. 4).

4. IFC Performance Standard 6 has been reviewed and will be relaunched in 2018 (I. Bray, personal communication, 2018).

5. James Cook University – https://www.jcu.edu.au/

Photo: Infrastructure development fuels deforestation, affecting these ecosystems and the diversity of species that dwell within them, including human communities. © Jabruson (www.jabruson.photoshelter.com)

CHAPTER 2

Impacts of Infrastructure on Apes, Indigenous Peoples and Other Local Communities

Introduction

Infrastructure is a common and expanding feature of the anthropocene, with human-altered landscapes across every part of the world (Laurance, Goosem and Laurance, 2009). Roads, bridges and railways, as well as hydroelectric dams, mining and processing plants, and electrification projects cover much of the earth's surface and infringe on even the most remote landscapes. Collectively, roads cover a distance of more than 83 round trips between the earth and the moon (van der Ree, Smith and Grilo, 2015, p. 3).

Fifteen years ago, an assessment of infrastructure using the GLOBIO tool—which models human impacts on biodiversity—revealed that up to 70% of tropical forest habitat in Africa and in Asia had been affected by infrastructure development and the associated human exploitation of the forests around it. Projections based on the GLOBIO tool and more recent assessments indicate that less than 10% of the habitat in African great ape ranges and probably closer to 1% of the habitat in orangutan ranges in Asia will be left untouched by 2030, as a result of infrastructure development and the associated habitat disturbance (Junker *et al.*, 2012; Nellemann and Newton, 2002). For apes and the majority of other animal and plant species, infrastructure development represents a major conservation threat.

Infrastructure also affects human populations living in or near tropical forest habitats, and not only in the intended positive manner. Infrastructure development fuels deforestation, affecting the complex dynamic of these ever-changing ecosystems and the diversity of species that dwell within them. Human communities are among those that depend on the forests and their resources. Forest peoples are part of the dynamic ecosystems of forests, living in them, adapted to them and shaping them—in stark constrast to the forces that are destroying forests. Strategies to mitigate damage to these ecosystems are most effective when they take into consideration both the potential social impacts of proposed infrastructure projects and forest peoples' capacity to help mitigate such damage. This approach serves not only to ensure the well-being of forest-dwelling and other local communities, but also to garner their support for proposed conservation measures, which are likely to fail without local backing.[1]

This chapter explores the ecological and behavioral impacts of infrastructure on apes in the forest, as well as social impacts of infrastructure development on forest peoples and communities dependent on forest resources. The first section considers the ecological impacts on apes and other species of fauna and flora across a range of infrastructure types; the second section explores the social impacts of infrastructure via examples from Cameroon. The chapter then offers some lessons learned and steps that can be taken to minimize the deleterious effects of infrastructure development.

With respect to the ecological impact of infrastructure, this chapter's key findings are:

- Infrastructure development is a major conservation threat for apes and for the majority of other animal and plant species.
- The major negative direct impacts of infrastructure development are habitat loss, road kills, and noise pollution and disturbance; indirect impacts include increased human access to previously remote areas, poaching, and the introduction of disease and invasive species. Some of these impacts are immediate, such as road kills, while others can have pernicious long-term and far-reaching consequences for wildlife populations.
- The anticipation of project implementation alone can exacerbate habitat loss and disturbance to wildlife in a locality, particularly through the development of roads to prospect areas and small-scale encroachment by local people, even if the project is not taken to completion.
- Industry-specific certification bodies already exist, such as the Forest Stewardship Council (FSC) and the Roundtable on Sustainable Palm Oil (RSPO), which require standards to be met for certification to take place, including those relating to associated infrastructure. There is thus scope to develop and implement standards for other large scale infrastructure development in relation to both the ecological and social impacts of such developments; and to monitor, maintain

> "Strategies to mitigate damage to forest ecosystems are most effective when they consider both the potential social impacts of proposed infrastructure projects and forest peoples' capacity to help mitigate such damage."

and promote the uptake of these standards through the development of additional certification requirements.

- In designing appropriate responses to infrastructure development, it is important to factor in direct and indirect impacts at both the local and landscape levels for all projects, be they expansive, such as roads, railways and transmission lines, or characterized by relatively small footprints.

With respect to the social impact of infrastructure, the chapter's key findings are:

- Infrastructure development in the traditional lands of indigenous peoples has a negative impact on their livelihoods, cultural practices and norms.
- Indigenous peoples traditionally manage and utilize natural resources from forests sustainably, but they can also become part of the cycle of destruction that is exacerbated by infrastructure development.
- Conservation efforts designed to mitigate and offset the impact of infrastructure development on biodiversity can further exacerbate negative impacts on indigenous peoples.

Ecological Impacts of Infrastructure on Apes

Impacts of different types of infrastructure can vary in intensity on several scales. Impacts can be direct or indirect; they can occur during the construction, utilization, production or decommissioning phases; they can be felt in the short or long term. The main direct impacts of infrastructure include habitat loss and fragmentation, behavioral disturbance and the creation of artificial barriers, which in turn disrupt movement patterns and affect habitat use, increase mortality rates, and hamper gene flow. Indirect impacts and threats, such as hunting or the risk of disease transmission, are often linked to the presence of people (see Table 2.1).

This section outlines the impacts of different types of infrastructure on apes. It covers transportation-related projects, such as roads, railways and ports; broader development infrastructure, such as dams, power lines, processing plants and human settlements (including temporary or permanent housing developments for workers); and other types of infrastructure, such as tourist lodges.[2]

Compared to industrial-scale agriculture and logging, which typically result in the conversion of thousands of hectares of forest or more, infrastructure such as roads or tourist lodges may be expected to have a relatively small impact on apes. Indeed, such linear and localized projects may pose a less significant immediate threat of habitat loss. Nevertheless, as forests are opened up for infrastructure development, people increasingly disturb previously intact ranges by hunting, capturing live animals, degrading and destroying the forest, producing noise, transmitting disease and polluting. In connection with infrastructure development, such human disturbance can have significant negative impacts on apes, affecting the landscape's structural connectivity (habitat type and composition) as well as its functional connectivity, which involves both the structure of the landscape and the ways in which animals interact with their environment (Kindlmann and Burel, 2008).

Various mitigating measures can be developed and implemented to prevent and respond to the negative impacts of infrastructure-related human disturbance in and around wildlife habitat. Designed to integrate conservation into infrastructure development, such measures can usefully be adapted to the characteristics of each

> When designing appropriate responses to infrastructure development, it is important to factor in direct and indirect impacts at both the local and landscape levels for all projects.

Photo: A common impact of all infrastructure development is the destruction or degradation of habitat wherever construction is taking place. Highway construction between Port-Gentil and Omboué, Gabon. © Julie Sherman

individual plan, be it managed exclusively by private companies, by a government or by a combination of stakeholders.[3]

Impacts of Infrastructure Development

Each type of infrastructure project can be expected to have a number of direct or indirect impacts on the local landscape. These impacts may differ in terms of their duration and extent, as well as in relation to the timescales required for the construction phase and the longevity of the infrastructure (see Table 2.1).

Three phases can be distinguished for infrastructure projects: their construction, their use and, in some cases, their decommissioning (as for dams, logging concessions and mines). These phases require separate consideration when it comes to assessing their impact on wildlife in general, and apes in particular.

Construction Phase

The overall impacts of infrastructure construction on apes are similar across development projects, but the scale of any impact depends primarily on the type of infrastructure being built. For example, setting up infrastructure that affects small areas of land, such as a power line or a pipeline, and that is mostly left alone after being established in the middle of a rainforest is likely to cause less disturbance than erecting a major structure, such as a dam, power plant or highway, in a similar area.

A common impact of the construction of any type of infrastructure is the human presence and the influx of workers to the construction site. The arrival of people increases indirect threats to wildlife, such as hunting, physical and noise pollution, risks of disease transmission and an influx of invasive species (Burgess *et al.*, 2007).

The noise of heavy machinery during construction is also likely to affect and possibly displace animals (see Box 2.1). In Uganda, for instance, mountain gorillas in Bwindi National Park reportedly shifted their range when the park service was building new office premises. In general, apes move away and shift their range in response to human disturbance.[4]

BOX 2.1

Impacts of Roads on Chimpanzees

Chimpanzees show flexible behavior that enables them to exploit anthropogenic landscapes; they may use human-made paths and cross large roads to access different areas of their home range (Cibot et al., 2015; Hockings, Anderson and Matsuzawa, 2006; Hockings and Sousa, 2013). At the same time, roads and paths can provide hunters with access to previously unreachable areas, where they can set traps and hunt chimpanzees and other animals for local consumption or for commercial trade (Blake et al., 2007; Poulsen et al., 2009; Robinson et al., 1999). When hunters use indiscriminate devices, such as snares or traps, they are also likely to capture non-target species.

Roads are generally risky areas for wildlife due to the increased human presence and the danger of collisions with vehicles (Jaeger et al., 2005). Research has shed some light on the risks associated with road development and utilization and how chimpanzees in particular manage road crossings.[5]

There is growing evidence that road crossing can cause injury or the death of individual chimpanzees (Krief et al., 2008; McLennan and Asiimwe, 2016). The danger is high although chimpanzees appear to assess the risks by looking left and right before and during road crossings, and despite the fact that they check on and wait for group members, especially more vulnerable ones (Cibot et al., 2015). Adult males are particularly at risk because they often take up the more dangerous positions at the front or rear of a group progression when crossing (Hockings, 2011). As shown in Figure 2.1, Bossou chimpanzees in Guinea spend more time waiting before crossing a large road than a small road. The large road had been widened prior to the period under review; between early 2005 and the end of that year, the chimpanzees reduced their waiting time at the road, most likely because they became habituated to its greater width.

Interestingly, Sebitoli chimpanzees in Uganda appear to maintain the old pathways they used prior to road construction, regardless of risk (Cibot et al., 2015). This finding highlights the need for road developers to identify chimpanzee paths and trails and to integrate such knowledge into road design and development plans.

FIGURE 2.1

Chimpanzee Waiting Time before Road Crossing, Bossou, Guinea, 2005

Key: ▪ Small road (3 m width) ▪ Large road (12 m width)

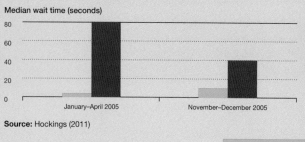

Source: Hockings (2011)

Another common impact of all infrastructure development is the destruction or degradation of habitat wherever construction is taking place. More often than not, these impacts result in habitat and population fragmentation and isolation, with possible long-term consequences (see Table 2.1).

Utilization or Production Phase

Apes generally prefer areas with lower levels of human disturbance.[6] The overall response of apes—and other mammals—to established infrastructure is avoidance of the built area, which results in reduced densities of the animals (Benitez-Lopez, Alkemade and Verwej, 2010). Several types of infrastructure can kill apes directly, such as through electrocution or collisions with vehicles on roads (McLennan and Asiimwe, 2016; see Box 2.1). Asian apes and other arboreal mammals are regularly electrocuted in the Kinabatangan region of Malaysian Borneo when they use power lines to move across the landscape. Apes and other animals sometimes recover from electric shock, but many die of electrocution; they may also drown near dams or in drains (see Annex I).

Causes of ape mortality that are indirectly linked to infrastructure typically involve hunting, most of which takes place fewer than 10 km from any roads (Laurance et al., 2009). Mortality rates are also affected by the transmission of emerging diseases due to close proximity to people or domestic animals, as well as reduced food availability due to habitat loss (see Table 2.1).

A primary concern associated with all types of infrastructure is the increased mortality rate among apes whose habitat has been destroyed and who are thus pushed away from their original home range or concentrated in small patches of forest. Mining and dam development have particularly significant effects on apes, especially if human settlements, be they temporary or permanent, are established alongside the infrastructure.

Habitat fragmentation results primarily from linear infrastructure with a pronounced edge effect, such as roads, railways, power lines, drains and canals. In the long term, fragmented and isolated ape populations become more prone to extinction due to genetic isolation, stochastic events (such as fire, flooding or the outbreak of disease) and reduced resilience to the impact of climate change (Gillespie and Chapman, 2008).

The impact of roads also depends on their size and the frequency with which they are used. Dirt and gravel roads that receive relatively little use may not be much of a barrier for apes, even for some arboreal species, such as orangutans. As their level of use increases, such roads can become a greater barrier and may ultimately prevent passage by apes.

The longevity of infrastructure is also of importance. For example, a dirt road that is not well maintained or is closed after the cessation of activities (such as a logging road) may be recolonized by the forest over time, unless it continues to be used. In contrast, the decommissioning of a dam is not likely to result in the full reversion of the dam site and associated flooded forest to their previous natural, functional state, even if the local river system recovers in part (see Annex VII).

Decommissioning Infrastructure

The decommissioning process involves the rehabilitation of areas after infrastructure is no longer in use. At that stage, mitigating the impacts of infrastructure can include the following steps:

- *A clean-up of the exploitation site*: removing machinery and equipment; destroying buildings and other infrastructure that is no longer used and cannot be recycled; removing chemicals and other toxic waste.

- *Habitat rehabilitation*: replanting trees; reforesting degraded areas; filling in a landfill or a mine. In areas that are known to harbor important gorilla populations, it should be noted that gorillas consume large quantities of terrestrial herbaceous vegetation (THV)—particularly *Marantaceae* and *Zingiberaceae*—and are likely to be attracted to these resources in areas with an open canopy. Rehabilitation in such habitats requires careful planning, as focusing on tree planting alone can be detrimental to the establishment of THV (Morgan and Sanz, 2007).

- *Habitat protection*: closing or monitoring of paths, roads and bridges to decrease the opportunities for access, illegal hunting and other encroachment. The costs of effective control of access can be prohibitive (Elkan *et al.*, 2006). If successfully applied, however, habitat protection can help to promote the natural regeneration of vegetation, which can complement habitat rehabilitation efforts.

General Impacts on Apes

Apes vary in their socioecological traits, such that infrastructure affects each species differently (see the Socioecology section, p. xvii). Nevertheless, all apes share social and behavioral characteristics that limit their ability to adapt to infrastructure development. Most notably, these include:

- No ape species can swim: a dam, canal or a wide drain without any natural bridge (such as overarching tree branches) represents an impassable barrier to any individual or group.

- All ape species have low reproductive rates; due to their long period of maturation, individuals do not begin to reproduce until they are at least ten years

Photo: No ape species can swim: a dam, canal or a wide drain without any natural bridge (such as overarching tree branches) represents an impassable barrier to any individual or group. Grand Poubara Dam, Gabon. © Steve Jordan/AFP/GettyImages

old. They typically have one offspring every 4–9 years depending on the species. As a result, apes characteristically experience very slow population growth rates. Increased mortality rates can thus have severely detrimental effects on population size. Populations may take a very long time to recover to their original size, if they ever do.

- Apes are susceptible to many diseases that affect humans. As apes come into increasingly close contact with people, the risk of disease transmission is heightened, along with the risk of infection and subsequent death among apes (Carne *et al.*, 2014; Köndgen *et al.*, 2008; Muehlenbein and Ancrenaz, 2009).

- All apes are highly adaptable: many of them will use new food resources planted by people. Crop owners may identify such apes as "pests" (Humle, 2015; Seiler and Robbins, 2016); in this scenario, it is not only difficult to harness these people's support for conservation initiatives, but the likelihood of retaliation and killing of apes also increases (Ancrenaz, Dabek and O'Neil, 2007; Humle, 2015).

- All ape species depend on forests for all or a significant part of their behavioral ecology. Even chimpanzees and some bonobo populations that occur in savannah-dominated landscapes need forest for nesting sites and food. Gibbons are exclusively arboreal and cannot cross large distances on the ground. While chimpanzees and gorillas typically travel on the ground, and orangutans may also do so to a certain extent (Ancrenaz *et al.*, 2014), any barriers in their habitat may restrict their ranging patterns, depending on the size and the level of disturbance.

- Except for orangutans, most apes live in social groups and are either territorial or have overlapping home ranges, so that multiple groups occur in the same area. Therefore, as the construction of infrastructure leads to a loss of habitat and apes are compressed into smaller areas, it becomes difficult or impossible for them to establish new territories or shift their range. Greater density leads to increased intergroup aggression and possible death owing to attacks between individuals (especially among chimpanzees), increased social stress, as well as a reduction in food resources (Mitani, Watts and Amsler, 2010; Watts *et al.*, 2006).

Table 2.1 presents information about impacts of different types of infrastructure on apes. The list is not exhaustive; several impacts are not included due to a lack of data (for example, dust and airborne pollutants, and invasive species). The table also identifies to what extent apes tend to be able to adapt to such impacts.

The Consequences of Infrastructure Development

Increased Access, Immigration and Human Settlement

Infrastructure development nearly always leads to increased access, human influx and human settlement in areas that previously were not easily reached. Of all the types of infrastructure, new roads are the ones that result in the largest increase in access (Clements et al., 2014). Access roads are almost always needed for other types of infrastructure, which in turn open up areas to human settlement.

Research shows that the distance to roads, villages and cities is a strong predictor of the presence of apes; indeed, ape densities decrease as human presence increases, largely because of hunting pressure.[7] One study that compares the abundance of large mammals at varying distances from roads inside an oil concession (a non-hunted area that received extensive protection) and

TABLE 2.1

Impacts of Infrastructure on Apes and the Likelihood of Ape Adaptability

Impact of infrastructure	Impact type	Duration of impact	Roads and railways	Ports and dams	Power cables	Human settlements
Increased access, immigration and human settlement (villages; tour lodges; and buildings of any sort)	Indirect	Short to long term				
	Indirect	Long term			*	
Hunting (commercial and personal)	Direct	Short to long term				
Habitat loss, degradation and fragmentation	Direct	Short to long term				
Creation of artificial barriers (which disturb movement patterns and affect habitat use, increasing mortality and/or hampering gene flow)	Direct	Short to long term				
Behavioral change	Direct	Short to long term				
Disease (or pathogen) transmission	Direct	Short to long term				
Mortality and injury associated with vehicle and equipment collisions	Direct	Short term				
Disturbance associated with noise and vibration (including blasting), project lighting, and presence of workers	Direct	Short to long term				
Hydrological impacts, including flooding and fragmentation	Direct	Long term				

Note: * Chances of ape adaptability are good if local settlements do not have access to electricity, limited or moderate if they do.

Likelihood of ape adaptability

■ Limited ■ Moderate ■ Good ■ Unknown

in the hunted territory beyond the concession demonstrates that *hunting*—rather than the roads themselves—leads to a decline in gorillas (Laurence *et al.*, 2006). Similarly, a recent study reveals that the distance from roads is the best predictor of bonobo nest occurrence; distance is an indicator of hunting of apes, rather than of the displacement of bonobos, as hunting intensity is greatest closer to roads (Hickey *et al.*, 2013; Laurance *et al.*, 2009).

As people settle into an area, land use practices change and subsistence agriculture generally expands, as does the extent of land under cultivation. These shifts can cause apes to forage on cultivars with greater frequency and can lead to an increase in encounters between apes and people, which may result in increased conflict and aggression (Bryson-Morrison *et al.*, 2017; Campbell-Smith *et al.*, 2011b; McLennan and Hill, 2012; McLennan and Hockings, 2016). Crop foraging may be driven either by necessity, due to the loss of natural foods, or by opportunities linked to agricultural expansion of palatable crops.[8] It leads to a loss of income for local community members, stoking negative reactions and behavior towards apes (Ancrenaz *et al.*, 2007; Naughton-Treves, 1997).

Close cohabitation may be particularly problematic if the people in question have no previous experience of living near apes. They may be afraid of the apes—due to their lack of experience or based on urban myths about apes—and may therefore be more antagonistic towards apes. Even among people who have traditionally lived near apes, increased encounters with them may erode traditional or religious taboos and beliefs that favor local ape conservation or tolerance of apes (Humle and Hill, 2016).

In addition, employment insecurity associated with a significant influx of people into an area can exacerbate people's engagement in alternative revenue-generating enterprises that can have significant negative impacts on apes. Such activities include artisanal mining, small-scale logging and subsistence or commercial hunting, which can be facilitated by increased access to ape habitat.

Habitat Loss, Degradation and Fragmentation

All types of infrastructure development lead to some level of habitat loss, degradation and fragmentation. While infrastructure itself can be relatively "small" compared to large tracts of forest, some types, especially roads, can transect extensive areas, and all types will have impacts at both the local and the landscape level. In some cases, roads can limit apes' access to food and nesting trees (Bortolamiol *et al.*, 2016). Such infrastructure may lead apes to shift their range or territory, thereby increasing intra- or interspecific competition for food and nesting, which causes social disruption and stress, as well as a heightened risk of intergroup aggression. This kind of aggression can significantly raise the mortality rate, especially among chimpanzees (Mitani *et al.*, 2010; Watts *et al.*, 2006).

For the more arboreal Asian ape species, disruption to canopy connectivity can compel apes to travel on the ground and thus heighten their exposure to pathogenic agents, including viruses, bacteria and parasites, which may be transmitted from humans and domestic animals, such as via attacks by dogs (Das *et al.*, 2009). In addition to limiting the spatial distribution of apes, the loss of canopy connectivity also increases the risk of predation and food shortage, particularly among gibbons (Channa and Gray, 2009; Cheyne *et al.*, 2013, 2016; Hamard, Cheyne and Nijman, 2010; Turvey *et al.*, 2015).

While more terrestrial apes are less constrained by the presence of railways and

Photo: Chimpanzees show flexible behaviour that enables them to exploit anthropogenic landscapes, which puts them at risk of injury or death when crossing roads. © Matt McLennan

roads, the latter may nevertheless act as barriers, depending on the intensity of traffic, road or rail width, travel speed and visibility (see Box 2.1). In Uganda's Bwindi Impenetrable National Park, three groups of gorillas tend to cross a 15-km-long gravel road a few times per year. There are plans to pave the road, which is expected to increase vehicular traffic and, in turn, heighten the risk of vehicle collisions. If the gorillas stop crossing the road once it is paved, their habitat will be fragmented, as about 10% of the 330 km² (33,000-ha) park would effectively be eliminated as suitable habitat. Plans to pave a road through the already fragmented habitat of Cross River gorillas in Nigeria would have similar detrimental effects (see Case Study 5.1).

In estimating or assessing the impact of infrastructure on great apes and other wildlife, it is crucial to consider the anticipated or sustained disruption of habitat connectivity and relationships among patches across the affected landscape. A study that compared the amount of structural and functional connectivity for the critically endangered Cross River gorilla showed that the decline in functional connectivity was double that in structural connectivity over a 23-year period (Imong *et al.*, 2014).

Disease and Pathogen Transmission

Apes are susceptible to many human diseases. Disease epidemics or parasitic infections can negatively affect reproduction and kill apes, thereby changing demographic patterns (Gilardi *et al.*, 2015). An increased risk of disease and pathogen transmission is likely in areas where there is garbage, such as tourist lodges, villages and roadsides. Artisanal mines, camps used by construction workers, and satellite communities typically have unsanitary conditions that pose a large health risk to apes (Plumptre *et al.*, 2016b). Habituated chimpanzees, gorillas and orangutans may range very close to tourist lodges and may even come into very close contact with humans in unregulated settings, such as those not monitored by park staff, which can lead to an increased risk of transmission

of respiratory and other diseases (Gilardi *et al.*, 2015; Macfie and Williamson 2010; Matsuzawa, Humle and Sugiyama, 2011). Such contact puts both the apes and people, including tourists and staff, at risk of injury and pathogen infection in case of attack.

Injury and Death Due to Vehicle and Equipment Collisions

Terrestrial apes are at risk of injury or death when crossing roads. There are reports of chimpanzees being injured or killed in vehicle collisions (McLennan and Asiimwe, 2016; see Box 2.1). Encounters with infrastructure can also be life-threatening for arboreal apes, and poorly insulated and bare power lines pose a risk of electrocution for all species (see Annex I). In Kinabatangan, Malaysia, and in Assam, India, several cases of gibbons and orangutans being electrocuted have been recorded, some of them fatal. In 2011 and 2014, two adult orangutans were electrocuted when they used a power line to access a fruiting durian tree in the village of Sukau, Kinabatangan. In both cases, the orangutan fell to the ground and was unconscious for several minutes before recovering from the electrical shock and fleeing to a nearby tree. The hands of the animals showed marks of burning. Although neither orangutan died at the time, it is unknown whether they survived in the longer term. Local villagers have reported that gibbons and monkeys have died after similar shocks (Das *et al.*, 2009).

Disturbance Associated with Noise and Vibration (including Blasting), Project Lighting and the Presence of Workers

The construction phase of all types of infrastructure is accompanied by noise and human activity, both of which tend to be reduced once the infrastructure is built. This additional noise and disruption can cause apes to avoid affected areas, leading to temporary displacement that can affect individual and group ranging, access to food and shelter, and dispersal. The disturbances can also cause heightened stress levels, with possible impacts on health and reproduction.

Rabanal *et al.* (2010) measured the impact of dynamite blasts for oil exploration on gorillas and chimpanzees and found that both avoided the area where the explosions had occurred for months after the exploration work, even though there were strict regulations in place to minimize disturbance (for example, chainsaws and mechanized vehicles were not allowed, and transects were very narrow). The dynamite blasts and increased human presence presumably caused the apes to keep their distance. In Borneo, noise linked to timber extraction—such as from the use of machinery and chainsaws—drives orangutans away from disturbance areas, although animals may recolonize the same areas after the disturbance is over (Ancrenaz *et al.*, 2010; MacKinnon, 1974).

Hydrological Impacts

In both intact and degraded landscapes, gallery, riparian and swamp forests often represent critical habitats for apes, be it for food or nesting (McLennan, 2008; Mulavwa *et al.*, 2010). Riparian habitats are also vital to healthy freshwater ecosystems, fisheries, clean water and other essential functions that support local people and agricultural productivity (Chase *et al.*, 2016). It is therefore crucial to preserve these habitat types.

Chimpanzee and bonobo populations that occur in more arid landscapes dominated by savannah can be severely constrained by water availability (McGrew, Baldwin and Tutin, 1981; Ogawa, Yoshikawa and Idani, 2014). In such water-stressed

Photo: Strategic road planning can reduce the number of roads that apes must cross in their home range, decreasing stress and risks. Road construction in Guinea. © Morgan and Sanz, Goualougo Triangle Ape Project, Nouabale Ndoki National Park

landscapes, it is particularly critical that infrastructure development not prevent access to or otherwise affect water sources.

Infrastructure such as roads and dams typically affects hydrological systems, for instance by changing water levels and flow. Infrastructure development can also cause erosion or indirect impacts on the local or regional climate, which can modify vegetation composition. How such changes affect apes largely depends on the impact of infrastructure on three main factors:

- land use patterns, such as agricultural activities (whose expansion may cause additional habitat loss for apes);
- the degree to which water acts as a constraint on local apes; and
- local vegetation species, some of which may be critical to apes for shelter (nesting) and food.

Steps Forward

Learning from Environmental Impact Assessments

Environmental impact assessments (EIAs) are designed to identify measures to prevent or reduce the negative impacts of infrastructure development on biodiversity. Appraisals that also consider impacts on people are known as environmental and social impact assessments (ESIAs). Chapter 1 discusses best practice in impact assessments (see Box 1.6, p. 36).

Unfortunately, not all infrastructure development projects require EIAs or ESIAs. Whether an assessment is obligatory depends primarily on a country's laws and policies; which, if any, lending or investment agencies are involved (such as the International Finance Corporation, the World Bank and development banks); and what type of infrastructure is being considered. In many

countries, assessments are not required for road or bridge construction. When they are requested, EIAs and ESIAs often consider only the impact that infrastructure is likely to have on the immediate vicinity of the specific project, although the impact typically extends far beyond the area under review and may contribute to cumulative impacts, depending on surrounding land use and the proximity of other projects. Furthermore, EIAs and ESIAs are often carried out too late to influence the decision-making process; in such cases, they become tools for mitigating—as opposed to preventing—environmental degradation (see Box 1.6).

In addition to being undertaken late in the process, the vast majority of EIAs and ESIAs are conducted over extremely short periods of time. A short time frame precludes a surveyor's ability to establish a proper understanding of the distribution and conservation status of impacted ape populations, as well as the potential seasonal or long-term impacts of any infrastructure development on these animals. Indeed, surveying apes properly is time-consuming and requires significant effort and resources, both of which are often lacking (Kühl et al., 2008). Companies have to secure resources in advance to be able to hire qualified experts in ape population surveys to carry out thorough assessments. To capture seasonal variations, such assessments require data collection periods of at least one full year, as well as sufficient time to analyze and report on the findings (see Box 1.6). In practice, these vital conditions are rarely met.

To avoid adverse effects on local people and to help to manage their expectations, ESIAs for any infrastructure project need to consider the expected impact on their lives and estimate how many external people are likely to be attracted to the area prior to and during implementation. The process is most effective when such aspects are considered early on in the planning stages. Activities associated with infrastructure projects can otherwise have aggravating consequences, as was recently the case with the Bumbuna dam expansion project in Sierra Leone. Small-scale logging activities increased in the dam's potential inundation zone as local people sought to exploit timber resources that they anticipated would be lost (R. Garriga, personal communication, 2016). Such activities, which are generally based on the assumption that a project will go ahead, thus have a negative impact on local wildlife even if a project is not taken forward. If such a project is indeed abandoned, the prospect of its implementation alone will have exacerbated habitat loss and disturbance to wildlife in the locality. By providing an accurate assessment of anticipated social impacts in the early phases of a project, an ESIA can highlight these risks and inform the development of effective mitigating measures, typically more comprehensively than an EIA.

Mitigation Measures That Can Reduce Negative Impacts on Apes

The following approaches can serve to mitigate the impact of infrastructure development on apes. While some are not applicable in all circumstances, others are used by several certification bodies, including the FSC and the RSPO.

- **Applying strategic land use planning.** Integrated, well-informed land use planning is the most effective way to minimize the negative impact of infrastructure development while enabling social and economic development. There is an urgent need for conservationists to identify key priority ape ranges on maps and

> " Integrated, well-informed land use planning is the most effective way to minimize the negative impact of infrastructure development while enabling social and economic development. "

to use these maps in efforts to prevent infrastructure development in those areas. Just as development takes place at the international, national and local levels, so too does effective land use planning. Such planning considers the different stakeholders involved in various types of infrastructure development: local private industry may support planning for a tourist lodge, while governments may drive efforts to develop road networks, and multinational corporations may back bids for hydropower projects, mining concessions, processing mills and industrial agricultural activities.

- **Minimizing the length of road networks.** Efforts to restrict the growth of road networks help to limit impacts on habitat and wildlife populations overall, even if restrictions are only applied on a temporary basis (Wilkie *et al.*, 2000). Strategic road planning can also reduce the number of roads that apes must cross in their home range, decreasing stress and risks. To minimize the impact of road development on apes, stakeholders can apply best-practice measures, such as by:
    - undertaking road construction at least 5 km from protected areas, and ideally 10–20 km (Morgan and Sanz, 2007);
    - avoiding the construction of roads in areas that are important to apes, such as the core of their habitat or areas with high densities of fruiting trees, bearing in mind that construction in open or monodominant forest will cause less disturbance and minimize the loss of tree species that are important to apes for food and nesting (Morgan and Sanz, 2007);
    - reusing old logging and similar road networks instead of opening up new road networks, as long as such "recycling" does not lead to increased damage to forest canopy (Morgan and Sanz, 2007);
    - constructing well-designed and -located wildlife crossing sites, speed bumps and other structures (whether arboreal or terrestrial) to allow safer passage for animals (Cibot *et al.*, 2015; McLennan and Asiimwe, 2016; see Box 2.2);
    - keeping road width to a minimum since apes perceive wider roads as posing higher crossing-related risks (Hockings *et al.*, 2006; see Box 2.1); and
    - installing signs to alert drivers to the presence of apes.

- **Avoiding fragmentation.** In landscapes that are already fragmented and deforested, infrastructure—such as roads and power lines—may become additional filters or barriers to wildlife movements. The construction of wildlife passages as linear corridors can serve to minimize mortality rates and restore connectivity.

- **Controlling domestic animals and invasive species.** In areas adjacent to infrastructure and ape habitat, strict controls and policies can be effective in preventing the introduction of domestic animals and invasive species, and associated risks of disease transmission to apes.

- **Dismantling temporary infrastructure.** The dismantling and destruction of temporary infrastructure—such as access roads, provisional camps and bridges—prevents its further use by people after a project has been completed. The FSC and other certification bodies already encourage such dismantling as best practice (FSC, 2015; Rainer, 2014). Any relocation of people from temporary camps requires careful assessment

> Just as development takes place at the international, national and local levels, so too does effective land use planning.

BOX 2.2

Apes and Wildlife Bridges: Examples from Asia

Infrastructure can act as an articifical barrier, preventing apes from moving freely within their habitat. No apes can swim; even small rivers or drains can become impassable barriers to them. Gibbons rarely come to the ground, so the construction of a road dissects their habitat and results in intense fragmentation.

Wildlife bridges allow animals to cross artificial barriers. Bridges that have multiple access points at various heights can provide different routes across a gap; by allowing several animals to cross at different points at the same time, they help to avoid bottlenecks in which conflict can occur between family groups or individuals. In the absence of such bridges, single-strand rope bridges can also be effective. Canopy bridges are an inexpensive, minimally disruptive way of manipulating the habitat to provide primates (and other animals) with access to a larger area of habitat and food sources while minimizing the need for the animals to behave in stress-inducing or dangerous ways, such as descending to the ground to cross gaps (Das et al., 2009).

In Sabah, the removal of large riparian trees along major tributaries of the Kinabatangan River resulted in the destruction of all natural bridges that were used by orangutans (and probably gibbons) to move across the landscape. As a result, these populations experienced further fragmentation (Jalil et al., 2008). The HUTAN–Kinabatangan Orang-utan Conservation Programme in Sabah decided to erect bridges that would enable these species to cross small tributaries or drains. The first bridges were built with used fire hoses, but these ropes degraded after a few years and needed regular monitoring and maintenance to prevent any fatal falls. The second bridge generation used weather-resistant ropes that do not decay under tropical weather conditions. Several types of bridges were erected: from single lines to a web-like design using up to five different intertwined lines. The widest gap between the two riverbanks was about 30 m and the height of the bridges was about 10 m above water level.

A major challenge was identifying suitable trees on both sides of the river that would be tall and strong enough to sustain the weight of these bridges. A total of eight bridges were erected and are now constantly monitored via direct observation and camera trapping. Monkeys and other small mammals started to use these bridges in a matter of hours or days, sometimes even before a bridge was fully established. It took several years for gibbons and orangutans to start using the bridges, however. Once they did, the frequency of passage by these two species increased steadily.

These bridges have proved to be effective ways to alleviate artificial travel bottlenecks for apes. They have also become a major attraction for tourists, who come to watch macaques (*Macaca* spp.) and proboscis monkeys (*Nasalis larvatus*) cross them. Regular monitoring is needed for maintenance purposes and to make sure that poachers do not ambush wildlife on or near the bridges.

of relocation areas to minimize the potential impact on apes. Following dismantling and destruction, rehabilitation activities to promote natural regeneration help to support repopulation by apes and other wildlife.

- **Developing and implementing ecological and social standards for large-scale infrastructure development and establishing certification criteria**. Certification can boost credibility, not only by satisfying legal or contractual requirements, but also by enhancing transparency and maintaining high standards. The infrastructure sector could take the lead from other industry-specific certification bodies, such as the FSC and the RSPO, which require adherence to sustainable practices to mitigate threats posed by industry and associated infrastructure. Other certification bodies—including future ones that might be focused on the large-scale infrastructure sector—could adopt similar ecological and social standards as part of their certification processes. By requiring such certification for large-scale infrastructure projects, lenders and donors would contribute to sustainable development.

Systematic monitoring of ape populations and people is a valuable means of assessing and demonstrating the usefulness of applied mitigation measures; it is also a reliable method for gathering evidence to inform management decisions. For details on the mitigation hierarchy, see Chapter 4 (pp. 119–128).

Reducing Knowledge Gaps

To date, there is a paucity of longitudinal data that could allow for a more comprehensive evaluation of the impacts of infrastructure on ape survival. At best, snap-shot data are available, but they are rarely published or easily accessible. Even when baseline data have been collected, they often become available only after the infrastructure has been put in place. The lack of data is an impediment to informing infrastructure development.

There is a clear need to undertake more longitudinal research into the impacts of infrastructure development on apes. Studies will be possible and relevant only if there is collaboration among those who are involved in the development, financing and use of infrastructure, namely private companies, governments and all other stakeholders. A first step in the promotion of studies that assess clear, scientific data gathered before, during and after infrastructure development is dialog between those who plan, finance and develop infrastructure and ape conservationists. Such collaboration can benefit both sides (see Box 2.3).

Some information is available about the correlation between roads on the one hand, and poaching and the decline of ape density in the vicinity of large-scale infrastructure on the other. Overall, however, there is a dearth of monitoring data on the short- and long-term impact of infrastructure development on ape survival. In view of such knowledge gaps and the issues highlighted in Table 2.1, urgent research questions include the following:

- How are apes using roads in relation to traffic intensity and road width?
- What are the best road and rail crossing mitigation strategies?
- At what point does traffic density on roads turn them into impermeable barriers for African and Asian great apes and gibbons?
- Are canopy and rope bridges effective tools for ape conservation? How many individuals or groups use them and for how long? What would be the ideal design for these bridges (see Box 2.2)?

Photo: Gibbons rarely come to the ground, so the construction of a road dissects their habitat and results in intense fragmentation. Wildlife bridges allow animals to cross artificial barriers. © Marc Ancrenaz/HUTAN–Kinabatangan Orang-utan Conservation Project

- What patterns emerge from short- and long-term monitoring data on road kills and injuries; health patterns (including human sanitary conditions); dust and airborne pollutants; and noise levels?
- What is the impact of power-line electrocution on gibbons and other apes? What devices could be effective in the prevention of electrocution (see Annex I)?
- How are great apes and gibbons affected by water-dependent infrastructure projects, such as hydropower dams and geothermal plants, given that rivers and large bodies of water can be significant natural barriers?
- To what extent do satellite communities that develop in proximity to infrastructure projects affect the local environment and biodiversity?

In the absence of data needed to evaluate the possible impact of infrastructure on ape survival, a cautious and preventive approach is necessary. It is difficult to predict the impact of some types of infrastructure due to the limited occurrence of certain structures, such as cable cars, in ape habitat. In the Virunga Volcanoes in East Africa, a proposed cable car would run through an area that was only recently re-inhabited by gorillas, members of one of the few ape populations that are currently increasing in size (Gray et al., 2013). With such a small population living in such a small habitat—about 500 gorillas in 450 km² (45,000 ha)—it seems too risky to assume that the impacts will not be great, in the absence of firm data to the contrary.

Social Impacts of Infrastructure

Introduction

Wildlife conservation and human welfare cannot be considered in isolation from each other; both rely on the well-being of tropical forests as dynamic, ever-changing ecosystems. Such systems include human communities that depend on and are part of forests. To be fully effective, wildlife conservation initiatives also rely on the support of local communities. The consideration of potential social impacts of infrastructure development and the formulation of associated mitigation measures are key steps in

---

BOX 2.3

Private Industry and Ape Conservation

In 2006 a private company, the China Petroleum & Chemical Corporation, or SINOPEC, began an oil exploration concession in Loango National Park, Gabon. Initially, the company was conducting exploration work (using dynamite explosions along a grid of transects cut through the forest) without any environmental regulations, even though work was being carried out in a national park. Following discussions with the Gabonese Ministry of the Environment, non-governmental organizations (NGOs) and researchers, an environmental impact assessment was conducted to inform the second phase of exploration in 2007. The assessment resulted in guidelines that:

- forbade the use of chainsaws and mechanized vehicles;
- called for narrow transects and allowed only trees with a diameter of less than 10 cm at breast height to be cut down;
- forbade hunting; and
- stipulated that a bridge providing access to a large area of the park had to be destroyed after the exploration work was finished (Rabanal et al., 2010).

With the help of routine monitoring of the area, SINOPEC followed these guidelines. Nevertheless, the disturbance caused by the noise of dynamite explosions resulted in displacement of chimpanzees and gorillas from the area for several months after the work. The exploration did not result in further exploitation of the area for petroleum extraction and, ten years after the exploration, the main access road is greatly reduced in width as the forest is slowly regenerating.

In some cases, a company's interest in maintaining infrastructure may be compatible with conservation goals. One example involves the oil giant Shell, which, until mid-2017, operated one of the highest-producing onshore oilfields in sub-Saharan Africa—Rabi, located between two national parks in Gabon. The company strictly limited access to this area; it also forbade hunting and implemented other regulations that reduced incentives for staff to hunt. These rules were in place largely to protect the infrastructure of the petroleum concession, but they resulted in higher densities of large mammals in this area, as compared to nearby landscapes that do not receive such high levels of protection (Laurance et al., 2006).

designing more effective strategies to prevent and minimize damage to these communities. At the same time, these steps can help to secure local support for efforts to protect wildlife and the environment.

Rather than attempting to cover the vast range of human societies affected by infrastructure development within ape range states, this section focuses on some forest-dwelling communities that retain an intimate knowledge of, and interaction with, complex tropical forests. By drawing on examples of oil pipelines, roads and railways in southern Cameroon, it examines the way industrial infrastructure development fuels deforestation. Analysing the impacts not only of infrastructure, but also of conservation-oriented attempts to offset the adverse effects such infrastructure has on indigenous peoples is critical to developing strategies to protect the forests on which both apes and such peoples depend.

Photo: Wildlife conservation and human welfare cannot be considered in isolation from each other; both rely on the well-being of tropical forests as dynamic, ever-changing ecosystems.
© Jabruson (www.jabruson.photoshelter.com)

> Wildlife conservation and human welfare cannot be considered in isolation from each other; both rely on the well-being of tropical forests as dynamic, ever-changing ecosystems.

Africa and Asia are home to relatively few indigenous hunter-gatherer populations that depend completely on forest resources, yet these continents are the most affected by activities that impact forests, including infrastructure. The prospecting, developing and operating of infrastructure have more extreme impacts on forest-dwelling peoples than on other communities that live near forest boundaries.

Forest peoples themselves have analyzed the dynamics involved in infrastructure development. In the Palangka Raya Declaration on Deforestation and the Rights of Forest Peoples of 2014, representatives of forest peoples from Asia, Africa and Latin America describe the situation as follows:

> Global efforts to curb deforestation are failing as forests are cleared faster than ever for agribusiness, timber and other land development schemes. We, forest peoples, are being pushed to the limits of our endurance just to survive. [...] Deforestation is unleashed when our rights are not protected and our lands and forests are taken over by industrial interests without our consent. The evidence is compelling that when our peoples' rights are secured then deforestation can be halted and even reversed (FPP, Pusaka and Pokker SHK, 2014, p. 117).

The Declaration goes on to highlight how the international bodies that are charged with halting deforestation are very often the same ones that are driving it:

> Global efforts promoted by agencies like the United Nations Framework Convention on Climate Change (UNFCCC), the United Nations Collaborative Programme on Reducing Emissions from Deforestation and Forest Degradation (UN-REDD) and the World Bank to address deforestation through market mechanisms are failing, not just because viable markets have not emerged, but because these efforts fail to take account of the multiple values of forests and, despite standards to the contrary, in practice are failing to respect our internationally recognised human rights. Contradictorily, many of these same agencies are promoting the take-over of our peoples' land and territories through their support for imposed development schemes, thereby further undermining national and global initiatives aimed at protecting forests (FPP et al., 2014, pp. 117–18).

Numerous examples from around the world, along with multiple studies that highlight the role of indigenous peoples and other local communities in forest conservation, indicate that conservation can succeed if it is based on securing forest peoples' rights to their lands and supporting them in conserving their lands. The opposite approach to forest conservation—one that destroys indigenous peoples' forests for "development" or evicts them from their forest for "conservation"—has been shown to fail (Seymour, La Vina and Hite, 2014). A survey undertaken by the Center for International Forestry Research compared 40 protected areas and 33 community-managed forests in 16 countries and found that community-managed forests were more than 6 times better at avoiding deforestation than protected areas (Porter-Bolland et al., 2012).[9]

Drivers and Impacts of Infrastructure in Cameroon

In relation to Cameroon, the Palangka Raya Declaration highlights that:

> logging, oil palm plantations and new infrastructure schemes are causing galloping deforestation, aided by colonial laws which deny our rights to our lands and forests and corrupt government officials who allocate our lands to other interests without regard for our welfare. Evictions are common and impoverishment results. Even protected areas set

aside to compensate for forest loss restrict our livelihoods and deny our rights (FPP *et al.*, 2014, p. 118).

The major direct causes of deforestation and forest degradation in Cameroon are commercial logging, cultivation of cash crops (mainly cacao and coffee), agro-industrial plantations (rubber and oil palm) and the exploitation of minerals (FPP *et al.*, 2014, p. 42). More recently, forests have been opened up and destroyed by infrastructure projects such as roads, railways and oil pipelines, and hydroelectric power, including the aluminum smelter at Edéa (Dkamela, 2011, pp. 32–5). This section identifies the overall drivers and consequences of such infrastructure development and presents specific examples from the rainforest areas of southern Cameroon.

Southern Cameroon is dominated by equatorial rainforest and is relatively sparsely inhabited by indigenous Bagyeli and Baka forest hunter-gatherer communities (the minority) and Bantu farming communities (the majority) (Kidd and Kenrick, 2009, p. 17; Nguiffo, Kenfack and Mballa, 2009; Owono, 2001, p. 249). Although many Bantu are also long-term inhabitants of the forest, they nevertheless acknowledge the Bagyeli and Baka hunter-gatherers as the first inhabitants of the forest (Dkamela, 2011, p. 27; Kidd and Kenrick, 2009, p. 16; van den Berg and Biesbrouck, 2000).

Between 1990 and 2010, Cameroon lost close to 20% of its forest cover, largely as a result of commercial logging, the expansion of medium- and large-scale commercial agriculture, and a major infrastructure project, the Chad–Cameroon pipeline (de Wasseige *et al.*, 2013; Freudenthal, Nnah and Kenrick, 2011; Ndobe and Mantzel, 2014, p. 5).

In 2009, the government of Cameroon set out its ambitious "Vision 2035" for becoming an emerging economy within 25 years through major growth in export agribusiness, mining, commercial logging and infrastructure development. Much of this economic activity is geared to export-led growth, which entails the supply of international markets with timber, rubber, palm oil, minerals and commodities (Dkamela, 2011, pp. 32–6; Republic of Cameroon, 2009a). To date, the resulting impacts on forests, wildlife and forest-dependent communities have often been exacerbated by poor governance and corruption, as well as by smaller companies and local elites who use the infrastructure opened up by export-led economic activity to encroach on forests and generate income on domestic markets, often at the expense of customary communities.[10]

The government's development plans do not make provisions for legal reform of outdated land laws, nor for addressing governance and corruption issues. As stipulated in ordinances issued in 1974, land that is not registered as private property (including any non-registered forest land) is under the administration of the state, a continuation of the colonial *terra nullius* principle, under which lands owned by local communities were appropriated by colonial administrations (Alden Wily, 2011b, pp. 50–51).[11] In practice, this means that communities are denied any collective property rights to forests and lands that they have customarily occupied and used for their livelihoods.

Cameroonian government officials generally grant forest concessions to private interests without consulting or compensating impacted communities (Alden Wily, 2011b; Perram, 2015). Based on the 1994 Forestry Law, which allows for community forests of up to 50 km² (5,000 ha), some groups have been granted community forests, or temporary access or use rights in protected areas and logging concessions. Community forests can be granted to and managed by customary communities, but, counterintuitively, they can also be granted

> **"** Numerous examples and studies indicate that conservation can succeed if it is based on securing forest peoples' rights to their lands and supporting them in conserving their lands. **"**

to and controlled by elite interests. Either way, communities generally gain little from these processes, since they are granted management but not tenure or property rights, and because they typically encounter widespread corruption and administrative barriers (Alden Wily, 2011b, pp. 66–83; Cuny, 2011).

At the international level, the principle of free, prior and informed consent (FPIC) is enshrined in the UN Declaration on the Rights of Indigenous Peoples (2007) and in International Labour Organization (ILO) Convention 169 (1989),[12] among other treaties. FPIC is embedded in the universal right to self-determination, which is itself embodied in legally binding instruments to which Cameroon is a party, such as the International Covenant on Economic, Social and Cultural Rights; the International Covenant on Civil and Political Rights; and the African Charter on Human and Peoples' Rights. Moreover, under Article 45 of its constitution, Cameroon is required to let its international law obligations take precedence over its national legislation (FAO et al., 2016, pp. 12–13; Franco, 2014, p. 5; Perram, 2016, pp. 6–7).

Although the government is thus legally required to consult communities about any project that may affect their customary lands, indigenous peoples typically learn that their forest has been allocated to a concession or infrastructure project via the sudden arrival of survey teams. Such teams may proceed to install concrete waymarkers to delimit a concession boundary, cut trails to make a new roadline, or dig pits and remove cores for mineral exploration.

Regulatory and administrative ambiguities and challenges currently prevent local people from accessing adequate, reliable information about development projects on customary land and from asserting their rights with developers or the government (Perram, 2016). The Mining Code, for example, makes provision for mining companies to pay compensation to customary land rights holders, but it does not identify how these rights should be determined (Nguiffo, 2016; Republic of Cameroon, 2001, art. 89).

Meanwhile, permits for mineral exploration frequently overlap with protected land and established logging or commercial agriculture concessions (principally oil palm and rubber), reflecting not only disregard for legally binding conservation commitments and community rights to FPIC,[13] but also a lack of coordination between the ministries responsible for issuing different permits. Mining permits now reportedly cover almost 100,000 km² (10 million ha), or about 20% of the country's total land area (Nguiffo, 2016); many overlap with forested areas and designated permanent forest estates, and 20% coincide with protected areas, including national parks (Dkamela, 2011; Mitchard, 2012; see Figure 2.2). Mining companies that have begun extracting or that are currently prospecting include:

- Caminex, a former Cameroonian subsidiary of Afferro Mining, which was taken over by the UK-based International Mining and Infrastructure Corporation;
- Cam Iron S.A., a Cameroonian subsidiary 90% owned by the Australian company Sundance Resources Ltd.;
- Civil Mining & Construction Pty Ltd. of Australia;
- Geovic Cameroon PLC (GeoCam), based in the United States; and
- G-Stones Resources S.A. of Canada (KPMG, 2014; Meehan, 2013; Profundo, 2016; Sundance, 2016).

For some forest-dependent individuals, the impact of Cameroon's development trajectory is not entirely negative in the short term, even if the longer-term consequences for families, communities and the forest

FIGURE 2.2

Bagyeli Customary Land, Forests, and the Chad–Cameroon Oil Pipeline and Proposed Railway in Southwestern Cameroon, as of November 2016

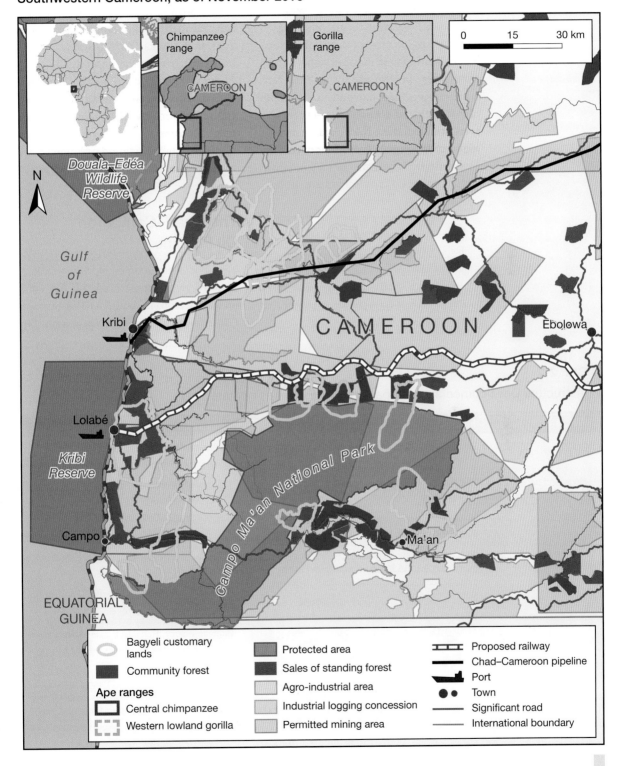

Chapter 2 Impacts of Infrastructure

itself often far outweigh immediate individual benefits. Those benefits can involve paid (but often short-term) employment opportunities, improved access to services and markets (as forest roads are often maintained by logging companies) and the arrival of mobile telephone masts in remote areas of rainforest. In some instances, developers promise to provide communities with healthcare facilities or school buildings on the basis of a "social contract" and, in principle, logging companies pay forest taxes. As observed by a Bagyeli man in 2014, however, such promises do not always materialize:

> We were promised 3 million CFA francs [US$5,000] as compensation for our land but so far we have received nothing. They told us this is development, yet we have no schools, no hospital and no transportation. The government did not respect its promise (FPP *et al.*, 2014, p. 44).

Cameroon's forest communities depend on the forest to provide them with food, clean water, shelter and medicinal plants. Forests are also the basis of the social and cultural identity and spirituality of the Bagyeli and Baka. Their customary practices are based on low-intensity hunting, freshwater fishing, gathering of wild honey and other forest products, and small-scale cultivation. For these communities, the negative consequences of large-scale deforestation and infrastructure development are varied and far-reaching (see Table 2.2).

The Chad–Cameroon Oil Pipeline

The Chad–Cameroon oil pipeline was constructed to transport crude oil from the oilfields of Doba in southern Chad, through Cameroon, and on to the coast at Kribi. On

TABLE 2.2
Infrastructure Developments and Impacts in Cameroon as of June 2017

Development	Impacts	Examples
Roads	In-migration, construction camps, poaching, artisanal logging, displacement	Djoum–Mbalam international road
Railway and port	Construction camps, displacement	Mbalam–Kribi proposed railway; Kribi deepwater port
Pipeline	In-migration, construction camps, commercial poaching, artisanal logging, displacement	Chad–Cameroon pipeline
Mining	Pollution and siltation of watercourses, loss of customary forests, destruction of sacred sites and medicinal trees, displacement, in-migration, commercial poaching, mining camps	G-Stones/BOCOM/MME Inc. mining Tsia sacred hill; Cam Iron mining sacred hill for iron ore at Mbalam
Commercial agriculture	Loss of customary forests, displacement, destruction of sacred sites and medicinal trees, extreme poverty	Oil palm and rubber by companies such as BioPalm Energy; Herakles Farms oil palm plantations; SOCAPALM; Sud-Cameroun Hévéa
Logging concession	Road construction facilitating poaching, loss of customary forests, destruction of sacred sites and medicinal trees, siltation of watercourses, in-migration, commercial poaching, mining camps	Logging concessions and standing sales such as 625,253 ha attributed to French timber group Rougier and 388,949 ha to Pallisco from the Pasquet group

Sources: Corridor Partnership (n.d.); Environmental Justice Atlas (n.d.); FPP *et al.* (2014); MME (n.d.)

the coast, the pipeline enters the ocean and, since 2003, the oil has been pumped to a stationary floating storage unit, from where it is offloaded onto tankers bound for the United States and Europe (IFC, n.d.).

Estimated at US$6.5 billion, the cost of construction was covered by the U.S. multi-nationals Exxon-Mobil and Chevron Texaco, Petronas of Malaysia and the International Finance Corporation of the World Bank. The southern portion of the pipeline, between Lolodorf and Kribi, traverses more than 100 km of rich biodiverse forest lands used by indigenous Bagyeli hunter-gatherer communities as well as local Bantu farming communities (Nelson, 2007, p. 2).

An 890-km stretch of the pipeline's total length of 1,070 km is on Cameroonian territory, where the route is 30 m wide. Its final 100 km have had a particularly destructive impact, especially on the Bagyeli hunter-gatherers and on the forest itself, including apes (Planet Survey/CED, 2003). Research has documented the adverse effects on the Bagyeli:

> Hunting is the most important Bagyeli activity, although they are also gatherers and increasingly farmers [...]. Construction of the pipeline brought large numbers of trucks, heavy equipment, workers, and work camp followers, including poachers, into the region, negatively impacting this form of livelihood. The pipeline has resulted in making hunting increasingly difficult for the Bagyeli. They say they now need to walk for at least three days in the forest before finding animals. Before the pipeline, they say, the animals were right next door and easy to hunt. Poachers are one of the problems, increasing competition for game, while not respecting the traditional methods of hunting without irreparably damaging the balance of the ecosystem (Horta, 2012, p. 221).

While World Bank policy required the development of an indigenous peoples plan to counteract any adverse impact on the Bagyeli, a study conducted in 2001 found that the Bank itself had failed to provide adequate and culturally meaningful space to enable Bagyeli participation in the design of the indigenous peoples plan (Nelson, Kenrick and Jackson, 2001, p. 3). Specifically, the plan did not address the Bagyeli's main priorities, but instead focused solely on supporting Bagyeli agricultural, health and education programmes. These programmes rarely reached their intended beneficiaries and ignored the fundamental need the Bagyeli had expressed, namely the protection of customary rights to their forests, which would have helped to secure their access to the forest itself and to agricultural land (Nelson, 2007, p. 15).

For the Bagyeli, the destruction of the forest by the pipeline has had very direct and devastating consequences. As one Bagyeli healer explained:

> When the pipeline destroys the medicinal trees, it will destroy everything. I am a healer; I don't use the medicines of the hospital. I was born in the forest, I live in the forest, I will die in the forest. I live from the forest—the pipeline destroys the forest by which I live (Nelson *et al.*, 2001, p. 12).

Another Bagyeli described how the process of constructing the pipeline intensified the exploitation of the Bagyeli by their dominant Bantu neighbors (referred to as the Myi):

> The Bagyeli work on the pipeline and the Myi take the wages. The monkey travels on high, but the chimpanzee takes what the monkey finds. I don't want to talk of the pipeline, because the pipeline makes the Myi take from us (Nelson *et al.*, 2001, p. 12).

Meanwhile, the pipeline opened up the forest not only for poachers, but also for loggers. Together they combined to destroy the rich biological diversity as well as the specific paths and places that made up the ecological and cultural richness that the Bagyeli always depended on and that had been sustained by their presence. A leading Bagyeli spokesperson, Madame Nouah, observed:

> The forest is very rich for us Pygmies, for us to nourish ourselves. Now we are afraid that things will be destroyed in the forest that are necessary and useful for us (Horta, 2012, p. 221).

Logging also removes non-timber products, such as honey and seeds, as well as other points of orientation. As a result of such losses, the Bagyeli are facing increasing poverty and "are now more frequently losing their orientation in the forest they used to know so well" (Horta, 2012, p. 221). In interviews, some Baka suggested that as the forest habitat became unrecognizable and filled with noise, humans, apes and other species most probably experienced disorientation and related disturbances in comparable ways.[14]

When "development" leads to the destruction of forests, the international community's standard response is to try to balance the damage with forest protection in the name of "conservation." This is exactly what happened in southern Cameroon:

> Since construction of the pipeline has led to the loss of important biodiversity in Cameroon's coastal forest, the World Bank's operational policy on Natural Habitats (OD 4.04) required the establishment of protected areas or national parks to compensate for these losses (Horta, 2012, p. 221).

The pipeline project gave the final justification and impetus for the establishment of Campo Ma'an National Park near Cameroon's coast (see Figure 2.2). The Campo Reserve had existed since 1932, but now funding for the national park came from the global fund managed by the World Bank's Global Environment Facility, which described the park "as part of the environmental compensation for the Chad–Cameroon pipeline project" (Owono, 2001, p. 248). As a result, hundreds of local Bagyeli communities were banned from hunting and gathering in many forest areas on which they had always relied, and so their livelihoods and ways of life became seriously threatened. The impact of this "green land grab" on the Bagyeli was severe:

> Previously, life within the Wildlife Reserve had been regulated, but with the creation of the park and the new funding which enabled the imposition of rules prohibiting access to the protected area and the use of any of the natural resources, the lives of the resident populations, especially the hunter-gathering Bagyeli Pygmies, have worsened. This is all the more paradoxical because the park was created as part of the environmental compensation for the Chad–Cameroon pipeline which, according to the World Bank, would help alleviate poverty. However, the establishment of the [park] will instead worsen the already precarious living conditions of the local hunter-gathering population (Owono, 2001, pp. 246–7).

As a case study on the implementation of the Chad–Cameroon pipeline notes, for peoples such as the Bagyeli, the forest is not so much a resource to be exploited or a wilderness to be protected; it is a place that is home, the source of livelihood and well-being. The Bagyeli have experienced the construction of the pipeline and the setting aside of land for conservation to compensate for the destruction of forests as a twofold existential threat. First, the Bagyeli—along with the rest of their complex forest ecosystem—were severely impacted by the pipeline construction and concomitant disruption; second, the "compensation" for this disruption further marginalized the community, impoverishing them and disrupting their lives (Planet Survey/CED, 2003, p. 12).

Like other forest peoples throughout the Congo Basin, the Bagyeli have been resilient despite centuries of discrimination by their more powerful neighbors and outsiders. As long as they have been able to move between the forests and the roadside Bantu villages, the Bagyeli have traded with their neighbors from a position of autonomy and resilience (Kenrick, 2006; Kenrick and Lewis, 2004; Kidd and Kenrick, 2011). Once they are no longer able to sustain their lives in the forest, however, the structural discrimination will become as permanent a feature of their lives as the poverty and sociocultural dislocation that resulted from having their forests destroyed by the pipeline. A conservation regime that excludes the Bagyeli from familiar places and from their hunting grounds effectively ignores their needs, their rights and their ability to sustain and be sustained by their forests (Kidd and Kenrick, 2011, pp. 16–21).

Road and Rail: Impacts of Extraction in South Cameroon

It has long been pointed out that Africa and Latin America are not intrinsically poor, but that so many of their inhabitants are poor

FIGURE 2.3

Ape Ranges and Road and Rail Impacts in Southern Cameroon, as of November 2016

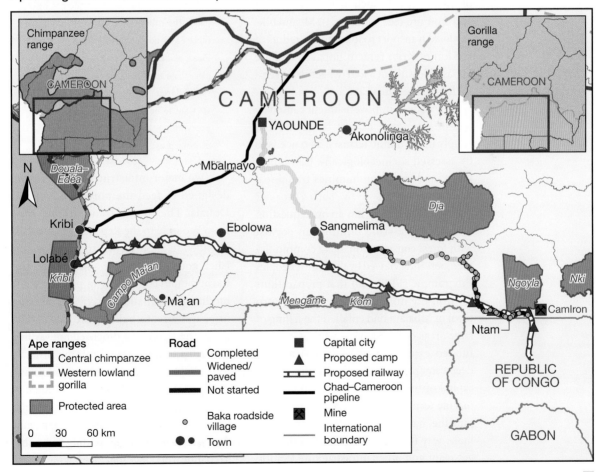

because far more powerful outsiders, along with national elites, have sought to extract the plentiful resources of both continents (Cotula, 2016).

The map of road and rail infrastructure is a clear indicator of whether the wealth of a country is being used to benefit its inhabitants. Uruguayan writer Eduardo Galeano points out that his continent's infrastructure was developed in order to suck its wealth into the ports, and thence into the colonial and neo-colonial economy; that infrastructure, he argues, was designed to leave as little wealth behind as possible (Galeano, 2009).

Similarly, in southern Cameroon, the proposed and developing roads and railways—and the Chad–Cameroon oil pipeline discussed above—very clearly run to the coast at Kribi so as to facilitate the extraction of inland wealth, such as tropical timber and iron ore (see Figure 2.3). Meanwhile, key local transport roads within a radius of 100 km of Kribi remain unpaved and are unpassable without a four-wheel-drive vehicle for parts of the year.

The issue of impoverishment caused by wealth extraction cannot be considered simply in economic terms; it also needs to be assessed socioecologically. Can biodiversity and forest communities' traditional livelihood patterns survive such a process?

More specifically, it is an open question whether large-scale mining can coexist with forest conservation. Baka community members interviewed by the Forest Peoples Programme (FPP) said that preparations for iron ore mining in the south-eastern town of Mbalam had entailed the felling of large areas of forest. Meanwhile, Chinese-funded expansion of West Africa's first deepwater port at Kribi, the administrative capital of the Océan Department and the marine terminal for the Chad–Cameroon pipeline, has involved forest clearance to make way for roads, mineral terminals, a gas plant and other infrastructure (Smith, 2013). This activity has had a severe impact on the local Bagyeli, who were relocated and have since experienced reduced access to the forest, an increasing scarcity of forest products, and noise and pollution from nearby construction work (FPP et al., 2014; Tucker, 2011).[15]

In the words of an older Bagyeli man named Bibera:

> The forest where we usually hunt and collect medicinal plants and non-timber forest products is disappearing, especially as the deep sea port, gas plant and roads are being constructed. The government has shown us a resettlement area, which has no forest, not even where you could find a tree to scratch the bark for medicine or hunt even a rat. We shall now be in the centre of the town; the railway line will be passing by us; roads are there; there is a gas plant. The calmness of the forest has been replaced by noise of vehicles and machines. Please tell the government to reserve us a place to go and collect medicines to heal our sick children. No one allows us to decide if we want to be resettled or not, and where. Everything is being imposed on us (FPP et al., 2014, p. 45).

Two major infrastructure projects are designed to feed the ports at Kribi and Douala. The first, a transnational road from Yaoundé to the Republic of Congo, is intended to allow for the transport of finished goods to Yaoundé and Douala, and the outbound conveyance of primary commodities. International civil engineering firms are currently building the road (AfDB, 2015). The second is a proposed railway line that aims to link several mining projects throughout southern Cameroon and deliver their resources to Kribi on the coast. Although this project is currently on hold due to the low price of iron ore, Cameroonians and Australians are seeking funding to be able to resume work once the price has increased.

Sundance Resources continues to request support from China and other international financial markets (Mining Review Africa, 2016).

Figures 2.2, 2.4 and 2.5 highlight the impact—and the potential impact—of these two projects by overlaying the road and railway line onto the community forests and customary lands of the Bagyeli and Baka. Of particular concern is the area around Ntam in the far southeast, close to the Congolese border and the Cam Iron mine at Mbalam. In this part of Cameroon, the concentration of Baka roadside villages is high and the road and rerouted railway run alongside each other.

The settlement of Ntam is on a road that has yet to be upgraded, more than 100 km away from the part of the road that is being improved (see Figure 2.4). Nevertheless, in anticipation of the road's completion and arrival at Ntam, the settlement is gearing up to become a big trading post. A large customs building has already been constructed in the town; moreover, local sources indicate that significant tracts of nearby community forest land have already been "sold" to incoming state functionaries, their families

FIGURE 2.4

Community Forests, Protected Areas, and Road and Rail Impacts in Southern Cameroon, as of November 2016

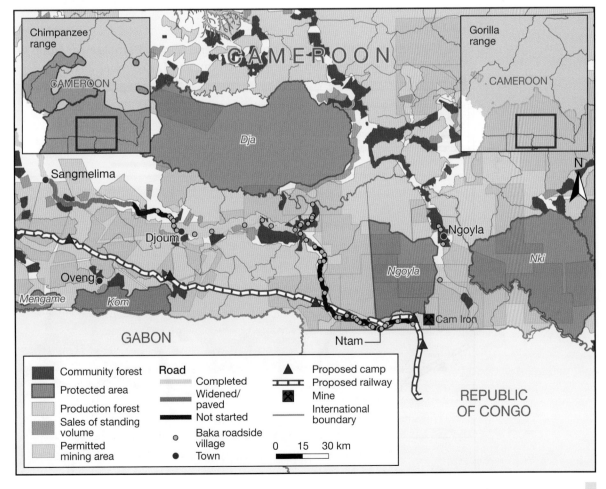

Chapter 2 Impacts of Infrastructure

and others—not always lawfully (J. Willis, personal communication, 2016). Ntam's transformation shows that impacts of infrastructure projects also precede—rather than simply follow on from—development. Indeed, the mere anticipation of infrastructure development opens up the forest for exploitation by major players. Small traders, poachers, small-scale loggers and others also make their way into the area to start exploiting the forest in the expectation of an exponential increase in opportunities and in the demand for various services and products.

The dynamics involved in the railway are similar to those of the transnational road but even more destructive because the railway opens up swathes of forest far from the road. A key point to notice in the environmental and social impact assessment of the railway is the effect of the construction camps (Cam Iron and Rainbow Environment Consult, 2010). The space cleared to build such camps and the number of people expected to populate the sites are indicators of the likely impacts on the area, not least in terms of unsustainable wild meat extraction. While the proposed line of the railway was rerouted to avoid the forest ranges of gorillas and elephants, it was consequently positioned to run through a series of villages along the road corridor, which is certain to exacerbate disruptions to communities' livelihoods and increase conflict over resources (see Figure 2.5).

Both the local communities and the forest are extremely vulnerable in the face

FIGURE 2.5

Baka Customary Lands, Production Forests and Mining Permits near Ntam on the Road–Rail Corridor in Southern Cameroon, as of November 2016

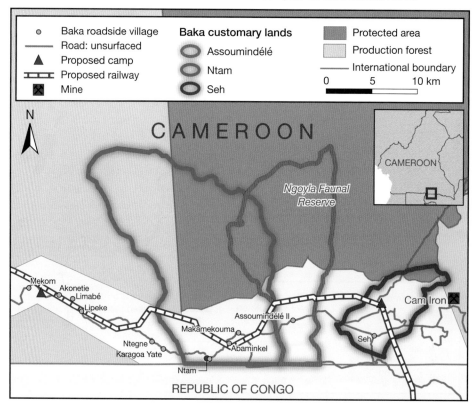

of these developments. Affected communities are rarely consulted with respect to such projects; if they are informed, project proponents tend to focus exclusively on the positive aspects—substantially easier transport options, opportunities to sell forest products to those in transit, and reduced costs for goods coming from outside the community. Community members thus have a limited understanding of the negative impacts, including increases in criminal activity and significant pressures that development-related activities will place on the lands and forest in which they live.

During a recent meeting of Baka community representatives in Assoumindélé, 12 km from Ntam, a Baka NGO staff member raised the issue of Djoum, where the road had already been paved, noting:

> Djoum is already full—there is no land left, and now it is starting to cause disputes within families.[16]

The social impacts of the destruction of their socioecological context include growing rates of alcoholism and suicide among the Baka, increased conflict within and between communities, displacement of whole communities along development corridors and elite capture of community forest concessions by influential Bantu.[17]

The Baka communities along prime transport routes targeted for "improvement" are in an extremely precarious position, as are the Bantu. The Baka, and the Bagyeli in the west, still rely significantly on the forest for their livelihoods. They generally cannot claim possession of their forests under national law, and their customary use rights are frequently violated in practice, particularly if more powerful people stand to profit financially. For the Baka and Bagyeli, the loss of forest areas translates into a loss of livelihood. No suitable compensation can restore that livelihood, nor can they expect any economic benefits from the road, since its construction and associated activities lead to the disappearance of the habitat on which they depend.

Without the possibility of obtaining land titles, the Baka and Bagyeli recognize that moneyed and authoritative outsiders can put pressure on them that is hard to resist, especially if the benefits they are promised sound appealing.

Conclusions and Strategic Approaches

Forest communities in southern Cameroon, particularly indigenous Bagyeli and Baka communities, are unprepared for the radical changes that large-scale road and rail infrastructure projects impose on them. The direct impacts include a reduction in livelihood opportunities; an increase in commercial poaching; and restricted access to land that has been allocated to different concessions (including conservation offsets). The social impacts outlined above, including disorientation, displacement, depression and substance abuse, and intra-communal conflict, compound the situation.

In Cameroon, the meaningful and effective inclusion of indigenous communities in economic development planning is extremely rare. The country's ten-year Growth and Employment Strategy, the cornerstone of Vision 2035, is focused solely on building infrastructure for national and regional resource extraction. In the same vein, financial observers predict that "recent developments in Cameroon's road and rail networks are set to drive the region's economic growth" (Williams, 2015). Efforts to promote such infrastructure expansion—through economic policy and land use planning—are being shaped at the national level, among government and business elites, international development banks and international private capital.

These efforts aim to develop infrastructure networks that will facilitate national and regional resource extraction. Put differently, the infrastructure is not designed to enable farmers and forest communities to bring renewable resources to market, or to allow them to access social provisions. Such planning is arguably based on a model of economic growth that has failed to protect the environment, and that has been unable to create the conditions for secure and stable societies (Blaser, Feit and McRae, 2004; Edelman and Haugerud, 2005; Martinez-Alier, 2002; Mosse, 2005).

The need to support indigenous communities faced with such a bleak future is as urgent and challenging as the need to support non-human forest communities. Neither is likely to be supported by an approach that focuses on economic extraction alongside aggressive conservation tactics, rather than one focused on securing communities' ability first to retain their lands and then, on that basis, to pursue development that is compatible with their well-being.

Below are some current and potential strategies that can enable government, conservation organizations and industry to support communities to challenge and adapt to infrastructure development. More fundamentally, these steps can help them to reclaim their self-determination and an ability to sustain and be sustained by socio-ecologies on which all living beings ultimately depend:

- **Securing community tenure**: This step is critical to enabling recognition in the national legal system of indigenous peoples' and local communities' rights to self-determination, self-governance, FPIC and participation in decision-making processes that affect them. As mentioned above, Cameroon is party to a number of conventions that recognize such rights; enabling them to be recognized in national law and practice may also require an acknowledgment that such communities are the ones best placed to secure the forests. The Cameroon-based Centre pour l'Environnement et le Développement (Centre for Environment and Development), FPP, the Rights and Resources Initiative (RRI) and many other organizations support communities in the use of mapping, the identification of legal strategies and the development of the capacity needed to sustain community lands to advance their goals. Central among these goals is the inclusion of communities in infrastructure-related decision-making processes that are likely to affect them, particularly in view of the fact that indigenous peoples are rarely, if ever, consulted about infrastructure development.

- **Participatory mapping of customary territories**: In Cameroon and other countries that do not recognize customary land tenure as representing legal land title, presenting evidence of such tenure can help to persuade developers to recognize land rights. Participatory mapping is a tool developed by international NGOs and communities to provide georeferenced maps of customary land use boundaries and key resources and sites within those boundaries (using GPS and GIS tools). Maps and supporting information can be used by a community and its NGO allies to challenge a project (for example, to oppose a development or reroute a roadline); to protect key resources and sacred sites; and to make a case for compensation. In Cameroon, a project is under way to develop a common set of protocols for identifying and mapping community land use and tenure across the country's diverse social and ecological landscapes. The project, part of the RRI Tenure Facility, is starting to garner support for

> " The impacts of infrastructure development include a reduction in livelihood opportunities; an increase in commercial poaching; and restricted access to land that has been allocated to different concessions (including conservation offsets). "

the adoption of common mapping protocols by government agencies responsible for the application of relevant land laws and ordinances, as well as the potential support of the land holders themselves, key private sector operators, civil society actors and donor agencies.

- **Capacity building**: One way to support communities is to provide them with information about infrastructure projects and their human rights in relation to infrastructure projects, as defined in national and international laws.
- **Development of indigenous peoples' representative structures**: Combined with capacity building, support for the development of networks of forest communities (such as federations, local associations or advocacy platforms) enables indigenous community voices to reach the elites, government officials and company shareholders. In Cameroon, the development of Bagyeli and Baka associations and their convergence into the Gbabandi platform in 2016 is starting to open political space for their issues at the national and regional levels.
- **Safeguard monitoring and complaints procedures**: With training and appropriate legal support, communities and community-based organizations are monitoring safeguards that developers and funders, such as the World Bank and the African Development Bank, are obliged to observe. They are also lodging formal evidence-based complaints to their grievance procedures whenever systemic or repeated failures to implement safeguards are documented, including the right to FPIC.
- **Advocacy**: Opposition to large-scale infrastructure development can take many forms, from direct mediation between communities or community-based organizations and developers (using legal texts, participatory maps and monitoring evidence); coalitions of national and international NGOs with social and environmental agendas that place pressure on government agencies and donors; and Internet-based campaigns (such as Avaaz, Survival International and various rainforest action networks) that raise the profile of an issue and apply pressure through petitions and letter-writing campaigns.
- **Compensation**: It is important to monitor social agreements and other forms of compensation (such as logging taxes) that developers and concessionaires have agreed to pay to communities, as they often fail to deliver on their part of the bargain.
- **Adaptation**: Steps can be taken to support agriculture-based livelihoods in order to compensate for the loss of forest resources; to develop microcredit and savings schemes; and to encourage added-value processing and market development. These measures generally require partnerships between rights-based organizations that work on community self-determination and development NGOs and international agencies that are more focused on meeting the United Nations Sustainable Development Goals.

The protection of land rights is often a prerequisite for the protection of the environment, and community-based forest management works best when it is rooted in communities that are recognized as legitimate owners of forest ecosystems.

In contrast to Asia and Latin America, Africa provides limited evidence on how forest communities' customary tenure can slow and reverse the loss of indigenous forest. This poor performance reflects many African governments' reluctance to recognize such customary rights, as well as the fact that community forestry has largely been limited to co-management regimes (Blomley, 2013). As the international land

> *The protection of community land rights is often a prerequisite for the protection of the environment.*

Photo: Cameroon is focused solely on building infrastructure for national and regional resource extraction; not to enable farmers and forest communities to bring renewable resources to market, or to allow them to access social provisions.
© Jabruson (www.jabruson.photoshelter.com)

tenure and resource governance specialist Liz Alden Wily has pointed out, however, "several states stand out as having purposely pursued democratic devolution of forest tenure, as well as management, in a bid to radically improve conservation" (Alden Wily, 2016, p. 11).

Alden Wily goes on to list Gambia, Liberia, Namibia and South Africa as countries in Africa that have advanced this process. She also notes: "Multiplication of community owned forests is especially advanced in Tanzania where, by 2012, 480 communities owned and managed their own forest reserves totaling 2.36 million hectares" (Alden Wily, 2016, p. 11).[18]

There is clear progress on community tenure of lands and forests in Africa, even though governments there remain far more reluctant to recognize such customary rights than those in Asia and Latin America (Alden Wily, 2011a, 2016; Nguiffo and Djeukam, 2008). In Asia, around a quarter of all forests had been brought under community ownership by 2009, and that percentage has been rising since (Alden Wily, 2016, p. 2).[19]

This rise in the proportion of community ownership of the world's natural forests reflects the growing recognition that community tenure is a prerequisite for sustainable forest management.[20] This shift is not only a result of acknowledgment that granting such community title is key to effective forest protection, but also a consequence of the fact that forests not owned by communities are more vulnerable to deforestation and are therefore vanishing.

The route to securing such rich and important forests is clear. As demonstrated in the discussed examples from Cameroon and in the relevant literature, however, the roadblocks are many. Concerted urgent action is required to remove those blocks, pursue that path and secure the forests that so many human, and non-human, communities experience as home.

Chapter 2 Impacts of Infrastructure

Overall Conclusion

Infrastructure development in ape range states can disturb forest landscapes in ways that have significant, long-term effects on both people and wildlife. Such effects may involve the removal of important species, structural changes that affect the use of the forest, noise pollution, and increased traffic and movement. This chapter's review of the ecological and social impacts of infrastructure development shows that there is an urgent, widespread need to ensure that infrastructure planning processes include effective measures to protect apes, their habitat and local populations.

Specific recommendations to mitigate the negative direct and indirect impacts of infrastructure development before, during and after project construction include conducting thorough environmental and social impact assessments, as well as ongoing monitoring and data collection (see Chapter 1, pp. 31–38 and Box 1.6); enabling and prioritizing participation through free, prior and informed consent of local forest-dependent populations; and developing appropriate mitigation and adaptation measures to counter any construction-related damage (for more information on The Mitigation Hierarchy see Chapter 4, p. 119). In the application of mitigation measures, particular care must be taken to avoid exacerbating any adverse impacts on indigenous peoples. As discussed in this chapter, deforestation is more likely to be halted if stakeholders recognize forest peoples' land rights and support their age-old approaches to sustaining and being sustained by their ecosystems, than if they evict these communities from their lands in the name of "development" or "conservation."

Countless examples exist of how infrastructure projects have severely affected ape populations and pushed local communities further into poverty, yet counterexamples are hard to find. Unless more effective measures are put in place, governments and private industry will continue to enter forests and exploit natural resources without adequately consulting local communities, without understanding the risks and likely impacts, and without considering the survival or well-being of affected people and wildlife. It follows that unless the environmental, social and economic impacts of infrastructure development are considered in a more holistic way, indigenous communities and endangered species will continue to suffer. This chapter has outlined the main impacts of the business-as-usual model, as well as some of the key measures that can help to prevent and mitigate the harm.

Acknowledgments

Principal authors of the ecological section: Marc Ancrenaz,[21] Susan M. Cheyne,[22] Tatyana Humle[23] and Martha M. Robbins[24]

Principal authors of the social section: Justin Kenrick, Jake Willis, Anouska Perram and Chris Phillips[25]

Annex I: Pamela Cunneyworth[26]

Reviewers: Cheryl Knott, Susan Lappan, Freddie Weyman and Elizabeth Williamson

Endnotes

1   For the purposes of this chapter, the term *indigenous peoples* is used interchangeably with *forest peoples*, *forest-dwelling peoples* and *forest-dependent peoples*. The term *local communities* is broader: it also includes farming populations that are local by proximity but tend to see the forest as a resource to be exploited or cleared for agriculture, rather than something that sustains them.

2   For detailed information on the impact of tourism on apes, see Macfie and Williamson (2010).

3   For examples of long-term government plans for infrastructure development, see ETP (n.d.), Indonesia CMEA (2011), SEDIA (2008).

4   For details on the shifting of ranges among bonobos, see Hickey *et al.* (2013); among chimpanzees, see Fawcett (2000), Plumptre and Johns (2001), Plumptre, Reynolds and Bakuneeta (1997) and

Reynolds (2005); among chimpanzees and gorillas, see Rabanal *et al.* (2010); among gibbons, see Cheyne *et al.* (2016); and among orangutans, see Ancrenaz *et al.* (2010).

5   For information on chimpanzee road crossings in Bossou, Guinea, see Hockings (2011) and Hockings *et al.* (2006); in Bulindi, Uganda, see McLennan and Asiimwe (2016); and in Sebitoli, Uganda, see Cibot *et al.* (2015).

6   For details on the impact of human disturbance on African apes, see Junker *et al.* (2012); on bonobos, see Hickey *et al.* (2013); on chimpanzees, see Brncic *et al.* (2015) and Plumptre *et al.* (2010); on Grauer's gorillas, see Plumptre *et al.* (2016b); on mountain gorillas, see Van Gils and Kayijamahe (2010); on western gorillas, see Laurance *et al.* (2006); and on orangutans, see Wich *et al.* (2012b).

7   See, for example, Blake *et al.* (2007), Brncic *et al.* (2015), Geist and Lambin (2002), Hickey *et al.* (2013), Junker *et al.* (2012), Marshall *et al.* (2006), Murai *et al.* (2013), Plumptre *et al.* (2016b), Poulsen *et al.* (2009), Robinson *et al.* (1999), Wilkie *et al.* (2000).

8   For details on crop foraging by chimpanzees, see Hockings, Anderson and Matsuzawa (2009), Krief *et al.* (2014), McLennan and Ganzhorn (2017); by mountain gorillas, see Seiler and Robbins (2016); and by orangutans, see Ancrenaz *et al.* (2015b), Campbell-Smith *et al.* (2011b).

9   See also Chhatre and Agrawal (2009); Nelson and Chomitz (2011).

10   Unpublished FPP trip reports, 2006–17.

11   See, for example, Ordinance No. 74-1 of 6 July 1974 on establishing the rules governing land tenure (especially articles 1, 2, 14, 16) and Ordinance No. 74-2 of the same date on establishing the rules governing state lands (Alden Wily, 2011b, pp. 50–1).

12   Cameroon has not ratified ILO Convention 169; doing so would help cement FPIC as a right. To date, the Central African Republic is the only African country to have ratified the convention, and the island of Fiji the only Asian one (ILO, n.d.).

13   That FPIC is a legally enforceable right is apparent in key regional rulings. Those who seek to override community rights must prove that such action is necessary, proportionate and in the public interest. In a very practical sense, they have the right to have their claim heard and judged in relation to other rights claims. To justify non-consensual conservation measures such as the establishment of protected areas, states must demonstrate that such actions are "strictly necessary" and that "they have chosen the least restrictive option from a human rights perspective to satisfy the stated public interest" (MacKay, 2017).

14   Author interviews with Baka community members, Lomie, Cameroon, February 2010

15   FPP interviews with Bagyeli community members, Cameroon, 2014.

16   FPP staff member observation during a Baka community meeting, Assoumindélé, Cameroon, 2016.

17   FPP staff member observations during field trips to the region, Cameroon, 2016.

18   See also Kigula (2015) and MNRT (2012).

19   See also Oxfam, ILC and RRI (2016) and RRI (2016, 2017).

20   For examples of the growth of community tenure, see FPP, IIFB and CBD (2016).

21   HUTAN–Kinabatangan Orang-utan Conservation Programme (www.hutan.org.my) and IUCN SSC PSG Section on Great Apes.

22   Borneo Nature Foundation (www.borneonaturefoundation.org) and IUCN SSC PSG Section on Small Apes.

23   Durrell Institute of Conservation and Ecology (DICE), School of Anthropology and Conservation, University of Kent (www.kent.ac.uk/sac) and IUCN SSC PSG Section on Great Apes.

24   Max Planck Institute for Evolutionary Anthropology (www.eva.mpg.de) and IUCN SSC PSG Section on Great Apes.

25   All at Forest Peoples Programme (www.forestpeoples.org) at time of writing.

26   Colobus Conservation (www.colobusconservation.org).

Photo: In Africa, the projected annual cost of infrastructure is around $93 billion. Highway construction between Port-Gentil and Omboué, Gabon. © Julie Sherman

State of the Apes Infrastructure Development and Ape Conservation

CHAPTER 3

Deforestation Along Roads: Monitoring Threats to Ape Habitat

Introduction

The International Energy Agency predicts that governments and development agencies will invest more than US$33 trillion to build 25 million km of new paved roads through 2050, a 60% increase over levels in 2010. Nearly 90% of new roadway infrastructure is expected to be built in developing nations (Dulac, 2013). The Asian Development Bank estimates that climate-adjusted infrastructure "investment needs" from 2016 to 2030 will reach about US$16 trillion in East Asia and US$3 trillion in Southeast Asia (ADB, 2017, p. 43). Transportation, the second largest sector, accounts for 32% of expected climate-adjusted infrastructure investments in Asia over the same period. In Africa, the projected annual cost of infrastructure is around US$93 billion, about a third of which is for maintenance, leading to US$1.4 trillion in expenditures over the next 15 years (AfDB, 2011a, p. 28).

The chronic failure of governments to avoid degrading critical wildlife habitat when planning and building infrastructure suggests that this massive investment in transportation networks will have devastating effects on remaining forests (Quintero *et al.*, 2010).

More than other types of infrastructure, roads facilitate forest access that enables logging, settlement, hunting and other resource extraction, beyond direct damage to ecosystems (Trombulak and Frissell, 2000). In fact, many road networks in forested areas in the tropics are sited explicitly to extract natural resources (Nellemann and Newton, 2002). By providing access to forested areas, roads also catalyze various indirect disturbances to remaining habitat —including charcoal production and overhunting—that imperil apes and other arboreal mammals (Coffin, 2007; Wilkie *et al.*, 2000). Greater contact between apes and humans also facilitates the spread of disease between them (Köndgen *et al.*, 2008; Leroy *et al.*, 2004).

The World Bank proposed the concept of "smart green infrastructure" to minimize harm to tigers and their habitat, which are facing a similar crisis (Quintero *et al.*, 2010). The tenets of the Bank's mitigation hierarchy—avoidance, minimization, restoration and offsetting of negative effects—could be applied to reduce the damage caused to ape habitat by infrastructure development (see Table 3.3 and Chapter 4, p. 119). The specializations of many forest-dependent species, including most apes, to the forest environment's stable, moist, shaded conditions and complex architecture make them particularly vulnerable to the damage associated with roads (Laurance, Goosem and Laurance, 2009; Pohlman, Turton and Goosem, 2009;

see Chapter 2). Of particular importance to apes, then, is determining whether "greener" infrastructure development can help limit the secondary clearing and resource extraction associated with roads built through forest.

In considering the role of roads in the deforestation of ape habitat, this chapter presents four original case studies and draws on the authors' extensive experience in monitoring forest cover loss, placing particular emphasis on recent technological developments that have allowed for unprecedented access to high-resolution satellite imagery. The research conducted for this chapter reveals the following key findings:

- The construction of new roads in intact forest landscapes is frequently followed by major episodes of deforestation, leading to negative consequences for forest-dependent species such as apes. Deforestation occurs in forest along roads regardless of the protection status of the surrounding area.

- Three case studies presented in this chapter show that drivers of deforestation vary by location, but that road construction is consistently associated with a spike in forest loss, followed by elevated deforestation rates and a progression of forest loss outward from the road over time.

- In the case studies, illegal logging and smallholder agriculture occurred in small clearings close to roads. These activities are more strongly associated with incremental expansion outward from the road and the growth of settlement enclaves than the more organized, and often legal, conversion of larger patches of forest to plantation.

- Planning to avoid critical areas, regular monitoring of forest status and additional conservation action are needed to reduce the negative effects of roads on wildlife habitat. Simple but powerful approaches to detecting and measuring forest loss can help resource managers monitor the construction and land use change associated with legal roads and halt the building of illegal road clearings in contiguous forest tracts.

- Road design must address the access provided to natural areas by roads that cannot be rerouted. Even if a road does not hinder the movement of apes, the associated conversion of formerly inaccessible forest to other land uses can decimate resident ape populations, as has been the case for western Tanzania's chimpanzee populations.

- In Peru's primate-rich Amazon forests, deforestation tracking combines weekly alerts of tree cover loss with verification using high-resolution satellite imagery. This useful model for combating illegal road building and associated land clearing activities could easily be adapted to ape habitats.

Proposed New Approaches to Road Monitoring

Roads and other forms of transportation infrastructure can bring rural communities much-needed social and economic benefits, including access to markets and resources; however, this is not always the case (see Chapter 2, p. 60). Ideally, these arteries connect people to markets and resources while avoiding primary forest, sensitive habitats, animal dispersal and migratory routes, and unique natural communities. However, recent road planning often fails to consider these factors. Without proper planning and post-construction monitoring, roads can incur tremendous costs of time and money while devastating the surrounding environment and creating public health issues (Clements, 2013; Laurance *et al.*, 2009; see Chapter 1).

This chapter provides three examples of road construction projects that have affected

Photo: More than other types of infrastructure, roads facilitate forest access that enables logging, settlement, hunting and other resource extraction. © HUTAN–Kinabatangan Orang-utan Conservation Project

surrounding ape forest habitat. A fourth case study was conducted outside an ape range but is relevant to monitoring primate habitat; it shows how new data and tools available to the ape conservation community can help detect, monitor, predict and minimize forest loss. Specifically, satellite imagery and associated spatial data analysis tools now permit resource managers to more effectively monitor changes in canopy cover of ape habitat surrounding infrastructure and other developments (see Annex II). This approach has already been used to assess remaining habitat for tigers and to influence landscape-level planning to ensure their survival (see Box 3.1). It can be applied to ape habitat in the same way.

Data and maps of expected tree cover loss associated with proposed infrastructure routes can inform road siting and suggest preventive actions to minimize deforestation, assuming that high-level decisions incorporate environmental information. These tools can also help to reduce damage from roadways by:

- estimating potential impact within the area surrounding a proposed road;
- detecting tree cover loss along a new roadway before it expands;

BOX 3.1

Applying Lessons from Tiger Habitat Analysis to Ape Habitat Monitoring and Conservation

Like apes, tigers need large areas to survive. But loss of habitat, combined with overhunting of both tigers and their prey, has diminished the global wild population to fewer than 3,500 individuals (Joshi et al., 2016). Nevertheless, sufficient forested tiger habitat still remains across the species' range to bring the tiger back from the brink of extinction.

A recent assessment of critical tiger habitat utilized a new satellite-based monitoring system to analyze 14 years of forest loss data within the 76 landscapes that have been prioritized for the conservation of wild tigers (Joshi et al., 2016). Published in 2016, the study identifies enough forest habitat within tigers' geographic range to achieve the international commitment of doubling the wild tiger population by 2022—an initiative known as Tx2 (World Bank, 2016b)—with additional conservation investment.

The researchers systematically examined forest cover change across globally recognized tiger conservation landscapes (TCLs), which have a median area of 2,904 km^2 (290,400 hectares (ha)) (Joshi et al., 2016; Wikramanayake et al., 2011). They used high- and medium-resolution satellite data provided by Global Forest Watch and Google Earth Engine, along with analysis from the University of Maryland (GFW, 2014; Google Earth Engine Team, n.d.).

The open-access GFW platform provides tools that forest managers and others can use to measure and monitor critical habitats, analyze risk and prioritize conservation efforts. The research team used annually updated GFW tree cover data at a resolution of 30 m × 30 m to detect and locate forest loss.

The researchers estimate that forest clearing between 2000 and 2014—an area equivalent to nearly 80,000 km^2 (8 million ha) or 7.7% of the tigers' remaining habitat—resulted in the loss of habitat that could have supported an estimated 400 tigers, more than one-tenth of the global population (Walston et al., 2010). Across the 76 TCLs, forest loss was actually much lower than expected, given the region's rapid economic growth and high population densities.

Loss was also unevenly distributed: 98% of tiger forest habitat loss across the 29 most critical TCLs for increasing tiger populations occurred within just ten of these landscapes, primarily in Indonesia and Malaysia, where oil palm plantations are driving deforestation. Many of these TCLs, especially in Sumatra, are also home to critically important ape populations (IUCN, 2016c; see Chapter 7).

The results of the habitat assessment allow scientists and tiger range authorities to improve their understanding of the spatial distribution of intact forest, tree cover loss and human development within the TCLs so that conservation resources can be applied where they are most needed to avert further damage.

In Indonesia, more than 4,000 km^2 (400,000 ha) of unbroken forest expanses in TCLs have been allocated for oil palm concessions. Conversion of these forests would fragment forest corridors and deplete habitat in protected areas. If this faster rate of habitat loss is to be addressed, conservation investment in these TCLs will need to be particularly intensive and targeted at commodity production practices.

The tiger habitat assessment introduces tools that, had they been part of the toolkit of forest and wildlife managers, could have helped detect and address forest change even at the landscape level. Forest regrowth in Khata, one of Nepal's tiger corridors, coincided with a community-managed forestry program to restore forests for tiger dispersal in this region

- identifying trends in tree cover loss over time and effectiveness of various conservation actions (Clements *et al.*, 2014);
- helping decision-makers understand the patterns of loss and potential mitigation options; and
- highlighting best-practice examples of road construction that are followed by conservation action to contribute to the growing trend towards smart green infrastructure (Quintero *et al.*, 2010).

Until recently, the use of satellite data required substantial expertise and funding for the acquisition, processing, verification and interpretation of the raw information (Curran *et al.*, 2004; Gaveau *et al.*, 2009b; LaPorte *et al.*, 2007; see Annex II). Assessing deforestation at the landscape scale provided valuable evidence of the effects of human activity on forests, but the cost and effort needed to obtain the satellite data prevented widespread use of such approaches.

Global Forest Watch (GFW), a new forest change analysis platform, has transformed the process and increased access to the power of satellite imagery. It provides free access to spatially explicit tree cover change data, derived from thousands of

(Joshi *et al.*, 2016). Community-based anti-poaching teams now also patrol the forests to prevent wildlife poaching and habitat degradation. Timely knowledge of these positive results would have allowed forest managers to assist the Khata communities and focus official protection work elsewhere.

In contrast, the clearing of forests by people in search of land around Nepal's Basanta corridor has impeded tiger dispersal to the north, resulting in the absence of previously seen tigers from recent surveys. The human settlement process was identified by regional experts, whereas near-real-time forest loss alerts would have notified managers far sooner and enabled them to try to guide the settlement so as to reduce forest loss (Joshi *et al.*, 2016).

Updated forest cover information would also have helped small, isolated reserves, such as India's Panna National Park, where tigers were wiped out by poachers and the lack of connectivity to other reserves precluded tiger recolonization (Wikramanayake *et al.*, 2011). The park's vegetation and prey base were left intact, but the government had to transfer five tigers from nearby reserves to catalyze population recovery to more than 35 adult tigers.

The tiger habitat assessment was presented at a meeting of environmental ministers from tiger range states in Delhi, India, in April 2016. In the Delhi Pledge for Tiger Conservation, the conference delegates vow "to protect the tiger and its wild habitat to ensure crucial ecological services for prosperity" (PIB, 2016b). Delegates from five countries asked to use the satellite-based monitoring tools presented in the assessment to conduct and update their annual national tiger habitat analyses, and others described how this tool could help them to monitor habitat across the tiger range countries at the same scale (PIB, 2016a). The Global Tiger Initiative, an alliance of governments, international agencies, the private sector and civil society groups whose aim is to prevent the extinction of wild tigers, has also endorsed the approach (World Bank, 2016b).

Doubling the tiger population by 2022 will require moving beyond tracking annual changes in habitat. GFW's new forest loss alert system (at a spatial resolution of 30 m) will soon generate weekly alerts for forests across the tropics (M. Hansen, personal communication, 2017). Once the system is in place, forest managers in range states will be able to receive alerts of forest loss within a certain reserve, corridor or TCL in near-real time and take appropriate action. Tiger range state officials have expressed interest in integrating the weekly forest loss alerts into reserve managers' regular monitoring and reporting activities, as even rapid alerts still require immediate action on the ground to stop habitat degradation and loss.[1] For a relatively poor disperser such as tigers, community forestry programs, government initiatives and other stakeholders should also monitor the extent to which forest connectivity is regained. GFW's weekly updating can help track and even promote these interventions.

Tracking and detecting forest change through tree loss is even more relevant to arboreal animals, such as apes. GFW alerts allow for a weekly assessment of the level of risk posed by the fragmentation of thinly connected forest blocks, which is particularly important for the 20 species of gibbon (GFW, 2014). A continually updated, spatially explicit assessment of forest change will help identify and refine key areas for apes and evaluate the type and degree of threat to enable authorities and resource managers to take appropriate action. By making a population recovery commitment based on Tx2, but for great apes and gibbons, ape range states and conservation groups could jointly create an opportunity to facilitate the flow of attention and resources to key areas within ape habitat.

The maps of tiger habitat and tree cover change can be found online at globalforestwatch.org.

Landsat satellite images at a resolution of 30 m × 30 m, updated annually for the entire world (GFW, 2014; see Chapter 7). As of mid-2017, GFW began offering weekly updates of tree cover change for most ape range states to enable near-real-time habitat monitoring (GFW, 2014; M. Hansen, personal communication, 2017). Ape range stakeholders can use the online GFW tools to view and analyze tree cover loss data for a country or protected area, create custom maps or download data for their target region. GFW thus allows users with basic skills to monitor changes in habitat and generate critical information on forest change that can enhance their conservation efforts or monitor the effects of road building in near-real time.

Case Study Approach

This chapter presents past and expected change in ape forest habitat in three case study sites around roads that were substantially upgraded between 2001 and 2014 (see Annex III) and in one site outside the ape range, in the primate-rich tropical forests of Peru. The first three sites—two in northern Sumatra, Indonesia, and one in western Tanzania—are home to a total of four ape subspecies. The Sumatran sites lie in the Leuser Ecosystem and are home to the siamang (*Symphalangus syndactylus*), Sumatran lar gibbon (*Hylobates lar vestitus*) and Sumatran orangutan (*Pongo abelii*); the site in western Tanzania supports eastern chimpanzees (*Pan troglodytes schweinfurthii*). Peru's rainforests harbor more than 50 primate taxa, and the species numbers at several sites are among the world's highest (IUCN SSC Primate Specialist Group, 2006).

Specifically, the analysis applies the Global Forest Change 2000–14 data set to show the loss of ape forest habitat in areas up to 10 km from individual roads in the years before and after a road's construction or improvement (Hansen *et al.*, 2013). The quantification of tree cover loss over time at a fine scale allows for estimates of the location and scale of the effects of roads on forest habitat, the detection of patterns and the identification of areas where future loss is likely.

Further, the chapter examines aspects of road development associated with negative effects on ape habitat. It also assesses the potential of open-access GFW tools, such as forest loss alerts, and data for: a) fine-scale monitoring of forest surrounding roadways built or expanded between 2001 and 2014; b) quantifying forest loss from infrastructure and associated secondary development; and c) helping reserve managers and others do the same. A description of methods can be found in Annex III.

Recommendations for Road Infrastructure in Ape Habitat

Zoning Roads to Maximize Societal Benefit and Minimize Damage to Ape Habitat

Planning new roads to minimize environmental damage while maximizing societal benefits must include consideration of both their location and design. Of primary importance is the avoidance of new road construction through pristine habitat, where soils are commonly of marginal productivity and which are far away from markets (Laurance *et al.*, 2015b; Quintero *et al.*, 2010; see Table 3.3). Laurance and Balmford (2013) and Laurance *et al.* (2014a) propose global "road zoning" to identify and map the areas where roads would best connect people to markets and resources and where roads should not be built, including areas of primary forest, sensitive habitats, animal dispersal, ▸ p. 1

CASE STUDY 3.1

Roads Facilitate Industrial-Scale Agriculture Threatening the Leuser Ecosystem in Sumatra, Indonesia

Background

Over the past 50 years, human activity has reduced Sumatra's vast expanse of tropical rainforest to isolated remnants and a few large patches. Oil palm, pulpwood and other large-scale plantations have rapidly replaced the island's natural forest and now occupy 20% of the land area (Abood et al., 2015; De Koninck, Bernard and Girard, 2012). Forest clearing in the north of the island began in earnest in the 1980s and led to the loss of more than half of the formerly intact forests in Aceh province by 2000 (De Koninck et al., 2012).

The Leuser Ecosystem encompasses 25,000 km^2 (2.5 million hectares (ha)), including the Gunung Leuser National Park (GLNP), and is by far the largest and most significant forest remnant in Sumatra. It occupies the last remaining lowland forests and the largely mountainous, biodiverse rainforests of Aceh and North Sumatra province (De Koninck et al., 2012; GFW, n.d.-c). The Leuser Ecosystem comprises 78% of remaining habitat for the Sumatran orangutan and supports more than 90% of the remaining population—an estimated 14,600 individuals (Wich et al., 2008, 2016). It is most probably a critical refuge for the Sumatran lar gibbon and siamang (Campbell et al., 2008; Nijman and Geissmann, 2008). All three taxa are endangered by hunting and habitat loss and require intact forest canopy to survive (Brockelman and Geissmann, 2008; Nijman and Geissmann, 2008).

Established in 1995, the Leuser Ecosystem is a legal entity managed for conservation of the region's biological diversity and designed to contain viable populations of native species (Van Schaik, Monk and Robertson, 2001). Even in this protected area, however, people continue to clear forest and large-scale plantations have come to cover much of the historical ape habitat.

Hunting and the conversion of previously logged forests into monoculture plantations are the two principal threats to the three ape species in the Leuser Ecosystem (Geissmann, 2007; Wich et al., 2011, 2016). Quantifying local hunting pressure is beyond the scope of this analysis. Therefore, road buffer-zone distances were set to reflect the previous finding that wild meat hunting typically extends between 5 km and 10 km from roads, as reported by Laurance et al. (2009; see also Annex III).

Incursion of the Ladia Galaska Road Network

Ladia Galaska is a 1,650-km all-weather road expansion effort that is meant to link Aceh's west and east coasts through the province's mountainous interior (De Koninck et al., 2012). Since the mid-1990s, the mega-development project has upgraded and connected previously built roads, including routes that were passable only during the drier seasons. The Ladia Galaska road network cuts across the Leuser Ecosystem's northern section, fragmenting formerly intact forest and threatening forest biodiversity and water supply services for lowland communities.

Ladia Galaska has generated heated debate since it was proposed in the mid-1980s (Eddy, 2015). Aceh's governors have pushed to speed up construction, and many local communities support the project because it would improve their options for transporting palm oil and other commodities (Clements et al., 2014).

Critics have argued that Ladia Galaska threatens essential ecosystem services provided by the intact forest, including water supply for several million local residents, erosion and flood control, fire suppression and tourism (van Beukering, Cesar and Janssen, 2003; Wich et al., 2011). They also note the reduction and fragmentation of forest that is habitat to numerous iconic and threatened species, including critical orangutan and gibbon populations (Clements et al., 2014; IUCN, 2016a). Further, many of the roads are constructed in forested areas with steep slopes that are prone to earthquakes and landslides (Riesco, 2005). Finally, the partially completed project has met with opposition because it will expand access to the area's forests, including the GLNP. By facilitating illegal logging, it will continue to have a negative effect on critically important habitat of all three ape species as well as other unique Sumatran wildlife, including tigers and elephants (Gaveau et al., 2009b; Panaligan, 2005; Wich et al., 2008).

For this analysis, improvements to roads at two nearby sites served as case studies (see Figure 3.1):

- the Tamiang Hulu–Lokop (TH–L) road in the Leuser Ecosystem's eastern portion; and
- the Blangkejeren–Kutacane (B–K) road that runs through the Ecosystem's center, separating portions of Gunung Leuser National Park.

Roughly 54 km apart, they are two of roughly 16 sections that comprise the Ladia Galaska road improvement scheme (De Koninck et al., 2012).

Tamiang Hulu–Lokop Road Development

The east–west TH–L route near the village of Tampor Paloh was initially a logging road, visible in the 1980s. It was intensively developed during 2009–10 (see Figure 3.2).

Effects on the Surrounding Area, as Identified by GFW

Roughly 1,072 km^2 (107,200 ha) of forest remained within 10 km of the road in 2000 (see Table 3.1). Of this area, 243 km^2 were within agricultural concessions zoned for conversion to plantations. Prior to 2000, some lower-elevation forest where the road connected to a large oil palm plantation on its eastern edge had already been cleared. Between 2000

FIGURE 3.1

The Tamiang Hulu–Lokop and Blangkejeren–Kutacane Roads in the Leuser Ecosystem, Aceh, Sumatra, Indonesia, 2001–14

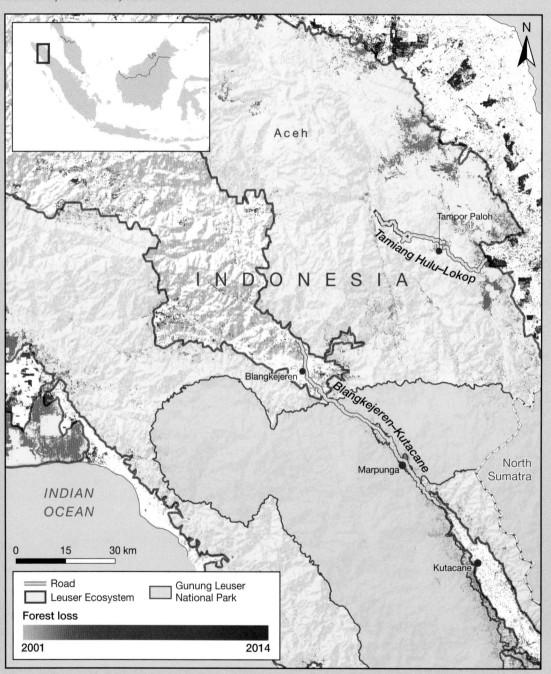

Notes: Forest loss is color-coded by year. Yellow–orange colors represent earlier years, and purple–blue colors represent later years.

Data sources: Google Earth Engine Team (n.d.); Hansen et al. (2013)[2]

FIGURE 3.2

Eastern Half of the Tamiang Hulu–Lokop Road with Forest Loss, Aceh, Sumatra, Indonesia, 2000–14

and 2014, additional natural forest was cleared within various concessions.

Most of the forest loss between 2000 and 2014 occurred within concessions that were still in natural forest in 2000 but cleared by 2014. This included 129 km² (12,900 ha) primarily in oil palm concessions within 0–5 km and another 114 km² (11,400 ha) within 5–10 km (see Table 3.1).

Outside of the concessions, the area along the road experienced scattered and limited deforestation before 2007. Between 2000 and 2006, the areas 0–5 km and 5–10 km from the road each lost less than 0.2% of the 2000 forest cover per year (see Figure 3.3). Prior to the road's improvement, most clearing took place immediately along the road or where it intersected rivers or prior clearings (roads, plantations). Much of the initial 2007 spike in deforestation occurred where the road intersected a river, due to an improved crossing and expansion of a local main road along the river's edge.

Improvement of the road in 2009 corresponded to a second surge in deforestation, as tree cover loss spiked again. The area within 5 km of the road lost nearly 0.8% per year for several years, after which the rate of loss declined (although plantation reclearing expanded).

Between 5 km and 10 km from the road, tree cover loss between 2009 and 2014 averaged 1.2% per year, a rate six times higher than the pre-2009 average. Despite improved access to interior forests all along the road, most loss took place in the lowland forests within previously designated concessions at the area's eastern edge or at intersections of the road with other roads or rivers. Deforestation within 10 km along much of the length of the road was limited to scattered small clearings that extended 100–200 m on either side of the roadway.

Notes: Forest loss is color-coded by year. Yellow–orange colors represent earlier years, and purple–blue colors represent later years. The large clearing at the eastern end of the road is an oil palm plantation established before the year 2000 and is excluded from the analysis.

Data sources: Google Earth Engine Team (n.d.); Hansen et al. (2013)[3]

TABLE 3.1

Tree Cover and Loss in Tamiang Hulu–Lokop Road Buffer Areas, Aceh, Sumatra, Indonesia, as Identified by Global Forest Watch

Buffer	Tree cover, 2000 (km²)	Tree cover loss, 2000–14 (km²)	Forest cover, 2000, excluding mature oil palm (km²)	Loss excluding reclearing (km²)	Total concession area (km²)
0–5 km	485	41	468	23	129
5–10 km	608	57	604	53	114
0–10 km	1,093	97	1,072	76	243

Notes: Values for tree cover in 2000 and tree cover loss in 2000–14 refer to the full extent of tree cover identified by GFW for those years. The values for forest cover in 2000 exclude a 17-km² mature stand of oil palm within 5 km and another 4-km² stand within 5–10 km that GFW mistakenly counted as forest (see Annex III). The reclearing of these areas between 2011 and 2014 was excluded from tree cover loss. Although nearly 25% (243 km² or 24,300 ha) of the total area with tree cover was within large-scale concessions, some of it was still natural forest in 2000.

Data sources: GFW (2014); Hansen et al. (2013)

FIGURE 3.3

Forest Loss Within the Buffer Zones of the Tamiang Hulu–Lokop Road, Aceh, Sumatra, Indonesia, 2000–14

Key:  0–5 km ■ 5–10 km

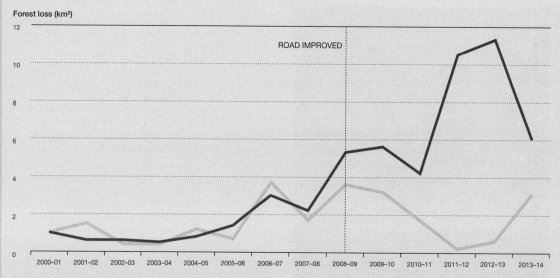

Notes: Road improvements took place in 2009. Loss values exclude the reclearing of a major oil palm plantation in the western edge of the buffer zones between 2010 and 2014 (see Figure 3.2).

Data sources: GFW (2014); Hansen et al. (2013)

Addressing Effects of Road Development

The findings suggest that, on its own, the upgrade of the TH–L road caused limited forest loss; however, it negatively affected ape populations because of its role in reducing key lowland forest habitat. Orangutans and lar gibbons favor lowland forest below 1,500 m (Brockelman and Geissmann, 2008; Campbell et al., 2008; Van Schaik et al., 2001; Wich et al., 2016). These species may persist at low densities within Leuser's remaining upland forest (Van Schaik et al., 2001; Wich et al., 2016). The improved TH–L road may have sped the conversion of lowland forest to oil palm within acknowledged plantation boundaries. Nevertheless, minimal settlement occurred along this route, concentrated along intersections with an existing road and river (see Figure 3.2). The narrow roadway clearing sits in a valley, and its hilly surroundings may have limited the establishment of side roads, which presumably would have led to additional forest clearing and hunting access.

Requiring agricultural concession holders to include in their management plans a series of maps of the forest types, endangered species, protected areas, roads and management activities could help identify where critical habitat may be in jeopardy. Accompanied by enforcement, such plans would encourage thoughtful concession design and enable independent and comprehensive review across a given region (Meijaard and Wich, 2014).

However, the power of Indonesian logging interests and a lack of capacity to control them have minimized restrictions placed on logging and conversion to plantations (De Koninck et al., 2012; Robertson, 2002). Most proposed improvements have either disregarded findings of their required environmental impact assessment or have ignored it altogether (Robertson, 2002; Singleton et al., 2004).

There is an urgent need to develop a systematic and refined land use monitoring system (De Koninck et al., 2012). The transparency afforded through regular monitoring by forest officials at various levels using tools such as GFW could greatly facilitate such efforts.

CASE STUDY 3.2

Roads Facilitate Small-Scale Agriculture and Encroachment into Gunung Leuser National Park, Sumatra, Indonesia

Blangkejeren–Kutacane Road Development

The Blangkejeren–Kutacane route, a section of the road that bisects both the Leuser Ecosystem and the Gunung Leuser National Park, also runs through a valley, but it differs from the TH–L road in that it does not provide access to large-scale plantations. Nevertheless, improved road access into the middle of the Leuser forest has invited serious encroachment and deforestation problems over time.

Historically, the B–K road served as a pathway between Blangkejeren and Kutacane. By providing access to the forest, it attracted settlers (Tsunokawa and Hoban, 1997; see Figure 3.4). The road was substantially improved in 2009, and illegal logging and agriculture have since widened the deforested strip along the road, which divides two large sections of the GLNP.

The road has provided transport and market access to two settlement enclaves, Gumpang and Marpunga, which were allowed to remain outside the boundaries of the GLNP (see Figures 3.4 and 3.5). These settlements have since expanded into National Park territory. The road has also provided forest access to loggers, who have illegally cleared sections alongside the adjacent Alas River and into the surrounding protected forest (McCarthy, 2002).

A lack of political will to enforce logging laws and collusion between powerful government officials and timber companies makes illegal logging in Leuser's protected forests especially hard to address (McCarthy, 2000; Wich et al., 2011).

FIGURE 3.4

The Blangkejeren–Kutacane Road, Aceh, Sumatra, Indonesia, Shown With Buffers of 5 km and 10 km, 2016

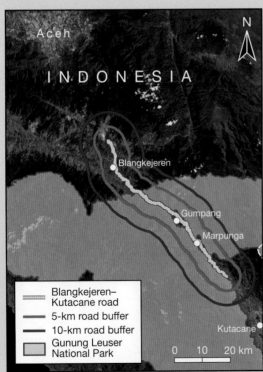

Notes: The road separates the forest blocks of Gunung Leuser National Park (in green). Two settlement enclaves along the road—Gumpang to the north and Marpunga to the south—are visible outside of the park boundary on the map.

Data source: Google Earth (n.d.)[4]

FIGURE 3.5

Section of the Blangkejeren–Kutacane Road with Protected Forest on Both Sides and Forest Loss, Aceh, Sumatra, Indonesia, 2000–14

Notes: Forest loss progresses over time outward from the road, including outward from the concentration of loss in the enclave of Marpunga. The clearing deep inside Gunung Leuser National Park at center left is a landslide.

Data sources: Google Earth (n.d.); Hansen et al. (2013)[5]

TABLE 3.2

Tree Cover and Loss in Blangkejeren–Kutacane Road Buffer Areas, Aceh, Sumatra, Indonesia, 2000–14, as Identified by Global Forest Watch

Buffer	Forest, 2000 (km^2)	Forest loss, 2000–14 (km^2)	Forest loss, 2000–14 (%)	Pre-2009 average annual loss (km^2)	Post-2009 average annual loss (km^2)
0–5 km	646	53	8.1	2.4	5.5
5–10 km	818	27	3.3	1.3	2.7
0–10 km	1,464	79	5.4	3.7	8.2

Note: No concessions occurred within the buffer areas of this road.

Data sources: GFW (2014); Hansen et al. (2013)

Effects on the Surrounding Area, as Identified by GFW

Roughly 1,464 km^2 (146,400 ha) of forest remained within 10 km of the road in 2000 despite decades of regular use (see Table 3.2). Forest loss between 2000 and 2006 was consistently greater along the B–K road than around the Tamiang Hulu–Lokop road, averaging 1–3 km^2 per year within 5 km of the road and 1.0–1.5 km^2 per year within 5–10 km.

The B–K road was upgraded in 2009. Forest loss tripled that year and remained high; the average area of forest lost annually between 2009 and 2014 was more than double that for the period between 2001 and 2008.

Between 2000 and 2008, roughly 3.7 km^2 (370 ha) of forest was lost each year within the entire 0–10-km buffer area. This rate more than doubled during the years after the road improvement (see Table 3.2 and Figure 3.6). Most of the loss occurred within 3 km of the road. Part of the improved road section runs through Blangkejeren, which was already well settled before 2000 and which lost relatively little additional

FIGURE 3.6

Forest Loss Within the Buffer Zones of the Blangkejeren–Kutacane Road, Aceh, Sumatra, Indonesia, 2000–14

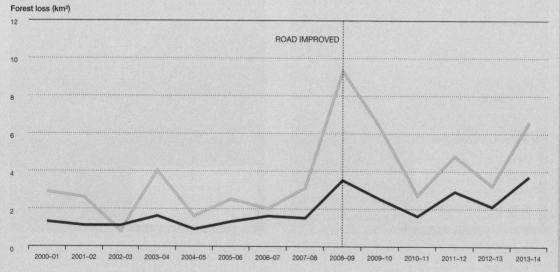

Note: The buffer zones did not contain plantations.

Data sources: GFW (2014); Hansen et al. (2013)

FIGURE 3.7
Progression of Forest Loss Along a Section of the Blangkejeren–Kutacane Road, Aceh, Sumatra, Indonesia, 2003–14

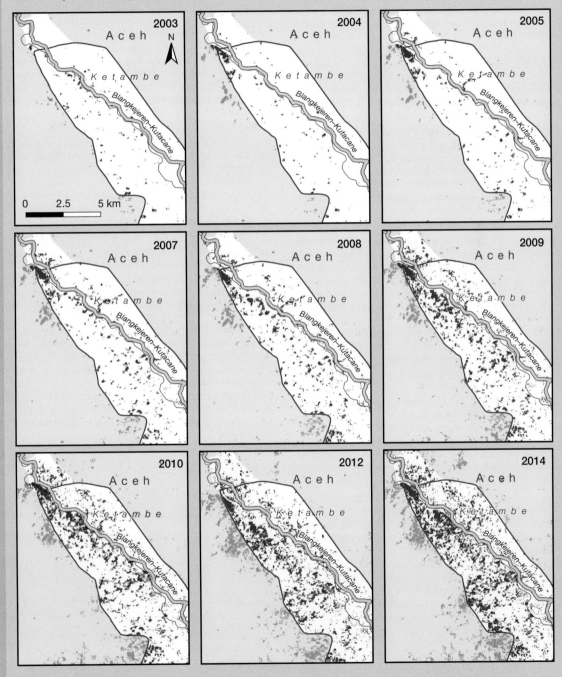

Note: Forest, shown in pale green, lies within Gunung Leuser National Park.

Data sources: GFW (2014); Hansen *et al.* (2013). All maps © OpenStreetMap and contributors (www.openstreetmap.org/copyright)

forest cover during the study period. Nonetheless, forest loss overall was more extensive on the two ends of the road section, near the established towns.

As with the TH–L road, the 2009 upgrade of the B–K road corresponded to a surge in deforestation (see Figure 3.6). The average rate of loss rose over the following years at both distances from the road, but particularly close to it. Within 5 km of the road, the 0.9% average annual rate of forest loss after the 2009 upgrade was more than double the 0.4% loss rate prior to the upgrade. At a distance of 5 km to 10 km from the road, forest loss between 2009 and 2014 averaged 0.3% per year, also double the pre-2009 average.

One explanation for the negative influence of roads on forest cover is the shifting of loggers' efforts once the road itself is no longer a barrier. As soon as a good road is available, loggers or settlers may be more willing to spend a day clearing forest from a point on the road than if they had already had to travel over 20–50 km of bad road that day. The upgrading of the road facilitated access to interior forests inside the GLNP, despite the hilly terrain that may have limited clearing in steeper areas.

Incursions into the GLNP thus accelerated over time (see Figure 3.7). Progression of forest loss within the park spiked in 2004, building on smaller incursions of the previous two years. Loss spiked again in 2008 and 2009, also after several years of smaller incursions. This pattern of smaller but consistent, incremental clearing along the B–K road stands in contrast to the clearing of larger blocks within the concessions of the TH–L road, where little settlement occurred. The images in Figure 3.7 show the spatiotemporal progression of deforestation inside the GLNP along the B–K road.

Predictive models have shown that forest areas near roads in Aceh are increasingly vulnerable to deforestation. Researchers expect the extent of orangutan habitat to decrease by another 16% between 2006 and 2030, which would cause major declines in the current global population (Clements et al., 2014; Gaveau et al., 2009b). Forest conversion and fires have followed logging activity along many Indonesian logging roads, heightening the vulnerability of resident ape populations (Clements et al., 2014; Laurance et al., 2009).

Addressing Effects of Road Development

The Ladia Galaska project is a case of poor land use planning, as exemplified by the B–K road (Wich et al., 2008). Whereas the TH–L road facilitated wide-scale conversion of lowland forest to oil palm inside designated plantations, the B–K road bisects the mountainous Gunung Leuser National Park. Aerial photographs taken before and after an earlier (1982) upgrade of the B–K route show that the improved access facilitated uncontrolled illegal settlements inside the park around the Gumpang and Marpunga enclaves (Singleton et al., 2004). The improved road enabled the settlers to enter the GLNP illegally, extract resources from the park and poach wildlife. The 2009 improvement further encouraged forest loss around these growing human enclaves, within an otherwise remote national park.

The Leuser Ecosystem is officially protected by presidential decree and provides water for millions of Aceh residents (Eddy, 2015; Singleton et al., 2004; van Beukering et al., 2003). Nevertheless, some Ladia Galaska roads traverse the region's steep slopes, cutting through protection forests, which have an average slope of 40% or more, as well as conservation forests, including the GLNP and water catchment areas. Scientists at the Center for International Forestry Research have recommended redirecting Aceh's road investment away from remote Leuser forests to existing roads that need improvements along the coast, where more agriculture and settlement occur and forests have been degraded. This shift would benefit more residents and incur lower environmental costs (CIFOR, 2015; Laurance and Balmford, 2013).

Projections based on economic and environmental data suggest that Aceh forests that are near roads have a higher risk of deforestation, leaving viable ape habitat only in the more remote sections of the Leuser Ecosystem (Gaveau et al., 2009b; Van Schaik et al., 2001). The indiscriminate spread of clearings along the B–K road and other roads within the Leuser Ecosystem will increasingly fragment the GLNP and two of the three largest remaining orangutan populations.

As the hills within the GLNP are quickly becoming a last refuge for apes on Sumatra, additional conservation action must address not only the access provided by this road and associated settlement enclaves, but also the lack of law enforcement capacity, as both factors enable illegal logging to continue within park boundaries (Eddy, 2015; Robertson, 2002; Wich et al., 2011). Along established roads, posts established by local non-governmental organizations (NGOs) and resource managers at road and river checkpoints could help to prevent loggers from entering the GLNP and to confiscate wildlife and logs that are being removed from the park illegally (Singleton et al., 2004). Planning new roads so that they avoid or minimize forest clearing will be crucial to apes' persistence in the Leuser Ecosystem (Jaeger, Fahrig and Ewald, 2006; Nijman, 2009).

CASE STUDY 3.3

Stepwise Road Construction through Chimpanzee Habitat in Western Tanzania

Background

The Ilagala–Rukoma–Kashagulu (I–R–K) road in the west of Tanzania has facilitated settlement of forests and woodlands east of Lake Tanganyika (see Figure 3.8). The region contains large tracts of intact woodland characterized by *Brachystegia* species (spp.) and *Julbernardia* spp. that provide high-quality habitat for a diversity of species, including the eastern chimpanzee (Piel *et al.*, 2015). Forested lands to the south of the Malagarasi River are under increasing threat from a human population growing at an annual rate of 2–5%, one of the highest rates in Tanzania.

The study area includes 20 villages, most of which lie along the lakeshore, and areas in six land tenure categories—village forest reserves, other demarcated village lands, Kungwe Bay Forest Reserve, local authority forest reserves, Mahale Mountains National Park (MMNP) and general land not reserved for a specific use or a particular village. Fishing and subsistence farming are the region's main economic activities; hunting is not a major economic venture in this region.

The road runs along the shores of Lake Tanganyika, from the Malagarasi River south to the southern border of the MMNP. Fewer than one-third of Tanzania's 2,500 chimpanzees live inside Gombe and Mahale Mountains National Parks, where they are well protected (Moyer *et al.*, 2006; Piel *et al.*, 2015; Plumptre *et al.*, 2010). Most chimpanzees in the region live at lower population densities outside protected areas. The most recent draft of the Tanzania National Chimpanzee Management Plan considers infrastructure, settlements and smallholder agriculture "very high" threats to chimpanzees and habitats at the national scale (TAWIRI, in preparation). A previous analysis, carried out in 2011 using the same methodology, ranked settlements and infrastructure as "high" (Lasch *et al.*, 2011); the reassessment suggests that the threat from infrastructure development increased from 2010 to 2016.

Incursion of the Ilagala–Rukoma–Kashagulu Road

The I–R–K road is the primary infrastructure development in the region. It is being built in sections. Section A of the road—between the Malagarasi and Lugufu rivers—connected villages along the lakeshore long before 2000 (see Figure 3.8). It was expanded during the main road construction phase in 2006–07, when a bridge was built across the Lugufu. The absence of a bridge prior to 2007 had limited travel between areas north and south of the river. Similarly, no road existed south of the Lugufu River before 2007, when extension of the road began. Subsequent sections were built over the next seven years, as funding became available. No road planning or impact assessment of its design or implementation was conducted for Sections A–E (K. Doody, personal communication, 2017).

Construction plans foresee an extension of the road southward, as a way of connecting Rukoma village, north of the MMNP, with remote villages south of the park. Narrow dirt roads of cleared vegetation already run from Rukoma for 20 km, connecting scattered settlements east and south of the MMNP (Sections E and G). As of 2017, a 13-km segment of Section F of the road, along the eastern boundary of the MMNP, was still at the proposal stage (see Figure 3.8).

Effects on the Surrounding Area, as Identified by GFW

Before 2006, areas across the region experienced moderate forest loss, even before road construction, as people were already living in the area and converting forests to farmland (see Figure 3.9). The building and upgrading of the I–R–K road, beginning in 2006–07, correlated with dramatic increases in forest loss, particularly within the 0–5-km buffer in the Lugufu–Ntakata area (5.5 km^2 or 554 ha), where the new road bisected large patches of pristine forest and miombo woodland. In the Masito area, a smaller 2007 spike in tree cover loss (1.2 km^2 or 121 ha) within the 0–5-km buffer reflected the area's already diminished forest cover, as deforestation along the existing dirt road there had begun prior to 2000. In contrast, no spike in forest loss occurred in 2007 in the Mahale East area, as the corresponding road section was not yet built. The increase in forest loss in Mahale East after 2011 is probably due to a gradual influx of settlers from shoreline villages north and south of the MMNP via dirt tracks.

Both high-resolution satellite images and community forest monitoring data indicate that the most important drivers of deforestation within areas up to 10 km from the road are the building of side roads and houses, farming, livestock grazing and charcoal production. The improved road in Section A and new road in Sections B–D facilitated residents' access to new agricultural and charcoal markets in Kigoma, north of the study area, and made it easier for people from villages north of the Malagarasi River to migrate south and settle in previously remote forest and woodland.

The road's construction in 2006–07 corresponded to a wave of forest loss reaching beyond the 10-km buffer in both the Masito and Lugufu–Ntakata areas (see Figure 3.10). In Lugufu–Ntakata, the largest forest loss in all years occurred within the 0–5-km buffer area, and forest loss decreased with distance from the road. In Masito, greater forest loss occurred in areas between 5 km and 10 km from the road. The road that existed in Masito before 2007 connected to an extensive network of footpaths. It is therefore likely that substantial forest within 5 km of the main road had already been lost in Masito before 2007.

An alarming trend in both Masito and Lugufu–Ntakata is the increase in forest loss 25 km to 30 km away from the I–R–K road—at levels significantly higher than before 2007. Most of these areas lack roads, enabling chimpanzees to range and disperse across the landscape. The Ntakata Forest east of Rukoma, the current terminus of the paved road, is critical habitat for chimpanzees, as it allows dispersal of individuals from and to the MMNP chimpanzee population (see Annex V).

FIGURE 3.8

Distribution of Forest and Woodland Vegetation with 5-km and 10-km Buffers Along the Ilagala–Rukoma–Kashagulu Road, Tanzania, 2000

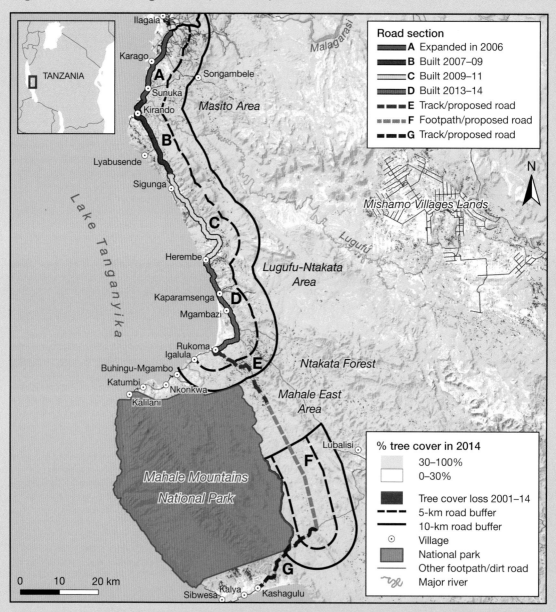

Notes: Letters refer to road sections built during different time periods. A dirt road in Masito (Section A) was improved and expanded in 2006. Between 2007 and 2013, Sections B–D in Lugufu–Ntakata were built, and a narrow dirt road cleared in Sections E and G. Section F surrounds a proposed future stretch of the road. The analysis excluded areas inside the MMNP because habitats within the park were relatively well protected during the study period. Forest and woodland vegetation were defined as areas with a tree cover density of more than 30% (see Annex III). ArcGIS Desktop (Esri, 2016) was used to digitize road construction based on DigitalGlobe satellite images from 2003–16, using the ImageConnect 5.1 plug-in; Google Earth was used to digitize road construction based on Landsat satellite images from 2000–16.

Data sources: Hansen *et al*. (2013); OpenStreetMap (n.d.)

FIGURE 3.9

Forest Loss in the Ilagala–Rukoma–Kashagulu Road's (a) 0–5-km and (b) 5–10-km Buffer Zones, Tanzania, 2000–14

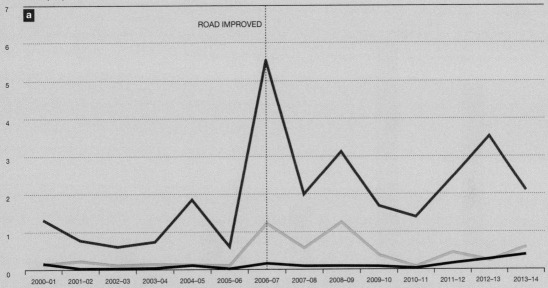

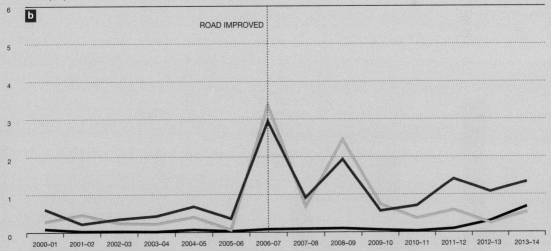

Notes: Lines correspond to the northern Masito, central Lugufu–Ntakata and southern Mahale East areas (Sections A, B–E and F, respectively, in Figure 3.8). The road development expanded a road in Masito and involved the construction of a new road to the Lugufu–Ntakata area. Tree cover loss spiked in 2007 in both Masito and Lugufu–Ntakata; deforestation continued at an elevated rate in Lugufu–Ntakata. The road has not yet reached the Mahale East area, to the south of the existing road.

Data sources: GFW (2014); Hansen *et al.* (2013)

Chapter 3 Deforestation Along Roads

FIGURE 3.10

Forest Loss Before and After Road Construction Within 5–30 km of the I–R–K Road in the (a) Masito and (b) Lugufu–Ntakata Areas, Tanzania, 2001–06 and 2007–14

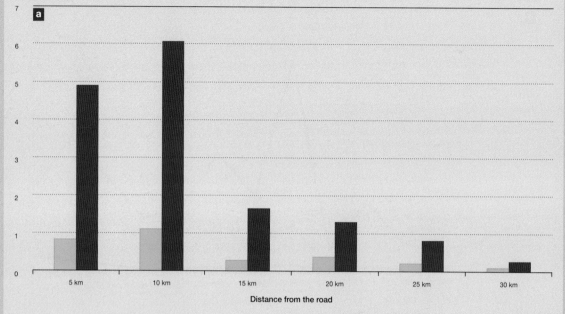

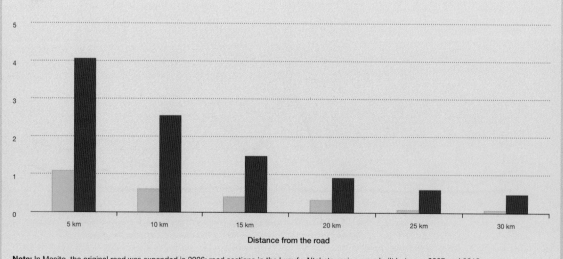

Note: In Masito, the original road was expanded in 2006; road sections in the Lugufu–Ntakata region were built between 2007 and 2013.

Data sources: GFW (2014); Hansen *et al.* (2013)

Tanzania's National Roads Agency (TANROADS) has permission and funding to clear an 18-km stretch of forest and woodland to build Section F, the next planned segment of the road (see Figure 3.8). The potential impact of building this section and upgrading existing footpaths and dirt roads along Section E is a cause of concern for chimpanzee conservationists. The increased access provided by these tracks has already hastened forest loss north and northeast of the MMNP. Unless it is properly planned and managed to restrict illegal settlements, the construction of the new road east of the park is expected to increase rural population density, intensify deforestation, and contribute to isolation of Tanzania's largest remaining and well-protected chimpanzee population in the MMNP—about 550–600 individuals. It also threatens the large numbers of chimpanzees living outside the park, partly because they depend on the zone of connectivity between MMNP and the Ntakata Forest.

The road itself will not stop chimpanzee movements; however, it will attract settlers who will clear adjacent forest to farm, graze livestock or burn charcoal in this otherwise remote region. Most of the area alongside the new road is general or village land and lacks any form of protection. Loss of intact, roadless areas for the most densely populated chimpanzee habitats in the region will have disastrous consequences for the overall health and viability of chimpanzees in Tanzania.

Addressing Effects of Road Development

As part of a conservation action planning (CAP) process, some communities along the road have developed village land use plans and established village forest reserves based on recommendations for mitigating habitat loss (Lasch et al., 2011). If they were to be granted protected status, these reserves could help maintain forest cover along the road and serve as buffers between the road and the core chimpanzee habitats.

Plans resulting from subsequent CAP processes have called for the identification of areas where roads are likely to expand into critical chimpanzee habitats and the application of a hierarchy of mitigation strategies for greening infrastructure (Plumptre et al., 2010; Quintero et al., 2010; TAWIRI, in preparation; see Table 3.3 and Annex V). The plan for the Mahale region advises against building the remaining sections of the road, suggesting that, at a minimum, the routes be moved farther away from the MMNP. If Section F of the road must be built, the plan urges development and implementation of a detailed land use plan to protect the forest on either side of the road, so that chimpanzees can cross the road safely and use the surrounding habitat.

Conservation groups have met with TANROADS to design the new section and to address the potential loss of chimpanzee habitat as people use the new road to move into the area (K. Doody, personal communication, 2017). In principle, TANROADS agreed to conduct an environmental impact assessment. Continued dialog between the TANROADS road developers, Uvinza district government, communities and conservation practitioners will be critical to the proper design of future road improvements and to the implementation of conservation strategies to avoid unplanned settlement and conversion of forests to other land uses.

One such strategy is to establish a new, locally administered protected area that serves as a buffer against future loss of forest and woodland along the road. Tanzania's ongoing conservation action planning processes, such as the chimpanzee management planning process, provide an opportunity to integrate road development, land use and other chimpanzee conservation efforts at the national level to maximize societal benefits from future roads while minimizing the effects on chimpanzees and biodiversity in general.

Photo: © Jabruson 2018 (www.jabruson.photoshelter.com)

CASE STUDY 3.4

Integrating Forest Loss Alerts with In-Depth Analysis to Tackle Deforestation in Near-Real Time

An innovative forest mapping effort in primate-rich Amazonian forests may provide a useful model for monitoring ape habitat at a fine scale. The Monitoring of the Andean Amazon Project (MAAP) integrates and applies a suite of remote sensing tools to detect and monitor the status of deforestation events (MAAP, 2016, n.d.). The project team combines Landsat satellite images (which are of medium resolution) with high-resolution images from DigitalGlobe and Planet, radar-based imagery and Global Land Analysis & Discovery (GLAD) forest loss alerts to identify patterns and drivers of deforestation in near-real time (GLAD, n.d.; see Annex IV).

The MAAP team's first step in identifying deforestation hotspots is receiving a GLAD alert in the area. On a weekly basis, the GLAD system accesses and analyzes Landsat imagery across the tropics. GLAD alerts are triggered when a threshold portion of a 30 m × 30 m pixel in a user's area of interest changes from forest to non-forest cover (Hansen et al., 2016). The team allows the alerts of tree cover loss to guide their investigations into deforestation events. Each of the thousands of GLAD alerts is presented as a pink spot on a map (see Figures 3.11 and 3.12). MAAP's area of interest is all of Peru, but the selected area could instead comprise a specific protected area, a road corridor or a multi-country region.

The MAAP team reviews the high-resolution imagery of a target spot from different time periods to confirm that the alert represents deforestation. The team can then bring the alert data into a geographic information system (GIS) to produce a detailed map or to investigate drivers of the forest loss (see Figure 3.12b–c).

At this writing, the MAAP team was improving its analysis of the distribution and intensity of alerts to identify overarching patterns and drivers of deforestation (M. Finer, personal communication, 2016). MAAP analyzed the average size of deforestation events in the Peruvian Amazon to help NGOs and national authorities understand deforestation patterns and prioritize response actions. The analysis found that large-scale deforestation (more than 50 ha)—mainly from cacao and oil palm plantations—accounted for just 8% of deforestation events, while small-scale deforestation (fewer than 5 ha) from clearings along roads made up more than 70% of deforestation events (MAAP, 2016). Since larger-scale clearing can expand rapidly, these monitoring activities need to remain a priority.

GLAD already operates in much of the Congo Basin, Indonesia and Malaysia, and it should be available to help managers easily and consistently monitor all tropical forests by late 2017 (GFW, 2014). By helping to detect habitat loss at the onset of road building, alerts will facilitate more timely, and therefore more effective and efficient, interventions (Hansen et al., 2016). Since forest loss alerts provide rapid updates, they can help guide associated development and enforcement, as they have in Peru, to ensure that no additional illegal development happens along roads where restrictions or planning regulations have been established.

FIGURE 3.11

Sample Set of GLAD Forest Loss Alerts Near Kisangani, Democratic Republic of Congo, January–March 2017

Notes: The image shows deforestation along roads and rivers, emphasizing the relationship between the access provided by these transport corridors and forest loss.

Data sources: GFW (2014); Hansen et al. (2013)

FIGURE 3.12

Sample Set of Images Showing the Process of Examining and Integrating GLAD Forest Loss Alerts into Forest Trend Mapping Near Cordillera Azul National Park, Peru, January–July 2016

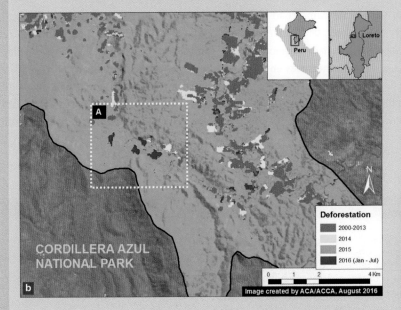

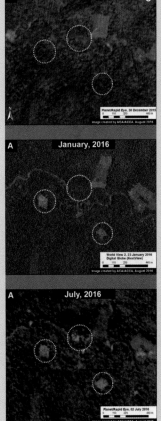

Notes: The images show illegal forest clearings in a protection forest. The initial (a) GFW alerts can be downloaded and combined with other data in a (b) geographic information system (GIS) and examined in greater detail with (c) high-resolution satellite imagery to help determine drivers of forest loss.

Data sources: (a) GFW (2014); Hansen *et al*. (2013); MAAP (2016); (b) and (c) DigitalGlobe (n.d.); MAAP (2016); Planet (n.d.).

Photo: Siting a new road in areas with substantial economic activity, such as northern Aceh, rather than through a large tract of intact forest, could improve farmers' market access and avoid a possible environmental disaster.
© Joerg Hartmann/TNC

migratory routes and unique natural communities. However, many decision-makers fail to consider these factors during road planning processes. The consequences can be devastating to natural environments while wasting time and money connecting areas in ways that help relatively few people (Laurance *et al.*, 2015b; see Chapter 1, p. 28).

Current road planning and mapping efforts inadequately assess environmental and socioeconomic impacts, and especially the indirect effects, such as unplanned colonization, hunting and secondary road building (Clements *et al.*, 2014; Laurance *et al.*, 2014a). Roads that stimulate uncontrolled immigration lead to greater satellite clearing and other forest damage by settlers (Angelsen and Kaimowitz, 1999; Liu, Iverson and Brown, 1993). Chimpanzees and orangutans appear to tolerate some road presence. However, the subsequent conversion of newly accessible forest to settlements, farming, charcoal and other uses encourages further forest clearing and hunting, a major threat to apes and other large-bodied animals (Laurance *et al.*, 2006, 2009).

If building new transportation infrastructure cannot be avoided, best practices can help to minimize negative consequences for the surrounding ecosystem (see Table 3.3). Monitoring logging roads and closing them after extraction ends can restrict access to illegal loggers and animal poachers (Laurance *et al.*, 2009). Following recommendations of environmental impact assessments that consider both roads and associated clearing and hunting and conducting enhanced patrolling and monitoring of forest on both sides of a road can further help to minimize the negative effects of infrastructure on forest ecosystems (Clements *et al.*, 2014; Quintero *et al.*, 2010).

Rerouting a proposed road may be the cheapest and most effective means of avoiding areas of critical wildlife habitat, but in poor countries covering this additional cost will probably require creative fundraising (Quintero *et al.*, 2010). Fees from ecotourism income and visitors, international payments for ecosystem services, public–private partnerships and sales of sustainably harvested timber within

TABLE 3.3

The Mitigation Hierarchy

Mitigation step	Description
Avoidance	Measures taken to avoid negative effects from the outset. These include careful spatial or temporal placement of elements of infrastructure in ways that completely avoid harming certain components of biodiversity.
Minimization	Measures taken to reduce the duration, intensity and/or extent of impacts that cannot be completely avoided, as far as is practically feasible.
Rehabilitation/ restoration	Measures taken to rehabilitate degraded ecosystems or restore cleared ecosystems following exposure to effects that could not be completely avoided and/or reduced.
Offset	Measures taken to achieve no net loss of biodiversity, by compensating for any significant adverse effects on biodiversity that could not be avoided or reduced, and/or by compensating for lost biodiversity that could not be rehabilitated or restored. Offsets can involve restoring degraded habitat, arresting degradation, averting risk or preventing at-risk areas from experiencing biodiversity loss.

Note: For more information, see Chapter 4, p. 119.

Source: Quintero *et al.* (2010)

production forests could help offset costs, pay for the rerouting of a road or allow for the mitigation of its environmental impacts (Dierkers and Mattingly, 2009; Laurance et al., 2014a). Park entry fees or impact fees for roads that transit protected areas can and should be used to minimize associated clearing in adjacent forest. Engaging lenders early in the process can help to guide funding to less damaging projects (Laurance et al., 2015a). Concentrating roads in already developed areas should make construction and maintenance costs, as well as use fee collection systems, more cost-effective. Such efficient use of funds could encourage international banks to support a project.

Applying Road Planning to the Local Context

Refining the global map of Laurance et al. (2014a) using local-scale data on distributions of natural resources and human communities for specific proposed roads could guide decision-makers in determining whether and where to site new roads. Siting a new road in areas with substantial economic activity, such as northern Aceh, rather than through a large tract of intact, unprotected forest, such as Gunung Leuser National Park, could improve farmers' market access and avoid a possible environmental disaster (Rhodes et al., 2014; Wich et al., 2011). In the case in western Tanzania, this protocol would call for avoiding a new road through the single remaining habitat corridor for chimpanzees and other woodland species in and out of Mahale Mountains National Park. In that context, integrating road planning with village land use planning and data collection, as recommended by Tanzania's CAP process, could help reduce local habitat loss (Clements *et al.*, 2014; see Annex V).

Laurance and Balmford (2013) suggest collaborative, multidisciplinary teams that combine satellite data on forest cover with information on transportation infrastructure, agricultural production, biodiversity distribution and other relevant factors to produce maps that can help government and other stakeholders plan roads in ways that achieve environmental and societal goals. Development banks and other major funding bodies have a key role in supporting efforts to harness the capacity of roads to improve local economies without damaging natural resources. Open-access monitoring tools would allow such integrated, cross-agency teams to analyze the effects of infrastructure-related development to improve monitoring and planning for future developments.

The dynamics of road infrastructure and human activity are complex and often case-specific. Roads not only respond to, but also stimulate, increased human population density. Some roads, such as those in western Tanzania, are built specifically to support existing settlements. Elsewhere, speculators are known to purchase and clear forested land to demonstrate ownership in anticipation of the progression of a new road into hitherto intact forest (Angelsen and Kaimowitz, 1999). Moreover, deforestation where roads are built to transport minerals, logs or palm oil from vast cleared areas that have few people may not depend directly on population density (Curran *et al.*, 2004; Kummer and Turner, 1994). Independent information sources are therefore essential to understanding the deforestation that accompanies various categories of roads.

> Development banks and other major funding bodies have a key role in supporting efforts to harness the capacity of roads to improve local economies without damaging natural resources.

The Potential of Remote Sensing Tools to Detect and Monitor Changes in Ape Habitat

Remote sensing imagery can serve as an independent source of information. At some point over the course of new infrastructure development, such imagery will capture the tree cover loss that results from construction and subsequent human activity. Through the above-mentioned weekly forest loss alerts, the detection of tree cover change will be sped up dramatically (see Annex IV). These data can be strengthened by undertaking ape habitat mapping and analysis using landscape metrics to assess habitat connectivity, fragmentation and patch size, shape and richness in relation to ape distribution and abundance (M. Coroi, personal communication, 2017).

Resource managers in ape range countries can verify the effects of infrastructure on forest cover by contrasting the status of surrounding forest before and after infrastructure projects in analyses similar to the ones presented in this chapter's case studies. Forest loss and land cover data can help predict where ape habitat and populations may already be degraded. Managers can complement data on a proposed road with lessons learned from previous case studies to inform the process of determining the new road's location and design. Proposed roads and other developments indicate where remaining ape populations will be most affected in the future (Laurance *et al.*, 2006). Detecting and monitoring loss of forest habitat in ape range countries through rapid analysis will also help managers to reduce the effects of infrastructure presence through targeted local action.

The drivers and patterns of deforestation in these cases vary by location, yet the spike from road-associated development is seen consistently in all cases and at various distances from the respective roads. GFW forest change analysis tools can help researchers, managers and policymakers to quantify changes in forest cover over time, from both road construction and the subsequent development associated with it. The accumulation of spatially explicit forest change data allows users to communicate these changes to policymakers and to maintain decision-making transparency.

The increasing fragmentation and conversion of ape habitat documented elsewhere in this volume underscore the importance of road construction as a proximate driver of that loss. Dealing with the underlying drivers of habitat loss is beyond the scope of this analysis, although they must also be addressed. In view of the ongoing expansion of road networks, the simplest solution is to focus on improving roads close to population centers; while at the same time avoiding the construction of new roads in intact forests and stopping the upkeep of roads that once were used for extraction purposes, so that forest access may be cut off (Clements *et al.*, 2014; Laurance and Balmford, 2013).

Numerous studies cited here and elsewhere suggest that roads and wildlife do not coexist well in any country unless stakeholders adopt principles of smart green infrastructure. A shift to a model that embraces these principles must become a prerequisite for development in all wildlife habitats, including in regions that harbor remnant wild ape populations.

Acknowledgments

Principal authors: Suzanne Palminteri[6], Eric Dinerstein[7], Lilian Pintea[8], Anup Joshi[9], Sanjiv Fernando[10], Agung Dwinurcahya[11], Serge Wich[12] and Christopher Stadler[13]

Annexes II, III, IV and V: The authors

Reviewers: Leo Bottrill, David Edwards and Wijnand de Wit

Endnotes

1. Author interviews with delegates at the Third Asian Ministerial Conference on Tiger Conservation, New Delhi, April 2016.
2. Map sources: Aerogrid, AEX, CNES/Airbus DS, DigitalGlobe, Earthstar Geographics, Esri, GeoEye, Getmapping, IGN, IGP, NOAA, swisstopo, USDA, USGS and the GIS User Community
3. See endnote 2.
4. See endnote 2.
5. See endnote 2.
6. Formerly RESOLVE, now wildlife researcher and writer/editor (wildtech.mongabay.com)
7. RESOLVE (www.resolv.org)
8. Jane Goodall Institute (JGI) (www.janegoodall.org.uk)
9. University of Minnesota (www.consssci.umn.edu)
10. RESOLVE (www.resolv.org)
11. Hutan, Alam dan Lingkungan Aceh (HAkA) (www.haka.or.id)
12. Liverpool John Moores University (www.ljmu.ac.uk)
13. McGill University (www.mcgill.ca)

Photo: The best way to limit the impact of human disturbance on wildlife and ecological processes is to ensure that core areas of parks remain road-free – Maiko National Park, DRC. © Jabruson 2018

CHAPTER 4

Apes, Protected Areas and Infrastructure in Africa

Introduction

Equatorial Africa sustains the continent's highest levels of biodiversity, especially in the wet and humid tropical forests that harbor Africa's apes. This equatorial region, like much of sub-Saharan Africa, is facing dramatic changes in the extent, number and environmental impact of large-scale infrastructure projects. A key concern is how such projects and the broader land use changes they promote will affect protected areas—a cornerstone of wildlife conservation efforts.

This chapter assesses the potential impact of new and planned infrastructure projects on protected areas in tropical Africa, particularly those harboring critical ape habitats. It focuses on Africa not because tropical Asia is any less important, but because analyses of comparable detail are available only for certain parts of the Asian tropics (Clements *et al.*, 2014; Meijaard and Wich, 2014; Wich *et al.*, 2016). Such knowledge gaps underscore the importance of future work on infrastructure impacts in Asia.

Ape range states in tropical Africa are encountering an array of important changes. These include an unprecedented expansion of industrial mining (Edwards *et al.*, 2014); more than 50,000 km of proposed "development corridors" that would crisscross much of the continent (Laurance *et al.*, 2015b; Weng *et al.*, 2013); the world's largest hydropower dam complex (International Rivers, n.d.-c); ambitious plans to expand industrial and smallholder agriculture (AgDevCo, n.d.; Laurance, Sayer and Cassman, 2014b); extensive industrial logging (Kleinschroth *et al.*, 2016; LaPorte *et al.*, 2007); and myriad other energy, irrigation and urban infrastructure projects (Seto, Güneralp and Hutyra, 2012).

Many of the largest infrastructure projects in Africa are being advocated because of concerns about the continent's booming population, which is expected to nearly quadruple this century (UN Population Division, 2017). This projection is creating apprehension about food security and human development, and broader anxieties about the potential for social and political instability (AgDevCo, n.d.; Weng *et al.*, 2013). Africa faces serious challenges revolving around:

1. effective design and assessments of new infrastructure projects to limit their environmental and social impacts;
2. good governance for nations experiencing unprecedented foreign investments for infrastructure and natural resource extraction; and
3. management of economic instabilities that can plague nations largely reliant on just a few natural resources or commodities for export income (see Chapter 1).

Key Findings

The main findings of this chapter are:

- Africa is experiencing an unprecedented proliferation of infrastructure projects and, consequently, dramatic changes in land use, the effects of which are likely to have an impact on many protected areas in critical ape habitats and beyond.

- Advances in remote sensing, computing power and databases are rapidly improving the quality and accessibility of information on the distribution of roads and other infrastructure, as well as on the attributes and threats affecting global protected areas.

- Foreign investment, in extractive industries in particular, is playing a key role in promoting infrastructure expansion in Africa.

- Protected areas in Africa are particularly vulnerable to reductions in size or downgrading of their protection status if they hinder exploitation of natural resources or limit infrastructure expansion.

- Growing pressures from infrastructure expansion and land use changes in the regions immediately surrounding protected areas can have adverse effects on ecological integrity, biodiversity and functional connectivity. Larger parks are generally less susceptible to such external pressures.

- While roads inside parks may foster ecotourism, the best way to limit the impact of human disturbance on sensitive wildlife and ecological processes is to ensure that core areas of parks remain road-free.

- There is an urgent need to implement considered land use and infrastructure planning, and to apply the "mitigation hierarchy" to avoid, minimize, restore and offset threats to endangered apes and other iconic species and critical habitats in equatorial Africa.

African Ape Ranges and Protected Areas

In Africa, several factors complicate efforts to conserve viable species and subspecies of apes. One concerns the limited geographic ranges of many apes (see the Apes Overview and Figures AO1 and AO2). Another is the imprecision of published range maps, which typically overestimate ape distributions, reflecting the fact that most species are patchily distributed as a result of natural habitat variability and spatially varying human pressures. When such patchiness is taken into account, many wildlife species are in fact more seriously imperiled than suggested by the International Union for Conservation of Nature (IUCN) Red List classifications (Ocampo-Peñuela *et al.*, 2016). Political conflicts, remoteness and limited scientific resources further hinder efforts to identify key threats and monitor ape populations.

Where reasonably robust data have been gathered, at least some ape taxa have been shown to suffer serious population declines. In the eastern Democratic Republic of Congo (DRC), for example, field surveys suggest that the critically endangered Grauer's gorilla (*Gorilla beringei graueri*), a locally endemic subspecies, has declined by 77% to 93% in abundance over the past two decades (Plumptre *et al.*, 2015).

Although more than 6,400 protected areas occur across sub-Saharan Africa, only a limited number are considered "large"—meaning that few cover more than 10,000 km^2 (1 million ha)—especially in the continent's equatorial regions that harbor ape populations (Laurance, 2005; Sloan, Bertzky and Laurance, 2016). In West and Central Africa, protected areas broadly coincide with ape ranges (see Figure 4.1 and Figure AO1). African apes are represented by five species and a number of restricted subspecies. They

FIGURE 4.1

Protected Areas in West and Central Africa

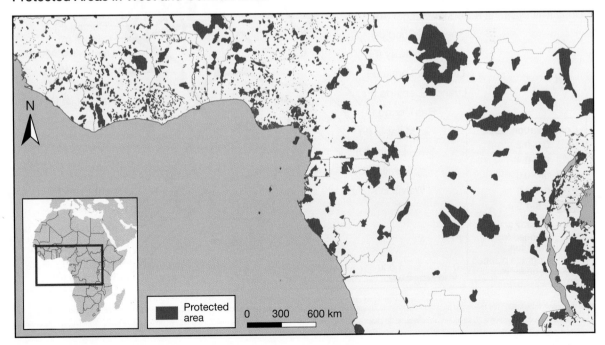

Data source: UNEP-WCMC and IUCN (n.d.)

are separated by geographic features such as the arid Dahomey Gap, which splits the West African rainforests and the extensive rainforests of Central Africa; major rivers, such as the Congo, which separates bonobos from other African apes; and two tall massifs that sustain populations of mountain gorillas (*Gorilla beringei beringei*).

FIGURE 4.2

Conservation Values of Habitats within 25 km of 33 Development Corridors in Sub-Saharan Africa

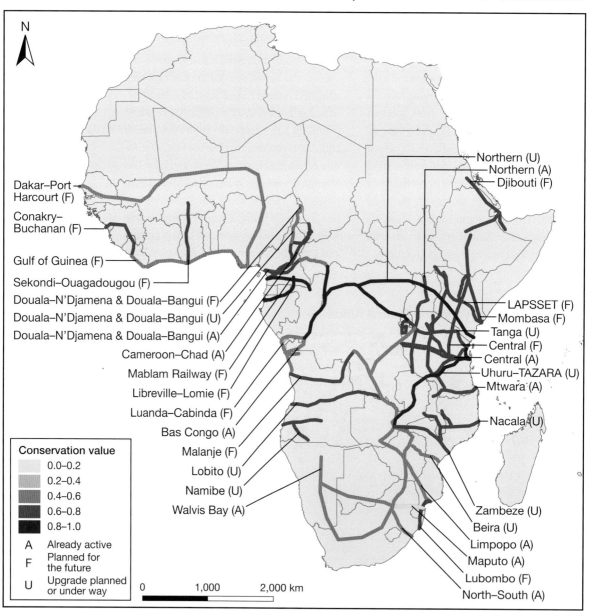

Notes: Conservation values are estimated based on biodiversity, threatened species, critical ecosystems, wilderness attributes, environmental services and human population densities of habitats within a 25 km buffer zone around 33 proposed or existing development corridors. Values are shown on a relative scale, from 0 (low conservation value) to 1 (high conservation value).

Data source: Laurance et al. (2015b)

Threats to Protected Areas from Infrastructure

Africa's "Development Corridors"

A true game-changer for African nature conservation is the proposed and ongoing construction of at least 35 development corridors. If completed in their entirety, the corridors will crisscross sub-Saharan Africa, spanning a length of more than 53,000 km in total (Laurance *et al.*, 2015b).

These corridors are likely to affect existing nature reserves in at least three ways:

- First, by bisecting reserves, fragmenting them and opening them up to illegal encroachment and poaching (Sloan *et al.*, 2016).
- Second, by promoting colonization, habitat loss and intensified land use around reserves, they could decrease the ecological connectivity of the reserves to other nearby habitats.
- Third, environmental changes in the lands immediately surrounding a nature reserve tend to infiltrate inside the reserve itself (Laurance *et al.*, 2012). To some degree, a reserve with extensive logging and hunting in its surrounding lands will be exposed to those same threats within its own borders.

A detailed analysis of 33 of the proposed and ongoing development corridors[1] indicates that:

- many corridors would occur in areas that have high conservation value and are only sparsely populated by people (see Figure 4.2);
- the corridors would bisect more than 400 existing nature reserves; and
- assuming that land use changes intensify within 25 km on either side of each corridor, more than 1,800 reserves could experience deterioration in their ecological integrity and connectivity, as well as additional human encroachment (Laurance *et al.*, 2015b).

In total, the 33 development corridors could bisect or degrade more than one-third of all existing protected areas in sub-Saharan Africa (Laurance *et al.*, 2015b). The 23 corridors that are still in the planning or initial upgrading phases would be especially dangerous for nature. These corridors would bisect a larger proportion of high-priority reserves—such as World Heritage sites, Ramsar wetlands and UNESCO Man and Biosphere Reserves—than would existing development corridors. Collectively, the 23 planned corridors would slice through more than 3,600 km of reserve habitat (Sloan *et al.*, 2016).

Of the approximately 2,200 African protected areas that could be affected by development corridors, a number include ape range habitats. For example, two epicenters of bisected reserves—the iron-rich belt spanning southern Cameroon and the northern Republic of Congo, and the Great Lakes region of East Africa (see Figure 4.2)—harbor vital ape habitats (Sloan *et al.*, 2016). There would also be considerable losses of important habitats outside of protected areas. A simulation model developed by the World Bank projects that in the Congo Basin, which is critical habitat for apes, expanding roads and transportation infrastructure will be the biggest driver of deforestation through 2030 (Megevand, 2013).

The Grand Inga Hydroelectric Project, DRC

While it is not possible here to describe the full range of infrastructure projects that could diminish African ape habitats, one cannot fail to mention the massive hydroelectric project under construction near Inga Falls on the lower Congo River. Should it proceed as planned, the Grand Inga dams will generate more electricity—40,000 megawatts (MW)—than any other single project on Earth. To achieve this level of output, however, the project will inundate more than 22,000 km² (2.2 million ha) of largely forested lands in the western DRC (Abernethy, Maisels and White, 2016). Dam projects in tropical regions often have deforestation footprints that markedly exceed the flooded reservoir itself, because road networks needed for dam and power line construction also provoke major forest disruption (Barreto *et al.*, 2014; Laurance, Goosem and Laurance, 2009; see Chapter 6).

Road Proliferation

One of the most serious effects of large-scale infrastructure projects—be they hydroelectric dams, mines, development corridors or nearly any other large development scheme—is that they provide a strong economic impetus for road building. Since they can open a Pandora's box of hunting, land colonization and other human activities, such roads often pose a greater threat to ecosystems and biodiversity than the original infrastructure project itself (Laurance *et al.*, 2015a). Moreover, many roads are constructed illegally; consequently, they do not appear on official road maps.

Hence, one of the most fundamental challenges facing those who seek to manage land use activities and limit their threat to nature is simply determining the locations of existing roads. The number of illegal and unmapped roads is generally much greater in developing nations, such as those that sustain ape populations, than in wealthier industrial nations (Ibisch *et al.*, 2016). For this reason, simply mapping existing roads is a major priority, one that is beset by some important technical challenges (see Box 4.1).

BOX 4.1

The Challenge of Mapping Roads

Key Uncertainties

A common misperception is that roads and other transportation infrastructure have been adequately mapped at the global scale, and that related data are readily available. In fact, they are not, and this lack of information creates serious challenges for nature conservation.

Road maps suffer from two key sources of uncertainty. First, the quality of road maps differs markedly across nations. In Switzerland, for instance, nearly every viable road is mapped, whereas in developing nations such as Indonesia or Nigeria, road maps are far from complete. Second, developing nations in particular have many illegal or unofficial roads that do not appear on any map. In the Brazilian Amazon, for example, a recent analysis found nearly three kilometers of illegal, unmapped roads for every kilometer of mapped, legal road; further, 95% of all deforestation occurred within 5.5 km of a legal or illegal road (Barber *et al.*, 2014). Since roads play such a dominant role in determining the pattern and pace of habitat disruption, it is vital to have a clear sense of where roads and other transportation infrastructure are located (Barber *et al.*, 2014; Laurance *et al.*, 2001, 2009).

For information on roads, the best freely available global data set is gROADS, the Global Roads Open Access Data Set, although it suffers from notable differences in accuracy and temporal coverage across nations (CIESIN and ITOS, 2013; Ibisch *et al.*, 2016; Laurance *et al.*, 2014a). gROADS staff manually digitized coarse-scale (1:1,000,000) hardcopy maps, often from the 1980s and 1990s. This process resulted in horizontal-accuracy limitations (±2 km) that restrict the use of gROADS to general comparisons, especially within, rather than across, nations.

Information Revolution

The late 1990s saw rapid growth in road mapping, driven by the rise of the in-car navigation industry. Often restricted to specific navigation devices and applications, the widespread use of global road data was revolutionized in 2005 with the launch of Google Maps (maps.google.com) and continued with subsequent data collection campaigns. These developments have generated detailed coverage for urban roads worldwide, although data for rural areas are much more patchy. Google Maps data have commercial applications (linked to advertising and location-based search results); their use for nonprofit websites and independent data analysis is thus restricted.

Despite their proprietary nature, Google Maps data are being used to help generate the Global Roadless Areas Map, a collaboration among Google, the Society for Conservation Biology and the European Parliament. This initiative began in 2012 under the aegis of RoadFree (www.roadfree.org), an initiative designed to highlight the importance of roadless wilderness areas for biodiversity conservation and the reduction of atmospheric carbon emissions. RoadFree has helped to spur interest in improving maps of transportation infrastructure, using a variety of data sources and techniques.

In parallel with commercial road data, an initiative known as OpenStreetMap (OSM) (www.openstreetmap.org) has grown dramatically. OSM aims to create a free and editable map of the world. Since its launch in 2004, it has grown into a community of more than 4 million registered members, around 2,000 of whom are making daily edits. Between late 2016 and mid-2017, the number of road features in the OSM database increased impressively from 376 million to 430 million, in addition to many other features, such as buildings.

Efforts are underway to focus OSM development on evolving environmental crises and to improve data for areas that are inadequately mapped. Notable among these are two programs aimed at mapping roads in tropical forests. The first, Roadless Forest (roadlessforest.eu), is a European Union initiative to assess the benefits of road-free forests, strongly linked to EU policies on reducing illegal logging and carbon emissions from forest disruption (FLEGT, 2016; REDD+, n.d.). The second is Logging Roads (loggingroads.org), which focuses on mapping logging roads in the Congo Basin. The good news is that all mapping improvements from these various initiatives are being placed immediately on the publicly available OSM database. An OSM Analytics platform (osm-analytics.org), released in 2016, enables tracking of this mapping activity for roads and buildings at the global level.

Technical Challenges and Advances

While the new road mapping initiatives are invaluable, many technical challenges remain (Laurance et al., 2016). For instance, the spatial resolution of available imagery can differ greatly across particular areas of interest, compromising efforts to create accurate and comparable infrastructure maps. Figure 4.3 illustrates that spatial resolution can vary across

FIGURE 4.3

Mapping Discrepancies in an Area of Rutshuru Reserve, Uganda, in OpenStreetMap

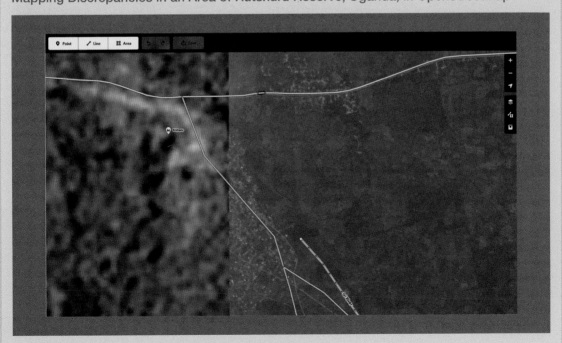

Source: © OpenStreetMap contributors – www.openstreetmap.org

FIGURE 4.4

Recent and Ongoing Logging-Road Activity in the Congo Basin, near Ntokou-Pikounda National Park, as Identified by Time-Series Analyses of Landsat Imagery

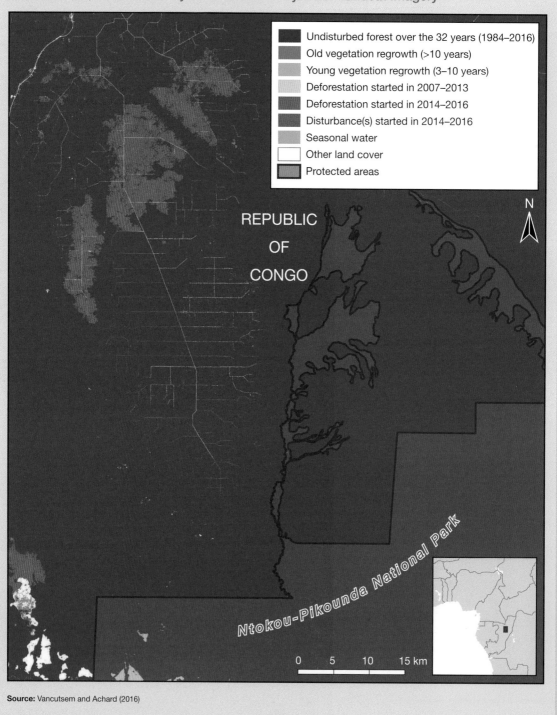

Source: Vancutsem and Achard (2016)

images; it also shows inaccurate road positions derived from older coarse-scale maps.

A common assumption is that increasingly higher-resolution satellite imagery is needed for better road mapping. However, spatial data from the Landsat and EU Sentinel satellites, and composite images produced by Google Earth, all have reasonably high resolution, sufficient for many road-mapping applications. Furthermore, with each satellite pass, higher-resolution sensors cover a narrower swath of land than do lower-resolution ones, and therefore they return to the same area less frequently. This slow return time can be a major constraint in the effort to find cloud-free images in the wet tropical regions that are key ape habitats. Fine-scale imagery (<1 m resolution) exists but is expensive, requires massive data storage capacity and is rarely available for the remote environments inhabited by apes. Finally, the long time period over which Landsat imagery has been available allows changes in land use and roads to be observed for intervals of up to several decades (given that Landsat commenced in 1972 and Landsat Thematic Mapper, with 30-m resolution sufficient for detecting roads in dense forests, began in 1982). This long-term coverage is extremely valuable for assessing the spatial patterns and drivers of land use change over time.

Until recently, high data costs, inadequate computing power and limited access to imagery precluded the systematic processing of remotely sensed data over periods exceeding 30 years. Prior to 2008, all Landsat data were provided on a commercial basis; as a result, the use of the data was meager. Once the data were made freely available, their use skyrocketed. This has fuelled numerous innovations, of which Google Earth Engine is perhaps the most notable. Launched in 2010, it has allowed global-scale analyses using the power of Google's own cloud-computing infrastructure.

With pricing and technical barriers for data and computing falling dramatically, opportunities for global-scale environmental analysis have grown rapidly. For example, researchers at the European Commission's Joint Research Centre have developed techniques to identify forest disturbances at a resolution of 30 m × 30 m as far back as 1982, using Google Earth Engine as the processing platform (Vancutsem and Achard, 2016; see Figure 4.4). Similarly, the rapid repeat time of Landsat has allowed researchers to find enough cloud-free images to effectively monitor the expansion of tropical logging roads. This technique can be used to highlight areas susceptible to road expansion and forest change (see Chapter 7), which in turn can feed into community-mapping programs such as OSM. The next step is to attempt to predict the environmental impacts of different road development scenarios on forests (Laurance et al., 2001).

Needed: Road-Detection Algorithm

For all the sophistication of modern remote-sensing technologies, researchers still lack an automated computer algorithm that can reliably detect and map roads under the hugely varying range of topographic, land use, sun-angle and road-surface conditions that one encounters in the real world. For this reason, actual road mapping is usually done with human eyes—by using the best available satellite imagery and manually tracing roads with a mouse onto a computer screen. Known as "armchair mapping," this method is still the most effective for mapping roads and determining whether they are paved or unpaved. Unfortunately, this is a very time-intensive process. Even with hundreds of active mappers, several years would be required to map all the roads on the planet. By the time the mappers had finished mapping Earth's roads, it would be necessary to start anew to identify the many new roads that would have been created since the project began. For such reasons, a holy grail for those studying roads is an automated system that can detect and map roads accurately in near-real time (Laurance et al., 2016).

Forest Monitoring

As a result of vastly improved data accessibility and computing power, forest monitoring by satellites has advanced impressively. In 2014, Global Forest Watch announced a revamped website (www.globalforestwatch.org), powered largely by Landsat satellite data (see Chapter 7). The next generation of Earth-observation satellites—the Sentinel-2 series from the European Space Agency—will have even higher spatial resolution (10 m), better spectral data (red, green, blue, near infrared), and faster return times (5 days) than does Landsat. The image characteristics of the Sentinel satellites will lend themselves to forest- and road-mapping applications (Verhegghen et al., 2016). The fact that their data are entirely free and open access should help to stimulate further innovations.

Next Steps

Finally, there is a need to go beyond simple maps of transportation infrastructure and look more broadly at accessibility. The World Bank and European Commission produced a Global Accessibility Map that estimates the travel time from any point on Earth to the nearest city exceeding 50,000 people (Nelson, 2008). Although focused on access to urban services, the map highlights the limited and shrinking extent of wilderness worldwide (Ibisch et al., 2016; Laurance et al., 2014a; Watson et al., 2016). With more and better roads, advances in vehicle technology and a rapid increase in the number of motorized vehicles, the globe is shrinking fast. Already, just one-tenth of the world's land surface is more than 48 hours' travel time from a major city (Nelson, 2008). Clearly, this is leading to increased pressure on ecosystems and biodiversity.

There is both enormous potential and an urgent need to devise better road-mapping tools, and to use these to assess road-related pressures to ape habitats. A logical next step is to identify critical areas that should remain free of roads to help ensure the long-term survival of apes and their habitats.

Photo: Bwindi's global importance was recognized in its designation as a UNESCO World Heritage site, particularly in view of the diversity of its habitats and its exceptional biodiversity. Bwindi hills.
© Martha M. Robbins/MPI-EVAN

Protected Area Downgrading, Downsizing and Degazettement (PADDD) in Africa

Documented PADDD Events

As development pressures increase, designated protected areas are sometimes diminished by legal means (Mascia and Pailler, 2011). In Africa, for instance, states have been known to reduce the size, contiguousness and protection status of reserves to allow new roads, mining, energy projects and other activities to expand. At least 23 African protected areas have been downsized or downgraded (Edwards *et al.*, 2014, table 1). Mining occurs more frequently in close proximity to protected areas in Africa than in either Asia or Latin America (Durán, Rauch and Gaston, 2013). Even natural World Heritage Sites, the global pinnacle of conservation, have been subjected to mining or fossil fuel exploration or development, with 30 sites in 18 African countries affected to date (WWF, 2015a). In the Republic of Guinea, for example, the Mount Nimba Biosphere Reserve, a World Heritage site, was downsized by 15.5 km² (1,550 ha) to allow for iron ore prospecting. An even greater concern is Zambia, where nearly 650 km² (65,000 ha) of land within 19 protected areas has been downgraded to permit mining activities (Edwards *et al.*, 2014).

A number of protected areas with key African ape habitats are under growing development pressures. In Nigeria, for example, a proposed "superhighway" would increase deforestation and other pressures on Cross River National Park, critical habitat for the endemic Cross River gorilla (*Gorilla gorilla diehli*) (see Case Study 5.1). Meanwhile, one of only two surviving populations of mountain gorillas, in Uganda's Bwindi Impenetrable National Park, could also be threatened by a major road-upgrading project inside the park (see Box 4.2).

BOX 4.2

Alternatives to Road Development in an Iconic African Park

Bwindi Impenetrable National Park in the southwest of Uganda supports a highly diverse range of plant and animal species, including the endangered eastern chimpanzee (*Pan troglodytes schweinfurthii*) and one of only two remaining populations of the critically endangered mountain gorilla (*Gorilla beringei beringei*) (Plumptre *et al.*, 2007, 2016a; Plumptre, Robbins and Williamson, 2016c).

Although it is relatively small (321 km²/32,100 ha), Bwindi contributes to local and national economies through Uganda's nature-based tourism industry and other ecosystem services the park provides. Its global importance was recognized in its designation as a UNESCO World Heritage site in 1994, particularly in view of the diversity of its habitats and its exceptional biodiversity, including Albertine Rift endemics (UNESCO WHC, n.d.).

In 1995 the aid agency CARE commissioned a study to assess the feasibility of diverting part of the Ikumba–Ruhija road, which cuts through Bwindi for 12.8 km, to land outside the park's boundaries. The study concluded that a road diversion was feasible, identified suitable alternative routes and indicated that a new route would promote long-term protection of the park while boosting economic activity in the area (Gubelman, 1995).

However, in 2012, the Ugandan government advertised a scheme to design and construct 1,900 km of new roads in the country, including an upgrade of the road inside Bwindi, whose clay surface was to be converted to a paved road as part of a much larger road circuit (Kampala, 2012). At the time of writing, an environmental impact assessment to identify the potential effects of the proposed road upgrade on the park's ecology and wildlife had yet to be conducted.[2]

Concerned that the proposed upgrade could harm the park's mountain gorillas and that it might provide few benefits for local villages outside Bwindi, the International Gorilla Conservation Programme

(IGCP)[3] partnered with the Conservation Strategy Fund and the National Environment Management Authority of Uganda to assess the upgrade scheme and to contrast it with the earlier plan to divert the road outside the park, as part of the Biodiversity Understanding in Landscape Development project, funded by the United States Agency for International Development.

This analysis showed that an alternative route, while costing more initially, would provide greater benefits for twice as many villages and would avoid the negative impacts on the park's gorillas. Furthermore, the study suggested that the government's plan would cost the economy upwards of US$214 million in tourism revenue losses over the 20-year life cycle of the road investment (Barr et al., 2015). These results were presented to the Uganda National Roads Authority and the Uganda Wildlife Authority.

Based on the results, representatives from the Uganda chapter of the Poverty and Conservation Learning Group conducted consultations with affected communities and prepared a position paper that supported diverting the road around the park (U-PCLG, 2015). During a meeting in March 2015, local stakeholders supported the view that road development around Bwindi is extremely important, and the government was urged to pursue the option of investing in diverting the road outside of Bwindi.

To date, however, the relevant government authorities have not changed their position. Government agencies claim they lack the funds needed to divert the route and compensate local land owners. Local and international stakeholders, including the IGCP, are continuing to urge the government to divert the road outside Bwindi and to take all steps necessary to protect Bwindi Impenetrable National Park and its iconic wildlife.

Prospects for PADDD

As infrastructure and resource-extraction projects proliferate across Africa, the potential for further PADDD events could increase dramatically. One tool that has considerable utility for monitoring threats to parks is a global database known as the Digital Observatory for Protected Areas (DOPA). DOPA provides a wide range of indicators of park features, habitats, species composition, irreplaceability and threats (see Box 4.3). These metrics could be used to monitor changes over time for a single park and to assess national trends in park protection. Comparisons of environmental threats across parks in different ecoregions or nations need to be conducted carefully because of potential differences in data quality and normalization procedures.

The research conducted for this chapter involved an evaluation of the practical utility of DOPA for assessing threats to parks. To that end, the effects of two factors that could influence the proliferation of roads inside parks were compared: park area and road pressure immediately outside the park. The study hypothesis held that larger parks would have fewer roads than smaller ones, and that parks with many surrounding roads would also have many internal roads.

For purposes of this research, road pressure inside the park was defined as the total number of kilometers of road length (km) divided by park area (km²). To quantify external road pressure, a 30-km buffer zone was defined around each park and an inverse distance–weight function was used to calculate pressure from all roads inside the buffer zone. This approach applies greater weight to roads near a park than to those farther away. In all cases, gROADS was used to generate data on roads (see Box 4.1).

The analysis generated data for 656 protected areas within ten countries in equatorial Africa:

- Cameroon;
- the Central African Republic;
- the DRC;
- Gabon;
- Ghana;
- Ivory Coast;
- Liberia;
- Nigeria;
- the Republic of Congo; and
- Sierra Leone.

Not all protected areas in these nations harbor apes or ape habitats, nor were all protected areas with African ape populations included in the analysis. Via a generalized linear mixed-effects model, "nation" served as a random variable, in order to reduce differences in road-map quality at the national level.[4]

Despite limitations in the available data sets, the results of the analysis appear clear: road pressure inside each park was strongly influenced by its external road pressure, but park size had a weaker and less consistent influence (see Figure 4.5).[5] These findings suggest that as roads proliferate across equatorial

---

BOX 4.3

Digital Observatory for Protected Areas (DOPA)

DOPA (dopa.jrc.ec.europa.eu) is an online system developed by the European Commission's Joint Research Centre to provide key indicators of the pressures facing more than 16,000 terrestrial and marine protected areas, each of which exceeds 100 km² (10,000 ha) (Dubois et al., 2015). DOPA uses freely available open data for its calculations.

DOPA provides a variety of information, including on the size, location, boundaries and protection status of each park; ecoregions, soils, topography, climatic and land cover data; and the number of threatened species of mammals, birds, amphibians and other selected taxa. It also features indices of species irreplaceability and measures of environmental pressures for five parameters, namely human population density around the park, the annual rate of change in the human population around the park, agriculture surrounding the park, roads inside the park and roads surrounding the park (Dubois et al., 2015).

FIGURE 4.5

Effects of External Road Pressure and Park Area on Internal Road Pressure for 656 Protected Areas in Ten Nations in Equatorial Africa

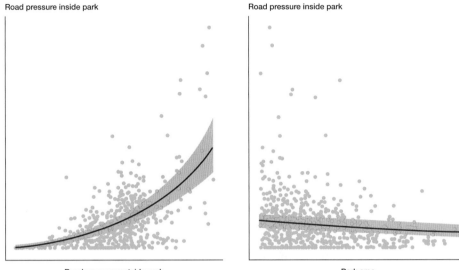

Notes: Curves show predicted values; shaded areas are 95% confidence intervals. Each curve shows the effect of the predictor variable on internal road pressure once the effects of the other predictor and cross-national differences were statistically removed.

Africa, protected areas could experience marked increases in internal road pressure. The effects of park size are variable, although the largest parks rarely suffered high internal road pressure.

The Mitigation Hierarchy: Reconciling Infrastructure and Ape Conservation

The Mitigation Hierarchy

Given the likelihood that many large-scale infrastructure projects will proceed, a major priority is to limit their various direct and indirect environmental impacts. The mitigation hierarchy can be applied throughout the life cycle of a project to aid the process of constructive engagement (see Figure 4.6 and Table 3.3). It aims to minimize negative impacts and to offset any significant impacts that remain (TBC and CSBI, 2015). A new report from Forest Trends identifies that "the

FIGURE 4.6

The Mitigation Hierarchy Applied to Infrastructure Projects within Ape Habitats

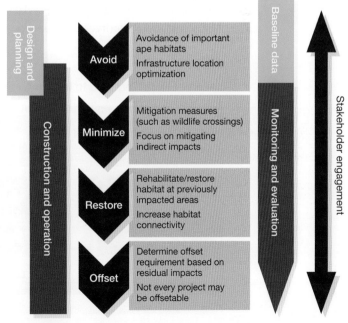

Source: © TBC, 2017

energy, transportation, and mining/minerals sectors were responsible for more than 97% of offsets and compensation measured by cumulative land area under management" (Bennett, Gallant and ten Kate, 2017, p. 5).

The mitigation hierarchy is increasingly required by project lenders, including the International Finance Corporation and World Bank (IFC, 2012c; World Bank, 2017). It is also becoming integrated into environmental legislation around the world, including in many ape range states (TBC, 2016). The hierarchy follows four sequential steps: avoid, minimize, restore and offset.

Step 1: Avoid

When operating in ape habitat, the first step, avoidance, is the most crucial and effective. It requires early data gathering and planning, ideally at the start of the design and planning phase (see Figure 4.7).

The consideration of alternative routes or project siting is an important early task as it may allow important ape habitat to be avoided. At this stage, projects are rarely able to finance extensive data collection and instead rely on readily available data. Available maps of priority areas for ape conservation, such as those produced by regional or national action planning processes, can be extremely useful (Golder Associates, 2015; Rio Tinto Simfer, 2012b). However, companies that design infrastructure projects may not be aware of such data, and therefore ape conservationists may need to take the initiative of sharing data in usable formats and of directing decision-makers to available resources, such as the A.P.E.S. Database (Max Planck Institute, n.d.-b).

Once a broad project option has been adopted, finer-scale optimization of infrastructure placement can further ensure that construction is avoided in sensitive ape habitat. This requires more detailed information on ape distribution and habitat use in relation to proposed infrastructure locations, as can be gathered via surveys conducted as part of environmental and social impact assessments (ESIAs). For example, the ESIA for the Simandou Iron Ore Project in Guinea revealed that chimpanzees principally used the western side of the mining concession. As a result, all mine-associated infrastructure was relocated to an economically suboptimal location in the east of the concession, to avoid important chimpanzee habitat (Rio Tinto Simfer, 2012a).

FIGURE 4.7

Levels of Data Required to Inform Avoidance Measures in the Mitigation Hierarchy

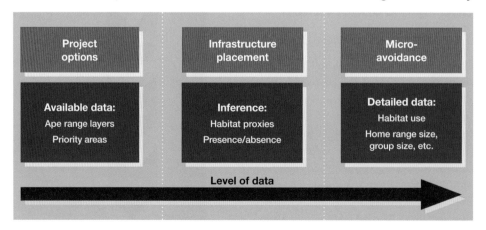

Source: © TBC, 2017

Step 2: Minimize

If it is not possible to avoid impacts on apes and their habitat entirely, minimization measures can often reduce the extent and intensity of remaining negative impacts. In addition to being good practice, minimization measures, such as noise and dust reduction and ape-specific measures, may be appropriate. Sufficient ecological data are required to guide informed planning of minimization actions for apes. If there is uncertainty, monitoring and adaptive management may be required.

For apes the indirect impacts of large infrastructure projects, particularly increased poaching and habitat loss due to induced access and in-migration, are usually the most serious (IUCN, 2014b; Vanthomme *et al.*, 2013). These impacts can occur on a large scale and thus effective minimization measures may also need to be implemented at large scales. Such minimization efforts were made in the context of the public–private partnership between the government of Cameroon and the private railway developer CAMRAIL, with the aim of reducing the illegal transport of wild meat, including chimpanzee, that could be facilitated by the railway (Chaléard, Chanson-Jabeur and Béranger, 2006).

Minimization measures can be capital-intensive while also requiring ongoing investment by infrastructure developers. It can therefore be difficult to demonstrate the business case for minimization if data or experience is limited. Such is the challenge regarding wildlife crossings, including artificial canopy bridges. While they have been shown to be effective at maintaining connectivity for the more arboreal gibbons and orangutans, these bridges have never been trialed with African great ape species (Das *et al.*, 2009; see Box 2.2). Hence, their effectiveness at facilitating movements, and their potential for making apes more vulnerable to poaching, are unknown. Other impacts of infrastructure projects that are poorly understood include tolerable noise levels and the potential barrier to ape dispersal caused by large-scale linear infrastructure projects.

Step 3: Restore

Complete restoration of ape habitat may not be possible or achievable within a project timeline, and thus it may be more suitable to consider habitat rehabilitation. Examples of rehabilitation measures include planting native tree species, preventing uncontrolled burning and removing damaging species (mainly non-native or invasive species).

Habitat rehabilitation is a long-term process. Apes use complex habitat and often rely on tree species that take many years to reach maturity. Native tree species used by apes may also require special conditions to grow that are challenging to re-create. It is thus nearly impossible to re-create original habitats and, as a result, it is not possible to rely on restoration actions to make a significant contribution to reducing the magnitude of project impacts on apes (Maron *et al.*, 2012). Nonetheless, targeted habitat rehabilitation can serve as a valuable means of increasing habitat connectivity in fragmented landscapes.

Step 4: Offset

Any negative impact that remains after the first three steps of the mitigation hierarchy have been applied is termed a "residual impact." Offsets of such impacts are measures of last resort; their use with respect to threatened and charismatic species such as apes is often seen as controversial (Kormos *et al.*, 2014). If they are poorly planned, large-scale infrastructure projects can have significant indirect impacts that are difficult or impossible to offset. This underscores the need to focus on avoidance and mitigation measures to minimize residual impacts.

FIGURE 4.8

Ape Range Countries with an Offset Policy (as of 2016) for (a) Bonobos, Chimpanzees and Gorillas; (b) Orangutans; and (c) Gibbons

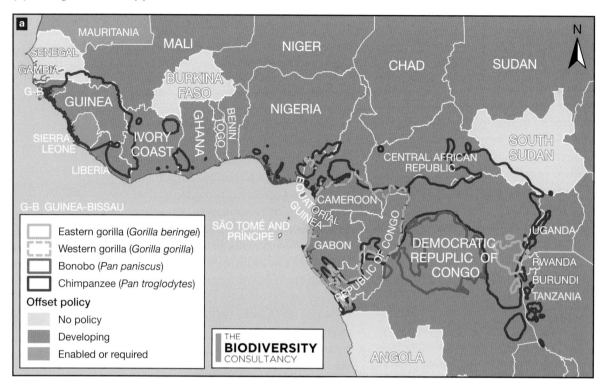

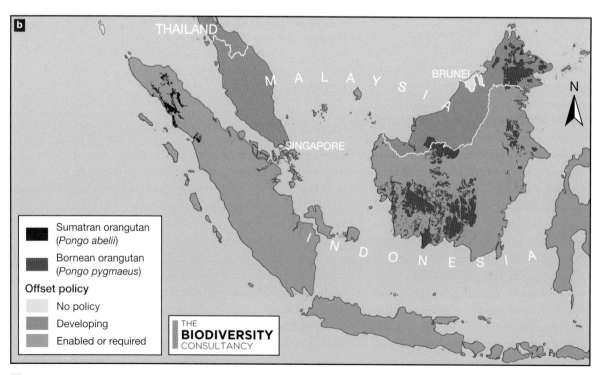

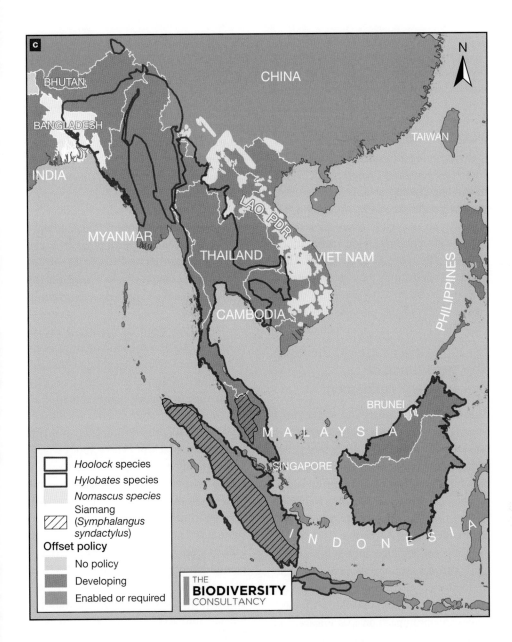

Several ape species and subspecies have very restricted geographic ranges (see the Apes Overview). A project that would have a negative impact over a significant extent of a species or subspecies' range would be difficult or impossible to offset, and thus would be unlikely to be supported by conservation stakeholders. Similarly, impacts that compromise the viability of identified regional priority areas for ape conservation may not be considered eligible for offsetting.

For projects associated with less serious residual impacts, the offset requirement is guided by aspects of the biology and behavior of apes, although it is also important to consider uncertainty in estimates of both the scale of impact and the scale of gains at the proposed offset site. Furthermore, the project would need to demonstrate that planned actions have an additional beneficial effect (over and above the status quo) and that they would contribute to an increase in the ape

Chapter 4 Protected Areas and Infrastructure

population in the long term (Kormos *et al.*, 2014). These requirements mean that the loss of even a few individual apes could translate into very significant offset requirements to meet "no net loss" definitions (IUCN, 2014a).

Offset requirements to compensate for impacts of development projects are increasingly becoming integrated in national legislation (ten Kate and Crowe, 2014). In Asia, most orangutan and gibbon range states have legislation either requiring or enabling biodiversity offsets, while many African ape range states are developing such national policies (TBC, 2016; see Figure 4.8). There is thus an opportunity for governments and ape conservationists to work together to ensure such policies provide appropriate protection for apes and their habitat.

The Importance of Stakeholder Engagement

Apes are iconic animals and any negative impacts on them or their habitats attract high interest and scrutiny from the general public, stakeholders and lenders. Therefore, infrastructure developers face potentially serious reputational risks when operating within ape habitat, which makes consultations with stakeholders and ape experts at an early stage advisable. Stakeholders such as universities and conservation groups can provide specialized knowledge that can be input into the project design and add credibility to a project, while reducing impacts on apes. Engagement with stakeholders is most effective when it begins in the early stages of a project and continues throughout its life span, through every step of the mitigation hierarchy.

Cumulative Impacts and the Mitigation Hierarchy

Cumulative impacts are defined as the incremental impacts of one project, combined with the past, present and foreseeable impacts arising from other developments (such as infrastructure, extractive or agricultural activities) within the same geographic and connected areas (IFC, 2012b). Cumulative impacts often arise when a country is undergoing rapid development, for example when multiple dams are planned for construction on the same river (Winemiller *et al.*, 2016). Environmental impact assessments for any single project often fail to adequately consider the wider or additive effects of other projects in the same vicinity (Laurance *et al.*, 2015a; see Chapter 1, p. 32). This can be severely detrimental to species such as apes, as numerous projects have large impacts across populations and reduce population connectivity.

There has been increasing pressure from stakeholders for individual projects to take cumulative impacts into consideration. Best-practice guidelines require cumulative impact assessments (CIAs); in practice, this step often receives insufficient attention or is omitted completely. A major barrier is the lack of clarity about whose responsibility it is to organize and pay for a CIA, particularly in a landscape that comprises multiple development projects with different timelines. However, if conducted rigorously and systematically, CIAs could greatly strengthen regional and national planning processes (IFC, 2013).

When adhering to the mitigation hierarchy, projects should take cumulative impacts into account (see Case Study 4.1). Ideally, neighboring projects would adopt coordinated mitigation measures and would be designed to share common infrastructure (such as railways and access roads) to reduce their footprint area. Governments can facilitate the management of cumulative impacts by carrying out strategic land use planning at the national or landscape scale, thereby preventing projects with competing interests (such as ape conservation and industrial development) from operating in the same area. Additional case studies

CASE STUDY 4.1

The Mitigation Hierarchy and Cumulative Impacts: A Case Study from Guinea

The Republic of Guinea in West Africa has large mineral deposits such as bauxite, gold and iron and its mining sector is undergoing rapid development. Major deposits can be found in different parts of the country, often inland, far from the coastline. Large infrastructure projects, such as railways and roads, are being planned to transport ore from mine sites to seaports for export to international markets (Republic of Guinea, n.d.).

Bauxite reserves in Guinea are concentrated in the northwest of the country, where they overlap with the range of the critically endangered western chimpanzee (*Pan troglodytes verus*) (Humle et al., 2016a). Several mining companies are active in this region and hold adjacent concessions. Most projects are operating independently and have not yet effectively tackled the issues related to cumulative impacts. Two neighboring companies, however, are working towards implementing international best-practice standards and addressing cumulative impacts. These companies—the Compagnie des Bauxites de Guinée (CBG) and Guinea Alumina Corporation (GAC)—need to develop or upgrade roads to transport their bauxite ore to a port site located about 140 km away.

They will share an existing railway so that they may reduce their cumulative impact (see Figure 4.9).

Following the mitigation hierarchy, both companies are considering the option of setting aside a portion of their concessions to avoid sensitive chimpanzee habitat. Extensive surveys for chimpanzees have been conducted to help inform mitigation planning. Mitigation measures, which were developed to minimize both direct and indirect impacts, are outlined in each company's biodiversity action plan.

GAC has also established a nursery with native tree species that are known to be used by chimpanzees for feeding and nesting. These species will be used to rehabilitate areas previously impacted by the project as well as other degraded areas that were cleared by the local population using slash-and-burn cultivation.

Despite the various measures, preliminary assessments show that both companies will have residual impacts on chimpanzees; offset requirements were thus estimated separately for each company. As Guinea lacks national offset planning and updated maps of priority areas for chimpanzees, GAC has supported a nationwide chimpanzee survey to find the most appropriate offset site. This site may be large enough to provide an aggregated offset, where other companies could also contribute towards protecting a large population of the western chimpanzee.

FIGURE 4.9

Locations of the CBG and GAC Mining Projects and the Railway to Be Shared, Guinea

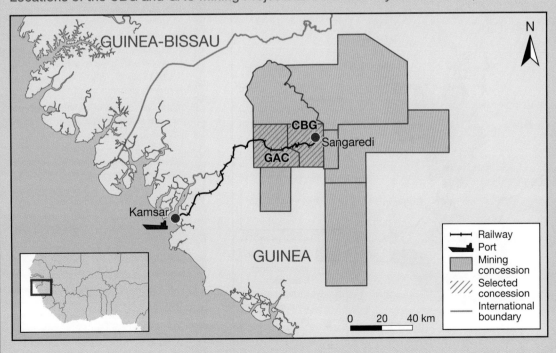

Sources: © TBC, 2017

and information on the mitigation hierarchy are available on the website of the Business and Biodiversity Offsets Programme (http://bbop.forest-trends.org/).

With its rapidly growing populations, a dire need for economic and social development, and exceptional natural riches, Africa represents serious challenges for environmental planners and managers. Unless these challenges can be addressed meaningfully, social instability and serious environmental damage will be unavoidable. The worst-case

FIGURE 4.10

Virunga National Park

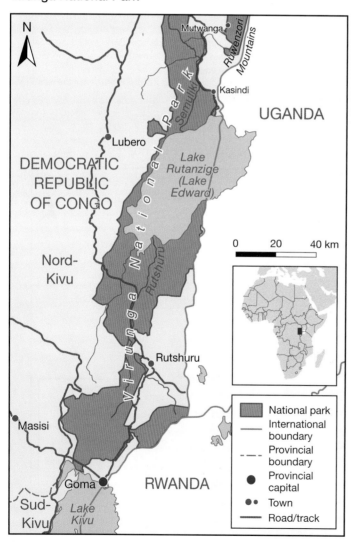

BOX 4.4

Virunga National Park: Promoting Socioeconomic Development alongside Conservation

The history of the Democratic Republic of Congo (DRC) is characterized by the exploitation of its vast natural resources. Yet despite the abundance of this natural wealth, extreme poverty has spread throughout the country. This paradox is exemplified by the DRC's water crisis: notwithstanding its immense freshwater resources, only 25% of the population has access to safe drinking water (and only 17% in rural areas), one of the lowest rates in sub-Saharan Africa (WSP, 2011). The legacy of colonialism, state collapse during the Mobutu years and recurring armed conflicts—of which the most significant followed the Rwandan genocide—have left the DRC with weak institutions and a chronically defective public infrastructure, particularly in the eastern provinces.

The catastrophic loss of life among civilians during the conflict years was principally attributable to indirect public-health effects, such as the dysfunction of water and sanitation infrastructure. Despite the international community's investments in peacekeeping, development aid and humanitarian relief (at an annual cost of up to US$15 billion), little has been achieved to prevent a resurgence of armed conflict.

In the face of overwhelming challenges, a community of institutions—the Institut Congolais pour la Conservation de la Nature (ICCN)—is working in partnership with the Congolese authority for conservation in Virunga National Park, in the eastern DRC (Figure 4.10). ICCN has invested more than US$60 million[6] to develop a holistic approach to social justice and conservation in this conflict-ridden region.

Virunga is Africa's oldest national park and a UN World Heritage Site, home to mountain gorillas and chimpanzees, as well as other endangered and endemic wildlife. It is plagued by ungoverned resource extraction as members of local communities hunt for food, clear forest for agriculture and gather fuelwood and charcoal for energy, lighting and heating.

TABLE 4.1
The Virunga Alliance's Hydropower Plan

	River/population center	Power	Users
Phase I	Butahu/Mutwanga	0.4 MW	1,200
Phase II	Volcano/Lubero	15.0 MW	160,000
	Rutshuru I/Rutshuru II	12.6 MW	140,000
Phase III	Various sites	80.0 MW	840,000

Source: Virunga National Park (n.d.)[7]

Along with the ICCN, a wider investment program known as the Virunga Alliance draws on the resources of the park to deliver broad-based services to the community in a way that is sensitive to the environment, focused on the needs of the poorest and most vulnerable, and supportive of stability in the region. Established in 2009, the Virunga Alliance was developed as three programs that may be visualized as concentric circles. The innermost circle is focused on conservation and protection of the park, as well as tourism. The second relates to socioeconomic development through the four main sectors of development: sustainable energy, tourism, agro-industry, sustainable fisheries, as well as measurable improvements in local infrastructure. These programs target the local population—principally the six million people in North Kivu (MONUSCO, 2015). The third circle targets private-sector investment to stimulate the local economy and to help bring people out of the cycle of poverty. Using a business approach to service delivery, the Alliance generates dividends from tourism and energy provision to industry, and reinvests these funds into the conservation and social infrastructure of the park.

Virunga's program of socioeconomic development—the second circle—focuses on renewable energy, sustainable fisheries, agro-industry and tourism. The region has vast natural wealth, including fertile soil, regular rainfall and abundant hydrological resources. The park's rivers feed Lake Edward, which flows into the Semliki River to form the source of the Nile. Millions of people depend on the park's healthy rivers and lake. There is very little infrastructure, however, to provide the local people with adequate water and energy supplies. The Virunga Alliance is working to supply hydroelectric power to nine towns in North Kivu on a build–operate–transfer basis. Eight hydropower plants, with the effective capacity of 108 megawatts (MW), and two interconnected networks will be built over 9 years, with the first completed in 2012 (see Figure 4.10 and Table 4.1). Two plants are already operational. Access to electricity is expected to provide a boost to local agriculture, thus helping to create 80,000 to 100,000 new jobs.

The hydropower feeds a grid, connecting consumers via a prepaid, smart metering system. Each megawatt of electricity is expected to produce as many as 1,000 jobs, based on the results of the Mutwanga hydroelectric pilot project in the north of the park, which was completed in 2013. The Matebe plant and Rutshuru grid are expected to create 13,000 permanent jobs, mostly in the small business sector.

There is a sizeable waiting list of consumers and small businesses that want to be connected to the grid, as grid electricity is substantially cheaper than the current power source—diesel generators. Indeed, a typical small business would save US$17 per month on power costs by connecting to the grid. This is a saving of US$204, which is more than half the average annual income ($394.25; Tasch, 2015). At present, the Mutwanga hydroelectric facility, managed by the park authority, provides electricity free of charge to schools and hospitals in the region.

The Virunga program assumes that increasing private-sector investment will accelerate economic development catalyzed by the hydroelectric program. Until now, Virunga has lacked a practical strategy to provide funding to small local businesses. Identifying a viable instrument for financing small Congolese-owned businesses is vital. The program is developing a Smart-Grid Small Business Loan Fund, capitalized with equity funding (grants or unsecured loans); the fund will approve, disburse, monitor and collect repayments on loans to small businesses that are also clients of the Virunga power grid.

The overall goal of the Virunga Alliance is to contribute to peace and prosperity via responsible economic development of natural resources for four million people who live within a day's walk of Virunga Park's borders. Economic opportunities and access to social services are an important factor in maintaining a long-term solution to violence. For the Virunga Alliance, a minimum of 30% of the park's revenues are invested in community development projects, which have been identified and defined by the local communities on the principle of free, prior and informed consent.

scenarios of natural-resource exploitation in Africa—driven by foreign capital, distorted by endemic corruption and resembling "feeding frenzies" of predatory behavior (Edwards et al., 2014)—are far too common. At the same time, innovative initiatives that are well planned and executed, attuned to social needs and sustainability outcomes, and long-term in nature are rare.

Africa does have a few examples of enlightened infrastructure projects—ones driven by visions of social and environmental betterment (see Box 4.4). Such endeavors, woven integrally into the surrounding cultural fabric, might be better described as "initiatives" rather than projects, in that their goals are less about generating profits than yielding broad-based social betterment and environmental sustainability.

Future Threats and Prospects

Narrow Window of Opportunity

The focus of this chapter is the potential effects of large-scale infrastructure expansion on ape habitats in equatorial Africa. The conclusions, by any measure, are alarming. Without determined efforts to modify, reconsider and mitigate the impact of current development schemes, apes and their biologically rich environments in Africa are likely to suffer irreparable harm.

The threats to African apes and their habitats are imminent, in the sense that many crucial changes will play out over the next 1–3 decades. However, the recent decline in global commodity prices, particularly for minerals and fossil fuels, provides a potential window of opportunity of a few years to employ direly needed land use planning and infrastructure-prioritization schemes (Hobbs and Kumah, 2015).

Two broad developments are critical to the promotion of strategic planning. The first is an expansion of the application of the mitigation hierarchy. The second is the implementation of viable financial strategies designed to help developing nations meet pressing economic and food-production needs while limiting the environmental impacts of rapid infrastructure development. For these nations, payments for ecosystem services, ecotourism and sustainable harvesting of native production forests, as well as strategic investments in natural capital could potentially help to balance economic and environmental priorities (Laurance and Edwards, 2014; see Box 4.5).

At a fundamental level, the challenges affecting Africa arise from its escalating population growth and serious needs for economic and human development, especially increased food security (AgDevCo, n.d.; Laurance et al., 2014b). As noted above, Africa's current population could almost quadruple this century, although such projections are not carved in stone (UN Population Division, 2017). Importantly, they can be altered by concerted efforts to promote family planning and, particularly, the education of young women. In demographic terms, educating young women has vital benefits, including delaying the age of first reproduction, which reduces average family sizes while increasing the mean generation time, thereby slowing the overall rate of population growth. Educated women with smaller families also enjoy greater marital stability, higher living standards and improved educational and employment opportunities for their children (Ehrlich, Ehrlich and Daily, 1997). Advocating for more sustainable infrastructure while ignoring rampant population growth in Africa is akin to plugging holes in a leaking dam while failing to notice rising floodwaters that threaten to spill over its top.[8]

BOX 4.5

Using Natural Capital to Promote Sustainable Infrastructure

The Idea

Healthy, intact ecosystems are essential for apes, gibbons and other wildlife. People also rely on these ecosystems for myriad benefits, including:

- medicinal plants;
- water supplies;
- areas of cultural and spiritual importance;
- carbon storage and sequestration; and
- pollination of crops (MEA, 2005).

Reflecting human dependence on nature, natural resources are increasingly viewed as "natural capital" that supplies "ecosystem services" (Kumar, 2011). These economic metaphors emphasize the importance of maintaining our stock of assets over time to ensure a long-term supply of benefits. The concepts can resonate with groups that have previously had limited interest in conservation, including ministries of finance and planning, private investors and business leaders (Guerry et al., 2015; Natural Capital Coalition, n.d.; NCFA, n.d.; Ruckelshaus et al., 2015).

The Challenge

It has been estimated that achieving the United Nations Sustainable Development Goals and realizing the climate commitments made in the Paris climate accord of 2016 will require approximately US$90 trillion in infrastructure investments, particularly in urban development, transportation and clean energy (Global Commission on the Economy and Climate, 2016). A majority of these investments will be in the developing world, including ape and gibbon range states.

This new infrastructure is essential for economic development, poverty reduction and human well-being. If the infrastructure is poorly planned, however, it not only imperils apes and gibbons, but also the benefits that are provided by nature to humans, undermining the very human development that the infrastructure was intended to support (Mandle et al., 2016a).

Environmental issues are typically considered late in the development planning process, and often when only marginal changes to project design can realistically be considered (Laurance et al., 2015a; see Box 1.6). Even though such gaps can be addressed through strategic environmental assessments, impacts on ecosystem services continue to be considered late or not at all, even when an infrastructure project's success depends directly on ecosystems, for example to reduce the risks of flooding or erosion (Alshuwaikhat, 2005; Mandle et al., 2016a). The transformation of this deeply flawed model of infrastructure planning and investment is a matter of critical urgency.

The Opportunity

Impacts on natural capital and those who depend on it can best be mitigated if they are centrally integrated into infrastructure planning, assessment and development processes from the outset. This early incorporation can build a pipeline of projects that genuinely take into account interlinked environmental, social and economic considerations. There is considerable demand for such projects: "patient" financial capital is invested to produce high-yielding, stable, long-term, income-oriented returns (Roberts, Patel and Minella, 2015).

Around the world, people are now developing, accessing and sharing information about natural capital to inform development planning (Brown et al., 2016; Guerry et al., 2015). These approaches identify the manifold benefits that nature currently provides and attempt to anticipate what might happen to those benefits in response to global climate change, and as resource management and human interactions with nature change (Ruckelshaus et al., 2015). Tools are being devised to help incorporate environmental priorities into real-world decision-making.[9] Governments and businesses can use this type of information to identify areas that are important sources of natural capital and that should be avoided or protected to minimize the negative impacts of built infrastructure (Laurance et al., 2015b). Such knowledge can also be used to identify positive impacts of ecological restoration—for example, investing in reforestation around rivers to enhance fisheries.

There is also demand from businesses and investors for help in determining the best locations for new infrastructure (Laurance et al., 2015a; Natural Capital Coalition, 2016). Environmental and social impact and risk assessments have often ignored companies' dependence on ecosystem services such as clean air, fertile soil and reliable water supplies. This puts companies at risk—for example, from flooding, drought and shortages that could affect their supply chains. To lessen these risks, companies can incorporate natural capital information in decision-making. The Natural Capital Protocol is a decision-making framework that provides guidance for businesses looking to manage risks and seize opportunities by integrating the value of nature into their internal decision-making (Natural Capital Coalition, 2016).

Some Examples

China provides an impressive example of strategic environmental planning at the national scale, one from which lessons can be drawn for ape conservation. In 1998, after decades of deforestation and overgrazing, China instituted major reforms in response to devastating floods that left more than 4,000 people dead and 13 million homeless in the Yangtze River Basin (Spignesi, 2004). Information on how nature benefits people is being used to design restoration and protection measures for ecosystems across almost half of the country. To date, about US$100 billion has been invested in ecosystems and to compensate 120 million people, with many millions of trees planted (Daily et al., 2013). China's first

Photo: Commitments made in the Paris climate accord of 2016 will require approximately US$90 trillion in infrastructure investments, particularly in urban development, transportation and clean energy, such as hydropower projects.
© Melanie Stetson Freeman/The Christian Science Monitor via Getty Images

national ecosystem assessment—carried out from 2000 to 2010—quantified and mapped changes in food production, carbon sequestration, soil retention, sandstorm prevention, water retention, flood mitigation and the provision of habitat for biodiversity. It showed significant improvements in most services, with the worrying exception of habitat for biodiversity conservation (Ouyang et al., 2016).

Incorporating natural values into project planning also has much potential to contribute to conservation in the ape range states of Africa and Asia, even where data and capacity are limited (Bhagabati et al., 2014; Mandle et al., 2016b; University of Cambridge, 2012; Watkins et al., 2016). In the Greater Virungas landscape, a key region for the conservation of gorillas and chimpanzees in Africa's Albertine Rift, a natural-capital assessment helped decision-makers in Rwanda and the DRC to identify the location and importance of areas for water yield, sediment retention, carbon storage and non-timber forest products (University of Cambridge, 2012). In Myanmar, a national assessment showed where and how natural capital contributes to clean and reliable drinking water, reduces risks from inland flooding and coastal storms, and maintains reservoir and dam functioning by greatly reducing erosion (Mandle et al., 2016b). In Indonesia, natural-capital tools were used to inform spatial planning in Sumatra and Borneo, and at the national level. The informed land use planning will be incorporated into efforts to build governance and financing that improve outcomes for people and biodiversity (Bhagabati et al., 2014; GEF, 2013; Sulistyawan et al., 2017).

BOX 4.6

The Bukavu–Kisangani Highway: A Threat to the Critically Endangered Grauer's Gorilla?

Extending over 6,000 km² (600,000 ha), the Kahuzi-Biega National Park (KBNP) in the eastern part of the Democratic Republic of Congo (DRC) comprises dense lowland as well as Afromontane rainforests. The protected area was originally created as a wildlife sanctuary to protect the small population of Grauer's gorilla (*Gorilla beringei graueri*) living in the mountain and bamboo forests between Mounts Kahuzi (3,308 m) and Biega (2,790 m). Having been upgraded to a national park in 1970, KBNP was extended in 1975 to comprise vast tracts of lowland forests, which make up more than 90% of its surface today (ICCN, 2009).

The park is one of the most important sites for biodiversity in the Albertine Rift and harbors 136 species of mammals, including 14 species of primate, 2 of which are great apes: the eastern chimpanzee (*Pan troglodytes schweinfurthii*) and Grauer's gorilla (ICCN, 2009). In view of this exceptional biodiversity, the park was designated a UNESCO World Heritage Site in 1980. KBNP suffered major impacts during the wars and civil conflicts in the DRC and has thus been on the list of World Heritage in Danger since 1997 (Debonnet and Vié, 2010).

KBNP harbors the largest surviving population of Grauer's gorilla that is endemic to the DRC. However, these apes are

FIGURE 4.11

The Bukavu–Kisangani Highway (RN3) and the Kahuzi-Biega National Park

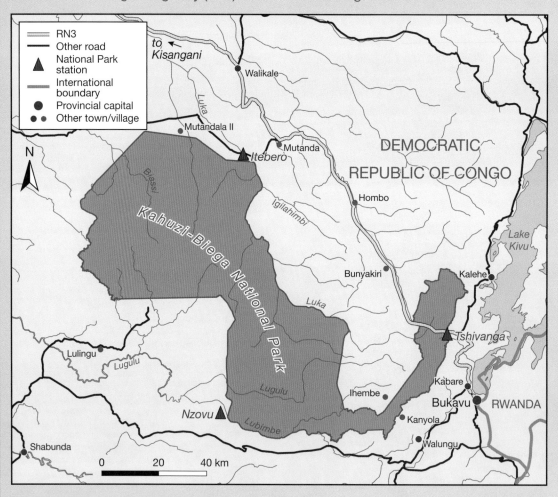

Sources: René Beyers; vector data from CARPE (n.d.); digital elevation model from USGS (n.d.)

increasingly threatened as a result of poaching for wild meat, an illegal activity that is linked to unlawful artisanal mining and the civil conflict. The population has declined by more than 77% since 1994 and is now critically endangered (Plumptre et al., 2016c).

Even before the outbreak of the civil conflict, the upgrade of the RN3, a major road connecting the cities of Bukavu and Kisangani, had raised concerns about possible adverse impacts on the park. The road bisects the highland sector of the park for 18.3 km, cutting through the habitat of several gorilla families (Bynens et al., 2007). After leaving the park, the road veers away from its boundaries before approaching it again in the vicinity of the village of Itebero, in the lowland sector (see Figure 4.11).

The road predates the creation of the national park. Traffic densities remained low until it fell into complete disrepair in the 1990s, when it became virtually impassable. Today, traffic is mostly local, transporting goods and people between Bukavu and the villages to the west of the highland sector. The road's poor condition beyond the village of Hombo has rendered through traffic to Kisangani virtually impossible since the early 1990s. To further mitigate the impacts, the protected area authority—the Institut Congolais pour la Conservation de la Nature (ICCN)—has erected checkpoints at the entry and exit of the park, where vehicles are registered and can be searched. The road is closed to all traffic between 6 pm and 6 am. Nevertheless, vehicles frequently stay in the park at night as a result of mechanical breakdowns or because they get stuck due to poor road conditions.

In spite of the inferior quality of the road, traffic on the stretch through the park has continued to increase. Data collected by the park show an upsurge from 1,485 motorized vehicles in 1999 to 47,489 vehicles in 2014—a 30-fold increase (Bynens et al., 2007; ICCN, 2015). Vehicle numbers vary widely between years, reflecting prevailing security conditions, but the past few years show a clear upward trend, paralleling gradual security improvements (ICCN, 2016). An upgrade of the road would allow vehicles to pass to Kisangani once again and would invite non-local traffic, which could result in a steep increase in traffic through the park.

The impacts of the road on the gorillas in the highland sector are not well understood. The road cuts through the territory of several gorilla families, which cross the road regularly, several times a week. The number of families living around the road and therefore needing to cross it has more than doubled over the years—from three in 2007 to eight in 2015 (ICCN, 2016). This increase may be partly linked to greater insecurity and human activities in the northern and southern parts of the highland sector, where illegal artisanal mining and farming are known to occur; these developments have led to a concentration of gorillas in the central region of the highland sector, which is safer.

Systematic follow-up of gorilla crossings in the early 1990s, when traffic flows were low, suggested that the number of crossings remained stable over time. However, the authors' field observations clearly indicate that road crossings are highly stressful for the animals. Ranger staff have documented that gorillas sometimes hide close to the roadside for long periods of time, waiting for humans to disappear before starting to cross. During the crossing, the silverback typically takes up a position in the middle of the road and waits for the family to cross safely[10]. It is therefore highly likely that a significant increase in traffic on the road would affect the current crossing patterns (Bynens et al., 2007).

The rehabilitation of the RN3 has been planned for a long time. At the end of the 1980s, rehabilitation started from Kisangani, with funding from the government of Germany. Following concerns raised by environmental experts and by the UNESCO World Heritage Committee, IUCN prepared an environmental impact assessment, which advised against rehabilitating the stretch through the park and recommended that it be rerouted around the northern boundary of the park (Doumenge and Heymer, 1992). Based on the results of the study, the German government informed UNESCO that it would not support the construction of the stretch through the park. As a result of the freezing of German aid to the DRC in 1990, the road was not built beyond the village of Walikale and thus never reached the village of Itebero (Bynens et al., 2007).

In 2007, the European Union undertook a new feasibility study for the rehabilitation of the road. Once again, the UNESCO World Heritage Committee expressed concerns that the measures proposed to lessen the adverse effects of the road in the park were insufficient and requested that the final report include clear proposals for mitigation measures to reduce the direct and indirect impacts (UNESCO, n.d.-a). The final study concluded that while the road would bring important socio-economic benefits to the local communities, the likely steep increase in traffic on the stretch through the park could have adverse impacts on the resident gorilla populations and the integrity of the World Heritage site. It thus recommended that the road be rehabilitated for through traffic to Kisangani only if the stretch through the highland sector of the park could be rerouted to avoid the park (Bynens et al., 2007). Kinshasa accepted this recommendation at the time.

To date, the RN3 remains impassable and no traffic is possible beyond the village of Hombo. A reopening of the road would bring important economic benefits to communities, which have lived in total isolation since the start of the civil conflict, at the mercy of the different armed groups and bandits who control the region. With the gradual return of peace and stability, the discussion regarding the road's rehabilitation will certainly be revived. A rehabilitated road would undoubtedly attract new threats to the lowland sector of KBNP, and it might increase illegal logging and stimulate the wild meat trade. At the same time, it would reintegrate this region into the modern world, allowing park authorities to exert better control over illegal activities. It would also persuade people who had settled inside the park after fleeing the violence to leave the park and resettle in the villages along the road; in this way, a revitalized road could garner conservation benefits. However, the rerouting of the stretch crossing the highland sector of the park remains an important condition that needs to be guaranteed before any rehabilitation is envisaged.

Priorities for Infrastructure and Protected Areas

Near-term priorities for limiting the environmental impacts of infrastructure expansion on African ape habitats and, more generally, vital protected areas include:

- Carefully scrutinizing plans for expanding "development corridors" in Africa in terms of their environmental costs and economic and social benefits (see Chapter 1). Taking this approach calls for the substantial modification or total abandonment of corridors that are likely to produce marginal benefits relative to their heavy costs, regardless of whether they are being planned or already being upgraded (Laurance et al., 2015a; Sloan et al., 2016).

- Limiting roads in and near protected areas. While protected areas need some road access for ecotourism, roads should avoid the core areas of parks whenever possible so as to limit human impacts. A variety of sensitive wildlife species shun areas with even modest levels of human activity (Blake et al., 2007; Griffiths and Van Shaik, 1993; Ngoprasert, Lynam and Gale, 2017; Reed and Merenlender, 2008; Rogala et al., 2011).

- Stemming the loss of buffering habitats and limiting infrastructure expansion in the habitats immediately surrounding protected areas. Unless they are curbed, these processes (1) reduce the ecological and demographic connectivity of reserves to nearby habitats, and (2) often "leak" into the interiors of protected areas themselves (see Figure 4.5). Both types of changes can have serious impacts on biodiversity (Laurance et al., 2012).

- Favoring large protected areas, which are superior to smaller protected areas because they typically (1) are less susceptible to human encroachment and external land use disturbances (Maiorano, Falcucci and Boitani, 2008; see Figure 4.5), (2) support larger wildlife populations that are less vulnerable to local extinction, and (3) provide a wider range of habitats, elevational and topographic diversity, as well as climatic regimes that can help buffer species against heat waves, droughts and other severe climatic events (Laurance, 2016b).

- Defending protected areas for African apes and designating new reserves in critical habitats. Two immediate priorities are Cross River National Park in Nigeria (see Case Study 5.1) and Kahuzi-Biega National Park (see Box 4.6) and its nearby critical habitats in the eastern DRC (Plumptre et al., 2015). Both parks harbor critically endangered subspecies of gorillas.

Acknowledgments

Principal author: William F. Laurance[11]

Contributors: Stephen Asuma, Ephrem Balole, The Biodiversity Consultancy (TBC), Neil David Burgess, Geneviève Campbell, Guy Debonnet, European Commission Joint Research Centre (JRC), Fauna and Flora International (FFI), International Gorilla Conservation Programme (IGCP), Annette Lanjouw, Anna Behm Masozera, Sivha Mbake, Emily McKenzie, Emmanuel de Merode, Stephen Peedell, UNESCO World Heritage Centre, United Nations Environment Programme World Conservation Monitoring Centre (UNEP-WCMC), Virunga National Park and World Wildlife Fund (WWF)

Mitigation Hierarchy and **Case Study 4.1:** Geneviève Campbell

Boxes 4.1 and 4.3: Stephen Peedell

Box 4.2: Anna Behm Masozera and Stephen Asuma

Box 4.4: Ephrem Balole and Emmanuel de Merode

Box 4.5: Emily McKenzie and Neil David Burgess

Box 4.6: Guy Debonnet and Sivha Mbake

Author acknowledgments: Mason Campbell provided helpful comments and assistance with statistical analysis, and Sean Sloan assisted with preparation of imagery.

Reviewers: Mark Cochrane and David Edwards

Endnotes

1. Two additional proposed corridors came to light during the study, making the total 35.
2. Author correspondence with Tom Okurut, executive director of the National Environment Management Authority, Uganda, 2016.
3. IGCP is a coalition program of Fauna and Flora International and the World Wide Fund for Nature, alongside the protected area authorities in the DRC, Rwanda and Uganda and local partners. http://igcp.org/
4. All variables were $\log_{10}$ transformed and then standardized prior to analysis.
5. Significance testing was carried out for external road pressure (t=13.72, df=651, P<0.000001) and park size (t=−2.65, df=651, P=0.008).
6. Internal calculation based on confidential ICCN documents reviewed by the author.
7. Some figures have been adjusted based on internal ICCN project update and assessment documents reviewed by the author.
8. The average woman in Africa had 4.72 children in 2010–15, exceeding the global fertility rate of 2.52 by about 87% (UN Population Division, n.d.).
9. See the Natural Capital Protocol Toolkit for information on a variety of available tools (WBCSD, n.d.).
10. In 1997, a soldier killed one of the park's most famous silverbacks, named Nindja, while he was standing in the middle of the road waiting for his family to cross.
11. James Cook University (www.jcu.edu.au)

Photo: KBNP harbors the largest surviving population of Grauer's gorilla. The population has declined by more than 77% since 1994 and is now critically endangered. © Jabruson 2018 (www.jabruson.photoshelter.com)

Photo: Increased vehicular movement produces air pollution, noise pollution, vibrations, light pollution and wildlife-vehicle collisions. © Jacqueline Rohen

CHAPTER 5

Roads, Apes and Biodiversity Conservation: Case Studies from the Democratic Republic of Congo, Myanmar and Nigeria

Introduction

As demonstrated throughout this volume, road construction is a leading cause of habitat fragmentation and loss. It reduces wildlife connectivity, threatening the survival of species by impeding their ability to move across a landscape in search of food and shelter and to mate. It also increases human access to, and the destruction and degradation of, previously remote and undisturbed areas, including essential forests (Laurance, Goosem and Laurance, 2009).

In addition to land use changes and loss of connectivity, road development alters the characteristics of habitats both close to and distant from the road, thereby changing the way wild animals use these habitats. Roads affect the movement of water and the patterns and severity of erosion, while increased vehicular movement produces air pollution, noise pollution, vibrations, light pollution and wildlife–vehicle collisions. By facilitating wildlife poaching, improved access has a particularly significant impact on species survival (Laurence et al., 2009).

Increased human encroachment into ape habitat exposes apes to greater hunting pressure and an increased risk of disease transmission, while also confronting them with a loss of habitat and connectivity. In 2002 the United Nations Environment Programme (UNEP) projected that by 2030 only 10% of the original gorilla range would be free of human impact, primarily as a result of infrastructure development, agricultural expansion and logging (Nellerman and Newton, 2002). This habitat destruction and fragmentation is one of the major threats to ape survival.

At the same time, roads can result in substantial economic and social benefits, which tend to form the cornerstone of national economic development plans, although these are not always realized (Berg et al., 2015; see Chapter 2, pp. 60–77). There are therefore trade-offs between improving human well-being and protecting the environment.

This chapter explores how advance planning that is evidence-based, inclusive and effectively implemented, monitored and evaluated can help to minimize the negative impacts of road development on biodiversity. To that end, it examines the interface between road development and the environment, focusing on the impact on apes in particular. The chapter presents three case studies on proposed and continuing road development in ape ranges in Africa and Asia:

- the Cross River superhighway of Cross River State, Nigeria;
- the Dawei road link between Thailand and Myanmar; and
- the High-Priority Roads Reopening and Maintenance (Pro-Routes) project of the Democratic Republic of Congo (DRC).

The first case study presents the context of the proposed Cross River superhighway, which is to connect a new deep seaport at Calabar in southeastern Nigeria to landlocked Chad and Niger. While the rationale behind the project appears to have some merit, the proposed highway will stop about 1,000 km short of Nigeria's northern border. Furthermore, Nigeria already has eight major seaports and experts doubt there is sufficient economic justification for constructing another one in Calabar (Shipping

Position Online, 2016). Moreover, the Calabar River is relatively shallow and prone to siltation, which is exacerbated by surrounding logging and deforestation, and consequently the "deep seaport" will require periodic and expensive dredging (Vanguard, 2015). In addition to considering the project's environmental and social impacts, the case study examines the role that local and international non-governmental organizations (NGOs) can play, especially in relation to drawing attention to the lack of adequate impact assessments, consultation and planning. It also highlights that thoroughly conducted environmental impact assessments (EIAs) are key tools for ensuring the integration of biodiversity conservation into all types of infrastructure planning (see Box 1.6).

The second case study focuses on the proposed 138 km road from the Thai border to the planned Dawei Special Economic Zone (DSEZ), an area that is to cover 250 km² (25,000 ha) in Myanmar's southernmost region, on the border with Thailand. The road's planned route bisects crucial ecological connectivity. Maintaining that connectivity in an area of weak governance, competing transnational interests and civil struggle urgently requires sustained, innovative approaches to infrastructure planning and design, as well as to conservation and environmental policy. In 2015 and 2016,

Photo: Road construction is a leading cause of habitat fragmentation and loss; one of the major threats to ape survival.
© WWF Myanmar/ Adam Oswell

Chapter 5 Roads

a multidisciplinary team from the World Wide Fund for Nature (WWF) and the University of Hong Kong (HKU) launched a campaign to promote ecological connectivity and sustainability in the region by increasing awareness and building capacity with stakeholders and decision-makers. In addition to several outreach strategies, the team released three reports: the first highlights the ecological systems at risk from the proposed road and argues for robust environmental policies; the second, a manual of sustainable road design, focuses on mitigating impacts on wildlife; and the last provides an explicit yet flexible method of locating wildlife mitigation measures and crossings, despite extremely limited biological and physical data for the area (Helsingen et al., 2015; Kelly et al., 2016; Tang and Kelly, 2016). In light of Myanmar's recent political shift, this case study explains these and other regional conservation initiatives in the context of decades of conflict and recent economic development.

The third case study traces the evolution of the Pro-Routes project, a major road rehabilitation project in the DRC, funded by the International Development Association and the United Kingdom's Department for International Development (DFID). It focuses specifically on the 523 km Kisangani–Bondo segment of this rehabilitation project and its anticipated impact on the Bili–Uélé Hunting Domain and the Bomu Faunal Reserve, referred to hereafter as the Bili–Uélé Protected Area Complex (BUPAC). At the outset, the project stakeholders aimed to consider the potential environmental and social impacts of the road's rehabilitation and planned to implement recommendations to mitigate projected negative impacts. As the case study reveals, however, there is almost no evidence that recommendations were implemented as planned. The study discusses the need for expertise in responsible infrastructure development, the critical role of external conservation specialists and the importance of timely and effective monitoring and evaluation.

Key findings of this chapter include:

- In the case of conflicting priorities, conservation organizations can play an important role in building relationships between various stakeholders by working with government agencies, local communities, industry, political actors and others who are sympathetic to conservation objectives.

- The fact that EIAs are required in relation to road development in all environmentally sensitive areas is useful, but not sufficient for ape conservation, as poorly conceived and conducted assessments can enable ill-advised or poorly designed infrastructure development in essential African and Asian ape habitats.

- Modeling is a valuable method for evaluating potential impacts of infrastructure, as it allows conservation actors to illustrate various scenarios and options to a wide range of stakeholders and decision-makers.

- By engaging with experts from relevant disciplines, project leaders can ensure that environmental factors are adequately addressed in project planning to allow for the development of effective mitigation measures.

- In the context of infrastructure development, integrated land use planning can serve to mitigate environmental and social impacts while also contributing to greater coordination, such as across ministries and within national agencies.

- Wherever landscapes do not have explicitly delineated areas for more traditional conservation planning, it is critical that conservation and environmental actors join forces, avoid overlaps in engagement and speak with one voice.

▶ p. 1

CASE STUDY 5.1

On a Road to Nowhere? The Proposed Calabar–Ikom–Katsina Ala Superhighway Project in Cross River State, Nigeria[1]

Introduction

With a population of more than 180 million people and massive oil reserves, Nigeria is Africa's giant, and, despite a recession, Africa's largest economy (*The Economist*, 2014). But the country has failed to live up to expectations for growth and development since independence in 1960 and now lags far behind comparable countries, such as Malaysia and Indonesia (Sanusi, 2012). The reasons behind this underdevelopment are complex, but endemic corruption and chronic mismanagement by a series of military and civilian governments are most likely to blame (Ojeme, 2011). Promising to tackle corruption, Nigeria elected a new leader, Muhammadu Buhari, in May 2015. New governors, who traditionally enjoy unrivalled autonomy in Nigeria, were elected at the same time in all 36 states of the federation.

The self-proclaimed environmentalist Benedict Ayade was appointed as the new governor of Cross River State. He soon announced a number of signature projects, including the construction of a six-lane, 20-km-wide, 260-km-long superhighway to connect a new deep seaport with northern Nigeria. The governor further boasted that this "digital superhighway" would be designed for the 21st century, with Internet connectivity along its entire length. Although Nigeria is in the grips of its biggest recession to date and Cross River is one of the most indebted states in the country—due to massive borrowing by previous governors to fund their own signature projects—an estimated US$2.5 billion has been budgeted for this ambitious project (Olawoyin, 2017; PGM Nigeria, 2016a, 2016b). Funding sources have not been disclosed, however, and although some potential investors reportedly pulled out, perhaps due to delays and controversy, it appears that a number of Chinese investors are still interested in the project (*This Day*, 2016). Designed to create jobs and sustainable revenue for Cross River State, the superhighway and deep seaport are to be developed and managed through a public–private partnership. At the time of writing, the superhighway was to pass through some of the state's most pristine remaining forests, including Cross River National Park, with catastrophic consequences for wildlife (Akpan, 2016a).

In September 2015, the initial groundbreaking ceremony for the superhighway was canceled at the last minute when the federal government realized that no EIA had been undertaken. In Nigeria, the law requires EIAs for all major development projects (Federal Republic of Nigeria, 1992). This was a huge political embarrassment for Governor Ayade. A compromise deal was soon reached, however, and the Federal Ministry of Environment issued an "interim EIA" to allow the groundbreaking ceremony to go ahead, on the understanding that no work would start until an EIA was submitted and approved.

FIGURE 5.1
The Proposed Cross River Superhighway

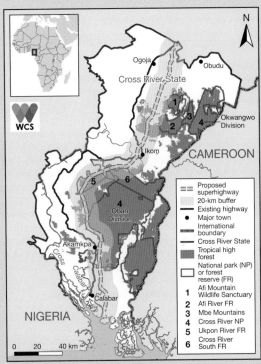

© WCS

Amid much pomp and ceremony, President Buhari arrived in Calabar on October 30, 2015, and performed the groundbreaking event. Through this act, Buhari tacitly gave the federal government's consent for the superhighway project, but Environment Minister Amina Mohammed would play a key role in ensuring that the state government had to produce an acceptable EIA (Akpan, 2016b).

Background

UNESCO has proposed that Cross River National Park—Nigeria's richest site for biodiversity—be listed as a Man and the Biosphere Reserve and potentially a World Heritage Site. WWF and the International Union for Conservation of Nature (IUCN) recognize the park as a Centre of Plant Diversity, and Birdlife International classifies it as an Important Bird and Biodiversity Area (Fishpool and Evans, 2001).

Within Cross River National Park lie the Oban Hills, whose biological importance was recognized as early as 1912, when a large part of the area was gazetted as a forest reserve (Oates, 1999). In 1991, the reserve was upgraded to create the Oban Division of Cross River National Park, through which the superhighway is now expected to pass (Oates, Bergl and Linder, 2004). Covering around 3,000 km² (300,000 ha) of lowland rainforest, the Oban Division is the largest remaining area of rainforest in Nigeria and is contiguous with the Korup National Park in Cameroon. With peaks reaching between 500 m and 1,000 m, the Oban Hills are also an extremely important watershed, giving rise to numerous rivers that guarantee a perennial supply of freshwater to hundreds of downstream communities in Cross River State (Caldecott, Bennett and Ruitenbeek, 1989).

In addition to apes, Oban contains a number of rare and endangered species, such as the Nigeria–Cameroon chimpanzee (*Pan troglodytes ellioti*), drill (*Mandrillus leucophaeus*), Preuss's red colobus monkey (*Procolobus preussi*), leopard (*Panthera pardus*), forest elephant (*Loxodonta cyclotis*), the slender-snouted crocodile (*Mecistops cataphractus*) and the gray-necked rockfowl (*Picathartes oreas*), as well as 75 plant species that are endemic to Nigeria (Oates *et al.*, 2004). The area is a center of species richness and endemism, particularly for primates, birds, amphibians, butterflies, fish and small mammals (Bergl, Oates and Fotso, 2007; Oates *et al.*, 2004). But the same area is also subject to intense hunting pressure to supply the wild meat trade, and rates of deforestation are among the highest in the world (Bassey, Nkonyu and Dunn, 2010; Fa *et al.*, 2006; FAO, 2015; Okeke, 2013). Given that it combines high levels of species richness and endemism with a high degree of threat, the area represents a biodiversity hotspot of global significance (Myers *et al.*, 2000).

Impact on Apes

Two different apes are found in Cross River State: the critically endangered Cross River gorilla (*Gorilla gorilla diehli*), the most endangered taxon of ape in Africa, and the endangered Nigeria–Cameroon chimpanzee (*Pan troglodytes ellioti*), the most threatened of four subspecies of chimpanzee (Morgan *et al.*, 2011). Due to hunting and habitat loss, these apes are restricted to two protected areas within the state—Cross River National Park and the Afi Mountain Wildlife Sanctuary—as well as a small area of community-managed land within the Mbe Mountains.

The Oban Division of Cross River National Park is expected to bear the brunt of the impact of the superhighway, while its Okwangwo Division will be relatively unaffected (see Figure 5.1). Although Oban supports an estimated 150–350 Nigeria–Cameroon chimpanzees, it does not contain any Cross River gorillas, a species found only in the Okwangwo Division, the Mbe Mountains and the Afi Mountain Wildlife Sanctuary (Dunn *et al.*, 2014; ellioti.org, n.d.). The superhighway only skirts the western edge of the sanctuary, however it directly threatens the Afi River Forest Reserve, a critically important corridor that links the sanctuary to the Mbe Mountains (Dunn *et al.*, 2014). The loss of such corridors in the landscape would be catastrophic for the Cross River gorilla and the Nigeria–Cameroon chimpanzee, as both survive in small isolated groups. The superhighway is expected to lead to massive deforestation along the entire route as farmers from neighboring states move into the area, and as improved access facilitates hunting (Laurance *et al.*, 2017a).

International Pressure Mounts

On October 20, 2015, ten days before the groundbreaking ceremony, a coalition of 13 international and national NGOs, including Birdlife International, the Wildlife Conservation Society (WCS) and the Zoological Society of London, submitted a letter to President Buhari expressing concern about the superhighway. In the letter, they conveyed support for the ongoing EIA but declared their outrage concerning plans for the superhighway to traverse Cross River National Park.[2] The route of the superhighway was subsequently adjusted, yet some argued that it was still too close to the edge of the national park and objected on the grounds that it would pass through some important community forests and forest reserves (Cannon, 2017b).

On January 22, 2016, the Cross River government published a notice of revocation of rights of occupancy within a 20-km-wide land corridor along the entire highway route (MLUD, 2016; see Figure 5.1). This single act dispossessed more than 185 communities within the corridor of their land, subjecting them to displacement at any time. With the notice, the state seized a land area of 5,200 km^2 (520,000 ha), or about 25% of the state's total area. Communities that had initially supported the superhighway rose up in revolt when they realized that they had been robbed of their ancestral lands overnight. Many people within the state began to call the superhighway project an elaborate land grab in disguise (Abutu and Charles, 2016).

Once freed of its occupants, this vast area of forest would represent an opportunity to generate significant revenue, first through the sale of the timber and then through conversion of the land to oil palm plantations. Even though the EIA had not yet been finalized, in February 2016 a number of bulldozers started the clearing and felling of trees along the proposed route. Some of the affected communities, such as Old and New Ekuri, blocked the bulldozers from entering their forest, but many more were powerless to prevent the destruction of their forests.

Direct intervention finally came in the form of a stop work order, issued by Environment Minister Mohammed in March 2016. The order forced the governor to suspend activities on the superhighway and await the outcome of the EIA (Ihua-Maduenyi, 2016). That same month five ambassadors of the UNEP–UNESCO Great Apes Survival Partnership sent a letter to the environment minister expressing concern regarding increasing threats to the integrity of the rainforests of Cross River National Park and requesting that the Nigerian government respect commitments made as part of the 2005 Kinshasa Declaration on Great Apes and UN-REDD (Reducing Emissions from Deforestation and Forest Degradation).[3]

Environmental Impact Assessment and Review Process

The EIA law in Nigeria exists to safeguard the population and environment with regard to any form of environmental degradation resulting from development projects. This legislation prohibits activities from being carried out in sensitive areas in the absence of mandatory studies.

Photo: The Cross River gorilla survives in small isolated groups in the Cross River National Park, Afi Mountain Wildlife Sanctuary and a small area of community-managed land in the Mbe Mountains. © WCS Nigeria

The environmental management consultancy PGM Nigeria Limited prepared an EIA of more than 400 pages on behalf of the government of Cross River State; in March 2016, it was submitted to the federal government for approval (PGM Nigeria, 2016a). Environment minister Mohammed appointed an independent review panel to assess the EIA and the document was circulated for public comments in April 2016. A professional review of the EIA, completed by the consultancy Environmental Resources Management on behalf of the international NGOs, identified 11 main flaws in the EIA. The review found that due to these flaws, the assessment could not be used as intended, namely to identify potential impacts of the project or to recommend adequate mitigation measures (ERM, 2016). The 11 main flaws were that:

- the scoping process was inadequate and provided no information on the rationale or analytical process that was adopted;
- baseline data were unclear, inconsistent, frequently contradictory and often incorrect;
- the project description was fundamentally flawed, most critically in that it failed to consider any impacts due to the 20-km-wide corridor of land acquired by the government of Cross River State along the entire route of the proposed superhighway;
- the EIA did not provide cost–benefit analyses for any of the proposed routes, a clear economic justification for the superhighway or reasons for building a new road as opposed to upgrading the existing highway;
- the EIA failed to consider the impacts of the superhighway on nearby protected areas, namely Cross River National Park, Afi Mountain Wildlife Sanctuary, Afi River Forest Reserve, Ukpon River Forest Reserve and Cross River South Forest Reserve;
- stakeholder engagement was extremely limited and failed to meet accepted standards as outlined by Nigerian legislation;
- the EIA failed to identify measures required to monitor effective mitigation of the impact of the superhighway;
- mitigation measures were described at a conceptual level only, with insufficient detail for implementation;
- the EIA failed to mention the presence of many rare and endangered species within the area, such as the critically endangered Preuss's red colobus and the slender-snouted crocodile;
- although more than 185 communities are likely to be affected by the proposed project, the socioeconomic study focused on only 21 communities and failed to assess the full impact on all affected communities, their livelihoods and vulnerability; and
- there was no consideration of any cultural heritage data (ERM, 2016).

NGOs Increase the Pressure

In May 2016 a second letter—this one from 13 international NGOs, including the Arcus Foundation, Fauna and Flora International (FFI), WCS and WWF—expressed further concern about the quality of the recently concluded EIA, requested that it be redone and called for compensation to be paid to affected communities.[4] In addition to these international NGOs, a number of national NGOs have also played a key role in the campaign against the superhighway (Uwaegbulam, 2016). Many local NGOs issued press releases or sent letters of protest, some on behalf of local communities, and a number of local NGOs brought lawsuits against the state government, although none was successful. Among the most active NGOs were the Ekuri Initiative, which has received international accolades for forest stewardship, the Nigeria office of the Heinrich Böll Foundation, the NGO Coalition for the Environment, and the Rainforest Resource and Development Centre (Akpan, 2017).

Rainforest Rescue in Germany organized an online petition against the superhighway, which drew more than 254,000 signatures—34,000 from Cross River State and 220,000 from concerned individuals worldwide. In September 2016, the petition was delivered to President Buhari through the Ministry of Environment in Abuja (Akpan, 2016c). Both the traditional press and social media have carried numerous stories and updates on the issue (Ingle, 2016). By April 2017, another 135,000 people had signed a separate WCS online campaign against the superhighway (WCS, n.d.).

A public meeting was held in Calabar in June 2016 to allow all stakeholders to present their views and opinions to the official review panel (Akpan, 2016b). The Federal Ministry of Environment, which eventually gave the EIA a "D" rating for gaping oversights and errors, ordered the assessment to be redone (Dunn, 2016). It subsequently rejected the revised EIA, a document of more than 600 pages submitted in September 2016, on the grounds that it still failed to meet basic international standards and that:

- there still had been no public consultation or dialog with important stakeholders, such as Cross River National Park;
- baseline data were still absent or weak;
- there was no consideration of the impacts of the 20-km-wide corridor;
- the economic justification for building a new superhighway, rather than simply upgrading the existing Calabar–Ogoja federal highway, had not been clearly demonstrated;
- there was insufficient consideration of the negative impact on local people;
- the EIA used the proposed national park boundary, which was never gazetted, rather than the current *legal* park boundary;
- the EIA failed to acknowledge the fact that the superhighway, as proposed, would pass through the national park;
- the EIA stated that there are no protected areas within the project area or within 50 km of the proposed area and that there are no protected areas within the sphere of influence of the proposed project, yet there are no fewer than five protected areas within the project area and the proposed route of the superhighway would pass directly through three different protected areas—Cross River National Park, Ukpon River Forest Reserve and Cross River South Forest Reserve—and the 20-km-wide corridor would also impact the Afi Mountain Wildlife Sanctuary and Afi River Forest Reserve (Dunn, 2016; Dunn and Imong, 2017; PGM Nigeria, 2016b).[5]

In the absence of an approved EIA, tensions mounted and the state government threatened to resume work on the superhighway, even without approval from the federal government (Vanguard, 2017). During preparation of a third version of the EIA, the Cross River State government finally became attentive to the environmental concerns and approached WCS for help. After a number of meetings with WCS, the state government announced in February 2017 that it was dropping all plans for the 10-km corridor on either side of the highway (Ihua-Maduenyi, 2017). However, since the route was still due to pass through some important Ekuri, Iko Esai and other community forests on the edge of Cross River National Park, as well as Ukpon River Forest Reserve and Cross River South Forest Reserve, conservation groups called on the government to do more (Cannon, 2017c).

Options for the superhighway were discussed, including rerouting it around the forests, even though such modifications would make the highway slightly longer and would increase the overall cost. In March 2017 in Calabar, at a stakeholder forum convened by the Federal Ministry of Environment to review the third version of the EIA, Governor Ayade announced the willingness of the Cross River State government to reroute the highway around the Ekuri community forest (Cannon, 2017a). While this was welcome news, stakeholders continued to demand that the highway be rerouted away from the Ukpon River Forest Reserve and Cross River South Forest Reserve. Finally, in April 2017, the state government agreed to reroute the highway away from most of the remaining forest (Cannon, 2017b; see Figure 5.1).

The fourth version of the EIA and a new biodiversity action plan were submitted to the Federal Ministry of Environment in May 2017 (PGM Nigeria, 2017). Significant improvements included the revocation of the 20-km-wide corridor and the rerouting of the superhighway to avoid important community forests and forest reserves on the edge of the national park. However, this version of the EIA also relied on inadequate data, and therefore its proposed mitigation measures could not be considered valid. Moreover, the EIA failed to assess indirect long-term impacts of hunting and habitat loss on Cross River National Park despite its proximity to the superhighway and improved access to the forest.[6]

Although WCS and others recommended that both the EIA and the biodiversity action plan be rejected, the Federal

Photo: The superhighway is expected to lead to massive deforestation along the entire route as farmers from neighboring states move into the area, and as improved access facilitates hunting. © WCS Nigeria

Ministry of Environment issued provisional approval of the EIA in July 2017. In so doing, the ministry specified no fewer than 23 conditions to be addressed and requested that the EIA be revised and resubmitted within two weeks. These conditions included the development of a biodiversity offset; a revised map on which the new route was to be clearly indicated; a resettlement action plan, including a list of affected communities; and compensation payments to affected communities.[7] At the time of writing, these conditions had not been met and, despite some misleading reports in the press, the ministry had not yet approved the EIA, nor had it issued an environmental impact statement or an EIA certificate.

REDD, Climate Change and Conflicting Policies

In September 2008, the UN Development Programme, UNEP and the Food and Agriculture Organization jointly established the REDD+ program in Nigeria, where it is being piloted in Cross River State. Three years later, Nigeria received a US$4 million REDD+ grant to realize the program's Readiness Project, which includes the preparation and implementation of REDD+ strategies with the active involvement of indigenous peoples, forest-dependent communities and other local stakeholders. In September 2016, the REDD+ program in Nigeria approved a new US$12 million strategy, one designed to deepen the initiative nationwide to combat climate change through improved forest governance (Uwaegbulam, 2016). That same month, President Buhari signed the Paris climate agreement and promised commitment from Nigeria as part of the global effort to reverse the negative effects of climate change. The construction of the superhighway as proposed would certainly conflict with the proposed REDD+ program under pilot in Cross River State, threatening the continuation of future funding from the UN.

Conclusion

Nigeria's Ministry of Environment has played an exemplary role in upholding the law, notably by insisting that the Cross River State government produce an EIA and by subjecting that EIA to critical review. In this respect, the leadership of Amina Mohammed, federal minister of environment at the time and currently Deputy Secretary-General of the United Nations, was instrumental. Without the strong leadership of the ministry, NGO concerns over the superhighway may have been brushed aside. The role of NGOs, both national and international, in opposing the superhighway has also been critical; NGOs were able to exploit social media and online petitions to generate publicity for their campaign.

Although the most recent EIA integrates an environmental and social management plan as well as a biodiversity action plan, it still fails to evaluate the long-term costs of the project. Given that every version of the EIA was paid for by the proponents of the very project it was meant to assess, it is not surprising that the analysis and results were unduly influenced. Despite significant environmental, social and financial concerns, the federal government is likely to succumb to political pressure and may eventually allow the superhighway to proceed without a comprehensive EIA and even though the construction of the deep seaport remains uncertain.

CASE STUDY 5.2

Engineering Conservation: Stories and Models of Infrastructure, Impact and Uncertainty in Southern Myanmar

Introduction

Tanintharyi, Myanmar's southernmost region, shares an extensive border with Thailand along the Dawna and Tenasserim mountain ranges and harbors some of the last remaining large forest areas in the Greater Mekong sub-region. This landscape is home to several endangered species, including the lar gibbon (*Hylobates lar*), Asian elephant (*Elephas maximus*), northern pig-tailed macaque (*Macaca leonina*), stump-tailed macaque (*Macaca arctoides*), langur (*Semnopithecus*) and tiger (*Panthera tigris*) (WCS, 2015a; WWF, 2016).

Isolated politically and economically due to more than half a century of civil war between ethnic groups and the Myanmar military regime, this region is now witness to intense pressure from domestic and transnational development proposals, weak land rights and large-scale exploitation of natural resources (Hunsberger *et al.*, 2015; Simpson, 2014). Since 2012, a ceasefire has been in effect between the Myanmar government and the Karen National Union (KNU), an opposition group that represents the Karen ethnic groups and still controls large areas of Tanintharyi Region (KNU, 2012).

New Conservation Efforts along the Road Corridor

Starting in 2008, the governments of Myanmar and Thailand agreed to collaborate on a series of projects, including the Dawei Special Economic Zone. Critical to the planned 250-km^2 (25,000-ha) DSEZ is a 138 km road link that will connect the economic zone to the Thai border (see Figure 5.2).

FIGURE 5.2
The Dawei road link and deforestation East of Myitta

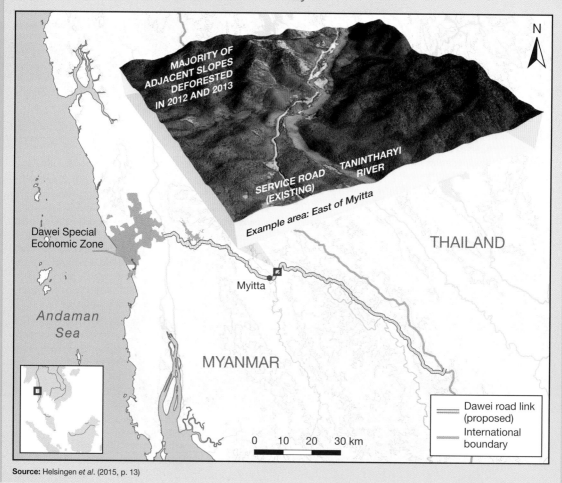

Source: Helsingen *et al.* (2015, p. 13)

This relatively short segment is the western end of the Greater Mekong Subregion's Southern Economic Corridor, a largely completed major trade route connecting Bangkok and Ho Chi Minh City (ITD, 2012). While the DSEZ and Dawei road link are key projects for renewed investment in Myanmar, political uncertainty related to the country's 2011 democratic transition, coupled with Thailand's 2014 military coup, KNU control of the border area and increasing civil society presence, limited investment. As a result, the project's scope has varied greatly over time, alternating between eight-, four-, and two-lane alignments, with and without rail, power lines and gas pipelines (ITD, 2011).

In view of these proposed development projects and the threats they pose to some of the most poorly documented yet biodiversity-rich forests in the Greater Mekong subregion, several international and domestic NGOs began to increase their presence throughout Tanintharyi in 2014. Their efforts have included:

- village and customary mapping by FFI and WCS;
- land cover change mapping by the Smithsonian Institution and a Myanmar-based NGO, Advancing Life and Regenerating Motherland, or ALARM;
- supporting forest management plans of the regional government; and
- biodiversity surveys completed by FFI and WCS, as well as by WWF in partnership with the Karen Environmental and Social Action Network and Karen Wildlife Conservation Initiative (Connette et al., 2016; WCS, 2015a; WWF, 2016).

While land use planning among local government offices, the KNU and NGOs is ongoing and has been somewhat effective in controlling the expansion of agroindustry and mining exploration, road development remains relatively unchecked, despite recent national environmental impact legislation (DDA, TYG and TripNet, 2015; METI, 2015).

Two Decades of Conservation and Ethnic Conflicts

Intense distrust between local civil society and both domestic and international institutions has long plagued conservation efforts in Tanintharyi. This distrust can be traced back to the mid-1990s, when multinational investment financed the precursors of today's DSEZ projects. In 1996, Thailand and the military-ruled Myanmar governments announced an industrial estate and road link, whose scopes and scales were similar to those of today's projects; the Industrial Estate Authority of Thailand completed a feasibility study and the Italian–Thai Development Company, which remains today's principal developer, carried out an initial survey (Arunmart, 1996).

Overlaid on these development proposals was the Myanmar government's controversial Myinmoletkat Nature Reserve, which was gazetted with the help of WCS and the Smithsonian Institution to include KNU-held protected areas, the proposed industrial estate and the road link, as well as the site of Total's Yadana gas pipelines (Mason, 1999; Noam, 2007).

The reserve was drawn predominantly on lands governed by the KNU ethnic armed group.

Between 1996 and 2004, the local villagers' landmark lawsuit and settlement against Total's partner Unocal in U.S. courts over the Yadana pipeline drew international attention (ERI, 2009). Given the Myinmoletkat Nature Reserve's link to the Myanmar military government, suspected endorsement by multinational oil companies, unjustifiably large expanse, and a record of forced relocations and disregard for human rights in the protected area, the Myinmoletkat Reserve was heavily criticized by the conservation community abroad (Brunner, Talbott and Elkin, 1998; Mason, 1999).

Within months of Myinmoletkat's establishment in 1997, the Myanmar military made a violent sweep of the planned transport corridor in KNU-controlled Tanintharyi. A Western aid worker noted that "bulldozers were flattening a broad swathe on the heels of the advancing army" (Moorthy, 1997). They destroyed at least eight Karen villages along the route and, in collusion with Thai logging companies, forced repatriation of refugees from Thailand to Myanmar, into an area of heavy fighting (Moorthy, 1997). In 1998, gas started flowing in the Yadana pipeline, which has since accounted for a significant portion of the national government's export income (Simpson, 2014).

In 2005, Myinmoletkat was turned into the substantially smaller Tanintharyi Nature Reserve Project, about 30 km north of the planned Dawei road link corridor; the reserve served as part of Total's contested corporate social responsibility program, itself funded by the lawsuit settlement and characterized by forced labor and other human rights abuses (ERI, 2009).

Current Status of the Road Corridor

The Dawei road link remains unpaved, despite an upgrade that was carried out between 2009 and 2012 (ITD, 2011, 2012).[8] At the time of writing, construction of the road had stalled due to a lack of investment; the developers were awaiting a final decision from the new civilian Myanmar government.[9] Meanwhile, the situation on the ground remained complicated by demands from villagers for appropriate compensation, competing land ownership claims among internally displaced persons and migrants, imminent refugee return from across the Thai–Myanmar border and military-sanctioned agroindustry land grabs (DDA, 2014). The democratization of land policies, notably in the 2012 Farmland Law and accompanying Vacant, Fallow and Virgin Lands Management Law, has opened up previously protected village lands to market interests and widespread land degradation (Oberndorf, 2012; Simpson, 2015).

In view of the complex situation of conservation and development in Tanintharyi, policy experts and conservation biologists from WWF teamed up with landscape planners, designers and civil engineers from the University of Hong Kong to construct a series of scenarios to predict possible outcomes, build capacity and provide tools for sustainable infrastructure development in southern Myanmar (Helsingen et al., 2015; Kelly et al., 2016; Tang and Kelly, 2016).

Photo: The forests along the Dawei road, east of Myitta, February 2016. © WWF-Myanmar/Adam Oswell

Predicting Impact on the Landscape

The best way to limit forest fragmentation as a result of road development is to avoid critical wildlife areas; if that step cannot be taken, it is possible to mitigate fragmentation by maintaining corridors through the construction of wildlife crossings and the management of vehicular traffic. Experience from infrastructure development in Europe and elsewhere has shown it is both more cost-effective and safer when wildlife and ecosystem services are taken into account early on in the planning process (Damarad and Bekker, 2003). Environmental and social considerations, supported by information on ecosystem services and wildlife, are effective when integrated further upstream in planning processes, well before road alignments are proposed.

Due to longstanding deforestation along the Thai border, the terrain running north–south in Tanintharyi is the last remaining link between two of the most significant forest conservation landscapes in tropical Asia: the Western Forest Complex and the Kaeng Krachan Forest Complex in Thailand. These landscapes are home to the lar gibbon and probably support the largest tiger population outside of India and Nepal (WCS, 2015a). Landscape connectivity is critical for both gibbons and tigers, especially as they require large home ranges and intact forest cover. The lar gibbon is a high-canopy species and is rarely found in the understory; loss of canopy connectivity and habitat isolates gibbons and leads to multiple negative effects on the population (Gron, 2010). Establishing and maintaining this ecological corridor would support the movement of gibbons, tigers and other wildlife along the trans-boundary landscape (Kelly *et al*., 2016). Without appropriate measures, the planned road link will lead to increased land cover change and threaten this corridor (Helsingen *et al*., 2015).

Land Change and Impacts on Wildlife

While the current access road for the Dawei road link has existed in some form since around 2000, deforestation has increased significantly in tandem with recent access road construction and upgrades over the past several years (BurmaNet News, 2000; Helsingen *et al*., 2015; see Figures 5.2 and 5.3). Construction of the road link has not yet formally begun, but the access road has been fortified and extended into new areas since 2010. These disturbances and the creation of isolated forest patches change the distribution of species. Unless urgent steps are taken to address deforestation, either through land use controls, infrastructure and investment regulations, or participatory forest management programs, significant habitat loss will continue to threaten Tanintharyi's remaining species.

Cases from Thailand bear witness to the increase in wildlife–vehicle collisions across the region. In one such incident, in 2014, a car crashed into three wild elephants on a road near Thailand's Khao Chamao–Khao Wong National Park, leaving six people and one of the elephants dead (Barbash, 2014). Without appropriate measures, the frequency of wildlife–vehicle accidents on the Dawei road link is likely to rise in

FIGURE 5.3
Deforestation within 5 km of the Planned Dawei Road Link, 2001–13

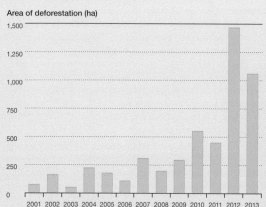

Source: Helsingen et al. (2015, p. 13)

line with the increase in traffic volume, speed and number of large vehicles. Gibbons are at high risk of car accidents, as they are uneasy travelling on the ground, while macaques and langur species tend to travel by and inhabit the ground more, which also exposes them to the risk of collisions (Baskaran and Boominathan, 2010). A further complication is that the proposed Dawei road link is meant for nighttime traffic,[10] which means that headlights from passing vehicles will pose particular risks to light-sensitive species such as leopards and other nocturnal wildlife.

Roads also enable poaching and promote the illegal wildlife trade by providing access to previously remote, undisturbed areas (Espinosa, Branch and Cueva, 2014; Clements et al., 2014; Laurance et al., 2009; Quintero et al., 2010). Myanmar is recognized as a major source of illegal animal parts to consumer and re-export markets in China and Thailand (TRAFFIC, 2014). As the road network in Myanmar's rural areas has essentially been unchanged for the past 50 years, options for trafficking wildlife are limited to the main transport corridors (Clements et al., 2014). Wildlife markets already exist in the area where the Dawei road link is planned. One wildlife market is held at Three Pagodas Pass, a border crossing between Myanmar and Thailand, just a few hours' drive north of Dawei (Shepherd and Nijman, 2008).

Once constructed, the Dawei road link will significantly shorten travel time to the Thai border. In turn, it is likely to contribute to the illegal wildlife trade—unless preventive measures, such as monitoring and enforcement, are put in place. During field visits in 2015 and 2016, the authors of this study observed numerous hunters and noted that wild meat, including gibbon and langur stew, was served in restaurants along the road. One restaurant owner reported that he bought primate meat from hunters from the surrounding forests for about US$1.50 per pound (US$3.30 per kg). As road traffic increases, wild

Photo: Deforestation along the access road for the Dawei road link, east of Myitta, February 2016.
© WWF-Myanmar/Adam Oswell

Photo: The Dawei road link currently remains unpaved, and already the majority of adjacent slopes have been deforested. The project's scope has varied greatly over time, alternating between eight-, four-, and two-lane alignments, with and without rail, power lines and gas pipelines.
© Atid Kiattisaksiri/LightRocket via Getty Images.

game is reportedly becoming scarcer and the price paid for primate meat is rising (WWF, 2014). There is a need for further studies in this area.

Applying Algorithms and Strategic Road Designs in Scenario Modeling

This section outlines how scenario modeling can be used to decide how and where on the planned Dawei road link to implement road mitigation measures specific to primate habitat and movement patterns.

Scenario modeling is a process often used in regulatory instruments such as EIAs to evaluate the potential impact of infrastructure on the environment. An EIA typically describes the proposed scenario and simulates the environmental, social and economic outcomes of a given project. It lays out the threats and possible mitigation measures required to encourage sustainable development. It also models alternative options such as a "no-build" scenario or "best-case" scenario, along with associated outcomes, in order to aid planners and governments in making informed decisions.

However, while a typical scenario modeling process provides options, it does not offer enough flexibility to support decision-making in rapidly changing contexts with poor legislation enforcement, such as in Myanmar. The evolving economic, social and political context of the Dawei road link requires an alternative approach to typical scenario modeling (Alcamo, 2008). As described below, WWF and HKU undertook several alternative approaches, both technical and story-based, to support the sustainable development of the transport corridor and raise community and government awareness of sound environmental and engineering choices.

In three reports on the planned Dawei road link, WWF and HKU utilized different yet complementary methods of modeling scenarios. The first predicts land use conversion due to development and the resulting environmental threats; it calls for a considered and transparent planning process that involves multiple stakeholders. The second offers a toolkit for sustainable infrastructure design, construction and maintenance possibilities; it constructs scenarios and their impacts for typical sites along the road link. While not scenario-based, the third model was pioneered and used to predict multispecies movement patterns and to identify locations for mitigating the road link's impact on wildlife.

For the first approach, land use conversion was modeled using Natural Capital Project's InVEST (Integrated Valuation of Ecosystem Services and Tradeoffs) Scenario Generator (see Figure 5.4). Three land use scenarios were produced using selected inputs, including the likelihood of change, the different kind of physical and environmental factors that influence change and the quantity of change under different scenarios (McKenzie et al., 2012). In the "limited" and "more" land use conversion scenarios, expanding frontiers of deforestation are primarily concentrated around existing and planned roads and settlements. In contrast, the "high" land use conversion scenario predicts a future with extensive forest conversion, at a rate similar to those of neighboring countries (Helsingen et al., 2015). Future steps for this work would include using additional participatory approaches to better understand the different inputs, including likelihood and quantity of change. However, for now they serve as a basis for decision-making and understanding of different possible futures and implications.

These land use conversion scenarios are complemented with a second approach, an illustrated design manual on sustainable road construction techniques and mitigation measures that provides tools for decision-making across a wide range of stakeholders. The manual outlines sustainable principles for the road's alignment, alternative engineering technologies and road design guidelines specifically for wildlife endemic to the landscape surrounding the road corridor. As part of this design manual, three typical sites were chosen along the road link. For each site, the following graphically illustrated scenarios were displayed:

FIGURE 5.4

Baseline Plus Three Conversion Scenarios for the Proposed Dawei Road Link

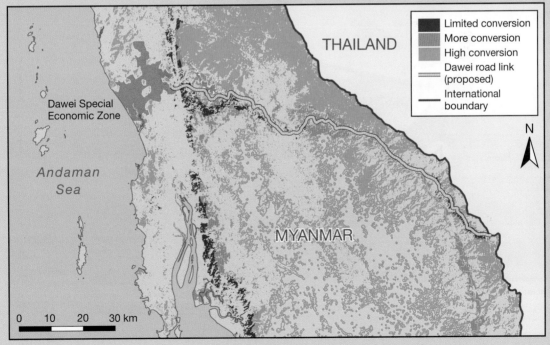

Source: Helsingen et al. (2015, p. 19).

- a business-as-usual engineering approach with no consideration for wildlife or ecological connectivity;
- an upgrade of the current access road using minimal construction standards; and
- an approach demonstrating the combination of "soft" engineering with vegetation (for slope retention), sustainable maintenance and mitigation measures for wildlife (Tang and Kelly, 2016).

These three scenarios were turned into 3D-printed models, which are much more effective for communicating the various options for alignment and mitigation measures to lay audiences in stakeholder meetings (see Figure 5.5).

Locating Wildlife Crossings for Many Species

Up to this point, WWF and HKU had built an argument for better planning processes and specified design guidelines to encourage and sustain wildlife habitat connectivity, but there were insufficient data on wildlife populations to identify crucial sites for mitigation measures that could link the landscapes north and south of the road link corridor (Kelly *et al*., 2016). Consequently, the team opted for a modeling method using techniques that simulate how electric current (as a proxy for wildlife) might flow—in this case, across a landscape (McRae *et al*., 2008). To that end, a multidisciplinary group of landscape planners, computation experts, conservation geographers and wildlife specialists from WWF and HKU's network compiled information about individual species' habitat preferences with regard to factors such as forest cover, human settlement, rivers and roads, so as to model individual species' rates of movement across a landscape.

However, while this technique for mapping critical areas for wildlife connectivity is well established for single species, it has frequently proved to be challenging to combine the movement preferences of multiple species and limited in its potential application to identifying sites for small-scale interventions, such as wildlife crossings (Brodie *et al*., 2015; McRae *et al*., 2008). To enable modeling multiple species and apply these methods to the specific landscapes along the road, the team's landscape designers and computation experts developed a framework for optimizing identification of wildlife crossing locations along the expected route of the road (Kelly *et al*., 2016).

Importantly, the final recommendations are flexible enough to accommodate pragmatic concerns such as alignment adjustments, engineering options and construction costs, while still providing enough crossings and maximizing the

FIGURE 5.5

Infrastructure Design Scenarios as 3D-Printed Models

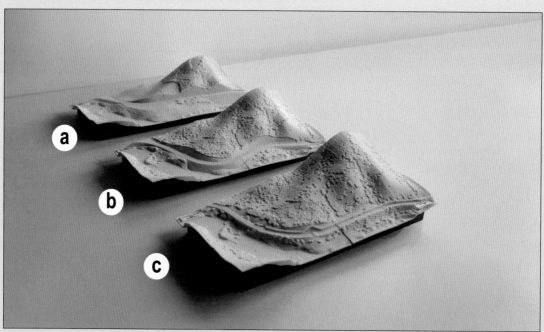

Notes: For a single site along the planned Dawei road link, these three models represent potential alignments, construction technologies, mitigation measures and impacts on surrounding land cover. The models show (a) the developer's likely alignment; (b) an upgrade of the existing access road; and (c) bioengineering and wildlife mitigation (Tang and Kelly, 2016).

Photo: © Ashley Scott Kelly, University of Hong Kong

FIGURE 5.6
Multispecies Movement Prediction Modeling

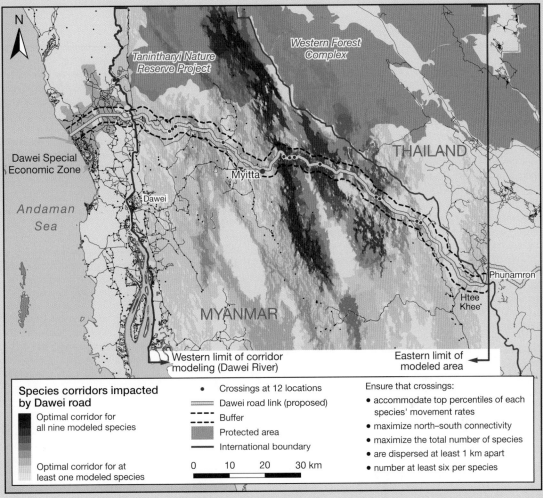

Source: Kelly et al. (2016, pp. 24–5)

number of included species. As shown in Figure 5.6, the crossing locations are not merely points, but rather segments measuring approximately 1 km in length that can take into account local cost engineering and a variety of mitigation measures. These measures are outlined for specific wildlife species in the accompanying design manual and are intended for the identified critical corridors, as well as mitigation strategies and sustainable construction technologies along the length of the Dawei road link.

Analytical modeling is most effective when decision-makers—who are often non-specialists—are able to understand the principles and factors involved. The Dawei road link combines "design thinking," which encourages scenario building with iterative approaches to problem solving, and the story-and-simulation approach, a hybrid of quantitative simulations and qualitative narratives (Alcamo, 2008). For instance, the creation of the design manual began with a series of specific example sites along the planned Dawei road; each of these sites was then used to develop potential sustainable engineering principles that could be useful along the entire route. In the end, these options were catalogued to provide a useful set of tools and recommendations. For the land use conversion scenarios, as an example of a story-and-simulation approach, technical modeling was coupled with narratives of environmental destruction and economic loss, each of which fed back into the other and demonstrated the decision-making processes—not necessarily the factors—that were critical to the desired outcome.

Wildlife and Ecosystem Services in the Infrastructure Development Process

In 2015, the government of Myanmar formally adopted EIA procedures (Thant, 2016). This was an important step towards better environmental management in the country. However, these procedures do not incorporate specific guidelines for different sectors, which would ensure that design, construction and mitigation measures are accounted for in both the EIA and the environmental management plan (ECD, 2016; MCRB, 2016). The Ministry of Construction recently formed an environmental safeguards division, a sign of more sector-level attention that may be able to mainstream ecosystem services and wildlife considerations at the national level. Moreover, public participation guidelines for consultations are under development, as is a system for the formal sharing of EIAs with the public.[11] Ideally, these efforts will improve consultations and access to EIAs, which currently lack transparency.

Nevertheless, in the EIA undertaken by the Dawei road link developer—ITALTHAI, Thailand's largest engineering and construction company (ITALTHAI, n.d.)—the sections on biodiversity and ecosystems are far from adequate. Perhaps most flagrantly, the EIA does not include biodiversity surveys for the area and only sets aside a very small amount of the budget for addressing negative environmental impacts. In response, WWF provided constructive criticism directly to the road developer and the EIA consultant. The three reports by WWF and HKU were also presented to the Myanmar EIA review committee and to the relevant ministries of the Myanmar government on several occasions, in efforts to encourage sector-specific guidelines for infrastructure nationally. At meetings and during capacity building initiatives, WWF presented Helsingen et al.'s *A Better Road to Dawei* and ongoing work on the design of mitigation measures to Dawei University and several government agencies, including the Ministries of Livestock, Fisheries and Rural Development; Environmental Conservation and Forestry; Construction; Agriculture; and Planning.

Building Capacity and Increasing Awareness

To support capacity building on how to plan, design and construct more sustainable roads, WWF facilitated attendance at conferences and organized a workshop for reviewers of EIAs from Myanmar's Ministry of Natural Resources and Environmental Conservation, as well as the Ministries of Transport and Construction. In addition, in September 2015, WWF, HKU and Stanford University's Natural Capital Project took 19 regional government officials from nine departments on a field visit to the project area to support their understanding of the connections between the environment, people and infrastructure. Government officials discussed what changes could be observed in the landscape, what factors were driving those changes, and how impacts might be addressed and mitigated to better protect forests and vegetation and prevent soil erosion and landslides along the road. This visit highlighted the need for integrated land use planning—especially with regard to infrastructure—and for greater coordination both horizontally among ministries and vertically within national bodies.

Last Resort: Offsetting Impact

Finally, as a last resort, options for offsetting or compensating for impact are under development. In April 2016, WWF showed the road developer an initial scoping study for one option concerning a financial mechanism that could support sustainable management of forests north and south of the proposed Dawei road link. The road developer subsequently requested a suite of options for a financial mechanism. According to WWF's initial assessments, the forests north and south of the road provide important sediment retention services that would protect planned bridges from damage and scouring.[12]

Considering the large amount of rainfall this region receives over short periods of time, the forests play a crucial role in regulating water and reducing the risk of floods and landslides. Erosion modeling undertaken by WWF in 2015 shows a number of sections at high risk from landslides (see Figure 5.7). Investing in the management of forest ecosystems adjacent to the road will help sustain the provision of services and reduce maintenance costs, while simultaneously reducing impacts from soil erosion and floods on surrounding communities and ensuring the long-term integrity of the landscape. At the time of writing, further studies to identify a set of design options for a financial mechanism were to be presented to the road developer. Until then, consultations with communities and civil society are necessary to understand the immediate needs of local people.

Conclusions and Lessons Learned

Having emerged in response to the Yadana pipeline case and gained experience in frequent cross-border exchanges with Thai counterparts, civil society in Tanintharyi has remained protest-oriented (ERI, 2009). Local groups seldom seek or accept collaborations with international NGOs. Their overall position regarding the DSEZ and Dawei road link frequently incorporates both rejection and acceptance, exemplifying Harvey and Knox's definition of an "impossible public" (Harvey and Knox, 2015).

In practice, Tanintharyi's civil society has claimed that much of WWF and HKU's work has aided the developer and argued "for the road"; however, this team did not see the comparative advantage of arguing from a singular or "protest"-oriented stance. A more suitable approach is to suggest alternative options and innovative solutions that would help mitigate and negotiate impacts. Opaque development plans, including a non-public EIA, also required more innovative approaches. Given this position, the team's efforts were developed to simultaneously offer toolkits in the form of future land change scenarios, design and construction scenarios, and wildlife prediction modeling to influence and build capacity with the national government, the local government,

FIGURE 5.7

Modeled Areas or "Servicesheds" that Impact the Proposed Dawei Road Link through Erosion and Landslides

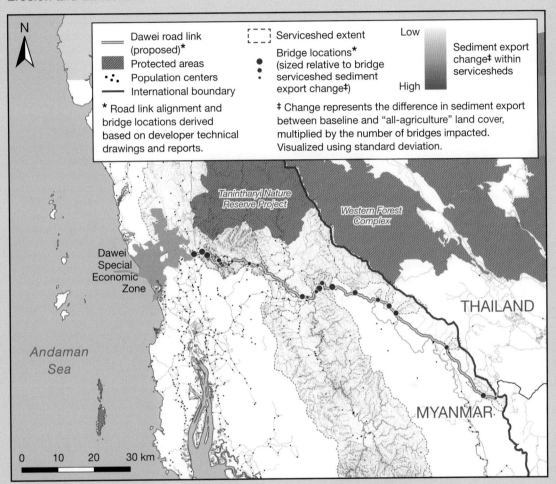

© WWF and HKU

civil society and the road developer. These tools are foremost intended to influence upstream planning, yet with enough geographic, physical and technical description and flexibility to negotiate infrastructural implementation in the absence of strong governance and environmental legislation.

Construction on the Dawei road link is anticipated to continue in 2018, as it has "continued" with or without necessary approvals, ambiguous land rights and tentative investment since agreements were signed between Myanmar and Thailand in 2008. While it is too early to tell whether WWF and HKU's spatially explicit strategies, designs and recommendations will be effective or implemented by the Thai road developer, in all likelihood they will suffice to inform civil society and the government of alternative and sustainable practices. However opportunistic, the scope of these efforts is also chosen to move beyond the uncoordinated and often competitive nature of NGO work in the region. Importantly, given many competing and overlapping interests, they have not explicitly delineated areas for more traditional conservation planning. Nor have they incorporated social and cultural knowledge into the process; the work remains largely within the technical and environmental silos. Nevertheless, these studies and toolkits help support a multitude of stakeholders in their different objectives. Critical to success for biodiversity conservation is flexibility, not only for land use and infrastructure planning, but also so that diverse stakeholders may appropriate these tools for their own use in securing ecological connectivity across the region.

CASE STUDY 5.3

Conservation in the DRC: Road Rehabilitation and the Bili–Uélé Protected Area Complex

Introduction

Aspiration 1 of the African Union's Agenda 2063 envisions a "prosperous Africa based on inclusive growth and sustainable development" (AU, 2015, p. 2). As part of that aspiration, the agenda pictures a continent on which "[c]ities and other settlements are hubs of cultural and economic activities, with modernized infrastructure, and people have access to [. . .] the basic necessities of life" (pp. 2–3). It goes on to visualize "Africa's unique natural endowments, its environment and ecosystems, including its wildlife and wild lands [as] healthy, valued and protected, with climate resilient economies and communities" (p. 3).

The continent is indeed experiencing a dramatic growth in infrastructure development, a process that is often accompanied by serious, irreversible environmental changes (Laurance et al., 2015b). Donors and policymakers are increasingly aware of the need to factor in environmental considerations at the onset of an infrastructure development project. In contrast, some current policies and guidelines appear to be lagging behind the growing intention to avoid causing a net loss to biodiversity, and perhaps to advance conservation goals in the process.

This case study examines the Pro-Routes project, a major road rehabilitation undertaking in the DRC that triggered the World Bank's strictest environmental safeguards (see Box 5.1 and Annex VI). In particular, this study considers the 523 km Kisangani–Bondo segment, the RN4, which is certain to have an impact on the Bili–Uélé Protected Area Complex (BUPAC) (see Figure 5.8).

A Brief Description of the BUPAC

For the purposes of this study, the BUPAC comprises the Bili–Uélé Hunting Domain (32,748 km²/3.3 million ha), a partial faunal reserve with low protection status, and the Bomu Faunal Reserve (10,667 km²/1.1 million ha).[13] With an area of more than 43,000 km² (4.3 million ha), this complex is the largest contiguous protected area in the DRC. Yet, very little is known about it and, until recently, no conservation organizations were working in the landscape and no protected area management was being undertaken.

The IUCN has identified the BUPAC as one of the most critical chimpanzee conservation units, as it harbors an estimated 20,000 endangered eastern chimpanzees (*Pan troglodytes schweinfurthii*). These individuals account for about half the DRC's population and one of Africa's largest contiguous populations (Hicks et al., 2010; Plumptre et al., 2010).

The BUPAC is remote and the few existing roads are barely accessible by car, if at all. Despite the near absence of infrastructure and low human densities, the threats to biodiversity are high; hunting and poaching have spread and illegal trade in wild meat and young orphaned chimpanzees is thriving locally, regionally and across the DRC's borders, in the Central African Republic and South Sudan. The situation is compounded by an increase in human encroachment, growing social conflict and small groups of presumed Lord's Resistance Army members terrorizing communities in the region (Gauvey Herbert, 2017; Hicks et al., 2010; LRA Crisis Tracker, 2016; Spittaels and Hilgert, 2010). The artisanal gold and diamond mining industries are also extensive, especially in the western area of the BUPAC (Hicks and van Boxel, 2010). While biodiversity in the complex previously seemed secured by the area's inaccessibility, this growing human encroachment—together with poor governance and law enforcement—has contributed to its depletion.

In 2014, the African Wildlife Foundation (AWF) and the Congolese Institute for Nature Conservation (ICCN) conducted a scoping mission in the region to support conservation action. The study resulted in a conservation and securitization program initiated by AWF, Maisha Consulting and ICCN in a core area of the BUPAC—the Bili–Mbomu Forest Savanna Mosaic—which covers about 11,000 km² (1.1 million ha) (AWF, 2015, 2016). Over the first year, 25 newly selected and trained rangers conducted reconnaissance walks covering more than 2,000 km. Having georeferenced and destroyed about 100 hunting camps, they were able to confirm that poachers had a substantial presence throughout the protected area.[14] In 2016, AWF and ICCN signed a co-management agreement to strengthen management of the protected area (AWF, 2016; Ondoua Ondoua et al., 2017). Without adequate protection and conservation action, further losses to biodiversity are inevitable.

The Need for Infrastructure and the Birth of the Pro-Routes Project

In early 2000, the transport sector of the DRC was in a very poor state. Following a decade of conflicts and a quasi-absence of management, the formerly operational, multi-modal transport network—which integrated roads, railways and waterways nationwide—had collapsed. The majority of roads were impassable, including more than 90% of the estimated 58,000 km national and provincial network (World Bank, 2008).

This situation has exacerbated rural poverty, particularly by impairing communities' access to social services and markets. More generally, it has hindered post-conflict economic recovery. In response, the government has strongly emphasized the critical importance of investing in transport infrastructure. It has presented a solid, well-maintained network as key to supporting the growth of the two pillars of the country's economy—the agriculture and extractive industry sectors—and to fostering trade at the national and regional levels (World Bank, 2008).

In 2004, the European Commission and the World Bank jointly created an infrastructure unit—the Cellule Infrastructures (CI)—as a financially autonomous body under the

FIGURE 5.8

The Pro-Routes Project and the Bili–Uélé Protected Area Complex (BUPAC)

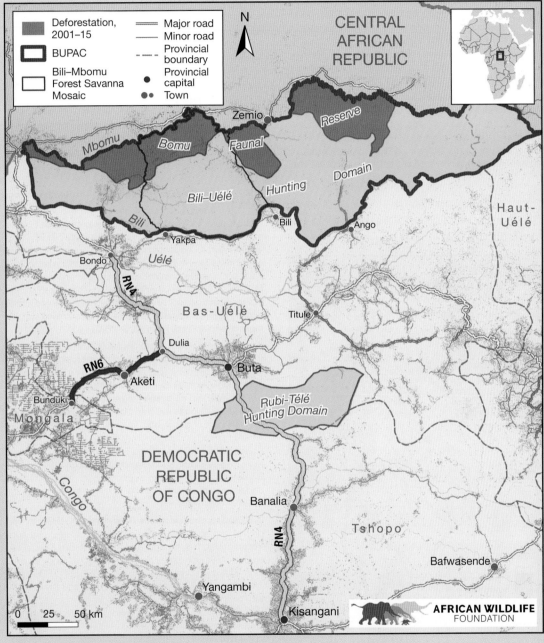

Data source: UNEP–WCM C and IUCN (2017)

authority responsible for infrastructure development, the DRC's Ministry of Infrastructure, Public Works and Reconstruction. CI provides institutional and technical support to the ministry, including capacity building. It also oversees the Pro-Routes project, which DFID initiated in 2005 (World Bank, 2008).

The main objective of the Pro-Routes project is to "re-establish lasting access between the provincial capital and districts, and districts and territories [. . .] in a way that is sustainable for people and the natural environment" (World Bank, 2008, p. 7). To support the project's implementation, DFID, together with the International Development Association, created a multi-donor trust fund administered by the World Bank. In 2008, agencies contributed US$123 million to this funding mechanism and to finance the rehabilitation of selected road segments (World Bank, 2008).

In the phase of upstream planning, the stakeholders concluded that the rehabilitation of existing roads would be the most economical and timesaving approach. The existing network had already reflected patterns of human activities, as corroborated by deforestation trends in 2001–15 (see Figure 5.8). Upgrading the network is expected to lead to a typical 10%–20% increase in deforestation, primarily within a 2-km radius of the targeted road segments, and largely close to urban centers such as Buta and Kisangani (Damania et al., 2016).

The national roads that were identified for rehabilitation in 2007—the RN4, the RN6 extension and the RN5—account for about 1,800 km within a 9,135-km-long target network (World Bank, 2008; see Figure 5.9). Importantly, the RN4 crosses the Rubi-Télé Hunting Domain; at its northern end, it stops at Bondo town, just before reaching the Bili–Uélé Hunting Domain of the BUPAC. The most severe negative impact on the environment is thus expected in the Rubi-Télé protected area, which is already severely degraded, with only 5–25 surviving elephants and virtually no ICCN presence (Hart, 2014; Thouless et al., 2016). As the BUPAC is considered the most biodiverse protected area in the region, it is the focus of this case study.

The Environmental Component of the Pro-Routes Project

Since the World Bank administers the Pro-Routes donor fund, its safeguard policies apply to the project (see Box 5.1 and Annex VI). Accordingly, under the auspices of CI, an environmental consultancy drew up an environmental and social management framework that identified the key potential impacts and recommended measures for managing them (AGRECO, 2007). Another consultancy then produced an environmental and social impact assessment (ESIA) to further explore potential negative impacts and recommend specific measures to address them (EDG, 2007).

Based on these studies, the project appraisal document (PAD), the design document of the Pro-Routes project, paved the way for the consideration of environmental and social impacts (World Bank, 2008). In assessing the critical risks to the environment as high, the PAD stresses the need for capacity building of ICCN and support to ICCN and the Ministry of Environment, Conservation of Nature and Tourism "in managing and protecting natural habitats, biodiversity, and forests and enforcing the pertaining laws" (World Bank, 2008, p. 36). Significant resources—US$18.7 million—were earmarked within the Pro-Routes project budget to support an environmental and social program, including US$8.18 million for environmental activities (pp. 62–66, 68).

In 2009, CI contracted a consultancy firm, SOFRECO, to lead project implementation as a delegated management contractor and to play the role of Bureau d'Études Spécialisés en Gestion Environnementale et Sociale (consultancy for environmental and social management, or BEGES) (DFID, 2010). The task of BEGES was to provide ICCN and the ministry with technical, operational and financial assistance to manage natural ecosystems and enforce related regulations and laws with regard to wildlife and protected areas, as outlined in the PAD (World Bank, 2008). In accordance with the project's classification under the World Bank's safeguard policy, CI recruited experts for an environmental and social advisory panel (ESAP), which was to provide guidance with respect to the management of environmental and social aspects (see Annex VI).

FIGURE 5.9

The Pro-Routes Project: Roads Selected for Rehabilitation

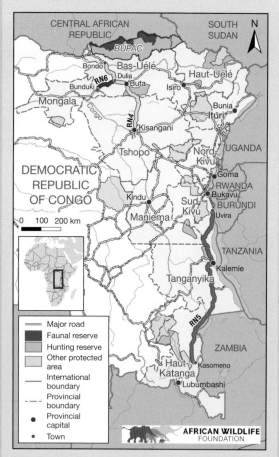

Data sources: UNEP–WCMC and IUCN, 2017; WRI and MECNT, 2010

BOX 5.1

The World Bank Infrastructure Development Imperative

A Low Infrastructure Baseline

When it comes to infrastructure, Africa lags behind the rest of the world on almost every development metric. The region has the lowest road density and levels of electrification, and few of its urban dwellers have access to piped drinking water or adequate sanitation (Foster and Briceño-Garmendia, 2010). At the same time, the infrastructure deficit is set to worsen with a burgeoning population that is expected to double by about 2050 (UN Population Division, 2017). Infrastructure development, including the provision of electricity, safe drinking water and transportation, is widely recognized as essential to reducing extreme poverty. It is also fundamental to achieving sustainable development and boosting shared prosperity.

The Challenge

In Africa, as elsewhere, infrastructure investments undertaken carelessly or without consideration of potential externalities can be counterproductive and undermine many of the sources of growth and livelihood in an economy. Evidence suggests that in Africa, where poverty is predominantly a rural phenomenon, the very poorest are the most dependent on forests for their livelihoods. In most cases, the poorest quintile derive more income from forests and the commons than from agriculture. An immediate implication is that forest income deserves at least as much attention from policymakers and at the project level as other sources of income. Neglect of such a significant component of economic value to the poor inevitably compromises the effectiveness of poverty reduction policies (Anderson et al., 2006b; Angelsen et al., 2014; Byron and Arnold, 1999; IUCN, 2016d).

Environmental and natural resources contribute to economic well-being and the ability to fight poverty sustainably. In that sense, they play a pivotal role in development, one that cannot be done justice if they are treated as mere afterthoughts in the development dialog (PROFOR, 2012; Sunderlin, Dewi and Puntodewo, 2007). Renewable natural resources in Africa merit particular scrutiny since the continent's poor are especially dependent on them.

Biodiversity Implications

With respect to biodiversity in general and ape conservation in particular, investments in two types of infrastructure—roads and dams—are especially relevant.[15]

Roads. In the process of connecting people—including the rural poor—with markets and services, roads are of fundamental importance. Ideally, they help to reduce poverty and stimulate economic development; in practice, however, these goals are not always achieved (see Chapter 2, p. 60). In sensitive locations, roads that are built or upgraded without adequate precautions can threaten apes and other biodiversity through their direct and induced (indirect) impacts. Direct impacts involve the footprint of the road itself, including forest fragmentation, altered drainage patterns and wildlife road kills. Induced impacts result from human activities that are made possible by new or improved roads, through improved access to remote areas; these impacts include new settlements, deforestation, logging and hunting of vulnerable species.

The most important planning decision available to address both direct and induced impacts of road development is careful site selection. In most cases, the World Bank requires that new roads—and major upgrades of existing ones—be located so as to avoid areas of high biodiversity value, including ape habitats. The one "special case" exception to this rule occurs when a road to a protected area might be supported by conservation authorities because it would allow for improved management or sustainable tourism. By avoiding remote forested areas where apes reside, new and improved roads are likely to benefit larger numbers of people by traversing more densely settled rural areas.

Approaches that consider potential road impacts at the very outset of the planning process enable decision-makers to steer development away from biodiversity hotspots towards areas where benefits can be maximized and adverse impacts largely avoided (see Box 1.6). The tools now exist to conduct detailed assessments of likely road impacts; some were pioneered in a recent analysis in the DRC (Barra et al., 2016). These tools offer a standardized and scientific way of assessing the environmental risks of an infrastructure investment, while also offering alternatives that may be equally beneficial, but less risky. A number of biodiversity-related databases—including the A.P.E.S. Portal, the Digital Observatory for Protected Areas (DOPA) and Integrated Biodiversity Assessment Tool (IBAT)[16]—provide easily accessible information on the locations of ape habitats and other important biodiversity areas. In planning roads and other infrastructure, a landscape approach is the most effective way to consider ape habitats within and outside of protected areas, as well as the potential connectivity between them.

Dams. In many African countries, hydroelectric and other dams are considered a key source of low-carbon electricity, potable water for cities and towns, and irrigation water to sustain agriculture (see Chapter 6). As with roads, site selection of dams is extremely important for avoiding and minimizing harm to apes and other biodiversity. A single proposed hydroelectric dam in Guinea, for instance, could adversely affect a major stronghold of the critically endangered western chimpanzee (Pan troglodytes verus), unlike other dams in the same river system.

In some cases, dam projects can further conservation goals through biodiversity offsets. For example, the World Bank-supported Lom Pangar Hydropower Project in Cameroon involved the establishment and on-the-ground strengthening of Deng Deng National Park, which protects an important population of the western lowland gorilla (Gorilla gorilla gorilla) (Ledec and Johnson, 2016; see Case Study 6.1). Many dams depend on the conservation of upper catchment areas for their long-term functioning; that dependence provides an important incentive for conserving upland forests and other natural habitats. Well-managed hydroelectric and water supply dams also generate annual revenues, a fraction of which can be devoted to recurrent costs of managing associated conservation areas.

Besides proper site selection and design, building infrastructure that is biodiversity-friendly means paying close attention to the construction practices used (see Box 6.1). The loss and degradation of natural habitats can be minimized through the establishment and enforcement of strong environmental rules for contractors (see Box 1.6), especially if these are reflected in the bidding documents and contracts for large infrastructure projects. Particularly important for apes and other wildlife are strict prohibitions on hunting, wildlife capture and the purchase of wild meat by all contractors and construction workers.

Getting It Right

Since much of Africa has yet to develop a basic infrastructure stock, there is potential for the process to be undertaken with due concern for the conservation of apes and other biodiversity, while avoiding many of the environmental mistakes that have often been made in other parts of the world. Getting it right will require more focused attention to biodiversity than has been the case to date in many countries.

The World Bank's commitment to biodiversity conservation as an integral part of infrastructure development is underpinned by its safeguard policies, particularly the Natural Habitats Operational Policy (OP) 4.04 and Forests OP 4.36 (World Bank, 2013b, 2013c). In July 2016, the Bank's board of executive directors approved a new Environmental and Social Framework, which will go into full effect in 2018; this framework includes Environmental and Social Standard 6 on biodiversity conservation and sustainable management of living natural resources (World Bank, 2017; World Bank, n.d.-d). The International Finance Corporation—the Bank's private sector affiliate—already operates under the very similar Performance Standard 6 on biodiversity conservation and sustainable management of living natural resources (IFC, 2012c). Beyond these mandatory environmental standards, the World Bank Group's Forest Action Plan for 2016–20 seeks to ensure that forests—including ape habitats—are effectively integrated within national development planning efforts and that new infrastructure investments follow a "forest-smart" approach to avoid or minimize any adverse impacts (World Bank, 2016a).

Balancing economic growth with environmental protection is a challenge faced by every nation on earth. There is growing recognition that degrading natural resources for short-term economic gain is ultimately a counter-productive strategy that can undermine development and growth. Recent technological advances have made available the information and analytical tools needed to avoid damage while harnessing and maximizing the net economic benefits of infrastructure development. The challenge lies in ensuring that governments, donors and policymakers use these tools to make better-informed and more effective decisions.

Assessments and Recommendations

By establishing four posts on the Buta–Kisangani road to control the illegal wild meat trade, BEGES immediately initiated implementation of the recommendations formulated by the environmental and social management framework and the PAD. Another ESIA for the 125 km section connecting Dulia to Bondo was carried out between 2012 and 2013. In addition, WWF and the consultancy TEREA released a study of Pro-Routes' impact on protected areas (WWF and TEREA, 2014). These studies resulted in the development of a two-fold approach.

The first element of the approach—the "emergency intervention package"—focused on poaching, which was expected to increase in the western part of the BUPAC due to the rehabilitation of the nearby RN4. The proposed wildlife conservation activities required technical and financial support to ICCN for improved anti-poaching measures in priority areas within the BUPAC, and support to communities to reduce dependence on the protected area. The latter component included the creation of a local development fund, awareness building and improved coordination between ICCN and communities living adjacent to the BUPAC's priority areas (WWF and TEREA, 2014).

The second element—the "priority action plan"—provided guidance on how to implement an ICCN-led participatory process to assess the BUPAC's status and revise the land use planning and management of the complex. The adjusted management objectives, governance mechanisms and spatial delimitation of the protected area complex would then be outlined in a management plan for the BUPAC. This design phase was established as a key step towards the effective management of the complex over time (WWF and TEREA, 2014).

Although WWF and TEREA strongly recommended the implementation of the full two-fold approach for the BUPAC, CI only prioritized the emergency intervention package. In author interviews, stakeholders suggested that BEGES had insufficient funding available for the implementation of the priority action plan, but this study was not able to corroborate this assessment.[17]

Implementation and Evaluation

From an economic perspective, the road rehabilitation project provided the expected benefits for users. Travel time between Kisangani and Buta was reduced from 3–4 weeks by bicycle to six hours by car, and corresponding travel costs plummeted. In towns along the road, the knock-on effects were immediate: the price of fuel fell by 50%, that of salt by 30% (World Bank, 2016d).[18]

Data are more elusive when it comes to evaluating the implementation of the mitigation measures designed to minimize environmental and social impacts of the Pro-Routes project on the BUPAC. The safeguard policies, recommendations and management approaches seem like a promising blueprint for the implementation of such measures. In the event, however, CI did not formally approve the approaches until after construction was well under way. In fact, the rehabilitation of the Kisangani–Buta and Buta–Dulia road sections was completed in 2013, six months before WWF and TEREA's recommendations were approved (Radio Okapi, 2013).[19]

Moreover, this study uncovered limited evidence that the mitigation measures were actually being applied. Road checkpoints are the only visible sign of such activity, but the staff does not appear to keep organized records. Beyond that, no reports or evidence is seemingly available on the implementation of the emergency intervention package. In author interviews, various stakeholders indicated that ongoing activities included anti-poaching patrols, meetings with local communities and collaboration with community-based organizations, yet none of these assertions is supported by verifiable reports, nor were such activities evident on the ground during this review.

In the absence of empirical evidence, it is difficult to confirm whether mitigation strategies are being implemented as intended and, if they are, whether they are effective. The project-wide lack of transparency may be partially attributable to the insular nature of the organizations in charge of overseeing the mitigation strategies. As discussed, CI delegated assessment and implementation responsibilities to a consultancy firm, which took on the role of BEGES. In turn, BEGES delegated the responsibility for implementation to government institutions, such as ICCN. BEGES was also tasked with contracting "an experienced and independent NGO of international renown" to work alongside the ESAP, in line with the PAD recommendation. This step was not taken for reasons that remain unclear but that may be linked to capacity constraints or conflicting priorities (World Bank, 2008, p. 12). As a result, BEGES was relegated to playing an intermediary role between government institutions, and was limited to facilitating the transfer of statements between the implementing and directing agencies, CI, ICCN and the World Bank.

A major weakness in the execution of this project, identified during the research for this case study, concerns the inertia exhibited by BEGES. The unit was charged with the implementation of the full array of policies and recommendations, both environmental and social. The wide diversity of expertise required to carry out this work would be difficult to gather in any single organization. Had BEGES solicited the input of a range of specialized organizations to implement specific aspects of the project, as initially envisioned, it could potentially have served as the linchpin of effective implementation (see Box 1.6).

Meanwhile, AWF, ICCN and Maisha Consulting successfully followed the two-fold approach recommended by WWF and TEREA in implementing their conservation and securitization program in the BUPAC's Bili–Mbomu Forest Savanna Mosaic. The project prioritized technical, operational and financial support to ICCN for improved anti-poaching measures in identified priority areas. Largely in line with the priority action plan, AWF and ICCN also conducted a participatory land use planning process for the affected region, including the BUPAC, in 2016. AWF provided the technical and financial support for staff selection, capacity strengthening, ecological monitoring and anti-poaching efforts, the creation and operation of a steering committee, and baseline data collection (AWF, 2016).[20] Although these activities overlapped with the Pro-Routes project recommendations and AWF requested that BEGES finance the implementation of local development plans and community-based management of natural resources, funding was not provided by the Pro-Routes project.[21]

Conclusion

Nowadays, the availability of economic data and georeferenced information on forest cover renders upstream planning both feasible and cost-effective (Damania et al., 2016). At its inception, the Pro-Routes project involved sound upstream planning that took account of potential environmental and social impacts of infrastructure development and identified options for the rehabilitation of habitat. Reinforcing this process, the World Bank's safeguard policies called for thorough environmental and social impact assessments and recommendations for the mitigation of adverse effects on the landscape.

In practice, however, these efforts have not yielded verifiable environmental mitigation measures within the reviewed aspects of the Pro-Routes project. On the whole, efforts to mitigate the impacts of the project lagged behind the roadwork, if they were undertaken at all. This study found no evidence that BEGES and ICCN actually implemented the emergency intervention package, which had initially been prioritized for action; nor did this study identify verifiable reasons that might explain why the priority action plan was not selected for implementation. In the end, neither component of the two-pronged approach was pursued even though the goals of each dovetailed with those of the Pro-Routes project. The road checkpoints remain the most visible concrete action, yet evidence as to their impact and effectiveness is limited. The results of this case study thus reveal that upstream planning alone is not sufficient to ensure effective, timely and coordinated implementation of mitigation measures.

The study also demonstrates that the input of external environmental experts can be invaluable. In this case, AWF and Maisha Consulting joined forces with ICCN, launching a conservation and securitization program that is contributing to the objectives of the Pro-Routes project—albeit without its financial support. If Pro-Routes had been developed as outlined in the PAD, BEGES—or a specialized conservation NGO contracted by BEGES—would have provided ICCN with technical, operational and financial assistance to manage natural ecosystems and to enforce related regulations and laws with regard to wildlife and protected areas. In reality, AWF played a role that BEGES should have played or facilitated, and financed.

This examination of the Pro-Routes project shows that the modernization of infrastructure and the protection of biodiversity in Africa—focal points of Aspiration 1 of Agenda 2063— require more than the establishment of goals and institutions, and more than upstream planning and donor funding. The implementation of recommendations to reduce the negative impacts of such development projects calls for relevant expertise and capacity, a clear allocation of tasks, continuous monitoring and recordkeeping, and the prioritization of environmental and social considerations by all stakeholders. In this context, the potential contributions of external conservation organizations cannot be overstated, regardless of whether they work in parallel or jointly with state structures.

Overall Conclusion

The construction of roads poses unique problems for environmental conservation. As the case studies illustrate, complex governance, technical and economic constraints can undermine the attainment of conservation goals, which may also compete with the need to ensure the welfare of affected communities. The studies demonstrate that the sustainable development of roads cannot be addressed by state or subnational governments alone. Active and sustained participation by various stakeholders is necessary to safeguard the environment and ensure equitable planning and implementation of large infrastructure projects.

Specifically, this chapter highlights the importance of advocacy by local and international NGOs in Nigeria, civil society engagement with industry and government actors in Myanmar, and the inclusion of specialized agencies in the planning and implementation of mitigation measures in the DRC. All case studies underscore the relevance of advocating for the integration of ecosystem and wildlife considerations into the planning and design of roads. In the case of Myanmar, the inclusion of civil society early in the planning process allowed for engagement with engineers and the production of multiple designs. This type of exploration may not have been fostered had conservationists not introduced environmental constraints prior to construction. The chapter also emphasizes that the building of relationships with local civil society groups requires respect and time, especially if there is a history of mistrust, as in Thanintharyi.

This chapter also demonstrates the various options for such advocacy, which ultimately relies on effective communication through a variety of channels. These include the media, direct engagement with government officials and developers, and the presentation of land use conversion scenarios to raise awareness of how infrastructure planning threatens to fragment or drastically alter remaining ape habitat and other areas of significant biodiversity. Only if decision-makers understand the various economic, social and environmental benefits and costs of a project can they take informed planning decisions. A first step in building that knowledge is conducting and disseminating state- and country-wide assessments of natural capital, biodiversity and ecosystem services needed by local people. Such analysis allows stakeholders to consider potential cumulative impacts of various projects, along with their viability.

A range of tools can be deployed to enhance our understanding of the risks and costs to the environment and society, including well-targeted scenario modeling. Also relevant is ongoing monitoring and evaluation of impacts and mitigation measures, as these activities permit stakeholders to respond to infrastructure development plans with suitable, evidence-based actions or adjustments. By presenting varied and cost-effective solutions, an evidence-based approach can help developers and policymakers plan and build more sustainable roads. Conservation actors therefore have a role to play in ensuring adequate scientific data are available to inform action. However, unless political actors and decision-makers prioritize environmental considerations, conservation organizations will be left to rely on financial institution safeguards, and regulations around impact assessments, to prevent biodiversity from being marginalized in large-scale infrastructure developments.

Acknowledgments

Principal authors: Andrew Dunn,[22] Jef Dupain,[23] Hanna Helsingen,[24] Ashley Scott Kelly,[25] Cyril Pélissier,[26] Helga Rainer[27] and Dorothy Tang[28]

Contributors: Hans Bekker, Nirmal Bhagabati, Ashley Brooks, Isaac Ho Wan Chiu, Grant Connette, Nicholas Cox, Richard Damania, IENE (Infra Eco Network

> "Only if decision-makers understand the various economic, social and environmental benefits and costs of a project can they take informed planning decisions."

Europe), Lazaros Georgiadis, Thomas Gray, Elke Hahn, HKU, George Ledec, Lisa Mandle, Natural Capital Project, Kity Tsz Yung Pang, Smithsonian Institution, Paing Soe, Robert Steinmetz, Amanda Ton, Joseph Vattakaven, A. Christy Williams, Stacie Wolny, World Bank and WWF

Case Study 5.1: Andrew Dunn

Case Study 5.2: Ashley Scott Kelly, Hanna Helsingen and Dorothy Tang

Case Study 5.3: Jef Dupain and Cyril Pélissier

Text Box 5.1: Richard Damania and George Ledec

Annex VI: Jef Dupain and Cyril Pélissier

Reviewers: Miriam Goosem, Ben Phalan and Kate Newman

Endnotes

1. This case study is adapted and updated from Dunn (2016) and Dunn and Imong (2017).
2. Copy of the letter reviewed by the author.
3. Copy of the letter reviewed by the author.
4. Copy of the letter reviewed by the author.
5. EIA reviewed by the author.
6. EIA reviewed by the author.
7. The letter from WCS to the Federal Ministry of Environment was written by the author and he reviewed the government's response.
8. Based on author observations of multispectral imagery and orthophotos acquired in 2013 and 2015.
9. Based on author meetings with authorities and the road developer, Bangkok, Thailand, 2015; Dawei, Myanmar, 2015; and Naypyidaw, Myanmar, 2015.
10. Author interviews with the road developer, Bangkok, Thailand, 2015; author review of unpublished technical documents.
11. Author interviews with authorities, Naypyidaw, Myanmar, September 2016.
12. The WWF assessments were not published but were presented to local stakeholders in September 2016.
13. Experts disagree as to the precise area encompassed by the BUPAC. This study relies most heavily on WRI and MECNT (2010).
14. Internal project reports and 2015 AWF project report to Global Forest Watch, all reviewed by the authors.
15. The extractive industries and industrial agriculture are also key drivers of habitat loss for apes and other species. These topics are examined in *State of the Apes* volumes 1 and 2.
16. For details on these databases, see European Commission (n.d.), IBAT (n.d.) and Max Planck Institute (n.d.-a).
17. Author interviews with CI, ICCN and World Bank representatives, DRC, 2016.
18. Author interviews with AWF field staff, CI and ICCN representatives, and community representatives, DRC, 2016–17.
19. Author interviews with ICCN and World Bank representatives, DRC, 2016.
20. Internal project reports and 2015 AWF project report to Global Forest Watch, reviewed by the authors.
21. Project correspondence and internal project reports reviewed by the authors.
22. WCS (www.wcs.org).
23. AWF (www.awf.org).
24. WWF Myanmar (www.wwf.org.mm/en/).
25. HKU (www.arch.hku.hk).
26. Independent consultant.
27. Arcus Foundation (www.arcusfoundation.org).
28. HKU (www.arch.hku.hk).

Photo: Dam construction tends to have substantial environmental and social ramifications. Grand Poubara dam, Gabon. © Marie-Claire Paiz/TNC

CHAPTER 6

Renewable Energy and the Conservation of Apes and Ape Habitat

Introduction

For thousands of years, humans around the world have been constructing dam-like structures to impound water for drinking and irrigation, to retain and control flood waters, to provide hydroelectric power, to allow for recreational amenities, and for various other purposes (Willems and Van Schaik, 2015). Yet, all too often, developers and regulators fail to consider the collective environmental, social and economic impacts of building dams, including the displacement of communities and the loss of ecosystem function and services (Babbitt, 2002; Poff *et al.*, 1997; Stanley and Doyle, 2003; WCD, 2000).

In 2000, the World Commission on Dams estimated that 40 to 80 million humans had

Photo: Direct impacts of dams include habitat fragmentation and loss due to the construction of dams, reservoirs and associated infrastructure, including new settlements for displaced communities. Construction of the new village of Ban Sam Sang, Lao PDR, for the relocation of four communities due to the construction of the Nam Ou Cascade Hydropower Project Dam 6. © In Pictures Ltd/Corbis via Getty Images

been displaced from their homes through the construction of dams (WCD, 2000). Dams can have major long-term consequences for river health, to the detriment of fish, wildlife and local communities that are reliant on the river system for drinking water, food, habitat and other uses (Brown et al., 2009; Tilt, Braun and He, 2009; WCD, 2000). Even small dams can have a major impact on fish migration and downstream fisheries, water quality, downstream water supply and overall stream flow, including the natural transportation of sediment and nutrients needed to replenish downstream forests and floodplains (Poff et al., 1997).

Hydropower, also known as hydroelectric power, generally provides low-carbon electricity and is often a primary source of energy for developing countries. Driven by the rising demand for electricity in developing economies, as well as a call for low-carbon energy as countries strive to meet emission goals, global hydropower capacity is projected to increase by 53%–77% between 2014 and 2040, and global electricity generation is expected to reach 6,000–6,900 terawatt hours (IEA, 2016, p. 249). This expansion is likely to entail the construction of thousands of large dams and tens of thousands of small dams.

Much of the hydropower potential is to be developed in the river valleys and mountainous areas of tropical regions in Africa and Asia. Since dam construction tends to have substantial environmental and social ramifications, the anticipated expansion of hydropower is certain to affect numerous communities and ecosystems, including great ape and gibbon habitats (Zarfl et al., 2015). Regardless of the projected deleterious effects—and despite the availability of alternatives that are more sustainable, more cost-effective and less likely to marginalize certain social groups economically—the green-lighting of large hydropower projects appears to be unavoidable (DSU, 2016).

Chapter 6 Renewable Energy

This chapter provides a review of the projected expansion of hydropower and the potential effects associated with the proliferation of dams, including the impact on apes and their habitat. It presents an initial estimate of the scope of this impact, assessed by overlaying projected dam build-out with the geographic range of great apes and gibbons. The chapter also features three case studies and a box that highlight best practices and strategies for avoiding and mitigating impacts.

With reference to the Lom Pangar Dam in Cameroon, the first case study considers the challenges of implementing best practices designed to protect apes once a project shifts from the planning to the construction phase. The second case study, which documents recent events in Sarawak, in Malaysian Borneo, explores how community activism and collaboration between communities and scientists can prevent the construction of destructive dams. These case studies are complemented by a box that focuses on a system-scale hydropower planning and design framework—"Hydropower by Design" —as a method to fuse planning for energy and water infrastructure with planning to maintain or restore environmental and social values. In recognition of the fact that hydropower is not the only form of renewable energy production associated with adverse impacts, this chapter features a final case study on the implications of a proposed geothermal plant in Sumatra's Leuser Ecosystem, alongside planned hydropower projects.

The chapter's key findings include:

- The negative impacts of dam construction on apes and their habitats across Africa and Asia are likely to increase over the coming years. Direct impacts include habitat fragmentation and loss due to the construction of dams and reservoirs, and of the roads and transmission lines associated with them; in turn, the roads facilitate access to habitats, thus enabling more widespread poaching and other indirect impacts.

- Hydropower development is likely to impact apes in Asia more significantly than in Africa, with gibbons identified as particularly vulnerable.

- Engagement, sharing knowledge and raising awareness of the potential adverse effects of large hydropower and other renewable energy projects can help at-risk communities avoid exposure to severe environmental and social impacts.

- Cost–benefit analysis is a key step in the planning phase of every large renewable energy project, particularly as it can reveal excessive environmental and social costs, issues related to carbon emissions and potential problems regarding delivery on economic objectives.

- The negative environmental and social impacts of dams and other large infrastructure projects are more likely to be minimized when their development planning incorporates a system-scale approach and draws on existing tools and processes, including the mitigation hierarchy.

- Once dam construction is in progress and mitigation measures have been implemented, ongoing monitoring and management of those measures are needed to verify that they remain effective. Given that both the life of a project and the attention of financiers tend to be finite, however, sustaining such activities represents a foreseeable and critical challenge to indefinite conservation.

Annex VII presents the reasons for, and the ramifications of, the decommissioning of dams.

Global Hydropower: Drivers and Trends

Hydropower accounts for approximately 16% of global electricity generation; it is the primary source of electricity in some countries, such as the Democratic Republic of Congo, Lao People's Democratic Republic (Lao PDR) and Uganda. As of 2014, hydropower represented more than 70% of all renewable electricity (IEA, 2016). Hydropower dams with storage capacity are essentially storing energy and are thus able to respond rapidly to changes in demand. Within an electrical grid, this storage function can facilitate a higher proportion of renewable sources with variable generation, such as wind and solar. Hydropower dams—both conventional and pumped storage—currently account for by far the greatest proportion of the world's electricity storage (Kumar et al., 2011).

Due to the rising demand for electricity in general—and for low-carbon and storable energy in particular—hydropower is drawing about US$50 billion in investments per year, although investment in wind and solar have eclipsed hydro in recent years (Frankfurt School-UNEP Centre/BNEF, 2017). In 2014, the International Energy Agency forecast that by 2040, global hydropower output would grow by about 3,000 TWh, particularly if the world were to transition away from fossil fuel sources of energy to achieve the reduction in emissions necessary to keep global temperature increases below 2° C above pre-industrial values (IEA, 2016, p. 250). Much of this development is expected to occur in Asia, although Africa will see the greatest growth rate in installed hydropower capacity (see Figure 6.1). The majority of hydropower expansion (70%) will occur in river basins that have the greatest freshwater biodiversity and where the well-being of people—including their food sources, livelihoods and cultural values—is most directly tied to healthy rivers and intact valleys (Opperman, Grill and Hartmann, 2015; see Figure 6.2).

Figure 6.2 indicates that the hotspots for hydropower expansion include the river basins of the Amazon, the southern Andes, the Balkan region of southeast Europe, and

FIGURE 6.1

Global Installed and Projected Hydropower Capacity

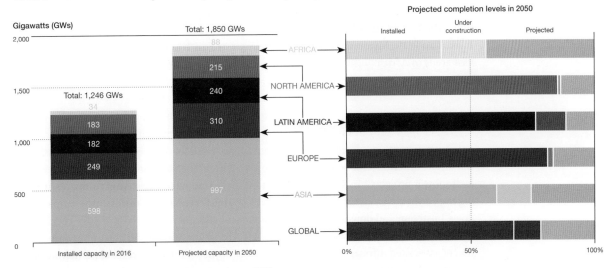

Source: Opperman, Hartmann and Raepple (2017, p. 21), courtesy of TNC

FIGURE 6.2

Hydropower Development in 2015: Dams Installed, under Construction and Planned

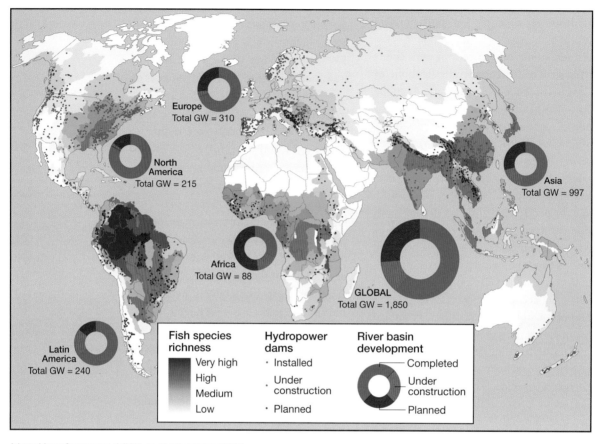

Adapted from: Opperman *et al.* (2015, pp. 16–17), courtesy of TNC
Data sources: Abell *et al.* (2008); IEA (2012); Lehner *et al.* (2011); Zarfl *et al.* (2015)

several regions that support ape populations: South and Southeast Asia (Cambodia, India, Lao PDR, Myanmar and Nepal) and vast areas of Africa.

Impacts of Hydropower

Extensive research has been undertaken on the environmental and social impacts of hydropower projects.[1] In addition to affecting the connectivity of organisms, nutrient flows, upstream and downstream resources, such projects typically involve the construction of associated infrastructure and significant greenhouse gas emissions, as follows:

Hydrological connectivity. Hydropower dams and reservoirs affect the downstream transport of wood, sediment and nutrients and disrupt the up- and downstream movement of organisms, including fish and invertebrates (March *et al.*, 2003). Declines in fish populations negatively affect human communities that rely on migratory fish for food, both up- and downstream (Richter *et al.*, 2010).

Impacts on upstream resources, including terrestrial habitats. The impacts on upstream resources typically receive the most attention in debates about dam development. For one, reservoirs behind large dams typically inundate agricultural land

and natural ecosystems, such as wetlands and forests (WCD, 2000). Perhaps more controversially, large dam development can displace human communities, raising serious social justice questions, as those who are displaced are often poor and lack political influence (Scudder, 2005). Terrestrial species, such as apes, are directly affected by impoundment; as reservoirs fill up and forests are replaced by open water, animals who are not killed in the process suffer a permanent loss of habitat. Further, hydropower reservoirs can convert previously passable river channels into impassable barriers for terrestrial apes and other species (WCD, 2000). Thus, hydropower dams and their reservoirs fragment ape habitat and affect dispersal.

Impacts on downstream resources. The impacts of dams on downstream environmental resources tend to be far greater than the upstream impacts, even if they attract less attention. As human livelihoods and communities are often directly tied to functioning river ecosystems, downstream environmental impacts can have considerable social costs (Richter *et al.*, 2010). Large reservoirs trap nearly all sediment, except for the smallest grain sizes, thereby disrupting the delivery of sediment and nutrients to downstream ecosystems, such as floodplains and deltas (Kondolf, Rubin and Minear, 2014). By altering river flows, dams also impair biological processes on which fish, floodplain forests, and other downstream species and ecosystems depend.

Impacts due to dam construction. In addition to a dam and a reservoir, hydropower development generally requires the construction of access roads and transmission lines, both of which can fragment forests and other habitats, affecting wildlife habitat and movement (Andrews, 1990). Roads, in particular, facilitate access to previously inaccessible areas, leading to an increase in settlement, forest clearing and hunting. During construction, major projects require thousands, or even tens of thousands of workers; in tropical forest regions of Southeast Asia and Africa, temporary settlements near dam sites have been associated with an increase in wild meat hunting (Laurance, Gooseman and Laurance, 2009).

Greenhouse gas emissions. Although hydroelectric dams are widely considered a low-carbon energy option, some reservoirs produce high emissions of greenhouse gases. Reservoirs produce significant amounts of methane, carbon dioxide and nitrous oxide when the land is flooded and organic matter rots and decays. Large dams[2] are the greatest single anthropogenic source of methane, responsible for roughly 30% of all anthropogenic methane emissions (Lima *et al.*, 2007, p. 201). The thermal, chemical and biological conditions in reservoirs in the tropics lead to higher methane emissions than those associated with reservoirs elsewhere (Fearnside, 2016a; Lima *et al.*, 2007). Other dam-related greenhouse gas emissions are linked to the use of fossil fuels during site excavation and building materials such as concrete in dam construction, land clearing for reservoirs, resettlement sites, transmission lines and access roads, and the expansion of irrigated agriculture (Houghton *et al.*, 2012; Pacca and Horvath, 2002).

Studies of the impact of hydropower projects around the world can be instructive with reference to mitigating effects on great apes and gibbons. As suggested above, the process of impounding a reservoir behind a hydropower dam involves the conversion of wildlife habitat, such as forest, into open water, and thus the direct loss of habitat. In addition, reservoirs fragment blocks of habitat and potentially obstruct dispersal routes, as has been the case for giant pandas (*Ailuropoda melanoleuca*) in China (Zhang *et al.*, 2007). A recent study of connectivity corridors in Brazil shows that roads and hydropower reservoirs are among the most

Photo: In addition to a dam and a reservoir, hydropower development generally requires the construction of access roads and transmission lines, both of which fragment forests and other habitats. An electric relay supplied by hydroelectric power from the Bang Dang dam, Thailand. © Thierry Falise/LightRocket via Getty Images

significant variables associated with the impairment of dispersal among jaguars (*Panthera onca*) (Silveira *et al.*, 2014). Similarly, in Costa Rica, the Reventazón hydropower project fragmented a jaguar dispersal corridor; to "offset" the negative impact of the reservoir, the developer funded reforestation of land adjacent to the inundated area to maintain a forested dispersal corridor (IDB, n.d.). Developers also used a biodiversity offset in Cameroon, where a forest reserve was elevated to a national park to compensate for the adverse environmental impacts of the Lom Pangar Dam (see Case Study 6.1). As noted above, the construction of roads and transmission lines linked to hydropower projects can also fragment wildlife habitat (Andrews, 1990; White and Fa, 2014). In discussing the various impacts of hydropower, the chapter highlights potential effects on apes and their habitat.

Hydropower and Apes

The academic literature provides limited information on how hydropower dams and reservoirs affect apes and their habitats (see Chapter 2, pp. 43–60). Since hundreds of dams are proposed within the habitats of great apes and gibbons, assessments of the impact of hydropower expansion are key to the conservation of these species and their habitats.

This section presents a simple spatial analysis that was conducted to assess the extent to which hydropower expansion could affect great apes and gibbons and their habitat. The analysis rests on two calculations: (1) the number of installed and planned hydropower dams in ape habitat; and (2) the potential length of new roads associated with planned hydropower dams. Given the lack of information on reservoirs and operations associated with potential future dams, this assessment does not evaluate the impacts of reservoirs, flow alteration, sediment delivery or greenhouse gas emissions, nor does it consider the impacts of resettlement areas, work camps, quarries or other associated infrastructure, or disturbances from transmission lines (see Annex I).

To identify installed and planned hydropower dams, this assessment draws on two sources: (1) the Global Reservoir and Dam (GRanD) Database for installed dams, and (2) a data set of future hydropower dams, which comprises dams that are either under construction or identified in planning documents (Lehner *et al.*, 2011; Zarfl *et al.*, 2015). The GRanD Database covers all types of dams, yet the majority of structures in ape ranges are hydropower dams, or multipurpose dams that include hydropower (Opperman *et al.*, 2015). The species ranges for great apes and gibbons were mapped based on information in the International Union for Conservation of Nature (IUCN) Red List of Threatened Species (IUCN, 2016b).

FIGURE 6.3

Number of Installed and Future Hydropower Dams in the Ranges of Great Apes and Gibbons

Key: ■ Installed dams
■ Dams planned or under construction

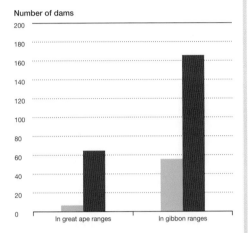

Data sources: IUCN (2016b); Lehner et al. (2011); Zarfl et al. (2015)

The number of hydropower dams in each ape range was quantified by identifying the intersection of dam locations with great ape and gibbon species ranges. The next step was to estimate the length of new roads associated with hydropower dams that are planned or under construction. It involved calculating the potential road distance between future dams and the roads closest to them based on a "least-cost path" or "path of least resistance," while also taking the local topography into consideration.

Importantly, both of the global data sets from which dam locations were derived—the GRanD Database and the data set of future dams—contain errors of omission and commission. Greater precision on dam locations may be available from finer-scale analyses that use data collected solely within the geographic range of ape species. Dam data collected at a finer scale may also include additional information that could be used to further improve the quantification of impacts on ape habitat. If, for example, dam data included the size of work camps at each dam, that information could be used to generate a more refined estimate of impacts. Further, the species range data may also contain errors. For instance, some proposed dams that are known to overlap with orangutan habitat are not included in the datasets used in this analysis (see Case Study 6.3). Nor does this study capture certain installed

FIGURE 6.4

Installed and Future Dams in the Ranges of Great Apes in Africa

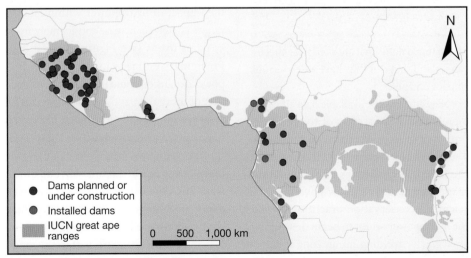

Sources: IUCN (2016b); Lehner et al. (2011); Zarfl et al. (2015)

FIGURE 6.5

Installed and Future Dams in the Ranges of Gibbon Species in Asia

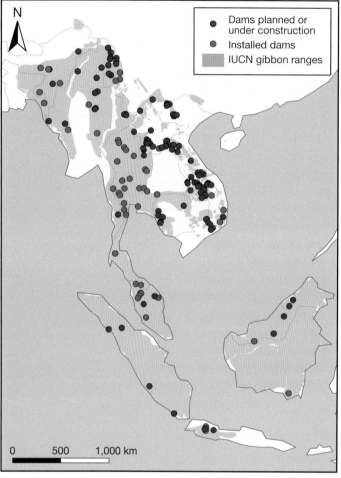

Sources: IUCN (2016b); Lehner et al. (2011); Zarfl et al. (2015)

FIGURE 6.6

Estimated Length of New Roads Associated with Construction of Future Hydropower Dams in Ape Ranges

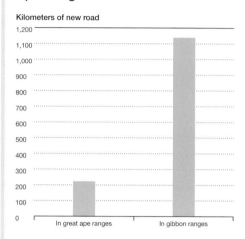

Data sources: IUCN (2016b); Lehner et al. (2011); Zarfl et al. (2015)

and planned dams that are sited near species ranges and that can thus have a deleterious impact on apes.

Nevertheless, the available data allow for a preliminary assessment of the potential impact of hydropower dams on great apes. The analysis can be used to call attention to the potential challenges of conservation management and to allow governments, scientists, conservation practitioners and the hydropower sector to begin developing strategies for avoiding, minimizing and mitigating impacts.

Results indicate that the impact of hydropower dams within great ape ranges will probably increase considerably in the coming decades (see Figures 6.4 and 6.5). Only six installed dams in the GRanD Database fall within the range of great apes, all in Africa. The number of dams affecting great apes could increase ten-fold, however, as 64 future dams are anticipated within the range of great apes—again, all in Africa. Similarly, the impact of hydropower within gibbon ranges is likely to increase considerably, from 55 dams to 165 (see Figures 6.3 and 6.5). Preliminary estimates indicate that hydropower expansion could lead to the construction of more than 200 km of new roads in great ape ranges and more than 1,100 km of new roads in gibbon ranges (see Figure 6.6).

As noted above, these data sets are known to include errors of commission and omission. The data set of future hydropower dams, for example, excludes a project that has been proposed in the Batang Toru ecosystem of North Sumatra, within the range of orangutans (Zarfl et al., 2015).

CASE STUDY 6.1

The Lom Pangar Hydropower Dam: Infrastructure and Ape Conservation in Cameroon

Introduction

Cameroon forms part of the Congo Basin rainforest and is home to some of the highest biodiversity on the continent. Its rich biodiversity, which represents 92% of Africa's ecosystems, includes significant populations of great apes, such as the western lowland gorilla (*Gorilla gorilla gorilla*) and the central chimpanzee (*Pan troglodytes troglodytes*), two endangered species whose habitats are in the rainforest (Republic of Cameroon, 2012). By dispersing seeds and maintaining forest health, these "forest gardeners" help to sustain the rich biodiversity in Cameroon.

Regardless of their role as keystone species, great ape populations are undergoing a dramatic decline, largely due to poaching, disease and habitat loss, which are driven by demands for wild meat, a lack of law enforcement, corruption and increased access to their once-remote habitat (Dinsi and Eyebe, 2016). Although Cameroon has made some effort to protect gorillas and chimpanzees—including by creating protected areas such as sanctuaries, reserves and national parks (Lambi *et al.*, 2012)—the ongoing expansion of industrial agriculture, logging, mining and infrastructure development projects will result in massive losses of habitat unless rapid, targeted action is taken.

In order to achieve its goal of becoming an emerging economy by 2035, Cameroon, a developing and still largely agrarian country, has prioritized infrastructure development. Part of the plan is to add 3,250 km of tarred roads between 2010 and 2020, alongside the construction of new railway lines. Meanwhile, the country aims to reduce the gap between the supply and demand for energy through the construction of several hydroelectric plants and dams, a heavy fuel thermal power plant and a natural gas power station (Republic of Cameroon, 2009b, pp. 59, 61–3). Expanding energy generation is central to the government's ambitions.

Cameroon's energy deficit is considered a serious impediment to its economic growth and development. In 2010, the

FIGURE 6.7

The Lom Pangar Hydropower Dam and Surrounding Area

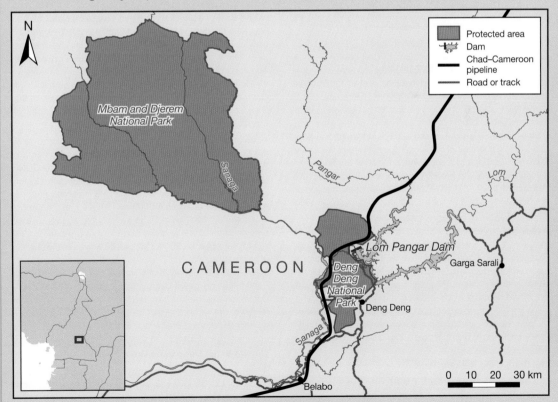

Sources: © OpenStreetMap contributors (www.openstreetmap.org); UNEP-WCMC and IUCN (n.d.)

country's total installed electricity capacity—comprising on-grid, self-generation and off-grid—stood below 2,000 MW. Hydropower plants accounted for about 73% of the total electricity produced in Cameroon in 2011, and thermal energy and solar sources made up some of the remainder. In order to increase installed hydropower capacity from about 719 MW in 2010 to 3,000 MW by 2020, the government intends to invest heavily in the energy sector (Africa–EU Energy Partnership, 2013). The Lom Pangar Hydropower Project (LPHP) was a critical first step in expanding Cameroon's energy production. This section explores the project's environmental impacts as well as efforts to mitigate them.

The Lom Pangar Dam

Cameroon relies on the LPHP as part of its efforts to provide a long-term solution to its energy supply gap. The primary purpose of the LPHP, which is designed to produce a modest 30 MW of electricity at the dam site, is to regulate the flow of the Sanaga River so as to increase and secure year-round power output for two existing downstream dams and an additional planned dam. While some estimates show that fewer than 20% of rural Cameroonians have access to electricity, the main purpose of the Lom Pangar scheme and the dams that it facilitates downstream will not significantly enhance rural electrification. Instead, the LPHP is geared towards expansion of the aluminum smelters owned by Rio Tinto, the world's largest mining company, which receives electricity at preferential rates (Ndobe and Klemm, 2014).

The management of the Lom Pangar Dam was handed over to the national Electricity Development Corporation in June 2017. A second phase that includes the construction of a 30-MW hydropower plant and electrification of 13 localities in the East Region is ongoing (BRM, 2017; ESI Africa, 2017; World Bank, 2012a). The dam is located in a remote part of eastern Cameroon, near the confluence of the Lom and Pangar rivers. Financing for the LPHP is drawn from a pool of donors, comprised of the African Development Bank, the Development Bank of Central African States, the European Investment Bank, the French Development Agency and the World Bank (ADF, 2011). The total cost of the construction of the dam and associated infrastructure is just under US$500 million (Ndobe and Klemm, 2014).

As the lead financier on the project, the World Bank assigned the project its highest environmental and social risk rating, Category A (see Box 5.1 and Annex VI). This categorization is reserved for projects that are likely to have significant adverse environmental impacts. The project received this rating in part since "the dam site is located next to portions of the Deng Deng Forest that are critical habitats, particularly because of the presence of a viable population of gorillas, and a significant population of chimpanzees" (World Bank, 2009, p. 5).

The Deng Deng National Park

The Deng Deng National Park (DDNP), which overlaps with the LPHP area, harbors a significant population of the northernmost population of the western lowland gorilla. In 2010, the Wildlife Conservation Society estimated that 300–500 gorillas lived inside the DDNP and in an adjacent logging concession (Live Science, 2011). The DDNP is also home to other threatened mammal species, including the central chimpanzee, black colobus (*Colobe satanas*), elephant (*Loxodonta africana*), hippopotamus (*Hippopotamus amphibius*) and giant pangolin (*Smutsia gigantea*) (Boutot *et al.*, 2005; EDC, 2011b).

When the World Bank agreed to finance the Chad–Cameroon oil pipeline in 1998, it insisted that the pipeline be rerouted to avoid any impacts on the Deng Deng Forest and its biodiversity (Dames and Moore, 1997; World Bank, n.d.-c). In fact, the potential impacts on the forest are among the reasons the Bank was reluctant to support the LPHP when Cameroon first sought financing in the early 2000s. At that time, the World Bank requested an environmental and social impact assessment (ESIA) to ensure that the LPHP would not have adverse effects on the Deng Deng Forest. In its review of the ESIA, the Bank cited concerns over potential impacts on great apes, especially during the construction phase, because of the large number of people expected to move to the area (EDC, 2011a, 2011b).

In 2012, in a sudden reversal of its earlier position, the World Bank decided to help finance the LPHP even though a portion of the Deng Deng Forest would be flooded by the dam's reservoir. To offset[3] impacts, the Bank required that the forest's status be upgraded from a forest reserve to a national park (World Bank, 2012a, 2012b). The Deng Deng National Park was thus created by decree on March 18, 2010; its surface area, which initially covered 523 km^2 (52,374 ha), was extended to 682 km^2 (68,200 ha) in 2013. The Wildlife Conservation Society provides technical assistance in the management of the DDNP, based on an ad hoc service contract with Cameroon's Ministry of Forests and Wildlife and its Electricity Development Corporation, with financial support from the French Development Agency (WCS, 2015b).

An enlarged Deng Deng functional ecosystem, referred to as the Deng Deng Technical Operations Unit, was created in 2010. Although it is yet to be gazetted, it includes the DDNP, two forest logging concessions, close to 20 community forests and two research forests. The Unit is spread out over a surface area of about 5,000 km^2 (500,000 ha); it harbors an estimated 990 gorillas who are roughly equally divided between the DDNP and the periphery of the park (IUCN, 2014c; Kormos *et al.*, 2014). One proposal involves the creation of an additional national park, the Lom Pangar National Park, to counteract hunting in the Mbam and Djerem National Park following development of the dam and the Chad–Cameroon pipeline. The proposed park would cover 1,775 km^2 (177,480 ha) within the dam project area and the pipeline corridor (Haskoning (Nederland B.V. Environment), 2011).

Threats to the Deng Deng Great Apes

While the creation and expansion of the DDNP were welcome conservation steps, significant threats to the great apes, as well as their habitat, remain. These include flooding, poaching, electrocution, and habitat degradation and

Photo: The Chad–Cameroon pipeline cuts through the Cameroonian rainforest. It was rerouted to avoid the Deng Deng forest, a portion of which will be flooded by the Lom Pangar Hydropower Project. © Gail Fisher/Los Angeles Times via Getty Images

loss, coupled with hunting pressures associated with artisanal mining.

Flooding

In September 2015, the contractors of the LPHP began a partial impoundment, or filling, of the dam's reservoir (EDC, n.d.-b). This step was highlighted in the project's ESIA (EDC, 2011b). Non-governmental organizations (NGOs) expressed significant concern that the full impoundment of the reservoir, which would cover approximately 590 km² (59,000 ha), about 320 km² (32,000 ha) of which is forest, would flood critical habitat of the gorillas, trapping them on islands or pushing them into populated areas (GVC, BIC and IRN, 2006). As a result, gorillas would be more exposed to poachers, the risk of disease transmission would grow due to more frequent contact with people, and human–wildlife conflict would increase in line with crop raiding (Kalpers et al., 2011). Many other, slower-moving species would likely drown during this phase.

Poaching

Large infrastructure projects tend to attract a huge influx of migrants in search of employment opportunities (WCS, 2011). In fact, the LPHP's own ESIA indicates that an estimated 7,000 to 10,000 people were expected to move to the area seeking jobs and secondary employment (Goufan and Adeline, 2005, p. 6). In a 2011 memorandum of understanding with the project contractor, China Water and Electricity Corporation, the Cameroon National Employment Fund agreed to facilitate the recruitment of an estimated 2,000 Cameroonians to work on the dam site (Agence Ecofin, 2012). Many others are likely to move to the project area without guaranteed employment, giving rise to a peripheral economy that would probably depend in part on poaching for wild meat and ivory trafficking, and that would also lead to further degradation of natural habitats.

In addition to permitting an influx of people during the construction phase of the dam, the Electricity Development Corporation intends to allow commercial fishing in the waters of the reservoir, anticipating an annual production of 1,500 tons of fish and an income of CFA 40 billion (US$65 million) (EDC, n.d.-a). Fishing opportunities are likely to draw even more people into the region, which is certain to increase pressure on biodiversity, including threats to the great apes (Goufan and Adeline, 2005; Mbodiam, 2016).

Transmission Lines

Although most tree species of high commercial value have already been exploited through illegal artisanal logging near the villages in the area, a further 5.28 km² (528 ha) of the Deng Deng Forest are to be cleared for the construction of transmission lines. Once the project goes live, it will present a risk of electrocution to wildlife (see Chapter 2 and Annex I).

Construction activities and noise pollution during the building of the transmission lines will also disrupt and temporarily displace local wildlife. A transmission corridor with a width of up to 50 m will cut into ape habitat along the eastern edge of the DDNP. Since this area represents a marginal strip of their habitat, the impact will probably be limited, depending in part on dispersal routes from the flooded land (AfDB, 2011b).

Artisanal Mining

Although the project area holds important gold reserves, the government abandoned its plan to ensure gold extraction from the reservoir area prior to impoundment as it would have delayed the project (Mbodiam, 2010). In view of the huge mining potential of Cameroon's East Region, however, the area is likely to attract artisanal and small-scale miners. Indeed, anecdotal evidence suggests that unauthorized mining operations are already under way in the DDNP itself (Charles-Innocent Memvi Abessolo, personal communication, 2016). Apart from disrupting behavior, altering habitat, reducing food resources and dispersing wildlife populations, such mining activities are associated with increased hunting pressures and disease transmission (ASM-PACE and Phillipson, 2014). Similar links between artisanal and small-scale mining and impacts on apes have been documented in the eastern Democratic Republic of Congo (Spira et al., 2017).

Mitigation Measures and Outcomes

In light of the adverse impacts identified through the ESIA process, the project developer and financiers set in place a number of mitigation measures. Nevertheless, environmental concerns persist with respect to the DDNP's staffing and viability.

Staffing of Deng Deng National Park

The LPHP relies on the deployment of rangers in and around the DDNP to control access to the park and to discourage and monitor poaching activities. The project put an emphasis on higher numbers of staff during the construction period of the dam, when the area would be most heavily populated. Once construction activities are complete, the number of rangers in the area is to be reduced to a base-case level.

The proposed number of guards to monitor the DDNP is one ranger per 10 km^2 (1,000 ha) within the park itself, and one per 25 km^2 (2,500 ha) in areas that are less vulnerable to poaching (EDC, 2011c; Charles-Innocent Memvi Abessolo, personal communication, 2016). Of the 58 managers and other staff members involved in securing and monitoring the DDNP and its periphery, only 17 are permanently assigned to the park; the rest are on temporary transfer from other services (MINFOF, 2015).

The number of permanent staff is modest for a protected area of more than 680 km^2 (68,000 ha), not including the periphery, especially given that the environmental and social management plan calls for 70 community guards and ecoguards (EDC, 2011c; MINFOF, 2015). Compounding the problem of insufficient personnel is the inadequate training of most of the staff.

There is clear evidence that poaching continues despite the presence of ecoguards at the DDNP. In 2015, 1,270 kg of wild meat was seized, including 20 kg of chimpanzee, and 290 kg of monkey and gorilla (MINFOF, 2015).

The Viability of Deng Deng National Park

In part to ensure the viability of the great ape populations, the World Bank established the DDNP as a biodiversity offset to be preserved in perpetuity. Yet, while the LPHP will facilitate access to the DDNP well beyond the period of construction, project financiers are expected to exit the project and thus cease monitoring by the end of 2018 (World Bank, 2012c). Therefore, a key question revolves around long-term viability and financial sustainability, including staffing and equipment for park surveillance.

The DDNP was expected to progress towards financial sustainability by drawing a growing number of ecotourists, but recent figures cast doubt on that assumption. In 2015, the park received only 23 visitors — 17 nationals and 6 foreigners — yielding a total fee of CFA 88,500 (US$150). In addition to park visits that year, auction sales of seized illegal forest products from poaching and illegal logging raised only CFA 1.1 million (US$1,891) (MINFOF, 2015). The lack of investment in the DDNP is evidenced by the ongoing absence of a dedicated DDNP office building. The park's temporary office is housed in one of the control posts.

Acknowledging that revenue from ecotourism is likely to be insufficient, the U.S. government insisted, as a condition of approving the project at the World Bank, that a portion of the water tariffs generated by hydropower installations be devoted to helping sustain the park financially. The installations are located downstream of Lom Pangar and payments are expected once the LPHP becomes operational. These details are included in the project appraisal document, which provides details on the World Bank's proposed credit to the government of Cameroon for the LPHP (World Bank, 2012c).

Arrangements to allocate a portion of the water tariffs to the DDNP have yet to be made, however. The matter is of relative urgency as construction activities are due to draw to a close over the course of 2018. These arrangements were to be finalized prior to full impoundment of the reservoir, which is also expected in 2018. Even once those arrangements are made, it is unclear what role the project's financiers will have in ensuring that the funds are utilized as intended, and what means they have to ensure compliance, should the agreement not be respected. The French Development Agency ceased payments to sustain the park in August 2016, the anticipated deadline.

Conclusions

The World Bank and other development financiers entered into the Lom Pangar Hydropower Project with full knowledge that such massive infrastructure in a remote and ecologically sensitive part of Cameroon was likely to have adverse

Photo: The number of permanent staff involved in securing and monitoring the DDNP is insufficient to ensure the protection of the western lowland gorillas and other species. © Chris Chaput

effects on important populations of great apes. Acknowledging the risks that the LPHP posed to these populations, the World Bank and other financiers stressed that instituting requirements to guarantee the preservation of the Deng Deng Forest, through the creation of an offset, was the only hope of ensuring the survival of the region's great apes (EDC, 2011a, 2011b, 2011c; World Bank, 2012a, 2012b, 2012c). However, evidence of the viability of these measures is clearly lacking, and the few reports from site visits point to deficiencies in efforts to guard the area against poaching. The lack of effective and regular monitoring means that the current status of great ape populations in the park is unclear.

Furthermore, the financial sustainability of Deng Deng National Park remains uncertain. The completion of dam construction means the World Bank will reduce its oversight of the project, and the project completion date at the end of 2018 will signal the termination of the World Bank's involvement, and that of the African Development Bank, the European Investment Bank, the French Development Agency and other financiers. Meanwhile, the lack of progress in developing the arrangements to ensure that a portion of the water tariffs derived from hydropower production will be devoted to Deng Deng National Park suggests that the park's viability is in danger.

In conclusion, the DDNP and its great ape population remain at risk of further degradation once the project concludes, unless urgent action is taken to ensure oversight beyond the project completion date and a secure revenue stream for the park. Given that the attention of financiers is typically finite, large infrastructure projects such as the LPHP can present critical, yet foreseeable, challenges to indefinite conservation. This case study demonstrates that even when the adverse impacts of an infrastructure project are acknowledged and assessed early on, they can nevertheless threaten the survival of endangered species such as gorillas and chimpanzees.

CASE STUDY 6.2

Community Resistance Against Infrastructure in Malaysian Borneo: The Case of the Baram Dam

Introduction

In 2006 the federal government of Malaysia embarked on a series of proposed economic corridors in an attempt to stimulate global and domestic investment in rural areas across the country. One of these corridors was the Sarawak Corridor of Renewable Energy (SCORE). It was to be established in Sarawak, one of two Malaysian states on the island of Borneo, and the largest of Malaysia's 13 states.

As part of SCORE, at least 12 dams were to be completed in Sarawak by 2030 (Shirley and Kammen, 2015). Two of these have already been completed: the Bakun and Murum dams (see Figure 6.8). Plans for the Baram Dam, which was next in line, were met with extensive community resistance from the indigenous communities in the Baram River Basin. Construction on the Baram Dam had been scheduled to start in 2014 but, by March 2016, after several years of community resistance, the state government legally withdrew its claim over the indigenous land earmarked for the dam site. This case study documents how a grassroots movement successfully prevented the realization of a large infrastructure project that had been backed by the government.

Background

The Bornean Rainforest

The third largest island in the world, Borneo is part of the Sunda Shelf, which extends from Vietnam to Borneo and Java. The rainforests of Borneo are a biodiversity hotspot acknowledged to be among the world's most species-rich ecosystems. At least 15,000 plants, of which 6,000 are found nowhere else in the world, grow in the swamps, mangroves and lowland and montane forests of the island. Borneo is home to an estimated 222 mammals (44 endemic), 420 birds (37 endemic), 100 amphibians and 394 fish species (19 endemic). Orangutans and gibbons share Borneo's forests with a number of other primate species, including langurs (*Semnopithecus*), macaques (*Macaca*), proboscis monkeys (*Nasalis larvatus*), slow lorises (*Nycticebus*) and tarsiers (*Tarsius*) (WWF, n.d.-a, n.d.-b).

The Baram River Basin lies in northeastern Sarawak (see Figure 6.8). The waters originate in the Kelabit Highlands along the border with Kalimantan (Indonesian Borneo), flow through the mountain highlands and hills for more than 400 km, and lead into the South China Sea (Encyclopaedia Britannica,

FIGURE 6.8

Baram River Basin and the Bakun and Murum Dams

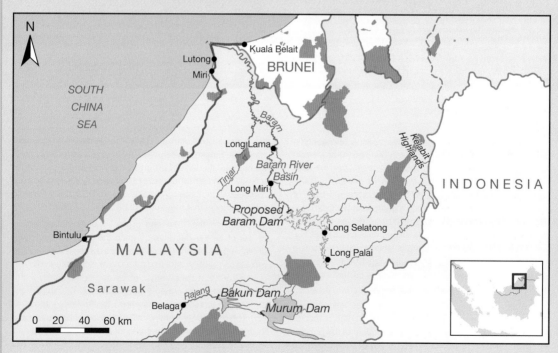

Sources: © OpenStreetMap contributors (www.openstreetmap.org); UNEP-WCMC and IUCN (n.d.)

1998). The forests of the Baram River Basin are home to a wide variety of fauna and flora, including gray gibbons.

Logging and Deforestation

In the past several decades logging has had a great impact on the forests of Sarawak; lush tropical rainforests are disappearing at an astonishing rate. Between 2005 and 2010, forest loss in Sarawak exceeded 2% per year, a rate higher than in any other major tropical forest territory. Between 2006 and 2010, about 9,000 km^2 (900,000 ha) of Sarawak's forest was lost—43% was converted to oil palm and 21% to timber plantations (Lawson, 2014).

From 1981 to 2014 Sarawak was governed by Abdul Taib Mahmud, who has been accused many times of gross environmental and human rights abuses for personal gain (Global Witness, 2012; Straumann, 2014). During his tenure, the state became one of the largest exporters of tropical timber in the world. In 2010, Sarawak accounted for 25% of the world's source-country exports of tropical logs, 15% of global tropical lumber and almost half of all tropical plywood—quite a feat for a forest estate that represents just 0.5% of the global total. Fewer than 5% of Sarawak's intact forests remain in a pristine state, unaffected by logging or plantations, with dire consequences for its wildlife and indigenous communities, which depend on the forests (Global Witness, 2012).

The Indigenous Population

The people of the Baram River are mainly indigenous Kayan, Kenyah and Penan, with a few Iban, Kelabit and Saban communities. They depend on healthy rivers and forests for their livelihoods. The native customary rights (NCR) of indigenous groups over their ancestral land are enshrined in the Sarawak Land Code and protected under the Malaysian Constitution (Colchester *et al.*, 2007). Nevertheless, the government has proceeded to license nearly the entirety of Sarawak, including land claimed for NCR, for logging and plantations, while simultaneously blocking attempts by communities to have their NCR land mapped, recognized and gazetted (Global Witness, 2012).

The people of Baram have a history of resisting deforestation in the area. Since the late 1980s, when logging and agricultural expansion began to change the landscape of Sarawak, indigenous communities resisted through protests and blockades against logging companies. Resistance has often led to arrests and political persecution, with the result that several prominent activists fled Malaysia in the 1990s. In the past several years the government has relaxed its approach to environmental and human rights activists; however, deadly conflicts are still occurring between indigenous activists and land developers.[4]

SCORE Hydroelectric Dams

The government of Sarawak and the dam builder, Sarawak Energy Berhad (SEB), have claimed that the energy produced by the SCORE dams would transform Sarawak into a developed state by the year 2020. Yet the project's 12 large hydropower dams were primarily designed to power the expansion of oil palm plantations and energy-intensive industries (Shirley and Kammen, 2015).

After five decades of delays, the Bakun Dam was opened in 2011, but since then it has only operated at half-capacity (Sarawak Report, 2014). This was the first of the SCORE dams to be built; looming at a height of 205 m, it is the largest dam in Asia outside of China (International Rivers, n.d.-a). The Murum Dam, the second in the SCORE series, officially opened in September 2016 (Then, 2016). The government began preliminary work on the Baram Dam in 2011 but officially canceled all works in March 2016, due to grassroots resistance. The Baleh Dam is next in line to be built, and while the government approved its environmental and social impact assessment in 2016, the details of the proposal and the ESIA have not been publicly released.[5]

The acronym SCORE stands for Sarawak Corridor of *Renewable* Energy, but the adjective "renewable" is inaccurate in this context, as the SCORE development plan entails the exploitation of coal reserves, the construction of new coal power plants and deforestation to accommodate the expansion of oil palm plantations (Shirley and Kammen, 2015). The power generated by the SCORE dams is intended to feed energy-intensive industries, such as aluminum and steel production. SEB, a state-owned electricity supplier under Sarawak's Finance Ministry, is responsible for the planning of all hydropower projects and coal plants in Sarawak. It is chaired by Abdul Hamed Sepawi, a cousin and one of the closest business allies of Sarawak's former chief minister, Taib Mahmud (Bruno Manser Fonds, 2012a, 2012b).

The Renewable and Appropriate Energy Laboratory (RAEL), an independent energy research facility at the University of California, Berkeley, recently conducted an in-depth analysis to explore the implications of building the SCORE dams, and the potential for clean energy solutions for Sarawak. The RAEL research agenda covered three main project areas: (a) modeling long-term, utility-scale electricity-generation alternatives for Sarawak to determine trade-offs across different technologies; (b) exploring to what extent rural communities in dam-affected areas would be able to satisfy energy access needs using local resources; and (c) demonstrating a rapid assessment method for estimating the impact of mega-projects on biodiversity. RAEL's research results call into question the necessity of building additional dams in view of potential lower-cost, lower-impact clean energy alternatives in the state (Shirley and Kammen, 2015).

The RAEL results show that the energy that would be produced by the SCORE dams is unreasonably excessive, even if the aims were to sustain aggressive growth in Sarawak. The SCORE initiative assumes an energy demand growth rate of more than 16% per year through 2030 (Shirley and Kammen, 2015). To put this in perspective, China's energy demand growth rate barely exceeded 10% for three years during the height of its industrial boom (Dai, 2013). The RAEL models show that there are a number of alternative choices to SCORE that meet future demand at an aggressive 7% energy demand growth rate and a very aggressive 10% energy demand growth rate at a lower cost than the SCORE plan. The Bakun Dam alone satisfies one-third of the demand by 2030 under a 10% growth assumption, and half of the demand under a 7% growth assumption. Two existing dams (Bakun in central and Batang Ai in southwestern Sarawak) and recently installed combined gas and coal-fired generators are sufficient to meet demand at a 10% growth rate if properly managed (Shirley and Kammen, 2015).

Social and Economic Impacts: Baram, Bakun and Murum

Although most of the dams are sited on native land, indigenous communities have not been properly consulted and are being forcefully relocated. The Baram Dam would have created a reservoir covering around 400 km² (40,000 ha) of forest and would have displaced up to 20,000 indigenous people (Lee, Jalong and Wong, 2014). Communities that were displaced because of construction of the Bakun and Murum dams have been severely impacted by relocation.

In 1998 the government of Sarawak relocated about 10,000 people to make way for the Bakun Dam. Two decades after resettlement, the displaced people are still struggling to eke out a living. The government required resettled communities to pay for their own housing, which forced many families into debt. Communities that had been able to catch fish in the river, hunt and gather forest products no longer have access to forests, and pollution from the dam has decimated fish stocks. Each family was promised 0.04 km² (4 ha/10 acres) of farmland but was only provided 0.01 km² (1.2 ha/3 acres), much of it a half-day's journey away from the resettlement sites; moreover, a large portion of the "farmland" is infertile, rocky and sandy land. This has not been enough to sustain a living (International Rivers, n.d.-a).

Similarly, communities displaced in 2013 by the Murum Dam are struggling in their resettlement sites. Construction of the dam began in 2008, even though neither the initial ESIA nor the resettlement action plan had been made public. The project developers did not begin an ESIA until after construction was already under way, and the resettlement plans were leaked in 2012 (International Rivers, n.d.-d).

The Sarawak government began resettling around 1,500 indigenous people from the Murum Dam area in July 2013. The resettlement sites are surrounded by vast expanses of oil palm and land earmarked for logging concessions to politically connected timber companies (International Rivers, n.d.-d). As of January 2018, the communities still had not been allocated land to cultivate. During a visit led by the Sarawak-based NGO Save Rivers to the Kenyah resettlement site at Tegulang in October 2016, the residents remarked that they felt as though they were "in jail."[6] Without land, they cannot grow food for their families or to sell at the market, and they are stranded without transportation to larger towns. The government has reduced their monthly rations twice, but the community still has no way of earning income or growing or gathering food to make up for the lost rations.

Photo: After five decades of delays, the Bakun Dam was opened in 2011, but since then it has only operated at half-capacity. Bakun Hydroelectric Dam, Sarawak, Malaysia. © MOHD RASFAN/AFP/Getty Images

Chapter 6 Renewable Energy

The dams also inflict considerable economic costs on the state. The Bakun Dam was built over the course of two decades at a total cost that was astronomically higher than projected. The dam was originally expected to cost MYR 2.5 billion (US$564 million), excluding transmission and all non-dam-related infrastructure. While the official expenditure figures have risen to MYR 7.4 billion (US$1.7 billion), researchers from the National University of Singapore put the cost of the Bakun Dam at MYR 15 billion (US$3.5 billion), six times the original estimate (Sovacool and Bulan, 2011). Construction began in 1994 and the dam was meant to be operational in 2003. It was not completed until 2011, but even today, it is not running at full capacity. The Murum Dam has also incurred significant cost overruns. It cost Sarawak MYR 530 million (US$120 million) more than the original price, according to the 2016 Auditor-General's report (Kallang, 2016).

Environmental Impacts

If the SCORE vision were to be realized as initially planned, 2,425 km² (242,500 ha) of rainforest would be destroyed to allow for the impoundment of reservoirs and the construction of dams, and additional land would be cleared for resettlement sites. The Bakun Dam reservoir alone covers 695 km² (69,500 ha) — about the size of Singapore (Kitzes and Shirley, 2015). Given that the rainforests of Borneo are among the most biodiverse terrestrial ecosystems in the world, it comes as no surprise that the three dams — Bakun, Baram and Murum — would have a tremendous impact on the rich biodiversity of the area.

The RAEL team conducted biodiversity impact studies for these three SCORE dams and uncovered alarming facts. Using global species range data, geographic information system (GIS) tools and species area scaling relationships, the team predicted three distinct measures of biodiversity impact: the total number of species affected by the dams, the number of individuals affected and the number of potential species extinctions that could result (Kitzes and Shirley, 2015).

The study found that the dams would have a negative impact on at least 57% of Bornean bird species and 68% of Bornean mammal species. The affected species include endangered and critically endangered birds and mammals, such as Abbott's gray gibbon (*Hylobates abbotti*), the Bornean bay cat (*Catopuma badia*), the Bornean peacock-pheasant (*Polyplectron schleiermacheri*), the flat-headed cat (*Prionailurus planiceps*), the smoky flying squirrel (*Pteromyscus pulverulentus*), Storm's stork (*Ciconia stormi*), the Sunda otter civet (*Cynogale bennettii*) and the Sunda pangolin (*Manis javanica*). In addition, the study found that two-thirds of all tree and arthropod species would be impacted, resulting in four tree and 35 arthropod species extinctions. The number of species extinctions does not take into account the potential extinction of subspecies or local populations, both of which may be critical to species' long-term viability (Kitzes and Shirley, 2015).

The study also provided numbers on individual organisms that would be lost — arthropods, birds, mammals and trees that would perish because of loss of habitat from clear-cutting and inundation. The three dams alone would cause the loss of an estimated 3.4 million individual birds and 110 million individual mammals. To put this into perspective, that is more individual birds than were counted in the North American Breeding Bird Survey in 2012 and more individual mammals than the entire inventory of cattle in the United States in 2012. A minimum of 900 million individual trees and 34 billion individual arthropods would also be lost (Kitzes and Shirley, 2015).

Community Resistance in Baram

The Formation of Save Rivers

In 2011 the state government began to hold briefing sessions about the proposed Baram Dam and started construction of the road to the dam site. In October of that year, eight Sarawak-based civil society organizations that were concerned about the implications for the people and the forests of Baram joined forces to form the Save Sarawak Rivers Network (Save Rivers), whose mission is to build broad-based support to educate and mobilize the public against the plans to build dams.

The first actions by Save Rivers were designed to raise awareness among the urban and rural populations about the dam and its implications. On February 16–18, 2012, the group organized an initial statewide conference in the city of Miri for representatives from the Bakun, Baram and Murum river basins. Following the conference, delegations from Save Rivers conducted roadshows, traveling by vehicle and boat to remote villages throughout the Baram river basin to inform communities about the proposed Baram Dam and its implications for them. At that point, the preliminary ESIA had been completed by Fichtner, a German consulting company employed by SEB; however, the full ESIA had not yet been initiated and the majority of impacted villages had not been informed about the plan to build the dam. The roadshows were conducted in all of the villages that were at risk of inundation; most villagers heard about the dam construction plans for the first time during these events.

Community Organizing, Nonviolent Direct Actions, Awareness Building

Since its formation, Save Rivers has continuously organized events and trips to build awareness and strengthen communities. Roadshows are conducted regularly to provide villagers with information and update them on the latest developments. One of the largest trips occurred in January 2013, through what is called the "Baram Wave." A delegation from Save Rivers travelled upriver in motorized canoes to distribute information and build solidarity. The group slowly made its way downstream, distributing information and encouraging canoes from each village to join. A flotilla of around 50 canoes arrived at Long Lama, the closest town to the access road for the dam site and, together with residents from around Baram, they held a rally to demonstrate their opposition to the dam. The Baram Wave fulfilled several vital functions, including raising awareness and solidarity among Baram

Photo: Long Lama blockade, the structure that blocks the access road to the Baram Dam site. © Jettie Word, The Borneo Project

communities and voicing the communities' concerns to government officials.

The next large event occurred in May 2013, alongside the International Hydropower Association (IHA) conference that was hosted by SEB in Kuching, in western Sarawak. Save Rivers brought together residents from Baram, local and international politicians, and local and international NGOs for an alternative conference on indigenous rights that included several protests and marches held outside of the IHA venue. The alternative conference drew supporters from around the state and around the country, greatly increasing local and national awareness about the issues and building solidarity.

In August 2013 the Sarawak government took the first steps to extinguish the land rights of indigenous communities near the Baram Dam site—without their consent (Lee et al., 2014). In response, Save Rivers traveled up and down the Baram River, helping the communities establish two blockades to prevent dam workers from accessing the proposed site of the Baram dam. One blockade was built centrally among Baram villages as a rally point. The second blockade was constructed at the beginning of the access road to the dam site near Long Lama. The blockades prevented construction, surveying work and logging at the proposed dam site, halting all progress. The blockades not only physically disrupted work on the dam, but also acted as community centers and observatories for monitoring illegal logging. In spite of numerous government attempts to dismantle the structures and disperse community members, the blockades have been continuously maintained and managed since October 2013. They are the longest-running blockades in the history of Sarawak, and their maintenance has required significant efforts. When the blockades were formed, Save Rivers also helped the communities file a suit against the government, in which they collectively demanded that their customary lands be returned.

In conjunction with blockades, rallies and roadshows, Save Rivers organized cross visits between Baram villages and communities that were forcefully relocated because of the Bakun Dam. During these visits the people of Baram were able to speak directly with individuals who had been evicted and witness for themselves what happens during resettlement. Save Rivers also organized several large conferences in Baram to distribute information and strategize among communities, as well as various acts of nonviolent direct action throughout the country. One of the larger events occurred in June 2015 during a visit by then chief minister Adenan Satem to the town of Long Lama for a bridge inauguration. Save Rivers rallied hundreds of Baram residents to line the streets of Long Lama and express their opposition to the dam. Their voices were heard loud and clear, and the chief minister acknowledged Save Rivers in his speech (Radio Free Sarawak, 2015).

Research and Publications

In addition to raising awareness and promoting community organization, the campaign against the Baram dam coordinated with local and international experts to produce several publications and studies about the situation.

A fact-finding mission to determine how SEB and the government had engaged with the communities of Baram was conducted by local experts and supported by several local and international groups. Based on detailed interviews in 13 villages along the Baram River, the mission report reveals how indigenous communities were denied information, prevented from participating in studies and decision-making, coerced into accepting the dam through threats and intimidation, and thus denied their rights, as ascribed under international agreements and treaties, to their lands and territories, self-determination, and to free, prior and informed consent (see Chapter 2). The report, entitled *No Consent to Proceed*, received significant media attention (Lee *et al.*, 2014).

Save Rivers also worked with experts from the University of California to increase transparency on energy development in Sarawak. As mentioned above, the RAEL team produced three studies that greatly informed the campaign. The studies show in detail that the energy that would be produced by SCORE is superfluous, and that the impacts on biodiversity would be severe. They also lay out a plan to increase rural energy through small renewable systems, such as solar and micro-hydro structures (Kitzes and Shirley, 2015; Shirley and Kammen, 2015; Shirley, Kammen and Wynn, 2014).

The RAEL studies were used to strengthen community resilience, as well as to increase awareness in the government. In March 2015 Save Rivers organized a trip to distribute the RAEL studies throughout Baram. The results reaffirmed and gave credence to the demands of the people. Three months later Save Rivers organized a meeting that brought together local activists, politicians, Dan Kammen, the founding director of RAEL, and Chief Minister Satem to discuss the energy options and the demands of the Baram people. Following the meeting, Satem, who has since died, asked for an alternative proposal to the SCORE dams, which was submitted in July 2015. In January 2018, the authorities had yet to respond to the submission. The campaign was working to resubmit the proposal and arrange a meeting with the new chief minister.

International Solidarity

In addition to networking among stakeholders, researchers and politicians, the campaign against the Baram Dam generated considerable international solidarity. International organizations have provided funding, strategy, media and networking support. In October 2015, Save Rivers, the Borneo Project and the Bruno Manser Fund organized the World Indigenous Summit on Environment and Rivers (WISER) to mark the second anniversary of the blockades. WISER brought together indigenous leaders who are fighting dams around the world, including the late Goldman Prize winner Berta Cáceres. Together, the WISER participants wrote the Baram 2015 Declaration on Dams and the Rights of Indigenous People. The Summit rallied more than 1,000 people from Baram at various events, building solidarity and drawing significant media attention.

Victory: Land Returned to Communities

On March 15, 2016, the Sarawak government revoked its claim to the land that would have been used for the Baram Dam, thereby legally restoring indigenous land rights and officially stopping all progress on the dam (Mongabay, 2016a). Stopping the Baram Dam was an unprecedented success for indigenous rights in Sarawak. This victory was won at a time when dams around the world were under increasing scrutiny. For Malaysia, where the space for civil society is constantly shrinking, the success of Baram gives hope to other struggles for rights and the environment (HRW, 2016).

Challenges and the Path Forward

The campaign experienced many challenges along the path to defeating the dam. One of the principal divisive tactics of the government was to divide the communities and label the people who opposed the dam "anti-development." The government also removed anti-dam village leaders, or headmen, and replaced them with pro-dam headmen.

In Sarawak activists often face exclusion from society. Many people choose to remain silent for fear that the government will withdraw support in the form of development projects and education grants. Leaders of the opposition to the Baram Dam have been socially ostracized by friends and family members who do not agree with the campaign.

Activists have also faced legal battles. SEB tried to sue 23 activists for tampering with samples and equipment at the dam site. As the land for the dam site has now been legally returned to the community, SEB has withdrawn the suit.

In spite of the victory against the dam, the blockades remain intact and running. The blockades now serve as a venue for community events instead of obstructing access to the dam site. Communities are wary that a new government may try to reinstate the project. To gear up for this possibility, Save Rivers is now focusing on campaigns to build long-term land rights protection measures in Baram through the Baram Conservation Initiative. The Initiative actively seeks to facilitate development systems that are chosen and managed by rural communities, in harmony with their environment. At the time of writing, the two main program aims were to establish a community-managed conservation zone and to build sustainable village-scale electrification systems, such as micro-hydro and solar systems.

A key lesson from the campaign against the dam is the importance of raising awareness in communities. Without proper knowledge of the situation, communities cannot act. Increasing awareness allows people to choose how to respond to projects.

Fostering community-based development models is key to avoiding the environmental and social destruction of large infrastructure projects. The promotion of community-based systems requires a paradigm shift away from top-down infrastructure projects that harm rural communities and forests.

BOX 6.1
Hydropower by Design

Introduction

The hydropower sector, governments, scientists and civil society groups have worked, often collaboratively, on finding ways to improve the sustainability of hydropower development and to achieve more balanced outcomes between energy development and other values. More balanced outcomes can occur at two scales:

- the planning and siting of new dams at the system scale (that is, at the scale of a river basin or a region); and
- the design and operation of individual dams.

Recognizing that the sustainability of hydropower is a function of the system and individual scales, The Nature Conservancy (TNC) developed an approach that integrates both scales: "Hydropower by Design." The approach encompasses a range of methods and tools to improve the planning, siting, design and operation of hydropower, as well as the mitigation of its adverse impacts (Opperman *et al.*, 2015, 2017; TNC, WWF and UoM, 2016). Hydropower by Design is a shorthand term for integrated, system-scale planning and management using a number of existing tools and processes, including the mitigation hierarchy (see Chapter 4, p. 119). By applying this approach, hydropower developers can:

- **avoid** building dams at the most damaging sites by directing development towards sites that result in less impact, specifically by identifying the spatial arrangement of dams that can produce optimal outcomes across social, environmental and economic values;
- **minimize** impacts, such as through best practices during construction;
- **restore** key processes and resources by adapting the design and operation of individual dams (such as building fish passage structures and managing the release of environmental flows to maintain or restore downstream floodplain fisheries); and
- **offset** adverse impacts that cannot be avoided, minimized or restored by investing in compensation to achieve no net loss of biodiversity.

Some progress has been made in the development of approaches that serve to improve the environmental and social performance of individual hydropower dams. Among these is a tool to measure the relative sustainability of projects—the Hydropower Sustainability Assessment Protocol ('the Protocol') (IHA, 2010). However, a number of major impacts from hydropower cannot be mitigated effectively at the scale of a single dam and project-level sustainability cannot address the complex issues posed by multiple hydropower developments across a river basin or region.

With respect to apes, certain impacts from hydropower can be addressed through best practices at the project scale, but some of the most important conservation objectives—such as the maintenance of large blocks of intact forest or connectivity between forests—can only be addressed through system-scale approaches that influence the spatial configuration of hydropower development, encompassing dams, reservoirs, roads and transmission lines.

When applied to ape conservation, the principles of Hydropower by Design can be organized to follow the mitigation hierarchy:

Photo: The negative environmental and social impacts of dams and other large infrastructure projects are more likely to be minimized when their development planning incorporates a system-scale approach and draws on existing tools and processes, including the mitigation hierarchy. Proposed site of a hydropower project, 'Chutes de l'Impératrice Eugénie' waterfalls, Ngounié River, Gabon. © Matthew McGrath

- **Avoid.** National parks and other formally protected areas should be maintained as no-go areas for dam development. System-scale planning can also be used to avoid siting or licensing projects that would impact high-value ape habitat outside of protected areas, such as dispersal corridors and large swaths of intact habitat. Multi-objective planning and analysis can identify investment options—combinations of project site, design and operation—that perform well across a range of metrics; such "win–win" or "close to win–close to win" outcomes can contribute towards energy targets while protecting the most important ape habitats. Ideally, areas that are "avoided" through a system-scale planning process would also receive formal protection from future development, potentially funded via mitigation or compensation measures, as described below. The most effective spatial planning for siting focuses not just on dams and reservoirs, but also on the siting of associated roads and transmission lines.

- **Minimize impacts during development and operation.** To protect apes, hydropower developers can implement management plans that minimize impacts during construction and operation. Construction management plans can include best practices to prevent workers from hunting for wild meat or engaging in other activities that harm wildlife. The environmental management plan for the Trung Son Hydropower Project in Vietnam, for example, includes a ban on hunting and possession of wild meat in the construction camps (Integrated Environments, 2010). During operation, a portion of revenue from a hydropower project could be dedicated to conserving intact forest in the watershed upstream of a project. This type of management fund can benefit projects by ensuring that upstream land cover promotes reliable flows of water and avoids excess sedimentation from land clearing and road construction. Wherever upstream watersheds also provide habitat for wildlife, including apes, this management fund can also be used to protect that habitat.

- **Compensate or offset.** Even if efforts are made to avoid and minimize impacts, the expansion of hydropower systems will almost certainly result in net negative impacts to natural resources such as ape habitat. For these "residual impacts," mitigation policies can promote compensation—investments in restoration or protection intended to "offset" residual impacts. For example, compensation funding could be used to support the durable protection of high-quality habitats that may be threatened by new development impacts by formally designating them protected areas and providing funding for their management. Compensation funding could also be dedicated to reforestation of migration corridors for apes; this type of funding was made available to reforest a corridor for jaguars with the Reventazón project in Costa Rica, for example (IDB, n.d.).

The outcomes of Hydropower by Design analysis and implementation are dependent on the participation and buy-in of all relevant stakeholders over the duration of the development process. In addition to governments, developers and financiers, a stakeholder group includes representatives from communities that may be directly or indirectly affected by the development of hydropower dams, as well as representatives with relevant expertise from academia and civil society. The stakeholder group is relied upon to identify social and environmental resources that may be impacted by the proposed hydropower development, determine whether the metrics used to assess those impacts are adequate through an iterative process, and participate in the decision-making process to select a hydropower build-out that best balances the trade-offs between development, conservation and social issues.

If the stakeholder group is not collaborative and transparent, the ultimate build-out of hydropower projects may not represent the best trade-off, with possible repercussions for environmental and social resources, including great ape and gibbon habitat. However, the process of identifying environmental and social resources and quantitatively measuring the impacts of a given hydropower build-out scenario on those resources inherently makes the planning process more transparent, even if the final decision is made in a political context that does not fully embrace the collaborative process that is at the heart of Hydropower by Design.

Implementing Hydropower by Design

In practice, Hydropower by Design is most effective when it is incorporated into the policies and practices of the key actors within the hydropower sector. Key actors are governments, financial institutions and hydropower companies, including developers and contractors.

Governments

Governments are generally in the best position to implement the concepts behind Hydropower by Design, particularly because they direct the planning of energy systems and license individual projects. A strong planning and site selection role by the government can identify both the river reaches or projects that should be developed, as well as the areas that should be protected, thereby reducing conflicts and increasing certainty for stakeholders, including hydropower developers, conservation organizations and local communities (Opperman *et al.*, 2017). For example, in the 1980s Norway conducted comprehensive studies of undeveloped rivers and river basins and identified a subset that would be eligible for hydropower development and another subset to be protected from future development, thereby reducing conflict and increasing certainty for energy development and other values (Wenstop and Carlsen, 1988).

In addition to planning, government licensing processes can be influential in determining which projects are built and which priority habitats are granted protection. Licensing agencies can identify areas for which licenses will not be granted (a categorization that is functionally equivalent to an "avoid" designation); they can also determine mitigation requirements for licenses, such as by setting compensation ratios based on impacts. However, such decisions are vulnerable to being overturned unless they are made durable through formal protected status. Particularly rare or important habitat types can have high compensation ratios (such as 5 ha of protection

Photo: Low water levels at the Mae Guang Udom Tara dam. In 2015, Thailand's key reservoirs fell to the lowest since 1987, and farmers were warned to delay planting their main rice crop. A number of major impacts from hydropower cannot be mitigated effectively at the scale of a single dam and project-level sustainability cannot address the complex issues posed by multiple hydropower developments across a river basin or region. © Dario Pignatelli/Bloomberg via Getty Images

or restoration per impacted hectare). Compensation funding generated for development that impacts habitats can then be used for acquisition or management of other high-value habitats. Colombia is integrating this approach into its licensing process for large infrastructure projects, including hydropower (Opperman et al., 2017).

Hydropower by Design does not necessarily require governments to adopt new policies or regulatory structures. Rather, existing policy or regulatory tools—such as energy master plans, strategic environmental assessments, environmental and social impact assessments, and licensing—can be updated or refined to move hydropower development away from a single-project focus and towards a system approach.

Financial Institutions and Developers

A variety of financial institutions fund hydropower projects, including private commercial banks and multilateral institutions such as the World Bank and the Asian Development Bank. Financial institutions can apply environmental and social policies to determine which projects they will fund and to attach conditions to their financing, such as mitigation requirements. Multilateral financial institutions have comprehensive environmental and social safeguards. However, these safeguards are generally applied at the project scale, and a review of hydropower standards by the International Institute for Environment and Development (IIED) concluded that few standards or safeguards address system planning or options assessments that can screen out detrimental projects (Skinner and Haas, 2014).

Specific hydropower-related risk screening tools can be used as a complement to general safeguards. The World Bank has acknowledged that, for its projects, the Hydropower Sustainability Assessment Protocol is a useful risk-screening tool that can be applied before its own safeguards (Liden and Lyon, 2014). The IIED review reported that only 10%–15% of new hydropower projects around the world were covered by international standards or safeguard processes. It concluded that the Protocol "represents the best currently available 'measuring stick' for respect for the [World Commission on Dams] provisions in individual projects" as it offers a set of principles that many civil society organizations see as the "gold standard" in terms of sustainability for dam development and operation (Skinner and Haas, 2014, pp. xi, 44, 75).

An "early planning facility" is an additional mechanism through which multilateral funders could help move hydropower development towards system-scale approaches (Opperman et al., 2017). Such a facility would combine funding and technical assistance to support governments in conducting system planning with the goal of developing a pipeline of projects. Projects that emerge through this process would represent low-risk opportunities for developers and investors that are consistent with objectives for the sustainable management of river basins or regions.

Developers generally do not have the ability to plan or manage at the scale of a system, with some exceptions (such as when a single company has multiple concessions or projects in a basin or when a company secures a contract to conduct a basin plan). However, companies can follow policies or practices that support sustainable hydropower, for example by adopting corporate sustainability standards or by using risk-screening tools such as the Protocol. Companies that recognize the value of reducing risk and uncertainty for hydropower development could signal their support for Hydropower by Design to governments and funders and find ways to contribute to its adoption.

CASE STUDY 6.3

Not All Renewable Energy is Sustainable: A Geothermal Project in the Leuser Ecosystem, Sumatra, Indonesia

On August 16, 2016, the governor of Aceh province wrote to the central government's Ministry of Environment and Forestry, requesting that a "core area" of the Gunung Leuser National Park (GLNP) be rezoned to allow development of a major new geothermal project. The location in question lies in the Kappi Plateau region of the park, in the northernmost province of the island of Sumatra, Indonesia (Hanafiah, 2016; see Figure 6.9).

Together, the Gunung Leuser, Bukit Barisan Selatan and Kerinci Seblat national parks comprise the Tropical Rainforest Heritage of Sumatra World Heritage site (UNESCO WHC, 2017). Covering 8,630 km² (862,975 ha), the GLNP itself is a UNESCO biosphere reserve and a Heritage Park of the Association of Southeast Asian Nations (ASEAN). It is contained within the confines of the 26,000-km² (2.6 million-ha)

FIGURE 6.9

Proposed Large-Scale Energy Infrastructure Projects in the Leuser Ecosystem and Beyond

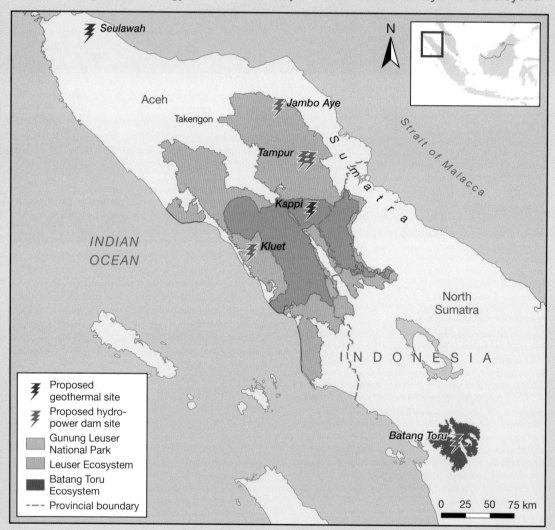

Map and Data Sources: © Rupabumi Digital Indonesia Map, Scale 1:50,000, BAKOSURTANAL, 1978; Ministry of Forestry Decree 190/Kpts-II/2001; About Demarcation of Leuser Ecosystem in Aceh Province; Leuser Ecosystem spatial plan draft; Aceh Spatial Plan; and Secondary Data. Courtesy of SOCP.

Leuser Ecosystem, which experts, including the IUCN, consider one of the world's "most irreplaceable protected areas"; it is ranked 33rd out of more than 173,000 protected areas worldwide (Le Saout et al., 2013). Protected under Indonesian law as a national strategic area for its environmental protection function, the Leuser Ecosystem is one of the largest contiguous intact rainforests in the whole of Southeast Asia, and the last place on Earth where orangutans, rhinos, elephants and tigers coexist in the wild (Rainforest Action Network, 2014).

The proposed project site lies at the very heart of the Leuser Ecosystem, in the Kappi Plateau. This area not only harbors some of the last remaining wild populations of all four of these iconic and critically endangered species, but is also the core of the only remaining major corridor between the eastern and western blocks of the ecosystem. Degrading this region would dramatically reduce the long-term survival prospects for these and a multitude of other species. Indeed, any major development within the Kappi Plateau will only serve to denigrate the Tropical Rainforest Heritage of Sumatra, which has been inscribed on the list of World Heritage in Danger since 2011. Given the extensive road and settlement infrastructure that would inevitably accompany construction, the ecosystem's Outstanding Universal Value would undoubtedly be severely depleted (UNESCO WHC, 2016). Destruction of the Leuser Ecosystem would also have far-reaching consequences for valuable ecological services, such as water supplies, carbon storage and disaster mitigation. A newly published study funded by the European Union determined that the forests of Aceh, more than 50% of which are in the Leuser Ecosystem, are worth approximately US$1 billion per year to Aceh's economy—if fully conserved (Baabud et al., 2016).

The Geothermal Project and Its Environmental Impact

Despite its critical importance in Southeast Asia, the Kappi Plateau is threatened by construction of a major new geothermal power plant by PT Hitay Panas Energy, an Indonesian subsidiary of the Turkish company Hitay Holdings (Hanafiah, 2016). This plan came to light after Indonesia's president publicly called for the country to become energy self-sufficient and to increase the use of geothermal energy to 23% by 2025 (Antara News, 2015; Tempo, 2017). Subsequently, the country's minister of energy and mineral resources announced, "I invite every stakeholder to study and make every effort to achieve these targets" (Antara News, 2015). In response to these policies and statements, numerous "renewable energy" projects are being planned and developed throughout Indonesia. The Kappi geothermal project is among the most pressing for those concerned about the continued conservation of the Leuser Ecosystem (Laurance, 2016c).

As of 2015, Indonesia had an installed production capacity of 1,345 MW, derived from ten geothermal plants (Mansoer and Idral, 2015). The PT Hitay Panas Energy project—one among several new sites under consideration in Aceh—is being proposed inside the Leuser Ecosystem. The governor of Aceh requested rezoning of an area covering 50 km^2 (5,000 ha) in Kappi for the purposes of geothermal development, even though a 25 MW site is only likely to require 10–40 ha for the power plant itself (Modus Aceh, 2016; T. Faisal, personal communication, 2017).

Interestingly, the company has not made details of its plans public, so it is difficult to ascertain the true potential environmental impact of the geothermal plant throughout the phases of exploration and drilling, construction, operation and maintenance, all of which will incur environmental impacts. During construction and drilling, transportation of heavy equipment is required, so an access road to the site will need to be built. Temporary workers will need access and housing. As an example, another geothermal plant of comparable size (20 MW) at Lahendong in North Sulawesi recruited more than 900 workers for the construction phase (Rambu Energy, 2016).

The target area in Kappi is forested and mountainous and has never had any form of prior road access. The nearest road is more than 10 km away at its nearest point and, due to the mountainous terrain, access to it would require a new road more than 10 km long. While such a new road could theoretically be removed after the construction phase, removal would not prevent severe damage from occurring to the forests, as roads open access for illegal logging, mining, encroachment and poaching of wildlife. Currently, the closest substation for transmitting electricity is more than 150 km away, in Takengon, and overhead transmission towers (150 kV) would therefore need to be built every 300 m from the plant to the substation, necessitating substantial clearing along the entire length of the route (T. Faisal, personal communication, 2017).

Land clearing, road construction, vehicle traffic and power plant construction can affect ecosystem services through increased erosion and runoff, increased risk of fire, toxic spills and disturbance of water, and interference with seed dispersal. These activities also pose a high risk to wildlife and species diversity. In addition, noise pollution threatens to disrupt breeding, migration and foraging behavior in this previously undisturbed area (Tribal Energy and Environmental Information, n.d.).

On September 15, 2016, the managing director of PT Hitay Panas Energy submitted a report requesting that the "core area" of the GLNP be rezoned as a "utilization area." Kappi is within a core area of the park by virtue of the fact that it meets stringent government criteria and regulations on biodiversity and habitat composition. As part of the core area, it cannot legally be exploited for geothermal development. In contrast, permits may be approved for geothermal energy developments in utilization areas, so long as the land does not harbor a concentration of priority biota communities (HAkA et al., 2016).

Indonesia's Ministry of Environment and Forestry, through its Directorate General for Conservation of Natural Resources and Ecosystems, publicly stated that the request to rezone the area, and thereby enable the geothermal plant to go ahead, would be rejected (Satriastanti, 2016). At the end of September 2016, the ministry informed the head of the GLNP that no rezoning of any part of the park's core area would be possible, regardless of recent Indonesian legislation, Law No. 21 of 2014 on Geothermal Energy, which allows for

Photo: Indonesia is pushing to become more energy independent and move away from traditional fossil fuels for electricity generation. New regulations opening up the possibility of geothermal energy projects in conservation areas highlight the pressure for new energy projects in areas that render them unsustainable and extremely damaging to the environment and conservation. Geothermal plant, Indonesia. © BAY ISMOYO/AFP/Getty Images

geothermal operations within the utilization area of conservation zones (Republik Indonesia, 2014; Satriastanti, 2016).

It later came to light that Hitay had previously commissioned an Indonesian university—Universitas Gadjah Mada (UGM)—to assess the feasibility of geothermal development on the site. Contrary to expectations, given the above-mentioned background, the assessment team made a "strong recommendation for changing the zoning in the Kappi area" in a report that was submitted to the Ministry of Environment and Forestry on December 1, 2016. One week later, at a meeting held at the GLNP headquarters in Medan, North Sumatra, the findings of the UGM study were shared with a group of NGOs and community members (PT Hitay and UGM, 2016). Subsequently, in a detailed review of the UGM surveys, a consortium of environmental NGOs identified poor survey design and other reasons why the UGM report was wholly inadequate, both for determining whether the requested rezoning was permissible within the Kappi Plateau and for arriving at the stated conclusions and recommendations. The review emphasized that the core area status should be maintained in view of comprehensive GLNP and other NGO data, which had been ignored or misrepresented by the UGM team, and based on current criteria and laws governing the zoning of conservation areas (Laurance, 2016a).

Yet, even though data from the GLNP and local NGO affiliates strongly support the rejection of the rezoning request, the matter is not yet fully settled (Satriastanti, 2016). Ongoing meetings and correspondence indicate that neither Hitay nor the GLNP considers the proposed project to be off the table, meaning that conservation NGOs and other civil society groups remain vigilant to ensure the development does not go ahead.[7]

A Chance for Change?

The Indonesian government's effort to move away from non-renewable energy sources, as part of its sustainable development strategy, is to be lauded. Such a pathway, however, should not include the destruction of one of Southeast Asia's most valuable conservation areas. The geothermal potential of the Seulawah and Takengon regions of Aceh have been thoroughly assessed and are already known. Both locations are also far closer to existing transmission networks and major population concentrations. As such, they could provide sustainable energy alternatives, meeting all of the president's goals, but without the destructive impact of development in the irreplaceable forests of the Leuser Ecosystem.

In addition to the proposed geothermal plant in the Kappi Plateau area, the Aceh government is also seeking approval and funding for several other large-scale infrastructure projects, including plans for mega-hydropower developments in the Jambo Aye, Kluet and Tampur water catchments (Gartland, 2017; see Figure 6.9).

Beyond the borders of Aceh province are additional sites of serious concern, in particular a major new hydropower project in the very fragile habitat of the recently identified Tapanuli orangutan (*Pongo tapanuliensis*)—the Batang Toru forests in North Sumatra province. The proposed project is especially worrying as this population of orangutans is genetically unique and among very few in Sumatra living outside of the Leuser Ecosystem. In fact, the new species immediately became the most endangered great ape species in the world, with fewer than 800 individuals remaining. The planned project would devastate a river catchment in which the highest densities of the Tapunuli orangutan are found. It would also sever an essential corridor linking two of the three main forest blocks that still harbor the new species, which could easily place the species on an irreversible path to extinction (Nater *et al*., 2017; Stokstad, 2017; Wich *et al*., 2016; see the Apes Overview).

With the push for Indonesia to become more energy independent and move away from traditional fossil fuels for electricity generation, and with the passing of new regulations that open up the possibility of geothermal energy projects in conservation areas, it is clear that there is strong pressure for new energy projects in areas that render them unsustainable and extremely damaging to the environment and conservation.

Instead of relying on unsustainable, large-scale energy generation schemes in unspoiled locations, Indonesia could significantly increase its electricity production by investing in smaller-scale 'run-of-river' hydropower schemes and other renewable resources. These would have a negligible environmental impact and provide a more stable and resilient power supply than would a few large, destructive schemes.

Conclusion

Hydropower represents a significant source of electricity for many countries and features in many economic development plans and projections. As this chapter shows, however, its negative impacts are concentrated in areas—river valleys and forested mountains—that have considerable environmental and social value, such as helping to buffer the effects of climate change, hosting river fisheries, encompassing habitat for apes and providing vital resources for local communities. Furthermore, as research has demonstrated, the oft-touted economic benefits of dams rarely materialize for the vulnerable sectors of society (see Annex VII).

Hydropower is expanding rapidly in remaining ape habitat, including in Southeast Asia and Central and West Africa. The preliminary assessment presented in this chapter suggests that the impacts of hydropower on apes and ape habitat will increase considerably in the coming decades. In this context, stakeholder engagement can serve to raise awareness, especially among indigenous and other local communities that are likely to be adversely affected by the construction of dams or geothermal plants. Such engagement can also help to identify opportunities for avoiding or mitigating negative impacts.

Some progress has been made in the development of tools that can serve to improve the environmental and social performance of individual hydropower dams. Nevertheless, many hydropower impacts are not effectively addressed at the scale of the system. This is particularly true for hydropower's impacts on apes, whose conservation requires large blocks of connected habitat. A system-scale approach to hydropower planning and management—including siting, licensing, mitigation and best practice during construction and operation—provides the best opportunity for hydropower expansion to be consistent with the conservation of environmental and social values, including the protection of apes and their habitat. To be successful, the application of such an approach requires collaboration among a range of actors in the hydropower development process, including governments, funders, developers and civil society.

Acknowledgments

Principal author: Helga Rainer[8]

Contributors: American Rivers, The Borneo Project, Emily Chapin, Emma Collier-Baker, Jessie Thomas-Blate, David Dellatore, Earth Island Institute, Joerg Hartmann, Erik Martin, The Nature Conservancy (TNC), Samuel Nnah Ndobe, Jeff Opperman, Ian Singleton, the Sumatran Orangutan Conservation Programme (SOCP) and Jettie Word

Hydropower and Apes: Emily Chapin, Erik Martin and Jeff Opperman

Case Study 6.1: Samuel Nnah Ndobe

Case Study 6.2: Jettie Word

Case Study 6.3: David Dellatore, Ian Singleton and Emma Collier-Baker

Box 6.1: Jeff Opperman, Joerg Hartmann, Emily Chapin and Erik Martin

Annex VII: Jessie Thomas-Blate

Reviewers: Josh Klemm and Kate Newman

Endnotes

1. See, for example, Richter *et al.* (2010) and WCD (2000).
2. The International Commission on Large Dams defines a "large dam" as one that has "a height of 15 metres or greater from lowest foundation to crest or […] between 5 metres and 15 metres impounding more than 3 million cubic metres" (ICOLD, n.d.).
3. Both the Campo Ma'an National Park and the Mbam and Djerem National Park were created to "offset" the adverse effects of the Chad–Cameroon oil pipeline. There is currently no evidence that these offsets were created with the aim of achieving "no net loss" as defined by the Business and Biodiversity Offsets Programme (BBOP, 2012).
4. In July, 2016 an indigenous land rights activist was killed in the city of Miri, purportedly in connection with his activism. In October of the same year, a clash between NCR landowners and people allegedly hired to intimidate them resulted in one death (Sarawak Report, 2016).
5. The ESIA was open for comments, as stipulated in Sarawak's procedures; however, it was not openly published or available to the public. Rather, a limited number of copies were available in a few government offices, where the public could read them. Comments had to be made within 30 days of publication. The ESIA was approved on March 13, 2015 (P. Kallang, personal communication, 2016).
6. Author interviews with residents, Tegulang, Sarawak, Malaysia, October 2016.
7. Confidential information and correspondence provided to the authors.
8. Arcus Foundation (www.arcusfoundation.org/what-we-support/great-apes).

SECTION 2

INTRODUCTION
Section 2: The Status and Welfare of Great Apes and Gibbons

Two chapters comprise this section of *State of the Apes*. **Chapter 7** focuses on *in situ* ape conservation in Africa and Asia. It presents the findings of a study of changes in ape habitats between 2000 and 2014, based on in-depth analysis of thousands of satellite images. By extrapolating current rates of deforestation, the chapter also projects future habitat loss, thus quantifying likely threats to the long-term survival of apes. **Chapter 8** reviews the status and welfare of apes in captivity around the world. It also considers the history and context of sanctuaries in ape range states, as well as the opportunities and challenges facing them and their role in broader conservation efforts.

The online Abundance Annex—available at www.stateoftheapes.com—presents updated population estimates for apes across their ranges. In combination with figures provided in the previous volumes in this series, the annex allows for the tracking of population trends and patterns over time

Chapter Highlights
Chapter 7: Mapping Change in Ape Habitats

This chapter examines the status of forested habitats used by apes by quantifying the rates of tropical forest destruction utilizing the Global Forest Watch platform. This is

Photo: © Jon Stryker and Ronda Stryker

Chapter 8: Sanctuaries and the Status of Captive Apes

Photo: © Jurek Wajdowicz / Arcus Foundation

Captive apes are held in a variety of contexts within ape range states. These include private homes, public displays for tourists, zoos and safari parks, and specialized non-commercial care facilities that are often referred to as rescue centers, rehabilitation centers or sanctuaries. This chapter presents the findings of a study of 56 sanctuaries in ape range states. It discusses their history and context, as well as opportunities and challenges linked to ongoing and emerging threats. Conditions varied across the facilities under review; only a small minority achieved independent accreditation based on their welfare and care standards.

A variety of drivers push apes into captivity. These include forest loss and degradation due to agricultural expansion, mining, logging and infrastructure development as well as the hunting and capture of apes for private collections and entertainment. The number of apes in need of captive care is growing and already far exceeds current capacity. At the same time, ape habitats are shrinking, meaning that options for reintroducing or translocating rescued apes are dwindling and that even rehabilitated apes are likely to spend their lives in captivity. Compounding the situation is the absence of legal consequences for many perpetrators of wildlife crimes, which places a double onus on sanctuaries if they are to contribute to conservation objectives. The first is to tie the intake of apes to appropriate legal consequences; the second is to raise the public's awareness of apes' protected status and the legal consequences of hunting or buying apes. In this context, increased collaboration between sanctuaries and governments, conservation NGOs, industry and other stakeholders needs to be strengthened.

the first in-depth analysis of forest loss to use spatially explicit, high-resolution forest change data across the entire ape range. Based on thousands of satellite images, the assessment quantifies annual loss of ape range forest in 2000–14 and projects future habitat loss rates for each ape subspecies. The results can serve as a measure of their long-term survival.

Protected areas are vital to the conservation of biodiversity, including apes, as they comprise 26% of African ape ranges and 21% of Asian ape ranges. Their "protected" status has not spared these areas from experiencing forest loss, however, even if the rate of loss was lower than it was beyond their borders. In total, a staggering 453,000 km² (45.3 million ha) of ape range was lost between 2000 and 2014. The findings also reveal that gibbon habitats were impacted to a far greater degree than those of great apes. Indonesia has been particularly affected, accounting for 63% of total habitat loss in Asia and 50% of the total loss of ape habitat globally. The extent of forest loss across all ape ranges indicates that ape conservation faces grave challenges regionally and globally. If forest loss continues at the same rate into the future, the consequences for both African and Asian apes will be significant, and particularly devastating for Asian apes.

Introduction Section 2

Photo: Until recently, quantifying rates of tropical forest destruction was challenging and laborious. © Jabruson 2017 (www.jabruson.photoshelter.com)

CHAPTER 7

Mapping Change in Ape Habitats: Forest Status, Loss, Protection and Future Risk

Introduction

This chapter examines the status of forested habitats used by apes, charismatic species that are almost exclusively forest-dependent. With one exception, the eastern hoolock, all ape species and their subspecies are classified as endangered or critically endangered by the International Union for Conservation of Nature (IUCN) (IUCN, 2016c). Since apes require access to forested or wooded landscapes, habitat loss represents a major cause of population decline, as does hunting in these settings (Geissmann, 2007; Hickey *et al.*, 2013; Plumptre *et al.*, 2016b; Stokes *et al.*, 2010; Wich *et al.*, 2008).

Until recently, quantifying rates of tropical forest destruction was challenging and laborious, requiring advanced technical skills and the analysis of hundreds of satellite images at a time (Gaveau, Wandono and Setiabudi, 2007; LaPorte et al., 2007). A new platform, Global Forest Watch (GFW), has revolutionized the use of satellite imagery, enabling the first in-depth analysis of changes in forest availability in the ranges of 22 great ape and gibbon species, totaling 38 subspecies (GFW, 2014; Hansen et al., 2013; IUCN, 2016c; Max Planck Institute, n.d.-b). Launched in 2014, GFW provides free access to spatially explicit, high-resolution forest change data derived from thousands of satellite images that are updated annually. The global forest change data set on GFW allows users to quantify annual change in forest cover within the geographic ranges of each ape subspecies and within protected and unprotected areas in those ranges (Hansen et al., 2013; see Figure 7.1).

This chapter presents the first assessment of the distribution of forest habitat in IUCN-defined ape ranges across Africa and Southeast Asia. It also quantifies yearly loss of ape-range forest from 2000 to 2014 in a spatially explicit manner. Abundance data are not available for all ape subspecies for this period. In future assessments, combining population and habitat data streams will be essential because hunting threatens ape population viability across taxa. Even so, the integrity of ape habitat can serve as a useful threshold for estimating ape occupancy until demographic information becomes available.

The chapter presents these data in combination with current protected area (PA) coverage to assess the adequacy of protection for each subspecies. Various lar gibbons (*Hylobates lar*) and western black-crested gibbons (*Nomascus concolor*), as well as Grauer's gorillas (*Gorilla beringei graueri*), are already confined mainly to PAs (IUCN, 2016c; Maldonado et al., 2012). Protected areas are increasingly important refuges for all ape subspecies (Geissmann, 2007; Tranquilli et al., 2012; Wich et al., 2008).

In addition, the chapter projects future habitat loss rates for each subspecies and uses these results as one measure of threat to their long-term survival. GFW's new online forest monitoring and alert system, entitled Global Land Analysis and Discovery (GLAD) alerts, combines cutting-edge algorithms, satellite technology and cloud computing to identify tree cover change in near-real time, thereby allowing those involved in ape conservation at the local level to monitor changes and generate critical information to enhance their conservation efforts.

The key findings show that gibbons are in crisis:

- Gibbons receive less attention in the public eye than African apes and orangutans, yet gibbon habitats have been degraded to a far greater degree. By 2000, ten taxa of gibbons had already lost more than 50% of their forest habitat, and five gibbon taxa native to the Asian mainland had each had their habitats reduced to less than 5,000 km² (500,000 ha).

- In Indonesia, three others—the agile gibbon, Malaysian lar gibbon and siamang—lost more than 30% of their forest cover between 2000 and 2014.

- During the period under review, the ranges of Asian apes lost up to 25% of their protected forests (median 5%), at a rate that must slow if apes are to persist over the next few decades. Eight gibbon subspecies lost more than 8% of their protected habitat. Two of them—the Malaysian lar gibbon and Abbott's gray gibbon—lost more than 13%.

- Plantations account for more than 75% of the loss of forest habitat of three gibbon subspecies—the agile gibbon (76%), Malaysian lar gibbon (87%) and the moloch gibbon (77%)—as well as more than 50% of habitat loss of nine other Asian gibbon and orangutan subspecies.

- Based on the trends of the period 2000–14, nine ape subspecies, all gibbons, are expected to lose all their habitat by 2050 unless decisive action is taken to stop or at least slow forest loss. Most of these species have enough area in legally defined conservation units to persist if reserves are managed effectively.

- Better protection of existing reserves within the ranges of 18 of the 25 gibbon subspecies should be able to support more than 1,000 family groups.

Ape conservation faces grave challenges:

- Between 2000 and 2014, Indonesia lost 226,000 km² (22.6 million ha) of forest cover, which constituted 63% of total habitat loss in Asia and 50% of the total loss of ape habitat globally. Large-scale agricultural plantations account for the majority of forest loss within ape ranges in Malaysia (84%) and Indonesia (82%), as well as nearly 30% of loss in Cambodia.

- Altogether, ape habitat around the world shrank by more than 10%—from nearly 4.4 million km² to under 4 million km² (440 million ha to under 400 million ha).

- Ape forest habitat in Asia shrank by 21% (357,500 km² or 35.8 million ha) between 2000 and 2014. African habitat fared relatively well, losing less than 4% (95,400 km² or 9.5 million ha) of forest cover in that period, despite increasing human population density, insurgencies and activities such as illegal logging.

- Africa was home to two-thirds of the remaining global ape habitat in 2014, but major transportation infrastructure has already begun to speed deforestation

FIGURE 7.1

Forest Cover and Loss in Ape Ranges and Protected Areas in Asia and Africa, 2000 and 2014[1]

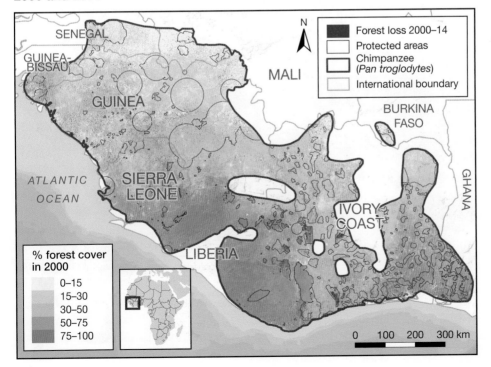

a. West Africa

b. Central Africa

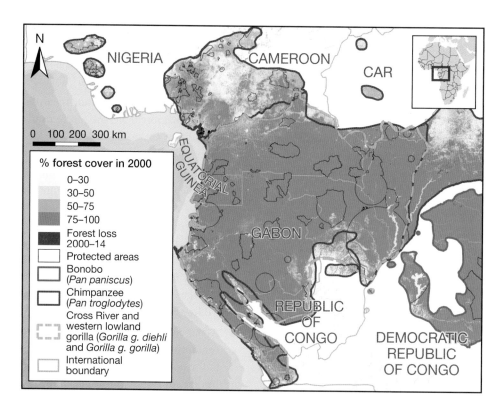

c. East Africa

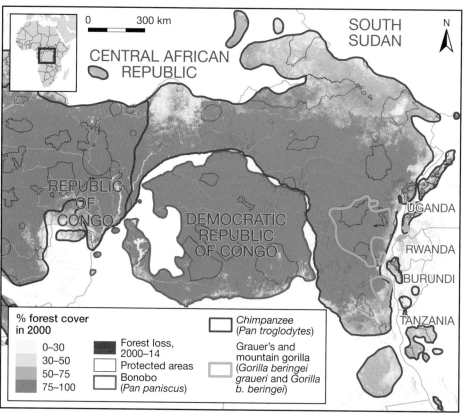

d. Northern Asia

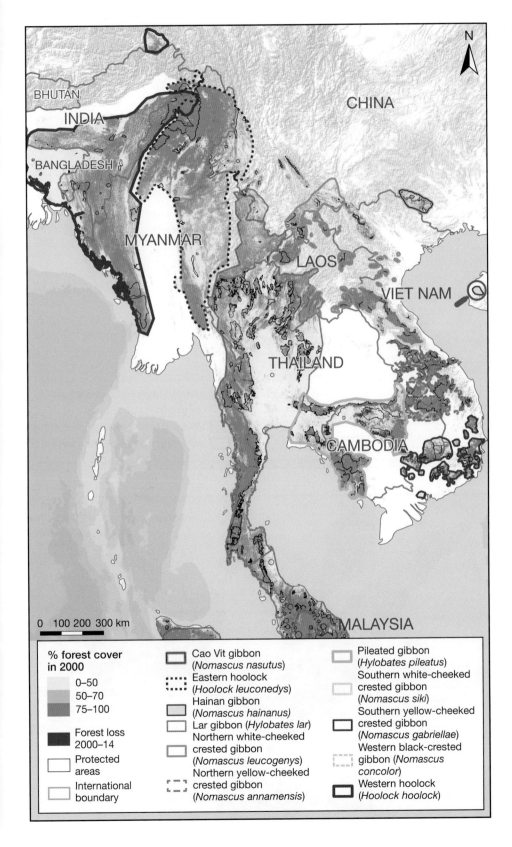

e. Southern Asia

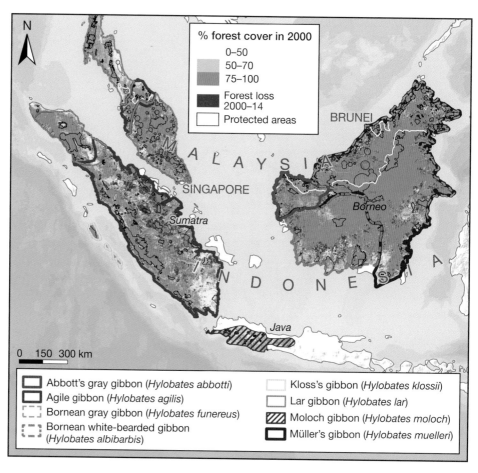

f. Southern Asia

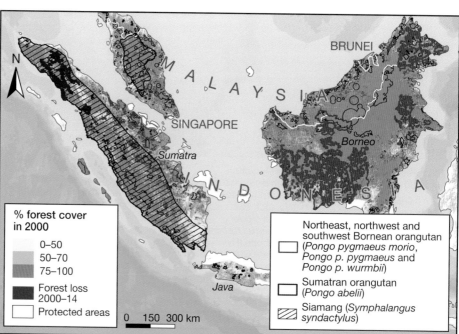

Data sources for Figure 7.1
a–f: GLAD (n.d.); Hansen et al. (2013); IUCN and UNEP-WCMC (2016)

and associated development (see Section 1).

- By 2014, individual African ape subspecies retained an average of 388,000 km² of forest habitat; Asian apes retained an average of just 41,000 km².

A Summary of the State of the Apes through the Lenses of Forest Cover and Protection, 2000–14

More so than other ape species, gibbons are in peril. Prior to 2000—the year used as a baseline for forest extent in this assessment —three gibbon taxa had each lost more than 60% of their historic habitat. The Cao Vit gibbon (*Nomascus nasutus*) retained just 26% of its forest habitat in China and Viet Nam; the Yunnan lar gibbon (*Hylobates lar yunnanensis*) had 27% in China; and the pileated gibbon (*Hylobates pileatus*) had 40% in Cambodia, Lao People's Democratic Republic and Thailand (Hansen *et al.*, 2013; IUCN, 2016c; see Table 7.1). Equally worrying are the situations of subspecies with highly restricted geographic ranges and limited forest cover, including the Hainan gibbon (*Nomascus hainanus*), with just 91 km² (9,100 ha) in 2000, and the Central Yunnan black-crested gibbon (*Nomascus concolor jingdongensis*), with just 672 km² (67,200 ha; see Figure 7.2).

Worldwide, ape ranges in 2000 contained 4.4 million km² (440 million ha) of forest habitat, about two-thirds of which was in Africa and the remaining one-third of which was in Southeast Asia (see Figure 7.1 and Box 7.1). In 2000, the median area of forest habitat within IUCN ranges of Asian apes (48,608 km² or 4.9 million ha) was one-tenth the area of forest habitat found in ranges of African apes (400,983 km² or 40 million ha; see Table 7.1). In 2000, eight countries each contained more than 200,000 km² (20 million ha) of potential ape habitat (see Figure 7.4). The Democratic Republic of Congo (DRC) and Indonesia, in particular, retained large expanses of tropical rainforest that supported multiple ape taxa. Most ape ranges in Sumatra and Borneo still contained high proportions of forest through 2000, despite high deforestation rates in the two previous decades (Gaveau *et al.*, 2016).

Forest Dynamics and Loss from 2000 to 2014

Forest Dynamics in the Geographic Ranges of Subspecies

In 2000, ranges of the 38 ape taxa contained a median of 78% forest habitat, based on a range of 26%–99% (see Table 7.1). Between 2000 and 2014, these ranges lost 1% to 44% of their forest habitat, with a median of 4.8%.

> **BOX 7.1**
>
> **Synopsis of Methods**
>
> The Global Forest Change 2000–14 data set, which is freely available on the Global Forest Watch (GFW) site, served as the basis for the habitat analysis (GLAD, n.d.; Hansen *et al.*, 2013; see Annex VIII). Tree canopy cover in the year 2000 served as a baseline forest cover; annual change in forest cover was calculated using tree cover data from Hansen *et al.* (2013), which is updated annually.
>
> Potential habitat (hereafter, *habitat*) for apes can be categorized by each subspecies' capacity to persist over time under varying degrees of canopy openness (see Table 7.1 and Annex IX). For example, eastern and western chimpanzees (*Pan troglodytes schweinfurthii* and *Pan t. verus*) have evolved in forests that are drier than those of their central African conspecifics and are believed to tolerate a more open canopy (L. Pintea and K. Abernethy, personal communication, 2016). To estimate forest change for each subspecies, this analysis applied values of "canopy density" that reflect the subspecies' tolerance of canopy openness and the overall vegetation cover in their respective ranges (IUCN, 2016c; see Annex IX). The GFW platform allows users to select canopy density values and thus recalculate the habitat assessment presented here with different estimates of canopy density. For more details on methods, see annexes VIII, IX and X.

FIGURE 7.2

Forest and Protected Areas in the Ranges of (a) Asian and (b) African Apes, by Subspecies, 2000 and 2014

Key: ■ Forest cover in 2000 ■ Forest cover in 2014 ■ Forest cover in PAs in 2000

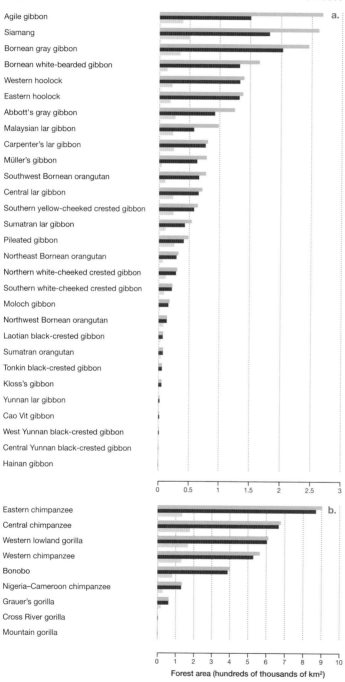

Notes: Subspecies are sorted by the amount of forest cover in the year 2000. PA-related data reflect area covered by PAs in 2016.

Data sources: GLAD (n.d.); Hansen *et al*. (2013); IUCN and UNEP-WCMC (2016)

Ranges of Asian apes lost more of their forest—from 2% to 44% (with a median of 8.3%)—than those of African apes, which lost anywhere from 2% to 6% (with a median of 2.1%).

The greatest recent loss of forest occurred in Southeast Asia within the ranges of orangutans and at least 11 gibbon subspecies (see Figure 7.1). The data reveal noteworthy variations. For example, the once-wide range of the agile gibbon (*Hylobates agilis*)—an area of 387,445 km² (38.7 million ha)—had already lost about 30% of its forest by 2000; it lost another 44% of its remaining forest cover in the following 14 years. In contrast, the extremely limited range of the Cross River gorilla (*Gorilla gorilla diehli*)—a mere 3,648 km² (364,800 ha) in Cameroon and Nigeria—was reduced by less than 1% over the same period.

The ranges of 15 Asian taxa overlap with mapped tree plantations, which have accounted for more than 50% of forest habitat loss in 12 of the ranges (see Box AX1 in Annex XI). Plantations correspond to more than 75% of the loss of forest habitat of three gibbon subspecies: the agile gibbon (76%), the Malaysian lar gibbon (*Hylobates lar lar*, 87%), and the moloch gibbon (*Hylobates moloch*, 77%). Plantations also overlap with distributions of all four orangutan subspecies (*Pongo* species (spp.)), representing 42%–59% of forest loss within their ranges.

Taxa of Conservation Concern

This analysis reveals that the forest cover in the ranges of 23 of 38 ape subspecies was reduced by almost 30% prior to 2000 (see Table 7.1). Forest loss before 2000 exceeded 50% in the ranges of ten gibbon subspecies, particularly those in mainland Southeast Asia (Bleisch and Geissman, 2008; Bleisch *et al*., 2008; Gaveau *et al*., 2016; Geissmann and Bleisch, 2008).

A closer examination of the data reveals several important findings for gibbons, Grauer's and Cross River gorillas, and both

TABLE 7.1

Ape Subspecies and Forest Cover Status and Loss, 2000 vs. 2014

Name	Range area (km²)	Forest cover, 2000* (km²)	% forest, 2000	Forest cover, 2014 (km²)	% forest lost, 2000–14	% PA forest, 2000	% PA forest lost, 2000–14
Bonobo (*Pan paniscus*)**	418,809	400,983	95.7	387,931	3.3	20.2	1.9
Central chimpanzee (*Pan troglodytes troglodytes*)**	710,681	676,693	95.2	666,152	1.6	26.2	0.8
Eastern chimpanzee (*Pan t. schweinfurthii*)**	961,246	902,867	93.9	869,160	3.7	14.9	1.2
Nigeria–Cameroon chimpanzee (*Pan t. ellioti*)**	168,393	133,806	79.5	130,257	2.7	21.4	2.6
Western chimpanzee (*Pan t. verus*)**	660,332	564,032	85.4	528,817	6.2	23.1	5.9
Cross River gorilla (*Gorilla gorilla diehli*)**	3,648	3,388	92.9	3,363	0.7	53.5	0.5
Grauer's gorilla (*Gorilla beringei graueri*)**	64,684	61,861	95.6	60,562	2.1	30.4	0.6
Mountain gorilla (*Gorilla b. beringei*)**	783	768	98.0	761	0.8	97.7	0.8
Western lowland gorilla (*Gorilla g. gorilla*)**	695,076	610,453	87.8	602,982	1.2	27.1	0.6
Northeast Bornean orangutan (*Pongo pygmaeus morio*)	32,931	32,149	97.6	29,163	9.3	19.9	7.1
Northwest Bornean orangutan (*Pongo p. pygmaeus*)	14,119	13,965	98.9	13,492	3.4	56.3	0.4
Southwest Bornean orangutan (*Pongo p. wurmbii*)	81,148	77,542	95.6	66,065	14.8	12.8	6.7
Sumatran orangutan (*Pongo abelii*)	7,848	7,783	99.2	7,452	4.3	46.8	2.0
Eastern hoolock (*Hoolock leuconedys*)	281,864	138,283	49.1	132,326	4.3	12.9	1.9
Western hoolock (*Hoolock hoolock*)	320,251	140,061	43.7	133,308	4.8	15.1	1.7
Abbott's gray gibbon (*Hylobates abbotti*)	147,330	124,499	84.5	92,208	25.9	21.2	13.3
Agile gibbon (*Hylobates agilis*)	387,445	267,607	69.1	150,787	43.7	14.4	8.5
Bornean gray gibbon (*Hylobates funereus*)	276,487	245,352	88.7	202,593	17.4	14.0	8.5
Bornean white-bearded gibbon (*Hylobates albibarbis*)	200,590	165,009	82.3	132,744	19.6	8.0	6.5
Carpenter's lar gibbon (*Hylobates lar carpenteri*)	265,446	80,531	30.3	76,918	4.5	29.9	1.1

Name	Range area (km²)	Forest cover, 2000* (km²)	% forest, 2000	Forest cover, 2014 (km²)	% forest lost, 2000–14	% PA forest, 2000	% PA forest lost, 2000–14
Central lar gibbon (*Hylobates l. entelloides*)	154,385	71,498	46.3	65,564	8.3	32.0	1.9
Kloss's gibbon (*Hylobates klossii*)	6,031	5,479	90.8	5,315	3.0	32.2	0.7
Malaysian lar gibbon (*Hylobates l. lar*)	137,898	98,344	71.3	57,445	41.6	22.7	25.0
Moloch gibbon (*Hylobates moloch*)	39,400	18,056	45.8	16,071	11.0	11.6	7.0
Müller's gibbon (*Hylobates muelleri*)	103,652	78,653	75.9	62,853	20.1	5.2	8.4
Pileated gibbon (*Hylobates pileatus*)	122,073	48,608	39.8	40,797	16.1	51.4	9.9
Sumatran lar gibbon (*Hylobates l. vestitus*)	73,254	53,886	73.6	42,519	21.1	19.9	2.6
Yunnan lar gibbon (*Hylobates l. yunnanensis*)	9,512	2,619	27.5	2,490	4.9	9.0	3.1
Cao Vit gibbon (*Nomascus nasutus*)	8,332	2,161	25.9	2,107	2.5	16.2	5.8
Central Yunnan black-crested gibbon (*Nomascus concolor jingdongensis*)	1,270	672	52.9	659	1.9	23.1	0.1
Hainan gibbon (*Nomascus hainanus*)	165	91	55.1	87	4.8	18.2	8.0
Laotian black-crested gibbon (*Nomascus c. lu*)	8,912	7,848	88.1	7,069	9.9	38.8	5.7
Northern white-cheeked crested gibbon (*Nomascus leucogenys*)	51,481	30,249	58.8	28,402	6.1	36.8	3.2
Southern white-cheeked crested gibbon (*Nomascus siki*)	26,634	22,674	85.1	21,817	3.8	39.4	1.6
Southern yellow-cheeked crested gibbon (*Nomascus gabriellae*)	95,205	64,243	67.5	57,912	9.9	37.3	5.0
Tonkin black-crested gibbon (*Nomascus c. concolor*)	13,097	6,149	47.0	6,012	2.2	25.0	0.8
West Yunnan black-crested gibbon (*Nomascus c. furvogaster*)	3,114	1,498	48.1	1,473	1.7	30.6	0.7
Siamang (*Symphalangus syndactylus*)	341,872	261,502	76.5	181,091	30.7	19.3	8.7

Notes: * Forest cover in 2000 is defined using the canopy density associated with each subspecies. ** African apes.

Data sources: GLAD (n.d.); Hansen *et al.* (2013); IUCN and UNEP-WCMC (2016)

FIGURE 7.3

Forest Cover, Protection and Loss between 2000 and 2014 in (a) Asian, (b) African and (c) All Ape Ranges, by Subspecies

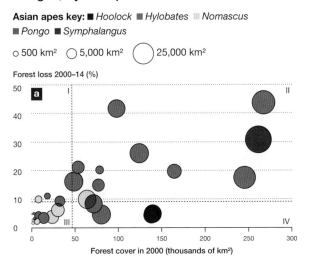

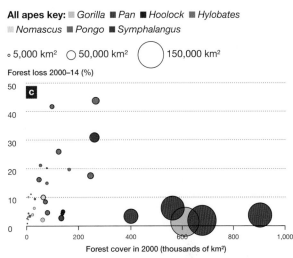

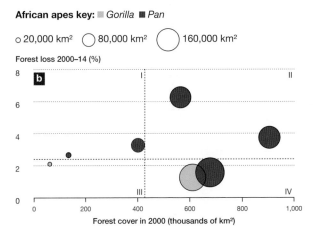

Notes:

The graphs show forest cover in 2000 and forest loss evident in 2014 in the ranges of (a) Asian, (b) African and (c) all ape subspecies.

The horizontal dotted lines in Figures 7.3(a)–(b) reflect the median percentage of forest loss for Asian (8.3%) and African (2.1%) apes.

The vertical dotted lines in Figures 7.3(a)–(b) show the median forest cover in Asian (48,600 km²) and African (401,000 km²) ape ranges in 2000.

The four resulting regions group subspecies according to the relative forest cover security of their ranges, from: (I) insecure (limited forest cover in 2000, high forest cover loss from 2000 to 2014) to (IV) secure (extensive forest cover, low forest cover loss).

Circle sizes in all graphs indicate the area of protected forest in each subspecies' range.

Data sources: GLAD (n.d.); Hansen *et al.* (2013); IUCN and UNEP-WCMC (2016)

orangutan species. Figure 7.3 combines effects of forest loss prior to the year 2000 and ongoing deforestation by dividing the data for each taxon into regions according to habitat remaining in 2000 and the percentage of habitat lost since then. The size of the circles in Figure 7.3 indicates the area of forest in PAs per range. In 2000, PAs covered 17 km²–50,470 km² (5%–56% of each range's forest cover) in Asia and 750 km²–177,300 km² (15%–98%) in Africa (see Table 7.1).

The subspecies in Region I are of greatest concern, as they have experienced the greatest forest loss in ranges with the most limited forest cover.

Habitat for several gibbons—the agile gibbon, the Bornean white-bearded gibbon (*Hylobates albibarbis*), the Bornean gray gibbon (*Hylobates funereus*) and the siamang (*Symphalangus syndactylus*)—was relatively extensive up to 2000 but decreased by 17%–44% from then until 2014 (see Figure 7.3a). These and other subspecies in Region II occur in areas where forest was relatively widespread in 2000 but was reduced substantially over the following 14 years.

Photo: Is there enough habitat for gibbons? The ranges of Asian apes lost up to 25% of their protected forests from 2000 to 2014. © Andrew Walmsley/Borneo Nature Foundation

The habitats of more than half of African and Asian ape taxa fall within Region III; these ranges had diminished forest cover in 2000 and experienced limited subsequent forest loss. On the whole, Asian apes lost roughly four times more of their forest habitat between 2000 and 2014 than did African apes (with a median loss of 8.3% vs. 2.1%, respectively).

The few African subspecies in Region IV have relatively large geographic ranges with more extensive forest cover (see Figure 7.3b). This group consists of the western lowland gorillas (*Gorilla gorilla gorilla*) and central chimpanzees (*Pan troglodytes troglodytes*). Of great conservation concern is the combination of limited forest cover and extensive forest loss within Asian ape ranges.

Forest Dynamics Inside vs. Outside Protected Areas

Protected areas are vital to the persistence of ape populations. Evidence indicates that areas that have undergone large-scale clearing of forest, such as for plantations, will not sustain viable ape populations over time, even though some ape species can make use of industrial plantations as supplemental food sources or corridors in the short term (Ancrenaz, Calaque and Lackman-Ancrenaz, 2004; Wich *et al.*, 2012b). Apes use agricultural habitats primarily in the absence of an alternative, if the natural forest in their range is cleared for agricultural and other uses, yet all need some natural tree canopy to find food and nesting substrate (Ancrenaz *et al.*, 2015a; Hernandez-Aguilar, 2009; Hockings *et al.*, 2015; IUCN, 2016c; W. Brockelman, personal communication, 2016).

Overall, about 26% of African ape habitat in 2000 was within PAs (median 81,152 km²/ 8.1 million ha of subspecies' geographic ranges). A slightly lower median proportion —21%, or 9,917 km² (991,700 ha)—of the habitat of Asian apes was protected that year. From 2000 to 2014, forest loss was detected within all PAs, although at lower rates than outside PAs. In African ape ranges, forest cover in PAs declined by less than 1%, which resulted in a median 79,573 km² (7.9 million ha) of protected habitat in their ranges in 2014 (see Table 7.2). Asian apes lost roughly 5% of protected forest during this period, which left their ranges with a median of 9,255 km² (925,500 ha) of protected habitat.

Median loss outside PAs in African ape ranges was three times higher than inside PAs. While it is encouraging that the mountain gorilla (*Gorilla beringei beringei*) experienced only a 0.3% decline in habitat outside PAs, such unprotected areas comprise less than 3% of this subspecies' entire, highly restricted range (see Table 7.1).

TABLE 7.2

Percentage of Forest Loss in Ranges of Asian and African Ape Subspecies, 2000 vs. 2014

	Asian ranges (n = 29)			African ranges (n = 9)		
	Lowest	Median	Highest	Lowest	Median	Highest
Inside protected areas	0.1	5.0	25.0	0.5	0.8	5.9
Outside protected areas	1.9	9.8	49.6	0.3	2.7	6.3
Overall range	1.7	8.3	43.7	0.7	2.1	6.2

Data sources: GLAD (n.d.); Hansen *et al.* (2013); IUCN and UNEP-WCMC (2016)

Among Asian apes, habitat loss inside PAs ranged from 0.1% (Central Yunnan black-crested gibbon) to 25% (Malaysian lar gibbon), with a median loss of 5%. Eight gibbon subspecies lost more than 8% of their protected habitat; two of them—the Malaysian lar gibbon and Abbott's gray gibbon (*Hylobates abbotti*)—lost more than 25% and 13%, respectively (see Table 7.1). Four gibbon subspecies and the northwest Bornean orangutan (*Pongo pygmaeus pygmaeus*) lost less than 1% of their habitat inside PAs. However, all five of these taxa have small ranges, with forest covering less than 15,000 km² (1.5 million ha) in 2000.

Not surprisingly, habitat loss was greater outside of PAs. Among the Asian ape ranges, median habitat loss outside PAs was nearly 10% and ranged from 1.9% (Cao Vit gibbon) to 50% (agile gibbon). Five subspecies, comprising four *Hylobates* gibbons and the siamang, lost more than 25% of their unprotected habitat. African ape ranges lost 2.7% (with losses reaching from 0.3% to 6.3%) of their unprotected 2000 habitat.

Given the rates of loss outside PAs, species may increasingly rely on the forest remaining inside PAs, where loss rates are lower. Yet, a relatively high proportion (more than 20%) of total annual loss of forest habitat of four mainland Asian gibbons and the Sumatran orangutan occurred in PAs.

Buffer zones, comprising habitats just outside parks, can play a critical role in preventing isolation of protected forests and enhancing their capacity to maintain healthy populations of apes and other wildlife (Hansen and DeFries, 2007; Laurance *et al.*, 2012). Forest loss between 2000 and 2014 within 10-km buffer zones did not differ statistically from loss outside of PAs overall (median = 8.7% vs. 6.1%, respectively), although it was substantially higher than loss within PAs (2.6%). Nevertheless, areas with greater forest loss in buffer zones also faced greater forest loss inside PAs.

Is There Enough Space for Gibbons to Persist in the Wild?

The results of this habitat assessment show that enough protected forested area may exist to support hundreds and even thousands of groups of most gibbon subspecies, if it is managed appropriately for native wildlife (see protection status in Table 7.1).

Gibbon densities range from 0.5–2.0 groups per square kilometer, such that a well-managed 5,000-km² park could technically support viable gibbon populations. This conclusion is based on the area of protected forest as calculated by this analysis and a conservative density estimate of one group per 2 km² (IUCN, 2016c).

In numerous ape-range countries, however, management of and law enforcement in parks has only been able to slow, rather than stop, the encroachment into and the loss of these forests (Curran *et al.*, 2004; Tranquilli *et al.*, 2014). Poor enforcement of laws against forest encroachment and poaching in PAs signals an urgent need for improved management, protection, patrolling and community involvement (Geissmann, 2007).

The ranges of Asian apes lost up to 25% of their protected forests (median 5%) from 2000 to 2014, a rate that must slow if apes are to persist over the next few decades (see Table 7.1). Other factors, such as hunting and disease, will intensify the effects of these projected habitat losses on population densities. In parts of Africa, habitat loss may be less of a concern than hunting (see Box 7.2). There is still enough time to prevent the decline seen in Asia from being replicated in Africa.

Based solely on the extremely limited amount of habitat remaining, it is clear that certain species will need more protected forest area to persist over time. The following gibbons are especially vulnerable:

- Abbott's gray gibbon;
- the Hainan gibbon;
- the pileated gibbon; and
- the southern yellow-cheeked crested gibbon (*Nomascus gabriellae*).

Gibbons and some great ape subspecies (mountain and Grauer's gorillas) persist primarily in protected conservation areas; they continue to face threats from hunting in PAs that are not well patrolled (Geissmann, 2007; IUCN, 2016c; Maldonado *et al.*, 2012). To be able to persist, the following species will, at a minimum, need better management of existing reserves within their ranges:

- both species of orangutan;
- the agile gibbon;
- the Malaysian lar gibbon;
- the West Yunnan black-crested gibbon (*Nomascus concolor furvogaster*);
- the Central Yunnan black-crested gibbon; and
- the mountain gorilla.

To remain viable in the face of reduced connectivity among populations, some species may need to be managed as metapopulations, linked by dispersal, by connecting reserves and buffer areas via forest corridors. However, results of this analysis also show that forest inside 10-km buffer zones around PAs, which would necessarily form the basis of dispersal corridors for apes, is as

BOX 7.2

Hunting May Wipe Out Ape Populations Sooner than Forest Loss

Assessing forest loss alone may greatly underestimate changes in ape population densities. Increased hunting associated with fragmenting and opening up of closed-canopy forest may, in fact, decimate ape populations before the loss of habitat quality does (Hicks *et al.*, 2010; Ripple *et al.*, 2016).

Deforestation facilitates access to previously intact forests, which, in turn, enables poaching for wild meat, participation in the wild animal trade, and disease transmission from humans (Köndgen *et al.*, 2008; Leendertz *et al.*, 2006; Poulsen *et al.*, 2009). Indeed, once people start cutting forest, they hunt game and target large mammals, including apes. While a substantial decrease in forest cover in an ape range — for example, from 90% to 30% — might not wipe out local species on its own, associated hunting may very well do so (Meijaard *et al.*, 2010b; Tranquilli *et al.*, 2014). Western lowland gorillas, for example, face a greater threat from hunting and disease than forest loss (Maisels *et al.*, 2016b; Walsh *et al.*, 2003).

Biologists are creating comprehensive layers of data on ape population densities and areas most affected by wild meat hunting (Max Planck Institute, n.d.-b). Once available, the data will be able to be used to complement information on forest change, thereby greatly improving our understanding of the trajectory of ape populations and assisting the conservation community in identifying and safeguarding the most vulnerable sites.

vulnerable to deforestation as other unprotected land. For some gibbon subspecies—such as the Hainan gibbon, whose habitat was reduced to less than 90 km² (9,000 ha) by 2014—remaining forest cover is insufficient in terms of both size and level of protection to enable metapopulation movements (see Table 7.1). The conservation community thus has only a few years to maintain or re-establish connectivity and to make sure that PAs are large enough and sufficiently protected to maintain viable populations of the subspecies.

Hunting is the other major threat. While the quantification of hunting within PAs is beyond the scope of this chapter, improved PA management will be needed to address this pressing concern (see Box 7.2).

Forest Dynamics by Country

Between 2000 and 2014, apes worldwide lost 453,000 km² (45.3 million ha) of forest, or more than 10% of the 2000 baseline. Of that loss, 79% took place in Asia. Asian ape-range countries lost 357,500 km² (35.8 million ha) of forest cover, or more than 20% of their forest habitat, an area nearly four times as large as that lost in African range states, which shrank by 95,400 km² (9.5 million ha) or 4% of African apes' total forest habitat (see Figure 7.4).

Destruction of ape habitat for agriculture has dramatically altered the forest landscape in some Asian states. From 2000 to 2014, Malaysia lost 33% of its forest, Indonesia lost 30% and Cambodia more than 20%; these rates significantly exceeded those of all other ape-range countries, each of which lost less than 10% of its forest cover. Forest loss in Indonesia (226,063 km² or 22.6 million ha) far surpassed even that of Malaysia (88,763 km² or 8.9 million ha), accounting for 63% of the total habitat loss in Asia and 50% of the total destruction of ape habitat globally.

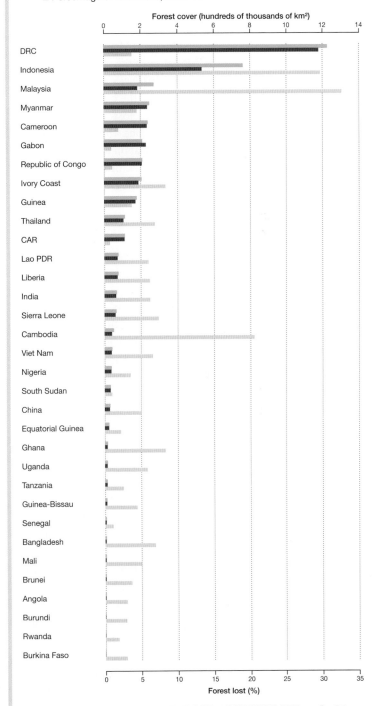

FIGURE 7.4

Forest Cover and Loss in Ape Range Countries, 2000 vs. 2014

Key: ■ Forest cover in 2000 ■ Forest cover in 2014
■ Percentage of forest loss, 2000–14

Data sources: GLAD (n.d.); Hansen *et al*. (2013); IUCN and UNEP-WCMC (2016); see Box 7.1

Large-scale agricultural plantations account for the majority of forest loss within ape ranges in both Malaysia (84%) and Indonesia (82%), as well as nearly 30% of loss in Cambodia. This expanding land use allocation affects at least ten gibbon taxa and all four orangutan taxa.

As noted above, Africa lost just 4% of its ape habitat over the same time frame. Much of that loss was concentrated in West Africa, where the highest percentage of forest base was lost in Ghana, Ivory Coast and Sierra Leone. The Central African Republic (CAR), Gabon and South Sudan each lost less than 1% of their ape habitat during this period. The DRC is home to the most ape habitat of any country—more than 1.2 million km² (120 million ha) or 28% of all ape habitat (see Figure 7.4)—and supports central and eastern chimpanzees (*Pan troglodytes schweinfurthii*), Grauer's gorillas and bonobos (*Pan paniscus*), of which the latter two taxa are endemic to the country. While the DRC lost more total forest cover (more than 46,000 km² or 4.6 million ha) between 2000 and 2014 than other African nations, this area represented less than 4% of its ape forest habitat, and the loss rate was only slightly higher than the median African rate of 2.9%.

Data indicate that the clearing of forest for plantations reduced the habitat of only one African ape subspecies, the western chimpanzee, between 2000 and 2014—by about 1% (GFW, 2014; Transparent World, 2015). The situation in Africa could rapidly change for the worse, however. Nearly 60% of oil palm concessions in Africa overlap with ape distributions, while 40% of unprotected ape habitat is in land suitable for oil palm (Wich *et al.*, 2014). Corporate demand to convert these concessions to palm is expected to increase sharply in Africa as land suitable for oil palm and other industrial-scale agriculture diminishes in Asia (Mongabay, 2016b).

Annual Forest Loss Trends in Ape Habitat

Cumulative Loss of Tree Cover

The availability of forest distribution data at 30-m resolution through the GFW platform allows for the tracking of annual forest loss for all ape taxa as of 2000. Annual data on cumulative forest loss over the study period reveals several worrisome trends (see Figure 7.5).

Ape taxa that lost the most forest habitat between 2000 and 2014 all live in tropical Asia (see Figure 7.5a). The period witnessed steady deforestation in previously extensive habitats of the agile gibbon, Malaysian lar gibbon and siamang, for example.

Figure 7.5b highlights ten subspecies that experienced the lowest cumulative forest habitat loss. Loss rates among the six African subspecies in this group have remained low but have increased, particularly since 2012, whereas those for the four Asian subspecies are tapering off. Absolute forest loss may be low in the habitats of these four subspecies, yet their forest cover was already restricted, ranging from less than 700 km² (70,000 ha) to just under 6,200 km² (620,000 ha) (see Table 7.1). In the limited forest that is left, each square kilometer lost is likely to have an outsized effect on the remaining population.

Data relating to the establishment of plantations were available only as single values for the period 2001–14, not on an annual basis. As a result, the cumulative annual loss values in Figure 7.5 exclude plantation data and so are only illustrative in their depiction of forest loss trends. Fifteen of the 38 ape subspecies, including the ten in Figure 7.5a, have faced substantially more extensive cumulative loss than is shown in Figure 7.5a, although the trends are indicative of the extent of their habitat loss (see Table 7.1). For example, the agile gibbon, Malaysian lar gibbon, Abbott's gray gibbon and siamang experienced the highest overall loss of habitat regardless of the inclusion of plantation data, and each showed even greater loss when plantations were fully included in the calculation (see Table 7.1 and Figure 7.5a). The amount of remaining habitat listed in Table 7.1 reflects the true 2014 habitat endpoint for subspecies whose ranges overlap with plantations.

FIGURE 7.5

Ape Ranges that Experienced the (a) Highest and the (b) Lowest Cumulative Annual Forest Loss, 2001–14

Key: ■ Sumatran lar gibbon ■ Bornean gray gibbon ■ Southwest Bornean orangutan ■ Pileated gibbon ■ Müller's gibbon ■ Bornean white-bearded gibbon ■ Siamang ■ Abbott's gray gibbon ■ Malaysian lar gibbon ■ Agile gibbon

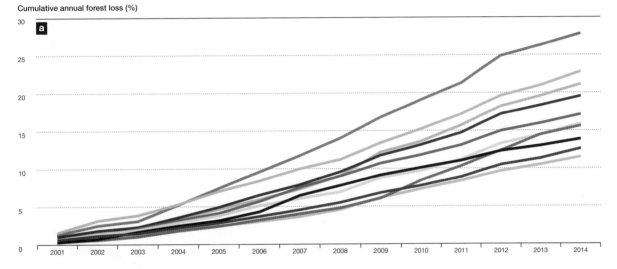

Key: ■ Cross River gorilla ■ Mountain gorilla ■ Western lowland gorilla ■ Central chimpanzee ■ West Yunnan black-crested gibbon ■ Central Yunnan black-crested gibbon ■ Grauer's gorilla ■ Tonkin black-crested gibbon ■ Cao Vit gibbon ■ Nigeria–Cameroon chimpanzee

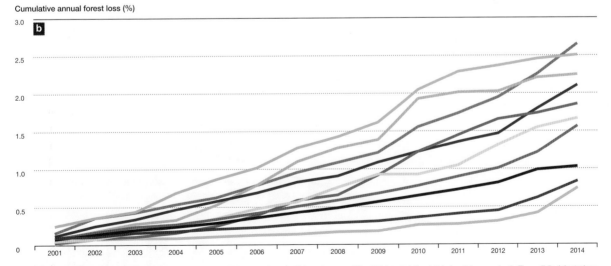

Notes: Plantation data were not available on an annual basis. Their inclusion would have increased the 2014 cumulative totals for all ten species in Figure 7.5a (plantations did not affect the subspecies in Figure 7.5b). For total cumulative loss values for all ape subspecies, see Table 7.1.

Data source: GLAD (n.d.); Hansen *et al*. (2013)

Projecting Forward

From 2000 to 2014, the annual rate of loss was relatively constant for most species, providing a rationale for projecting this same rate forward. Before future forest loss could be estimated, a regression line was fitted to the cumulative deforestation data; Figure 7.6 shows two examples. The resulting equations were then used to predict the amount of deforestation based on past trends, as discussed below.

The tight fit of the regression function to the data allowed future losses to be projected with a high degree of confidence (see Figure 7.7). The increasing loss rate for habitat of eastern chimpanzees stands in contrast to the decreasing loss rate of Hainan gibbon habitat (see Figure 7.6). The latter was severely diminished both before and during the study period, due to massive deforestation activities throughout Southeast Asia (Achard *et al.*, 2014). Hainan gibbons currently persist in a single island protected area.

The forest loss rates derived for each subspecies served as the basis for predicting remaining forest habitat in the medium term (2030) and longer term (2050), as shown in Figure 7.7. To avoid speculation about

FIGURE 7.6

Regression Lines Fitted to Cumulative Forest Loss for (a) the Eastern Chimpanzee and (b) the Hainan Gibbon, 2000–14

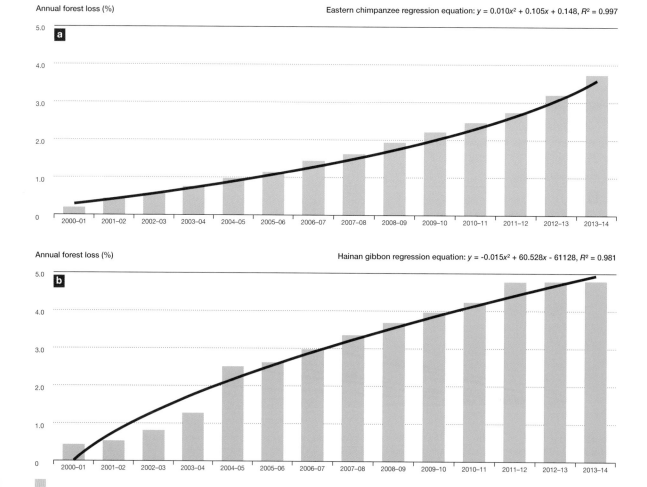

Eastern chimpanzee regression equation: $y = 0.010x^2 + 0.105x + 0.148$, $R^2 = 0.997$

Hainan gibbon regression equation: $y = -0.015x^2 + 60.528x - 61128$, $R^2 = 0.981$

changes in rates of forest loss, this assessment relies exclusively on forest loss data to make the projections.

If forest loss continues at the same rate into the future as it has since 2000, consequences for apes, particularly Asian taxa, will be severe. Five subspecies are predicted to lose half of the habitat present in 2000 by 2030 (see Figure 7.7). Nine subspecies, all gibbons, are projected to lose all their habitat by 2050, assuming the rate of habitat loss remains constant (see Figure 7.7).

In most cases, forest loss rates are projected to increase. In some cases, however, the rate of habitat loss slowed over time, potentially to the point of becoming negative, indicating possible regeneration. For the Hainan gibbon and Kloss's gibbon (*Hylobates klossii*), the calculations project a reduced amount of loss in 2050 compared to 2030, based on quadratic equations that best fit the loss data for 2000–14. When extrapolated, the tapering loss rate for the Hainan gibbon shown in Figure 7.6b predicts a negative loss rate for the coming decades—and possibly forest regeneration.

These forest loss projections are simplistic, and land use changes are dynamic within ape-range countries. Slower rates of forest loss within PAs, as shown in Table 7.3, suggest that as a higher percentage of a given taxon's range is under protection—either because more area is protected or less unprotected forest remains—the rate of loss will be slower in the future. As discussed throughout this volume, however, massive transportation infrastructure investments in Southeast Asia and central Africa are expected to speed deforestation and associated agriculture and development, at least along new roads and railways (Dulac, 2013; Quintero *et al.*, 2010). The discovery of minerals underneath reserves has led to the downgrading or even degazetting of PAs to facilitate extraction (Forrest *et al.*, 2015; see Chapter 4, pp. 116–119). Exploration

FIGURE 7.7

Projected Loss of Forest Habitat, by Subspecies, 2000 vs. 2030 and 2050

Key: ■ 2030 ■ 2050

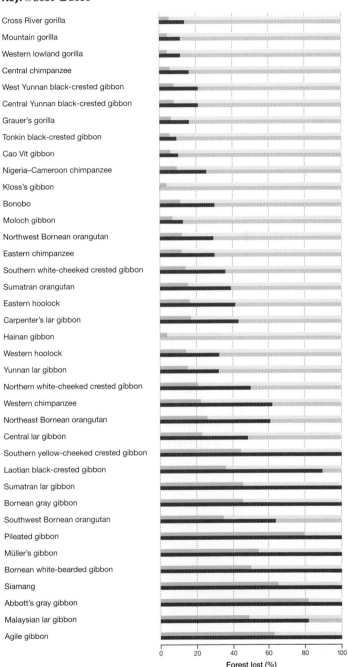

Notes: Projections reflect the percentage of total forest habitat in 2000 that is predicted to be lost by (a) 2030 and (b) 2050, using best-fit regression equations based on annual percentage loss from 2000 to 2014. Ape subspecies are ordered by their cumulative loss during 2000–14. Nine subspecies, all gibbons, are projected to lose all their habitat by 2050, assuming a constant rate of habitat loss.

and extraction could affect forest loss rates even in current reserves.

Regardless of the extent of forest cover, adverse impacts of human activities in ape habitats—such as hunting, forest degradation and disease transmission—are major conservation issues for apes. Even so, the availability of sufficient forest with adequate connectivity is a benchmark that must be planned against if these species are to persist into the future (Plumptre *et al.*, 2016b; Tranquilli *et al.*, 2012).

A critical finding of this is that gibbon subspecies with small geographic ranges face a particularly uncertain future. These taxa are little studied and poorly represented in conservation organization action plans; moreover, their plight is less recognized by the public and the media than that of chimpanzees or gorillas. Conserving remaining forest within gibbon ranges is possible, but only if the conservation community replaces this apparent complacency about the future of gibbons and dedicates the same attention and resources to gibbons that it does to the great apes.

Regular Monitoring of Forest Change

Forest loss in remote areas, including within and between PAs, often goes undetected until large areas have been cleared, as forest monitoring is typically limited to patrolling on the ground by park staff (Dudley, Stolton and Elliott, 2013). This chapter aims to help range-state institutions and conservation managers to:

(a) remain informed of habitat change in their areas of interest through frequent forest monitoring; and

(b) plan for enhanced ape protection by enabling them not only to identify areas of key forest habitat, but also to detect and respond to forest loss quickly.

Regular monitoring of remaining forest cover will be a critical conservation tool as surviving ape populations take refuge in increasingly isolated regions (IUCN, 2016c; Junker *et al.*, 2012). Early detection of the presence and location of forest loss can guide further investigation of a target area through higher-resolution aerial images or

by rangers on the ground (see annexes XII and XIII).

Repeating analyses in particular areas would allow managers to monitor key performance indicators of ape habitat over time. Updated forest cover data provide a tool for primatologists and conservationists to integrate current habitat status information into their analyses of population status and local threats. If PAs are losing forest, it is likely that they are losing apes directly to hunting as well (Walsh *et al.*, 2003; Wich *et al.*, 2012a). Regular monitoring of habitat change can lead to more rigorous assessments once population and wild meat hunting data become spatially explicit across all ape species and habitats.

Photo: Large-scale agricultural plantations account for 52%–87% of detected forest loss within the ranges of at least 12 ape subspecies in Malaysia and Indonesia.
© HUTAN–Kinabatangan Orang-utan Conservation Project

GFW now offers a new system of weekly tree cover loss alerts at 30-m resolution; for ape conservationists, this may be the most important tool released to date. GFW's online forest monitoring and alert system combines cutting-edge algorithms, satellite technology and cloud computing to identify where trees are growing and disappearing in near-real time. Having been piloted in a few countries in 2015, these GLAD alerts covered virtually all ape range countries by early 2017 and are to cover the entire tropics by the end of 2017 (M. Hansen, personal communication, 2017).

A new collaboration between GFW and RESOLVE will make GLAD alerts in critical ape regions easily accessible to the general public, along with a weekly feature called "places to watch," which highlights changes in tree cover that are of greatest concern to ape conservation. Alternatively, subscribers can receive these near-real-time alerts of detections of forest loss for whatever areas they select, be it a country, a forest reserve, a conservation landscape, a road buffer, or a hand-drawn polygon on the platform's interactive map.

Future habitat assessments could evaluate patterns of GLAD alerts as possible indicators of the intensity of imminent forest loss. In areas for which GLAD alerts have been set up, analyses could also track factors associated with forest loss, including slopes, distances to clearings, roads and towns (see annexes XI and XII).

Incorporating near-real-time GLAD alerts to improve the enforcement of existing PAs would go a long way towards conserving many ape populations, in particular the small gibbon populations and their remaining forest patches in both mainland and insular Southeast Asia. For these and other apes, the approach would allow managers to identify critical forest corridors and buffer zones that warrant conservation action and to enhance monitoring of forests within recognized corridors and buffer zones.

> Protected areas are becoming a last stronghold for remaining populations of a growing number of ape taxa, both in Asia, and, increasingly, in Africa.

Conclusion

The greatest recent forest loss has occurred within the ranges of at least 11 species and subspecies of gibbon and orangutan (see Table 7.1). Ape ranges in Sumatra and Borneo contained substantial forest through 2000 but lost it rapidly during the 2000–14 study period, as clearing for plantation agriculture in Indonesia and Malaysia triggered some of the world's highest rates of deforestation. Large-scale agricultural plantations account for the majority (52%–87%) of detected forest loss within the ranges of at least 12 ape subspecies in Malaysia and Indonesia, as well as nearly 30% of loss of ape habitat in Cambodia.

Available data reveal that plantations in Africa corresponded to just 1% of habitat loss for only one African ape subspecies, although nearly 60% of oil palm concessions occur within African ape ranges. Close to 40% of unprotected ape habitat in Africa is land suitable for oil palm (Wich *et al.*, 2014); as land available for expanding oil palm and other industrial-scale agriculture diminishes in Asia, corporate demand for undeveloped land is likely to increase in Africa. Such demand is likely to fuel a surge in both deforestation and degradation from associated infrastructure development (Barber *et al.*, 2014; Laurance *et al.*, 2015b).

In 2000, African ape ranges were 94% forested (see Table 7.1). By 2014, African apes still retained substantial forest cover in their ranges, but rates of loss had increased in the previous five years. In contrast, ape ranges in Asia were only 69% forested in 2000. While the overall rate of forest loss in Southeast Asia slowed somewhat in the following decade—particularly when compared to the extremely high rates caused by massive deforestation during the 1990s (Achard *et al.*, 2014)—apes there persist in isolated forest fragments and PAs.

Protected areas are becoming a last stronghold for remaining populations of a growing number of ape taxa, both in Asia, where forest loss continues to threaten ape populations, and, increasingly, in Africa. PAs experience lower rates of habitat loss than unprotected areas, but, as this analysis underscores, losses are still considerable (Gaveau *et al.*, 2009a; Geldmann *et al.*, 2013).

The need to act is most acute in Asia. If the frontier of deforestation is around PAs, where forest remains, and loss rates stay constant into the coming decades, forest connectivity will be lost, as will the chance to ensure that PAs are large enough and well protected enough to maintain viable populations of subspecies. Stabilizing expanses of protected forest and improving PA management effectiveness are priorities for ape conservation in the immediate future.

Acknowledgments

Principal authors: Suzanne Palminteri[2], Anup Joshi[3], Eric Dinerstein[4], Lilian Pintea[5], Sanjiv Fernando[6], Crystal Davis[7], Matthew Hansen[8]

Annexes VIII, IX, X, XI, XII and XIII: The authors

Reviewers: Leo Bottrill, Mark Cochrane, Mark Harrison and Fiona Maisels

Endnotes

1. Circular representation of protected areas: The authors used the April 2016 version of the World Database on Protected Areas (WDPA) for global spatial analysis. In the WDPA data layer, some protected areas are provided only as points, without boundaries. Circular buffers were created around these points, based on area information in the WDPA. These buffered points were then merged to create one visual layer for protected areas. This approach resulted in the circular representation of some protected areas in these maps.
2. Consultant
3. University of Minnesota (www.conssci.umn.edu)
4. RESOLVE (www.resolv.org)
5. Jane Goodall Institute (JGI) (www.janegoodall.org.uk)
6. RESOLVE (www.resolv.org)
7. World Resources Institute's Global Forest Watch initiative (WRI-GFW) (www.globalforestwatch.org)
8. University of Maryland (geog.umd.edu)

Photo: As a result of loss of natural habitats and wildlife, people will increasingly encounter apes only in captive settings. © Jurek Wajdowicz / Arcus Foundation

CHAPTER 8

The Status of Captive Apes

Introduction

As a result of human population growth and the attendant loss of natural habitats and wildlife, people will increasingly encounter apes only in captive settings. The contexts of these settings influence how viewers perceive the conservation status of apes (Leighty et al., 2015).

Apes in range states are held in a variety of captive settings: they are kept in private homes; publicly displayed as tourist attractions, in zoos, safari parks and by individuals; and taken in by specialized, non-profit care facilities. The latter facilities, which are dedicated to providing care for orphaned, confiscated and injured apes, are known as sanctuaries, rescue centers or rehabilitation centers. While rescue and rehabilitation centers typically focus on short-term care and treatment of injured animals, sanctuaries provide long-term or lifetime care (CITES, 2010a; Durham, 2015). Some zoos also hold orphaned or confiscated apes; since the provision of such care is not their primary function, however, zoos are not discussed in this study.

This chapter comprises two main sections. The first considers the history and context of range state sanctuaries, focusing on 56 such facilities identified by the authors. It examines the outlook for sanctuary apes and explores the opportunities and challenges for these sanctuaries in view of current and emerging threats. Unless otherwise cited, information is based on the authors' knowledge and observations; accounts and data provided by sanctuary practitioners and external experts; and unpublished data, as well as details provided on official and facility websites.[1] The key findings of this sanctuary review include the following:

- Conditions at range state sanctuaries vary widely. Many have exemplary programs, but few facilities have been independently inspected and accredited to verify their performance against welfare and care standards.
- Suitable habitat for reintroduction and translocation is increasingly limited, meaning that most of the thousands of apes already in sanctuaries and the thousands more still in need of captive care will spend their lives in captivity. If reintroduction or translocation is possible, careful site selection, proper rehabilitation, candidate selection and post-release monitoring are critical to prevent significant adverse effects on the welfare and conservation of both wild and rehabilitant apes.
- Overcrowding and resultant poor welfare lower the quality of life for sanctuary apes. Careful consideration is needed to determine whether and when new apes can be accepted without diminishing welfare standards for existing and new residents.
- In the absence of legal consequences for perpetrators of wildlife crimes, rescues and even confiscations do nothing to deter further illegal hunting of wild apes; in fact, they may contribute to illegal ape poaching and trade.
- Increased collaboration and collective efforts by sanctuaries, conservation-focused non-governmental organizations (NGOs), governments, industry and other parties are needed to address the habitat destruction, poaching and human–wildlife conflict that drive apes into sanctuaries.
- Sanctuaries can improve welfare and conservation impacts by: undergoing independent inspection, accreditation and evaluation against robust welfare and conservation standards; accepting external scientific review of reintroduction or translocation methodologies; committing to intake polices that support welfare standards, contribute to law enforcement and prevent corruption; and increasing engagement to address the root causes that lead apes to need captive care.

Section II updates captive ape population statistics and discusses the regulatory landscape affecting captive apes. The key findings of the statistics update are:

- While the United States is starting to witness a transfer of chimpanzees from laboratories to sanctuaries, the slow pace is of concern, in part because of the number of older chimpanzees.
- Ensuring transparency regarding the number, location and welfare of apes is an ongoing challenge. In the United States, the government recently removed considerable amounts of previously available data from online databases, raising concerns about accountability.

> In the absence of legal consequences for perpetrators of wildlife crimes, rescues and even confiscations do nothing to deter further illegal hunting of wild apes.

- Regulatory changes and actions by federal agencies in one country sometimes have an unexpected impact on sanctuaries within and beyond that jurisdiction. A recent case in point concerns a permit application for exportation of chimpanzees from the United States to the United Kingdom. The move raised issues regarding the international impact of the U.S. Endangered Species Act, the management of captive apes within Europe and the illegal international trade in wild animals—all of which affect sanctuaries and their missions.

I. Beyond Capacity: Sanctuaries and the Status of Captive Apes in Shrinking Natural Habitats

Background

History and Scope of Range State Sanctuaries

Ape sanctuaries have been operational in range states for several decades. They are a response to the specialized care needs of apes who have been confiscated from poachers or from the illegal trade, held as pets or retired from unsuitable zoos. The authors identified 56 range state sanctuaries that care for apes, based on personal knowledge, expert accounts, and online descriptions and photos. Most of these sanctuaries were founded and are run by dedicated individuals or NGOs with an interest in improving ape welfare and contributing to ape conservation. Eight of the 56 facilities (14%) are currently government-owned.

Many ape sanctuaries have evolved from an initial focus on individual rescues to a broader scope that includes local conservation and community projects, contributions to the understanding of species behavior, and the provision of behavioral enrichment and care centered on quality of life. A 2011–12 survey of 22 Pan African Sanctuary Alliance (PASA) centers—including three facilities that do not care for apes—demonstrated the breadth of sanctuary projects beyond ape rescue and welfare. Most PASA sanctuaries conducted conservation education programs: 86% were organizing on-site activities and 82% were running off-site conservation education. Cumulatively, these programs reached an average of 19,730 people per sanctuary per year. Most educational messaging was around wildlife laws and biodiversity (Ferrie *et al.*, 2014).

Other activities conducted by PASA sanctuaries included:

- staff development, including support to attend alliance workshops (at 86% of all surveyed facilities) and exchange with overseas zoos and sanctuaries (32%);
- supporting or assisting in the construction of roads, bridges and boreholes (46%) and health clinics and sanitation facilities (27%);
- supporting schools or education centers (87%) and community centers (27%);
- local grant programs or enterprise development assistance (36%);
- population and habitat viability analysis and other censuses (64%);
- research on ecology (55%) and social behavior (46%);
- funding or staffing anti-poaching patrols (73%);
- regular monitoring of primate habitats (46%);
- conducting anti-logging patrols (14%); and
- tree-planting (59%) (Ferrie *et al.*, 2014).

In addition to providing employment worth more than US$1.3 million per year for 21 sanctuaries, PASA sanctuary contributions to local economies totaled an average

> In the United States, the government recently removed considerable amounts of previously available captive apes data from online databases, raising concerns about accountability.

of more than US$78,000 annually (Ferrie *et al.*, 2014).

The authors' review of Asian ape sanctuary websites and interviews with Asian ape sanctuaries indicate a similarly broad scope of activities, with conservation programs including co-management of natural protected areas, acquisition of ape habitat to be designated as protected areas and collaboration with private land owners to protect habitat corridors for apes (Durham, 2015; Durham and Phillipson, 2014).

Sanctuary Standards

Conditions at ape sanctuaries vary widely. Importantly, standards of welfare, health care and facility management have improved over the past few decades alongside the expansion of captive facility activities. Relevant guidelines are now available for both great apes and gibbons (Farmer *et al.*, 2009; GFAS, 2013a, 2013b; PASA, 2016a). Through alliances, networks and advisory groups, sanctuary collaboration among facility directors, staff and outside experts has had a positive influence on the development and implementation of standards and the depth of expertise in sanctuaries, as described in Box 8.1 (Ferrie *et al.*, 2014; K. Farmer, personal communication, 2016).

The Global Federation of Animal Sanctuaries (GFAS), the Orangutan Veterinary Advisory Group (OVAG), PASA and the Wild Animal Rescue Network (WARN) have contributed to sanctuaries' recognition of ape captive care and welfare standards. PASA was formed in 2000, prior to the existence of published standards for in situ care of captive African apes. The African primate sanctuary community and outside experts jointly led the development of PASA's standards for African apes and other primates (Farmer *et al.*, 2009). PASA also published manuals to guide primate healthcare and conservation education practices (Cartwright, 2010; Unwin *et al.*, 2009). OVAG

Photo: Independent verification or accreditation of captive facility standards is critical to ensuring ape welfare in sanctuaries.
© Gorilla Rehabilitation and Conservation Education (GRACE) Center/Rick Barongi

publishes workshop reports with orangutan health care and welfare protocols (Commitante *et al.*, 2015).

GFAS formed in 2007 and developed international welfare standards for both great apes and gibbons. The Federation offers independent inspections to verify or accredit facilities' adherence to these standards. GFAS accreditation involves a more rigorous screening than verification, including operational as well as welfare standards (GFAS, n.d.-c). WARN has been collaborating with GFAS to encourage its members to seek GFAS verification or accreditation (GFAS, personal communication, 2016). Many PASA members are also seeking GFAS accreditation or verification.

At the time of writing, only 13% of sanctuaries considered in this chapter had been inspected and confirmed as complying with GFAS standards. One WARN member ape sanctuary, International Animal Rescue (IAR) Ketapang, was accredited by GFAS, and six PASA member ape facilities—the Chimpanzee Conservation Center, the Fernan-Vaz Gorilla Project, Jeunes Animaux Confisqués au Katanga (J.A.C.K. – 'young animals confiscated in Katanga'), Centre de Réhabilitation des Primates de Lwiro (Lwiro Primate Rehabilitation Centre), Sanaga-Yong Chimpanzee Rescue Center and Sweetwaters Chimpanzee Sanctuary—were verified by GFAS (GFAS, n.d.-b).

Between 2000 and 2014, PASA conducted on-site inspections of 13 of its 19 ape sanctuaries. The Alliance's revised standards no longer mandate regular on-site independent inspection of member sanctuaries, instead requiring sanctuaries to complete a questionnaire every five years; follow-up inspections are undertaken if deemed necessary by PASA (PASA, 2016a). In contrast, GFAS requires on-site inspections for every sanctuary verification or accreditation (GFAS, n.d.-a).

Independent verification or accreditation of captive facility standards is critical to ensuring ape welfare in sanctuaries. It is the only means for donors, governments, the public and partners to ensure that sanctuaries are meeting international welfare standards. While inspections reasonably focus on the essential questions of quality welfare and care, increased emphasis and clear standards around environmental practices,

---

BOX 8.1

The Role of Collaborations

Historically, it has not been easy for ape sanctuaries to communicate regularly with each other or with outside experts. Remote locations, a lack of Internet and phone connectivity, and an absence of travel funding can be barriers to communication. Collaborations among sanctuaries and with outside experts—including accredited zoos and zoo Species Survival Plan programs, field researchers, independent welfare experts and veterinarians—have helped to develop the capacity of sanctuary staff and interested experts. These collaborations continue to be an effective way to foster communication and learning.

Nearly three-fourths (71%) of the 56 sanctuaries considered in this chapter are part of collaborations—alliances, advisory groups or networks—and some participate in more than one. Sixteen are members of PASA; 9 are members of WARN; 10 have participated in OVAG; 5 are members of the Jakarta Animal Aid Network; and 3 are members of the Gabon Great Ape Alliance. One captive facility, a former zoo, is also a member of the South East Asian Zoos Association.

OVAG, PASA and WARN bring outside experts to sanctuaries and facilitate information exchange and reciprocal visits among facilities. These collaborations provide sanctuaries with access to experts on conservation education, strategic planning, reintroduction, and veterinary medicine and health care. Funding raised by alliances, networks and advisory groups has been used to pay for meeting space, accommodation and food to host sanctuary staff, travel costs for outside experts and travel of sanctuary staff to attend training.

conservation activities (including reintroduction) and collaboration in law enforcement efforts would improve verification and accreditation practices. The relevance of these issues to sanctuaries is discussed throughout the chapter. Developing and incorporating these standards could strengthen sanctuary and accreditation organization partnerships with conservation NGOs, governments, field researchers and donors.

Drivers of Intake at Ape Sanctuaries

Drivers and proximate reasons for apes' captive care needs differ across regions and range states. They include habitat loss and degradation, poaching and weak law enforcement.

National laws prohibit the hunting of and trade in apes in all range states.[2] With the exception of South Sudan, all ape range states are parties to CITES, the Convention on International Trade in Endangered Species of Wild Fauna and Flora (CITES, 2016a). All apes are listed in CITES Appendix I, which bans international commercial trade in listed species (CITES, 2017). However, enforcement of these laws and of CITES is inconsistent and transgressions are common (Bennett, 2011; Campbell *et al.*, 2008; Cotula *et al.*, 2015; Imong *et al.*, 2016).

Weak law enforcement facilitates the poaching of wild apes. In Africa, illegal hunting for wild meat (meat from wild animals, often referred to as "bushmeat") is a significant threat to apes in Angola, Cameroon, Central African Republic (CAR), the Democratic Republic of Congo (DRC), Equatorial Guinea, Ivory Coast, Liberia and the Republic of Congo (Fruth *et al.*, 2016; IUCN, 2014d; Maisels, Bergl and Williamson, 2016a; Plumptre *et al.*, 2010, 2015; Refisch and Koné, 2005). In some range states in Asia, including Bangladesh, India, Indonesia, Lao PDR and Viet Nam, orangutans and gibbons are regularly poached for wild meat. In addition, the demand for ape body parts for use in traditional medicine leads to poaching of chimpanzees (*Pan troglodytes*) and gibbons in some range states (Campbell *et al.*, 2008; Davis *et al.*, 2013; Geissmann *et al.*, 2013; Lao MAF, 2011; Molur *et al.*, 2005; Moutinho *et al.*, 2015; Rawson *et al.*, 2011). Infants captured by poachers are often opportunistically sold as pets. Poachers target some gibbon species in particular for sale as pets or to zoos and safari parks (Campbell *et al.*, 2008; Geissmann *et al.*, 2008; Molur *et al.*, 2005; Nijman and Geissmann, 2008; Rawson *et al.*, 2011). If confiscated or abandoned, these illegally captured apes are often delivered to sanctuaries.

The killing or capture of apes is also common in the context of human–wildlife conflict (Davis *et al.*, 2013; Rawson *et al.*, 2011; Williamson *et al.*, 2014). Sanctuaries are often called on to remove wild apes threatened by these conflicts, and to translocate them to other natural habitat or place them in captive care. If the apes are not removed, they are often killed or captured, and the infants sold or kept as pets (Ancrenaz *et al.*, 2015a; Durham, 2015).

Both poaching and human–wildlife conflict are associated with habitat destruction and fragmentation, which are direct consequences of human activities such as logging and forest clearance for the expansion of industrial, subsistence and small-scale agriculture, livestock grazing, extractive industries and infrastructure (see Chapters 1–6).[3] As their habitats shrink, these apes are exposed to a growing risk of being hunted, captured or killed. Examples of habitat destruction abound. Across Indonesia and Malaysia, forest conversions destroy and fragment ape habitats, often isolating apes in tiny patches of trees, where adults can easily be killed and their infants captured (Ancrenaz *et al.*, 2015a; Campbell *et al.*, 2008; Singleton *et al.*, 2016). In Indonesia in particular, fires set to clear land for agriculture

Photo: As their habitats shrink, apes are exposed to a growing risk of being hunted, captured or killed. © Jabruson 2017 (www.jabruson.photoshelter.com)

have exacerbated this habitat destruction (Tabuchi, 2016). In the DRC, chimpanzees, Grauer's gorillas (*Gorilla beringei graueri*) and mountain gorillas (*Gorilla b. beringei*) are imperiled by the illegal local charcoal trade and mining (Plumptre *et al.*, 2015; UNEP/CMS, 2009). Infrastructure such as roads provides access for poachers and a means to bring wild meat and live animals to market (Poulsen *et al.*, 2009). Roads threaten gibbons more than other apes, as these species rarely travel on the ground and can have difficulty crossing these barriers (Chan *et al.*, 2005).

Civil unrest presents threats to apes, particularly to chimpanzees, Grauer's gorillas and mountain gorillas, as they are subject to increased poaching and habitat destruction by displaced persons, armed militias and military forces (Plumptre *et al.*, 2015; UNEP/CMS, 2009). Several pet apes have been seized from military forces in the DRC over the past several years (Engel and Petropoulos, 2016).

As apes are increasingly captured or driven from their natural habitats, the demand for space in ape sanctuaries is certain to grow (Durham, 2015; Durham and Phillipson, 2014). Among the most at risk are Bornean orangutans (*Pongo pygmaeus*), as infrastructure-related projects are predicted to disrupt the vast majority of their habitat by 2030 (Gaveau *et al.*, 2013). Their situation is further compounded by climate change, which is projected to render much of their current habitat unsuitable (Grueter *et al.*, 2013; Struebig *et al.*, 2015). In fact, the catastrophic forest fires that are used to clear land for agriculture in orangutan range states play a role in exacerbating global warming and heightening the risk of larger, more frequent forest fires; as a consequence, more habitat is at risk of destruction and more orangutans are likely to sicken and need sanctuary care (Ancrenaz *et al.*, 2016; Tabuchi, 2016). Concurrently, climate change may impact food availability for other apes, such as the mountain gorillas (Grueter *et al.*, 2013; Struebig *et al.*, 2015).

Human population growth in ape range countries is also expected to cause increased demand for ape sanctuary capacity. Human populations in Angola, Burundi, the DRC, Tanzania and Uganda are projected to increase five-fold by 2100. Nine countries are projected to account for 50% of global human population growth between 2015 and 2050, among them five ape range countries: the DRC, India, Indonesia, Tanzania and Uganda

(UN, 2015). Since important ape populations occur outside of protected areas in these five countries and human population growth is sure to exacerbate illegal hunting and trade, apes will be placed at increasing risk (Indonesia MoF, 2009; IUCN, 2014d; Molur et al., 2005; Plumptre et al., 2010).

While improved enforcement of ape protection laws is urgently needed, it is also likely to increase demands on ape sanctuaries. In some African range states, better enforcement has entailed an increase in seizures and rescues, a trend that tends to persist unless law enforcement effectively deters poachers from further illegal activity (K. Farmer and D. Cox, personal communication, 2012). Meanwhile, international media coverage of CITES and wildlife laws has increased pressure on range states to enforce bans on hunting CITES-listed species, including apes (see Box 8.2). Ideally, such scrutiny will result in improved law enforcement and better protection of wild ape populations.

Chapter 8 Captive Apes

BOX 8.2

The Illegal Trade in Apes

The fact that the Chimpanzee Conservation Center and the Centre de Réhabilitation des Primates de Lwiro recently took in three chimpanzees confiscated from international trade indicates that trafficking in African apes continues, even if in relatively low numbers.

A recent study shows demand for wild-caught apes in Peninsular Malaysia and Thailand, two regions where apes continue to be acquired by zoos and for wildlife attractions such as safari parks, tourist photo props and performances (Beastall and Bouhuys, 2016; see Table 8.1). Interviews of facility staff indicate that most of the apes whose origin was known had been caught in the wild. The researchers found that Thai facilities held non-native apes in numbers far exceeding those recorded as legal imports, including a gorilla and gibbons for whom there were no legal import records. Zoo studbooks in Peninsular Malaysia and Thailand list dozens of orangutans as wild-caught or of unknown origin, although some wild-caught individuals arrived as a result of enforcement actions (Beastall and Bouhuys, 2016). The data indicate that illegal trade in Asian apes remains a concern and needs to be addressed through legislation, improved enforcement and public awareness campaigns.

Although prohibition of hunting and trade in apes is universal across range states, legal protections for apes vary widely. CITES depends on national laws for implementation. CITES has four requirements for each state party's national legislation:

1. designation of at least one management authority and one scientific authority;
2. prohibition of trade in species in violation of the Convention;
3. ability to penalize such trade; and
4. confiscation of specimens illegally traded or possessed (CITES, 2010b).

Only 10 of the 26 ape range states have laws that satisfy all four requirements: Cambodia, Cameroon, the DRC, Equatorial Guinea, Indonesia, Malaysia, Nigeria, Senegal, Thailand (see below) and Viet Nam. The remaining 16 range states do not meet the four requirements. Eight range states have laws meeting one to three of the four requirements: Bangladesh, Burundi, Gabon, Guinea, India, Mali, the Republic of Congo and Tanzania. Eight range states—Angola, Guinea-Bissau, Ivory Coast, Lao People's Democratic Republic (Lao PDR), Liberia, Myanmar, Sierra Leone and Uganda—do not have legislation meeting any of the four requirements. Required legislation is in development in all 16 of the above-listed range states (CITES, 2016a). Once passed and promulgated, this legislation is expected to improve the states' ability to confiscate illegally held apes and prosecute perpetrators. These steps, in turn, are certain to increase the number of apes in need of sanctuary care—and thus the demand for additional sanctuary capacity.

Notably, states can meet CITES requirements for national legislation while still providing insufficient protection for apes, as is the case in Thailand. A recent analysis of Thai wildlife laws highlights several significant shortcomings that imperil apes. The law currently places the burden of proof on the government to demonstrate that wildlife was obtained illegally, rather than requiring those possessing wildlife to prove they obtained it legally. In addition, current criminal penalties for illegally held or traded wildlife may not provide sufficient deterrence against wildlife crime. The study authors propose detailed recommendations for improving a draft amendment to Thailand's Wild Animal Preservation and Protection Act, B.E. 2535 of 1992, which is under consideration (Moore, Prompinchompoo and Beastall, 2016).

In Indonesia, the government is considering revisions to its Law for Conservation of Living Resources and Ecosystems, Law No. 5 of 1990, following government recognition that wildlife hunting and trade cases have typically resulted in short prison sentences (under one year) and fines of less than 100 million rupiah (US$7,500) (Jong, 2016).

Another issue undermining ape protection laws is fraudulent international trade of apes under CITES, often with the use of captive-bred source codes for wild-caught apes (CITES, 2014). Such fraud was particularly associated with trade cases from Guinea between 1999 and 2012. Guinea has no captive breeding facilities for apes; claims of captive-bred apes from this state are thus inevitably fraudulent, and the animals involved can be assumed to be wild-caught (CITES, 2012). CITES Trade Database records show that 122 chimpanzees and 10 gorillas were traded by Guinea as captive-bred (CITES, n.d.).

In 2016, the Conference of the Parties to CITES responded by approving a mechanism for CITES to review, investigate and enforce prohibitions on fraudulent uses of captive breeding codes (CITES, 2016b). This effort is intended to prevent further laundering of wild-caught animals.

While the illegal trade in apes persists and presents a threat to these species, it is typically a byproduct of illegal hunting, involving the opportunistic sale of infants for additional income. Among the threats to apes, the illegal trade is thus of a lower order of magnitude than the key drivers of population declines, namely habitat loss and fragmentation, illegal hunting and human–wildlife conflict, all of which can facilitate the capture and sale of apes.

The trade poses a proportionally greater threat to some gibbon species, however. Gibbon species that are specifically targeted are Kloss's gibbon (*Hylobates klossii*), the lar gibbon (*Hylobates lar*), Müller's gibbon (*Hylobates muelleri*), the Bornean gray gibbon (*Hylobates funereus*), the southern yellow-cheeked crested gibbon (*Nomascus gabriellae*) and the siamang (*Symphalangus syndactylus*) (Brockelman and Geissmann, 2008; Geissmann and Nijman, 2008a, 2008b; Geissmann et al., 2008; Nijman and Geissmann, 2008; Whittaker and Geissmann, 2008).

TABLE 8.1

Apes in Peninsular Malaysian and Thai Zoos and Wildlife Attractions, 2016

Ape species	Number of apes in zoos and wildlife attractions		
	Peninsular Malaysia	Thailand	Total
Chimpanzee (subspecies unknown)	14	36	50
Western lowland gorilla	–	1	1
Bornean orangutan	31	–	31
Sumatran orangutan	2	–	2
Orangutan (species unknown)	1	51	52
Agile gibbon	5	2	7
Lar gibbon	37	107	144
Moloch gibbon	1	–	1
Müller's gibbon (subspecies unknown)	1	–	1
Pileated gibbon	–	34	34
Hylobates gibbon (species unknown)	–	2	2
Nomascus gibbon (species unknown)	–	14	14
Siamang	7	3	10
Total	**99**	**250**	**349**

Notes: The agile gibbon, lar gibbon and siamang are native to Peninsular Malaysia and Thailand. The pileated gibbon is native to Thailand.

Data source: Beastall and Bouhuys (2016)

Apes in Range State Sanctuaries

Origins of Apes in Range State Sanctuaries

Most apes arrive at sanctuaries as a result of illegal wild meat hunting, habitat destruction and fragmentation, human–wildlife conflict or after they are abandoned by or rescued from individuals who kept them as pets. Far fewer apes are in sanctuaries because they were confiscated from the international wildlife trade.

Data from the Indonesian ape sanctuary IAR Ketapang show that 43% of its rescues were illegally held as pets, 31% came from oil palm plantations and 12% were found in local community agricultural landscapes, while only 1% were liberated from the international illegal wildlife trade (Durham, 2015). Similarly, in PASA range state sanctuaries, most apes became residents as a result of human actions within national borders, as opposed to the international trade. In the DRC, the Centre de Réhabilitation des Primates de Lwiro received 16 chimpanzees in 2015–16; all originated in the DRC. One was confiscated in Rwanda, after having been transported there by poachers (I. Vélez del Burgo, personal communication, 2016).

The number of trade-related confiscations is somewhat higher in Guinea, which has been a hotspot of international trade in African apes (CITES, 2014). One Guinean ape sanctuary, the Chimpanzee Conservation Center, accepted seven chimpanzees in 2015–16; the group included six who were native to Guinea and two confiscated from the international trade. The sanctuary took in one orphaned chimpanzee from Senegal, where there are no sanctuary facilities (C. Colin, personal communication, 2016).

The prevalence of hunting and local trade as proximate causes for the intake of apes in range state sanctuaries corroborates data showing that habitat destruction, poaching for wild meat and traditional medicine, and killing related to human–wildlife conflicts remain the most pressing threats to the majority of wild ape species (Brockelman and Geissmann, 2008; Campbell *et al.*, 2008; Davis *et al.*, 2013; Indonesia MoF, 2009; IUCN, 2014d; Plumptre *et al.*, 2015).

Status and Outlook for Apes in Range State Sanctuaries

Table 8.2 lists range states with ape sanctuaries and the species they hold. Except for Bangladesh and Myanmar, Asian ape range

TABLE 8.2

Captive Center Capacity in Ape Range States, 2016

	Ape range countries with sanctuaries	Species accepted
Africa	Cameroon	Central chimpanzee, Nigeria–Cameroon chimpanzee, Cross River gorilla, western lowland gorilla
	DRC	Bonobo, central chimpanzee, eastern chimpanzee, Grauer's gorilla
	Gabon	Central chimpanzee, western lowland gorilla
	Guinea	Western chimpanzee
	Liberia (facility in development)	Western chimpanzee
	Nigeria	Nigeria–Cameroon chimpanzee
	Republic of Congo	Central chimpanzee, western lowland gorilla
	Sierra Leone	Western chimpanzee
	Uganda	Eastern chimpanzee
Asia	Cambodia	Native gibbon species
	China (Hong Kong)	Lar gibbon, pileated gibbon
	India	Western hoolock
	Indonesia	Bornean orangutan, Sumatran orangutan, agile gibbon, Bornean white-bearded gibbon, Kloss's gibbon, moloch gibbon, Müller's gibbon, siamang
	Lao PDR	Northern and southern white-cheeked crested gibbon, other native gibbon species
	Malaysia	Bornean orangutan
	Thailand	Lar gibbon, pileated gibbon, other native gibbon species
	Viet Nam	Pileated gibbon, northern white-cheeked crested gibbon, northern yellow-cheeked crested gibbon, southern yellow-cheeked crested gibbon, other native gibbon species

Data sources: Wildlife Impact (2015, 2016); online and unpublished facility accounts, reviewed by the authors

states have sanctuaries that hold apes (Wildlife Impact, 2016). The rescue center at Kadoorie Farm & Botanic Garden in Hong Kong is not currently known to hold gibbons, but it is equipped to rescue and quarantine them (KFBG, n.d.).

Nine African ape range states—Cameroon, the DRC, Gabon, Guinea, Liberia, Nigeria, the Republic of Congo, Sierra Leone and Uganda—have sanctuaries that hold apes (Wildlife Impact, 2015, 2016). More than half of the African ape range countries—namely Angola, Burundi, CAR, Equatorial Guinea, Ghana, Guinea-Bissau, Mali, Rwanda, Senegal, South Sudan and Tanzania—do not have sanctuaries that are currently equipped to care for apes (Wildlife Impact, 2015, 2016). Ivory Coast does not have a sanctuary, but the Abidjan Zoo has accepted chimpanzees in need of rescue. In 2014 it was at full capacity due to high rates of intake, including of pet chimpanzees left at the zoo during the Ebola crisis (R. Champion, personal communication, 2014).

The number of apes in need far exceeds existing captive facility capacity. Many facilities are full and others have space for very limited numbers of additional apes. More than 6,000 gibbons and between 25 and 126 African apes are estimated to be illegally held in range countries (Durham, 2015; Wildlife Impact, 2015). These numbers exclude the 66 chimpanzees abandoned by the New York Blood Center in Liberia (Gorman, 2015a; see below). An estimate for orangutans was not available.

Many range state sanctuaries have the ultimate aim of reintroducing apes back into their natural habitats. In practice, however, reintroduction is not always feasible, as it may be inconsistent with conservation aims. As noted by Durham (2015), the reality is that many apes entering captive settings will become lifetime residents. Even apes at transit centers or other short-term facilities often spend many years, or the remainder of their lives, in these facilities. Many sanctuaries would need to invest heavily in infrastructure and staff to take on additional lifetime residents. Overall, overcrowding issues in sanctuaries are likely to worsen given the number of apes in need, apes' long life spans and current intake practices. Even now, sanctuaries would not be able to accommodate or provide minimum acceptable welfare standards to the thousands of apes held illegally, nor the newly captured ones.

Some countries without designated rescue centers have shown a reluctance to confiscate unlawfully held or traded live animals (André et al., 2008; Teleki, 2001). In personal communication with the authors in November 2016, zoologist Tamar Ron and Maiombe National Park administrator José Bizi describe recent ape confiscations in Angola, a gorilla and chimpanzee range state that lacks sanctuaries:

- Of five infant chimpanzees and two infant gorillas confiscated by the Maiombe National Park in the past two years or so, only one chimpanzee has survived. That one is being taken care of, together with a number of other chimpanzees of different ages, in a private facility of a person who has been trying to save infant chimpanzees and gorillas over several decades, with his own means, but unfortunately succeeds in providing them only with very substandard, inadequate conditions.

- The [Maiombe National] Park staff does not have adequate capacity, means and conditions to take care of confiscated apes over time. There are no adequate facilities in the country, and the transfer to facilities elsewhere would also require resources that are not available. In addition to the abovementioned private initiative, there is an unknown number (estimated at several dozens) of chimpanzees of different ages held privately, mostly in Cabinda and Luanda, all in quite inadequate and at times appalling conditions. The government has expressed strong interest in establishing an ape sanctuary as part of its strategic wildlife crime action effort, but would require substantial outside support to fund development, running costs and staff capacity building, and to create the enabling conditions required for this ambitious endeavor.

The creation of new sanctuaries might appear an obvious solution. In practice, however, they are very expensive and difficult to establish, requiring both specialized expertise and a commitment for the lifetime of long-lived, cost- and care-intensive rescued apes. Few are willing or able to take on this challenge, especially in range countries with great need but with high levels of civil unrest or other challenges.

Photo: Ape translocation or the release of captive animals into natural habitats can pose significant risks to the health and welfare of released and wild ape populations, other wildlife, ecosystems and human populations.
© Alejo Sabugo, IAR Indonesia

Further, the relationship between sanctuary presence or absence and the need for ape rescues remains unclear, particularly as ape seizures continue in states that have long had sanctuaries, such as Cameroon, DRC and Indonesia. Numerous factors influence seizures and intake of apes by sanctuaries, including the presence and effectiveness of law enforcement, corruption, public awareness of laws and their consequences, poverty and food availability, access to employment and livelihoods, the accessibility and ease of capture of wild ape populations, and demand for and access to markets for wild meat, ape body parts and live apes.

Certainly, the presence of sanctuaries in range states makes ape confiscation more practicable, in part because they can play a key role in facilitating law enforcement (Farmer, 2002; Teleki, 2001). Sanctuaries, particularly those accredited as maintaining high standards of care, also enable improved welfare, lifelong care and, potentially, reintroduction for rescued apes (Trayford and Farmer, 2013). Thorough analysis of need and feasibility, along with collaboration among organizations, individuals and governments, may be a more sustainable path to sanctuary development than the ad hoc approach often used to date. Integrating sanctuaries into broader efforts to address habitat destruction, ape killing and capture, and other factors that lead apes to require care would further improve sanctuary effectiveness.

Reintroduction and Translocation

Suitable Habitat in Range States

Suitable habitat is rapidly disappearing across ape range states (Funwi-Gaba *et al.*, 2014; Williamson *et al.*, 2014). Despite diminishing populations of wild apes, the size and carrying capacity of existing suitable habitat currently make it impossible to release all the captive apes in range states. In some areas there may simply be no suitable habitats that are not already occupied by viable populations of conspecifics or that do not first require forest restoration, protected area designation, sustained anti-poaching enforcement or other long-term conservation efforts.

Given the rapid rate of orangutan habitat conversion, experts have long concluded that suitable habitats that still support orangutans are already populated at or beyond carrying capacity (A. Russon, personal communication, 2016). The situation is similar for gibbons in Kalimantan, Indonesia, as discussed in the previous volume of *State of the Apes* (Durham, 2015). Cross River gorillas (*Gorilla gorilla diehli*) are limited by the extent of human encroachment and habitat use within their range (Imong *et al.*, 2014a). Under these circumstances, even habitat restoration is unlikely to enable gorilla reintroduction, as these human populations and activities would create risks for humans as well as released apes.

Reintroduction and Translocation Benefits and Risks

The release of captive animals into natural habitats can pose significant risks to the health and welfare of released and wild ape populations, other wildlife, ecosystems and human populations (IUCN/SSC, 2013). Nevertheless, reintroduction and translocation are the only ways to re-establish species in habitats from which they have been extirpated.

Used with appropriate precaution in suitable circumstances, reintroduction and translocation can thus be valuable tools. They can add genetic diversity, boost population numbers and provide a focus for species and habitat protection (IUCN/SSC, 2013). Another commonly recognized conservation value of release projects is an increased presence of both enforcement authorities (rangers or ecoguards) and wildlife monitors (including translocation project staff), which deters poaching and other illegal activities at the release site (Humle *et al.* 2011). Released animals can also act as a catalyst for ecosystem conservation (Humle *et al.*, 2011; King, Chamberlan and Courage, 2012).

> Where reintroduction or translocation are feasible options, monitoring of progress and impacts is essential to determine whether a project is achieving measures of conservation success.

Nevertheless, reintroduction and translocation can create myriad risks. One is the risk of spreading disease to conspecifics, other wildlife and humans, which can potentially undermine any positive conservation impacts (Beck *et al.*, 2007; Campbell, Cheyne and Rawson, 2015; IUCN/SSC, 2013; Jakob-Hoff *et al.*, 2014; Schaumberg *et al.*, 2012; Unwin *et al.*, 2012). Further, wild populations generally fill suitable habitats to carrying capacity unless conditions prevent their success (Moehrenschlager *et al.*, 2013). As a consequence, captive apes are often released into areas that are already inhabited by conspecifics and where conditions—such as hunting or deforestation—limit the size of wild populations.

Studies of wild chimpanzees and bonobos (*Pan paniscus*) indicate that individuals released into populations of wild conspecifics reduce the reproductive success of wild females (Wrangham, 2013). Other research suggests that male chimpanzees should not be released into wild chimpanzee ranges, as they are likely to be attacked or killed by wild conspecifics. Data from chimpanzee releases in the Republic of Congo, for instance, show that many released males were killed by wild conspecifics (Goossens *et al.*, 2005). For ex-captive female orangutans who have been translocated to habitats with wild orangutans, establishing a home range is extremely difficult because they are ostracized by resident females, who do not recognize them as part of their social network (M. Ancrenaz, personal communication, 2016). Indeed, the social pressure imposed on translocated animals by resident individuals is huge; it generates stressful situations that can be long-lasting and that may explain why many translocations fail (M. Ancrenaz, personal communication, 2016). Superimposing individuals onto viable conspecific populations is thus not sound conservation or welfare strategy, as it can diminish space and resources for wild apes while compromising the welfare of released apes.

Numerous factors determine appropriate reintroduction and translocation candidates, including sex ratios and social groupings among wild conspecifics, behavioral health and socialization, age, temperament, cognition and learning issues, human bonding and human-focused behaviors (Bashaw, Gullot and Gill, 2010; Russon, 2009). Not all individuals who do well in captivity are good release candidates. Once apes are past infancy, human-focused behaviors and human bonding pose serious safety risks and problems for individual welfare and successful release (Campbell et al., 2015; Riedler, Millesi and Pratje, 2010; Russon, Smith and Adams, 2016). Indeed, overly habituated apes are more likely to approach, harass or even attack humans, thereby increasing their own risk of being killed or captured (Macfie and Williamson, 2010; Russon, 2009).

As part of the feasibility assessment required by IUCN guidelines, reintroduction and translocation should be compared with other conservation measures to determine the most effective actions for species and habitat protection under the circumstances (Beck et al., 2007; Campbell et al., 2015; IUCN/SSC, 2013; Wilson et al., 2014). Wilson et al. (2014) found reintroduction and translocation to be significantly more costly and labor-intensive than other habitat conservation measures.

Where reintroduction or translocation are feasible options, monitoring of progress and impacts is essential to determine whether a project is achieving measures of conservation success, whether animals are surviving and adapting under differing seasonal conditions and whether breeding success is leading to population viability (Guy, Curnoe and Banks, 2014; Osterberg et al., 2014). Long-term monitoring also enables identification of animals who might need additional support through provisioning or even removal back to a captive setting (Farmer, Jamart and Goossens, 2010; Humle and Farmer, 2015). Although some reintroductions and translocations are carefully researched, monitored and documented, many are not, and overall there is little transparency regarding issues and outcomes (Guy et al., 2014). Unmonitored projects can overlook ape deaths and harm to wild conspecifics, released apes and humans. Conversely, even among well-monitored projects, some may intentionally avoid reporting on adverse outcomes for fear of losing funding or public trust.

Funders and governments can promote scientific evaluation and rigor in ape reintroductions and translocations by requesting or funding external scientific review of methodologies. Governments can also promote effective reintroduction and translocation efforts by providing administrative support, building law enforcement and monitoring capacity, and enabling habitat protections.

Captive Facility Sector Impact: Benefits and Risks to Ape Conservation and Welfare

Benefits to Ape Conservation and Welfare

The rising acceptance of GFAS verification and accreditation and increasing interest of funders in demonstrated impacts, coupled with the sincere desire of most sanctuaries to improve welfare and address conservation issues affecting apes, provide an environment ripe for positive change. Several sanctuaries are pursuing exemplary welfare standards, good governance and conservation programming that complement sanctuary operations. Some sanctuaries that have historically been run by expatriates have recently handed over leadership to local successors. Others are actively working to find and train local management-level staff. Many sanctuaries do an exceptional job of rescuing and caring for apes, while also providing opportunities for learning about rehabilitation, care and disease. Sanctuary education and outreach work is generally seen as filling an important role, particularly as sanctuaries are permanent fixtures in local communities.

Furthermore, as holders of rare ape species, range state sanctuaries are uniquely placed to be ambassadors for these species. Many people may never have seen these animals before, and seeing them rescued and in good care in a conservation-related context could make a compelling case for their protection.

Importantly, most of the 56 sanctuaries considered in this chapter participate in some form of anti-poaching patrol or ape tracking. Researchers have found that sensitization, community involvement and the presence of researchers and trackers or rangers can help deter ape poaching (Steinmetz et al., 2014; Sunderland-Groves et al., 2011; Tagg et al., 2015). Deterring poaching by prosecuting poachers is also expected to have a positive impact on ape protection, particularly when coupled with sanctuary care of captive apes. If anti-poaching efforts—such as education, the removal of snares and traps, and anti-poaching or tracking patrols—can decrease capture and deter poachers, there is hope of protecting apes in their natural habitats.

Sixteen African sanctuaries reviewed for this chapter disseminate public information on how their work benefits local communities. Two of them offer micro-credit schemes and ten have alternative livelihood programs, including artisan activities. Some of the sanctuaries provide local communities with services such as education development, medical care and infrastructure, as well as training or technical expertise in areas such as farming and livestock husbandry. Training for sanctuary staff, including in veterinary care, education and community development, has led to significant improvements in the skill levels—and thus the employability—of many staff members.

Challenges to Ape Conservation and Welfare

Standards and Quality of Care and Welfare

The quality of care and welfare in ape sanctuaries ranges from much-lauded accredited or verified facilities, to those that are known to be operating well below PASA or GFAS standards, and even to some that ape experts deem totally unacceptable by any standard. Many facilities have acceptable standards for short-term care but are not suitable for lifetime care for apes.

Issues at sanctuaries that operate below acceptable standards include overcrowding or insufficient suitable space; a lack of behavioral enrichment; and unsuitable social settings, such as solitary housing for social ape species and unsafe facilities from which apes can break out or where they can come into contact with visitors. Several sanctuaries allow some public contact with apes, increasing the risk of disease transmission for both visitors and apes and serious safety risks for humans (Macfie and Williamson, 2010). Further, this approach may perpetuate the concept that apes are suitable as pets.

Few sanctuaries across ape habitat regions have been independently inspected or accredited. Of the 56 sanctuaries considered in this chapter, only 7 (13%) have been inspected and accredited or verified as meeting GFAS standards. This number may understate ape facility engagement with independent inspection, as it does not include sanctuaries that are seeking GFAS verification or accreditation. Yet even when the latter group of sanctuaries is taken into account, it remains clear that an increase in independent inspection and verification of sanctuary standards is needed.

Government accountability in implementing animal welfare and captive care standards could also be improved. Both promulgating and enforcing national laws on welfare tied to GFAS standards would help to ensure good care and welfare for apes in all types of captive facilities.

Photos Depicting Contact with Apes

Studies by Leighty *et al.* (2015) and Ross *et al.* (2008) demonstrate that photos depicting apes in contact with humans promote the perception that these animals are good pets, and that they are not endangered.

A review of publicly available images on websites, Facebook and Twitter from 22 African ape sanctuaries from 2013 to 2015 shows that 19 sanctuaries (86%) publicly displayed photos of humans in direct contact with (touching) apes. Sixteen sanctuaries (73%) had Facebook photos showing this type of contact with primates. These 16 facilities posted 247 such photos between January 1 2013 and November 25 2015. Written context for these photos, such as explanations of veterinary care or rehabilitation, was present less than 70% of the time (Sherman, Brent and Farmer, 2016).

Photos of people hugging apes without safety gear (masks or gloves) elicited comments such as "Awhhh, I want one! They are so adorable!!" (Sherman *et al.*, 2016). Photos of new infants, particularly very young captive-born infants, being held and fed by humans drew similar responses, such as "I want!" (Sherman *et al.*, 2016).

These photos fuel arguments that sanctuary media messages may reinforce interest in apes as pets. Many sanctuaries have rules prohibiting volunteers and visitors from posting photos of themselves in contact with resident apes. Sanctuaries need to pay equally close attention to social media reactions to sanctuary-posted photos and should scrupulously avoid posting photos of staff interacting with apes in any manner that could create the impression of apes as pets.

Sanctuary Capacity

Breeding is a serious issue in many range state sanctuaries. Some sanctuaries purposefully breed apes, while others have what facility

Photo: Demand for sanctuary space puts significant pressure on facilities—many of which are underfunded, understaffed and operating in difficult settings.
© Sanaga-Yong Chimpanzee Rescue Center

> Apes born in range state sanctuaries occupy valuable space needed for victims of poaching and habitat destruction.

managers consider "accidental births." Captive births were confirmed at ten African ape sanctuaries between 2014 and 2016. Seven of these ten posted about the births on social media—on their websites, Facebook or Twitter—and in some cases used the births to fundraise. A review of social media posts from January 1 2013 to November 25 2015, shows at least 19 births at these seven sanctuaries (Wildlife Impact, 2015). Left unchecked, this level of breeding will overwhelm sanctuaries or at least necessitate significant expenses for facility expansion. Good information on preventing accidental births and technical assistance on contraception are readily available from zoo partners and veterinarians.

There is no conservation argument for the breeding of apes at range state sanctuaries, but there are clear arguments against it. Ape conservation action plans do not recommend captive breeding in range state sanctuaries, except in the context of the reintroduction of agile gibbons (*Hylobates agilis*) and in case of emergency management scenarios for Hainan black-crested gibbons (*Nomascus hainanus*).[4]

Apes born in sanctuaries occupy valuable space needed for victims of poaching and habitat destruction. Models of PASA chimpanzee sanctuary capacity show that even occasional sanctuary births have large impacts over time as they cause the total population and costs to swell (Faust *et al.*, 2011). These effects are of particular concern given the continued influx of confiscated apes and limited facility space. Current sanctuary populations already far exceed numbers that could be released. Similarly, there is no welfare argument for breeding apes in range state sanctuaries, many of which successfully manage non-breeding populations.

Demand for sanctuary space puts significant pressure on facilities—many of which are underfunded, understaffed and operating in difficult settings—to make painful choices. It is an unfortunate reality that sanctuaries cannot always rescue additional apes without diminishing the welfare of existing residents.

Sanctuaries should clearly define their maximum carrying capacity based on good welfare standards for resident apes, and then develop intake policies designed to maintain those standards. As part of their decision-making process, sanctuaries need a realistic understanding of their options to expand capacity, if any, and information on the capacities of other captive facilities with appropriate standards, ideally within the subspecies habitat region.

In the absence of such alternatives, a policy on euthanasia should be developed, as long as it is legal in the country. Such a policy can be designed to define circumstances under which a sanctuary may make a choice to end suffering and prevent consigning an ape to a poor-quality life. Ending a life is never easy and never without opponents; however, an ape in poorly operated and overcrowded facilities can suffer from increased aggression, increased stress (resulting in lower immunity and increased illness), poor diet and abnormal behavior, while also causing greater physical harm to low-ranking members of the group. Conversely, there can be social and conservation costs to euthanizing otherwise healthy apes, particularly if it perpetuates public perceptions that apes are less valuable alive than dead.

In these difficult circumstances, an important consideration is that apes and other native wildlife are the responsibility of the state, not of sanctuaries. Sanctuaries, together with conservation and welfare groups, need to ensure that governments are aware of the situations driving wild apes to need captive care, and to hold the state responsible for the ultimate outcomes for those apes. Periodic independent inspections and evaluations would also help sanctuaries assess viable options and make evidence-based decisions. Such analyses could be quite useful in helping sanctuaries ensure that their strategic focus contributes to concrete welfare and conservation objectives.

Intake Policies

Sanctuary intake policies differ primarily in whether they require confiscation and legal action to accept animals. Confiscation can denote anything from legal action and prosecution to a piece of paper saying the animal was confiscated—without consequences for the perpetrator. Some sanctuaries take only confiscated animals, while others accept all apes, regardless of how they were acquired. Some sanctuaries claim they must take every ape delivered by the government. Others have successfully negotiated agreements with governments to require law enforcement procedures as prerequisites for each new intake, or they have protocols to identify solutions for animals they do not have space to accept.

Unless sanctuaries address such intake-related concerns with governments, they may only perpetuate the failure of wildlife law enforcement. While intake issues can grow thorny and divisive in the difficult operating environments of range state sanctuaries—which are often compounded by corruption, as discussed below—they are nonetheless crucially important in defining sanctuary purpose and assessing the impact on ape conservation and welfare.

Community surveys undertaken in the Republic of Congo and Kalimantan, the Indonesian portion of Borneo, demonstrate that public awareness of apes' legally protected status is generally widespread. Surveys found that 90% of respondents in Congo and 73% of respondents in Kalimantan knew that apes were protected under national laws (Cox *et al.*, 2014; Meijaard *et al.*, 2011). In Kalimantan this knowledge was associated with a reduction in the killing of orangutans (Meijaard *et al.*, 2011).

These findings have two key implications for sanctuaries. First, public awareness of apes' protected status and the legal consequences of hunting or buying apes is critical in addressing poaching and local markets that sell ape meat and apes as pets. Sanctuaries can therefore play a valuable role in raising public awareness through targeted education campaigns.

Second, sanctuaries should generally not accept apes if they have not been legally confiscated or if there is no possibility for legal consequences for buyers or poachers, such as prosecution, fines or incarceration. In the absence of confiscation and legal consequences, buyers are likely to purchase another ape. However, if the person or persons who sold or bought the ape are arrested and sentenced, and the money is recovered, then the law has been enforced and a deterrent message has been sent to poachers, traffickers and buyers. To give the law teeth, the government must publicize the consequences of holding and selling apes and ensure that convicted offenders serve out their full sentences.

Unless their intake policies are tied to legal consequences, sanctuaries may undermine ape conservation efforts by implying that it is acceptable to buy, transport and house apes. Moreover, if they do nothing to promote the enforcement of wildlife legislation in cases where it is clear that government officials are ignoring the law or involved in the illegal ape trade, sanctuaries are essentially allowing the government to flout the law, thereby perpetuating the trade.

Tying the intake of animals to appropriate legal consequences is a protocol that the Eco Activists for Governance and Law Enforcement (EAGLE) Network, a coalition of law enforcement and conservation NGOs in Africa, has long urged sanctuaries to follow. The protocol is also in line with the procedures used by the Humane Society of the United States (HSUS) in rescuing illegally held animals. Prior to undertaking any such rescues, the HSUS works directly with law enforcement to ensure the perpetrators will be held accountable under the law

> To give the laws teeth, governments must publicize the consequences of holding and selling apes and ensure that offenders are convicted and serve out their full sentences.

and to prevent them from simply acquiring other animals and repeating their transgressions (K. Nienstedt, personal communication, 2016). While an analogous process in developing countries is clearly more challenging, the international community could do more to support governments, sanctuaries and NGOs in their efforts to increase transparency, reduce corruption and improve law enforcement effectiveness. Together, these changes would help to encourage sanctuaries to tie rescues to legal consequences.

Sanctuaries are rarely involved in prosecutorial aspects of wildlife law, but they can play a significant role in supporting enforcement through partnerships and outreach activities, as discussed below. Some sanctuaries are demonstrating good practice by ensuring that every animal they receive has a legal history that can be traced, thereby assisting law enforcement in holding suspects accountable and creating a deterrent for people who are considering wildlife crime.

Government Relations and Law Enforcement: A Path for Improved Transparency, Accountability and Deterrence

Historically, NGOs have carried the burden of supporting welfare-oriented projects such as building and maintaining animal sanctuaries to allow for the disposition and care of illegally held wildlife confiscated by governments. Many sanctuaries and related NGOs have come to accept that government partners are unwilling to make financial contributions to ensure the welfare of confiscated animals, and that they limit their involvement to allowing such facilities to operate within their boundaries. If governments value this capacity for humane care of confiscated wildlife, however, then they themselves should increasingly accept more of the financial burden involved in this costly process. To that end, sanctuaries should be taking stock of their role in the long-term conservation of apes and drawing up a division of responsibilities and financial commitments among all parties, including governments, in a written agreement.

Sanctuaries may benefit from being more assertive in requesting financial and operational support from government partners. Governments that authorize the establish-

ment of ape sanctuaries have historically neglected to assume these important responsibilities, although this step is ultimately needed to ensure the appropriate placement and long-term humane care of these animals. Moreover, range state governments have largely failed to enforce laws pertaining to illegal activities that support the live animal trade, resulting in near-total impunity for poachers, wildlife traders and influential individuals who participate in or facilitate the trade in protected species (Lawson and Vines, 2014; TRAFFIC, 2008; WWF and Dalberg, 2012). In this way, governments also fail to establish much-needed deterrents to wildlife crime. At the same time,

Photo: Sanctuaries, government partners and other stakeholders all must take additional action for the confiscation and rescue of apes to contribute to effective enforcement of wildlife legislation and to the maintenance of viable populations of great apes in the wild. © Jabruson 2017 (www.jabruson.photoshelter.com)

governments continue to appeal to the sympathetic nature of sanctuaries. By accepting the long-term financial burden that comes with caring for these animals, these facilities further remove government partners from any sense of responsibility. At the very least, a government's role as a partner in a sanctuary should include the capacity and willingness to ensure appropriate enforcement of wildlife laws.

The long-term financial burden on sanctuaries has become increasingly untenable as they become overcrowded. Moreover, acquiring necessary operational funds is becoming more difficult as the demand for sanctuary space continues to rise and funding sources become increasingly rare or competitive. Only when governments assume more responsibility and are obliged to become more involved will they begin to take a serious leadership role in enforcing national laws pertaining to protected species, as well as managing the operational and financial challenges faced by sanctuaries. This same scenario largely holds true for in situ conservation projects; however, governments have recently begun to assume some of the financial burden of implementing costly conservation activities, including law enforcement. Government partners may not become committed to conservation and welfare activities until they have made a considerable financial investment, which should simultaneously support programs that aim to reduce the number of apes in need of sanctuary care and provide better protection to wild ape populations.

Although it is difficult to gather data on instances of corruption due to their inherently clandestine nature, a wealth of anecdotal evidence suggests that high levels of corruption characterize most incidents through which apes are brought into captivity. In addition, multiple publications have linked poor governance and corruption with increases in illegal wildlife trafficking (Bennett, 2015; Smith *et al.*, 2015). In some cases, sanctuaries have prioritized animal welfare concerns over adherence to ape protection laws by skirting processes aimed at formally registering intakes and attempting to bring offenders to justice. A typical form of corruption is the willingness of government agents to accept bribes not to arrest perpetrators or, more passively, simply to allow an animal to be released or "dumped" at sanctuaries without legal consequences (especially if the animal belongs to a government official, influential businessperson or other prominent individual).

Indeed, corruption enters the picture long before an ape ever reaches a sanctuary. Infant apes are very recognizable; they are not likely to make their way from a distant forest block to an urban center without attracting the attention of a host of residents and civil servants, including wildlife rangers, police officers, and military and customs officials. It is quite common for traffickers to bribe authorities in order to avoid arrest and to gain free passage to transport an ape. In many cases, apes end up with high-level individuals in the government, the military, business or the expat community. These individuals or companies are often immune to arrest due to their strong connections or because they paid bribes to escape prosecution. Once they begin to see an ape as a long-term financial burden or physical risk, they typically attempt to transfer the animal to a sanctuary. Given their resolute concern for individual apes, the sanctuaries have historically been open to accepting such burdens, with few questions asked. If this cycle of impunity, corruption and crime is to be addressed, governments, sanctuaries and conservation NGOs must no longer turn a blind eye.

Prosecution, sentencing and effective deterrence against future crime are fundamental to successful law enforcement. Deterrence is in place if an established punishment for committing a crime is sufficient to discourage a potential offender from breaking the law. In corrupt legal systems, the deterrent effects are rarely sufficient, such that the motivation to break the law to obtain future benefits remains intact (Bennett, 2015). Prosecution and sentencing for wildlife crimes is still nascent in some ape range states, and even when perpetrators are convicted and incarcerated, they may simply pay bribes to be liberated (Martini, 2013; WWF and TRAFFIC, 2015; Wyatt and Ngoc Cao, 2015). In some cases, judicial personnel need training in the prosecution of crimes and the development of sentencing that will deter crime. To be effective, deterrents must also reflect national contexts. Punishments that would deter Indonesian villagers who might kill orangutans that raid their crops may not be effective to forestall wild meat traffickers in Africa. Prosecutors should establish deterrents that can be monitored and evaluated for effectiveness in their jurisdictional context. Those who break wildlife laws—be it companies, paid or traditional hunters, or pet traders—need to be prosecuted consistently, and their cases should be publicized to ensure deterrence.

By securing an appropriate and humane placement for animals confiscated by law enforcement officials, sanctuaries can play a vital role in contributing to in situ field conservation efforts. Conversely, if facilities accept animals from law enforcement officials based solely on a legal document that authorizes the transfer but lacks any information on the prosecution or sentencing of those responsible, they do little to deter future confiscations and may even serve to encourage the trade.

If sanctuaries are to play an important role in species conservation efforts, they must either be directly engaged in amplifying deterrence against future wildlife crime, or in assisting the government and other stakeholders in doing so. This does not imply that sanctuaries should undertake this work alone. Rather, it is incumbent upon sanctuaries to accept protected wildlife on the condition of enforcement follow-up, and to ensure that such follow-up is indeed taking place. To that end, they may decide to work more closely with government partners, NGOs that specialize in law enforcement efforts or local and international NGOs that support wildlife conservation efforts.

Many sanctuaries conduct educational outreach programs aimed primarily at younger audiences in order to discourage them from considering the illegal hunting of and trade in wildlife as a future occupation or source of additional income. Increased collaboration with stakeholders that are more closely linked to forests where apes are poached—such as conservation NGOs, government partners, development workers and industry—could ensure that these educational activities are delivered to targeted audiences for a more positive impact. Many sanctuaries are located near urban centers, which are not typically areas in which poachers reside. However, urban areas tend to be home to wealthier individuals who finance the trade; these people are important targets who may be responsive to information about wildlife laws and related court prosecutions. Consequently, it may be worthwhile to enhance collaboration with conservationists and researchers who are close to the rural origins and the urban centers of the illegal trade chain.

Equally important is the ability of sanctuaries, conservation NGOs and all others engaged in conservation education and awareness raising to monitor the extent to which these activities help to achieve conservation objectives. To date, despite millions of dollars spent on these seemingly important themes, data that demonstrate the value of conservation education remain surprisingly scarce.

> If sanctuaries are to play an important role in species conservation efforts, they must either be directly engaged in amplifying deterrence against future wildlife crime, or in assisting the government and other stakeholders in doing so.

It is challenging to demonstrate that any single program or campaign has influenced behavior in a way that has led to a decrease in the illegal hunting of apes, or in the destruction of ape habitat and habitat connectivity. Pre- and post-education campaign surveys can reveal increases in awareness, but they do not prove changes in behavior (Carleton-Hug and Hug, 2010). Survey responses can also indicate that people are consciously keeping quiet about illegal or unpleasant activities, or that they have learned the "right" answers to survey questions (Nuno and St John, 2015; L. Pintea, personal communication, 2015).

To demonstrate that a change in behavior has led to a decrease in the demand for apes, data on the behavior of people who buy and sell wild meat and apes are needed. Sanctuaries need to show that they have reached appropriate demographic groups—those comprising individuals who are most likely to kill, sell or buy apes—and that these audiences have not only gained relevant knowledge, but also modified the behaviors that led to ape poaching. To halt ape poaching behaviors, government partners must also actively deter illegal hunting by conducting effective anti-poaching patrols, ensuring that wildlife laws are properly enforced, and visibly prosecuting and sentencing offenders.

In summary, sanctuaries, government partners and other stakeholders all must take additional action for the confiscation and rescue of apes to contribute to effective enforcement of wildlife legislation and to the maintenance of viable populations of great apes in the wild. These steps would require that:

- sanctuaries do not accept apes that have been illegally held unless there is official documentation demonstrating that the government agency responsible for the confiscation has conducted a thorough investigation of the illegal act and has arrested, is actively seeking to arrest, or is planning to prosecute and sentence suspected individuals;

- sanctuary staff members request periodic meetings with the appropriate government enforcement agency to confirm that adequate follow-up of all ongoing cases with pending judgments has occurred or is in process;

- sanctuaries work in partnership with authorities and conservation organizations that pursue legal outcomes of wildlife cases to ensure that adequate sentencing guidelines exist and that sentences are indeed served by convicted perpetrators;

- governments enforce legal consequences consistently for all perpetrators of wildlife crime;

- sanctuaries periodically share critical data and intelligence information with partners that are strategically placed to help tackle the problem at the geographic origin of the confiscations, and to facilitate coordinated intervention efforts to prevent future poaching and trafficking incidents; and that

- sanctuaries regularly disseminate collected data to strategic conservation and advocacy partners and to media outlets, or to partners that specialize in public communications designed to deter audiences from involvement in the illegal ape trade.

Habitat Protection and Conservation Planning

Sanctuaries could further advance ape conservation by becoming more active partners in broader conservation action and planning efforts. At present, many sanctuaries do not work closely with conservation organizations, field researchers, businesses or governments on management planning for

> "Sanctuaries could further advance ape conservation by becoming more active partners in broader conservation action and planning efforts."

ape habitats (Wildlife Impact, 2016). These plans determine the management of lands that are the source of many sanctuary apes. Significant populations of some apes—such as Bornean orangutans, western lowland gorillas (*Gorilla gorilla gorilla*) and central chimpanzees (*Pan troglodytes troglodytes*)—are mainly located outside of protected areas (Ancrenaz *et al.*, 2015b; IUCN, 2014d). The importance of working closely with conservation NGOs, field researchers, businesses and governments to engage the agriculture and logging industries and traditional land owners within ape habitats thus cannot be overstated.

Further, sanctuaries and NGOs should press governments to ensure that national laws provide adequate protections for critical ape habitats. It is legal in some range states to destroy ape habitats, and in some cases conservation laws protecting apes may be overridden or ignored in favor of commercial concessions (Rainer and Lanjouw, 2015; Tata *et al.*, 2014; E. Meijaard, personal communication, 2017). Sanctuaries that do not have the capacity or time to focus on these broader conservation issues could collaborate with or help promote the work of conservation partners to deliver in situ projects aimed at ensuring the long-term survival of wild apes in their natural habitats.

One particular area of concern regarding habitat conservation relates to how sanctuaries and private companies address wild-to-wild Asian ape translocations. In Borneo, some translocations have actually led to additional forest clearing (M. Ancrenaz, personal communication, 2016). Companies have been known to ask sanctuaries or governments to remove what they call "problem" orangutans living in small patches of forest in mosaic landscapes. If sanctuaries agree to remove the orangutans, industry actors tend to clear the patches because they no longer contain species of high conservation value (M. Ancrenaz, personal communication, 2016). In these situations, it is not known whether the individual orangutans can adapt and survive after being translocated.

Scientists report that companies feel they have done a good thing and the issue is resolved once they contact a sanctuary to remove the "problem" ape (S. Cheyne, personal communication, 2016). While companies do take a positive step by notifying sanctuaries about apes, they typically lack awareness of the cost and long-term requirements of a translocated ape. Moreover, companies rarely contribute to translocation, post-release monitoring or long-term care costs. Many translocations simply displace a problem without addressing the reasons apes need to be translocated in the first place, such as poor land management by companies or plantation managers (S. Cheyne, personal communication, 2016).

Allowing industry actors to clear-cut forest patches within the landscape makes the overall landscape less and less suitable for orangutans and other wildlife. Research shows that where hunting is not an issue, orangutans can use oil palm and sustainably logged landscapes, but to do so they need corridors and forest patches (Ancrenaz *et al.*, 2015b; Wich *et al.*, 2012b). Once these small forest "islands" are removed, animals cannot use the landscape anymore and the population becomes extremely fragmented and not viable in the long term (M. Ancrenaz, personal communication, 2016). Sanctuaries, industry and governments need to collaborate on solutions that incorporate established oil palm plantations and logging concessions while also accommodating apes.

Efforts by sanctuaries, NGOs and industry are needed to promote sustainable management of these mosaic landscapes. Instead of removing individual animals at the expense of habitat for local wild apes, sanctuaries should encourage industries, government, and other stakeholders to focus on saving natural habitat—whatever size the patches—as a way to help support ape populations.

> Efforts by sanctuaries, NGOs and industry are needed to promote sustainable management of mosaic landscapes. Once small forest "islands" are removed, animals cannot use the landscape anymore.

Photo: Efforts by sanctuaries, NGOs and industry are needed to promote sustainable management of mosaic landscapes—whatever size the patches. © HUTAN–Kinabatangan Orang-utan Conservation Project/Marc Ancrenaz

Sustainability and Funding

Relatively few grant programs support range state sanctuaries. Many sanctuaries struggle with funding shortfalls, particularly for basic operations (administration and salaries), animal care and facility needs. Funders increasingly expect grantees to provide empirical evidence to demonstrate whether and how they are impacting long-term survival of the species in the wild. This presents a particular hurdle for sanctuary applicants, who rarely collect the type of data required to answer this question (Wildlife Impact, 2015).

Another issue is that many sanctuaries lack succession planning and are thus exposed to further sustainability risks. Since building management-level capacity of local staff to sustain sanctuaries in the long term is difficult and time-consuming, it is often overlooked. Ape sanctuaries, as well as many smaller conservation organizations, rarely undertake professionally led strategic planning, empirical monitoring of outcomes

or independent evaluation, even though these processes are essential to identifying successful actions and addressing shortfalls (Farmer, 2012; Ferraro and Pattanayak, 2006; MEA, 2005).

Transparency about governance and outcomes is likewise uncommon. Indeed, sanctuaries rarely document or share lessons learned from failure or near collapse of facilities with other actors in the sector, thus depriving the community of valuable insight and the chance to avoid known pitfalls. Collapse of peer facilities can create enormous pressure on other national or regional sanctuaries to find space for the failing facility's animals, which could, in turn, overwhelm these sanctuaries' capacity to accept orphans.

Sanctuaries that struggle with a lack of sustainability or risk complete failure are not likely to be able to address the root problems of their instability without changing their management structures and activities. Sanctuaries can increase transparency and share knowledge through captive facility alliances; they can also obtain fresh perspectives from outside experts, professionally led strategic planning, monitoring and independent evaluation. These processes can help sanctuaries to identify problems and potential solutions, focus efforts on project goals, inform good governance and sustainability, provide empirical evidence of impacts and guide the application of best practices. It is worth noting that planning, monitoring and evaluation require a continuous commitment, which can be difficult for sanctuaries in terms of time, funding and expertise. Funder recognition and support of these needs is thus important to their uptake, as is knowledge sharing and guidance from colleagues who have already gone through these processes.

Conclusion

Ape sanctuaries can be found in most ape range states in Asia, and in just under half of African range states. Collaborations have enabled information sharing and training among sanctuaries and with outside experts; they have also played a role in the evolution of these facilities into organizations with broad missions that encompass welfare, conservation and community development. Sanctuaries are currently under tremendous pressure to provide care for the many apes rescued from the wild meat trade, habitat destruction, human–wildlife conflict and the pet trade. Explosive human population growth, which is predicted in several African range states and in Indonesia, will exacerbate threats to wild apes and increase the need for confiscations of poached and trafficked apes.

Moreover, international attention on wildlife legislation is having a positive effect on promoting the enforcement of laws that forbid the capture of and trade in wild animals. With increased confiscations of apes, overcrowding and pressures on sanctuaries are likely to build. Sanctuaries, governments, donors, conservation NGOs and other partners need to collaborate to identify sustainable ways to ensure high standards of captive care for confiscated wildlife while simultaneously improving the protection of wild apes and their habitats.

The reintroduction or translocation of apes is often touted as a solution to captive facility overcrowding and ape welfare needs. In fact, they are high-risk options that can endanger the conservation of wild apes and other wildlife, as well as the welfare of both the wild ape populations and the released apes. The ongoing destruction of forests renders both options increasingly difficult, as little suitable habitat remains that is not already home to wild apes. Feasibility studies, comparisons of available conservation tools and a good understanding of the local ecological, political and community landscape can help sanctuaries determine whether reintroduction or translocation is appropriate, or whether other conservation tools would cost less and save more lives. Sanctuary accreditation organizations, independent evaluators and donors can play an important role in creating accountability on adherence to IUCN reintroduction and translocation guidelines and best practices. Granting foundations in particular can drive positive change by suggesting, or requiring, an independent scientific review of reintroduction methodologies or asking to see feedback from such efforts.

A significant number of apes currently in sanctuaries or in need of rescue will not be releasable and are thus likely to need lifetime captive care. For many sanctuaries, securing operational funding is a significant hurdle, as are recruiting skilled staff and ensuring that facility space can provide high welfare standards for increasing numbers of residents. As confiscation numbers increase, these issues will be compounded. It is thus ever more critical that sanctuaries ensure that their rescue and conservation

> " Explosive human population growth, which is predicted in several African range states and in Indonesia, will exacerbate threats to wild apes and increase the need for confiscations of poached and trafficked apes. "

activities are carefully coordinated, targeted and evaluated to facilitate law enforcement and to demonstrate progress in addressing the root causes driving apes to need sanctuary.

Sanctuaries that fail to hold authorities to account on the enforcement of wildlife law may further discourage effective enforcement, potentially exacerbating the illegal ape trade. Conversely, improved engagement with governments on confiscation and conservation planning and management activities, targeted education programs and partnerships with conservation NGOs offer diverse opportunities for sanctuaries to make a positive impact on these issues.

Many facilities have already taken the lead on these efforts. They adhere to transparent standards and accreditation, including non-breeding and no visitor–animal contact policies, demonstrated commitment to addressing the root causes of the need for sanctuary, the application of IUCN guidelines on reintroduction and translocation, and a willingness to undertake monitoring and independent evaluation. In so doing, they provide a pathway for all sanctuaries to demonstrate their successes, a critical step in attracting new funding and the support they require to improve ape welfare and conservation.

II. The Status of Captive Apes: A Statistical Update

The regulatory landscape continues to shift in a number of ways that impact how apes may be kept or used in captivity. Some of these changes have followed from legislation, petitions and other regulatory mechanisms, or activism (Durham, 2015). Other changes have stemmed from law enforcement or lawsuits. In Argentina, for example, a judge decreed that Cecilia, a chimpanzee living in isolation at a zoo, must be transferred to a specialized sanctuary in Brazil as a matter of protecting her rights (Tello, 2016). By contrast, the enforcement of the U.S. Endangered Species Act was the key issue in a lawsuit against an unaccredited Alabama zoo that held a chimpanzee named Joe (USFWS, 2015). After the case was filed, Joe was moved to the private sanctuary Save the Chimps, in Florida, and the U.S. authorities subsequently ordered the zoo to close (Brulliard, 2016; Sharp, 2016).

Captive Apes in the United States, Japan and Europe

While changes in the law and in law enforcement are important, the benefits for apes are not always delivered swiftly (Durham and Phillipson, 2014, p. 300). In the United States, growing restrictions on breeding, invasive biomedical testing, use in entertainment, private ownership and trade have led to a drop in the number of chimpanzees used in various commercial endeavors. While these changes have been accompanied by an increase in the number in sanctuaries, however, controversy surrounds delays in the transfer of chimpanzees to these facilities (Fears, 2016; see Table 8.3 and Figure 8.1). Given the age and health status of many chimpanzees used commercially in laboratories and entertainment, such delays can mean that some individuals will die before they reach a sanctuary or shortly after arrival. The ethical imperative when it comes to regulations, actions and practices designed to enhance apes' quality of life is to remove barriers and disincentives to change so that the apes themselves benefit.

The size and operations of chimpanzee sanctuaries in the United States vary considerably. Some care for just a few chimpanzees alongside hundreds of other animals ranging from chickens to tigers (Fund for Animals, n.d.); others specialize in chimpanzees, holding anywhere from seven to more than 250 (see Table 8.4). As of October

> "The ethical imperative when it comes to regulations, actions and practices designed to enhance apes' quality of life is to remove barriers and disincentives to change so that the apes themselves benefit."

TABLE 8.3

Number of Chimpanzees in Different Forms of Captivity in the United States as of October 2016

Captivity type	2011[a]	2014[b]	2016[c]	% change 2011–16
Biomedical labs	962	794	658	-32
GFAS sanctuaries	522	525	556	7
Zoo accredited by the Association of Zoos and Aquariums (AZA)	261	258	259	-1
Exhibition*	106	196	111	5
Dealer or pet owner	60	52	37	-38
Entertainment	20	18	13	-35
Total	**1,931**	**1,843**	**1,634**	**-15**

Notes: * Exhibition comprises non-AZA zoos and other facilities that may or may not be open to the public. The category includes apes in sanctuaries that were not accredited by GFAS or members of the North American Primate Sanctuary Alliance.

Data sources: (a) Durham and Phillipson (2014); (b) Durham (2015); (c) ChimpCARE (n.d.)

2016, Chimp Haven, the sanctuary for federally owned chimpanzees, and Save the Chimps accounted for 76.4% of chimpanzees in accredited sanctuaries; the remaining eight sanctuaries were housing 141 individuals (23.6%). A new facility called Project Chimps opened in 2016 and had nine chimpanzees in residence by October of that year (Baeckler Davis, 2016). While it was not yet accredited, the organization stated intentions to expand over a number of years to house more chimpanzees from a laboratory that is phasing out its operation (Milman, 2016).

In earlier volumes of *State of the Apes*, data extracted from U.S. government inspection reports were analyzed to determine (1) the number of apes in different forms of captivity, and (2) risks to ape welfare associated with violations of the Animal Welfare

FIGURE 8.1

Number of Chimpanzees in Different Forms of Captivity in the United States as of October 2016

Key: ■ 2011 ■ 2014 ■ 2016

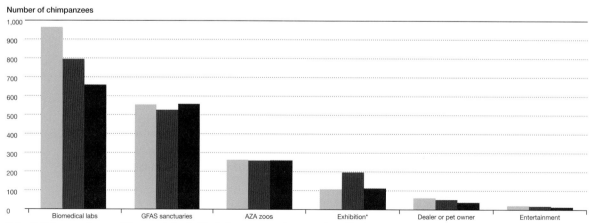

Notes: Exhibition comprises non-AZA zoos and other facilities that may or may not be open to the public. The category includes apes in sanctuaries that were not accredited by GFAS or members of the North American Primate Sanctuary Alliance.

Data sources: 2011: Durham and Phillipson (2014); 2014: Durham (2015); 2016: ChimpCARE (n.d.)

TABLE 8.4

Number of Chimpanzees in Selected U.S. Sanctuaries, October 2016

Sanctuary name	Number of apes	% of total
Center for Great Apes	28	4.7
Chimp Haven	204	34.2
Chimpanzee Sanctuary Northwest	7	1.2
Chimps Inc.	7	1.2
Cleveland Amory Black Beauty Ranch	2	0.3
Primarily Primates	38	6.4
Primate Rescue Center	9	1.5
Project Chimps	9	1.5
Save the Chimps	252	42.2
Wildlife Waystation	41	6.9
Total	**597**	**100.0**

Data source: ChimpCARE (n.d.)

Act (Durham and Phillipson, 2014). In 2017, however, the U.S. Department of Agriculture removed the taxon field from database search options and stopped providing animal counts in its search results; as a result, it was not possible to update key information on captive apes in the United States for this volume. Subsequently, the agency purged even more data, including information about violations and enforcement actions under the Animal Welfare Act, a move that spurred broad criticism and legal action (Brulliard, 2017c; Wadman, 2017b; see Box 8.3). The fact that U.S. government authorities are no longer making certain data available online raises concerns about transparency and accountability.

In contrast to the recent changes in the United States, Japan has a program of full transparency whereby the name, age and location of every ape in the country is openly reported through the Great Ape Information Network (GAIN, n.d.). Current ape figures for Japan are shown in Table 8.5.

Given that significant chunks of U.S. data have been made inaccessible, this update provides only figures for chimpanzees and other apes reported in captive breeding programs of the Species Survival Plans (SSPs) of the U.S. Association of Zoos and Aquariums. As shown in Figure 8.2, the numbers for most ape taxa in captivity in the United States have not changed dramatically since 2012, the year covered in the previous volume of *State of the Apes* (Durham, 2015). Data for gibbons exhibit a more conspicuous change:

BOX 8.3

Access Denied: The Disappearance of U.S. Animal Welfare Data

In early 2017, the federal agency that oversees the U.S. Animal Welfare Act (AWA), the U.S. Department of Agriculture, abruptly removed public access to online data and official AWA compliance documents (Wadman, 2017b). The agency terminated access to the searchable database and electronic annual reports; it also cut access to inspection reports, which provide details on inspections that find full compliance, new and repeated instances of non-compliance and associated terms of the agency's citations, such as the period allowed for correction (Daly and Bale, 2017).

A number of stakeholders—from animal rights organizations and industry organizations for zoos and laboratories, to members of Congress—expressed concerns about the overall impact for transparency and public perception (Wadman, 2017a). While the agency restored a small number of the deleted records, lawsuits under the Freedom of Information Act (FOIA) and Administrative Procedures Act are pending (Wadman 2017a, 2017b). There is no clear resolution in sight and new concerns continued to surface as recently as August 2017 (Brulliard, 2017a).

Although people may still file FOIA requests for records, responses are notoriously slow and the government may withhold or redact information, which can involve blacking out anywhere from a few characters (such as a name or dollar amount) to full pages (Winders, 2017). An attorney involved with a suit recently received nearly 1,800 pages that were entirely blacked out (Abel, 2017; Winders, 2017). Advocates for transparency have made efforts to fill the gap by posting records from archives on other websites (Chan, 2017).

As noted in this chapter, the number, species, locations and names of licensees who hold apes in captivity are no longer in the public database; such records were used in prior volumes of *State of the Apes* and were publicly available for several years prior (Brulliard, 2017a, 2017b). The impact on the figures provided in this volume are most significant for small apes, as they are more likely to be privately owned as household pets, or in private menageries and unaccredited roadside zoos.

numbers appear to have dropped from 624 to a significantly lower range of 374–97 (Gibbon SSP, unpublished data, 2016; Species360, 2016). However, while differences in taxonomy and species coverage in the cited sources may account for some of the disparity, the drop largely reflects the lack of "private ownership" data for pets, roadside zoos and entertainment. This information was available on government databases at the time of the previous review, but that is no longer the case (see Box 8.3).

While the quality and coverage of information available about apes and their welfare remain of concern for specific forms of captivity and certain jurisdictions, steps are being made towards improving standards and practice. In 2015, for instance, the European Commission released a good practices document for zoo compliance (European Commission, 2015). In countries of the European Union, the vast majority of apes in captivity are found in zoos, subject to regulation under Directive 1999/22/EC (Council of the European Union, 1999).

The number of apes in European zoos is significant when compared to the U.S. (see above), South American (33 apes) and Australian figures (158 apes) (Species360, 2016). Figure 8.3 shows numbers and the proportion of apes in each group in European zoos. In total, the European data set contains information on 2,354 apes in 215 member institutions, whose holdings range from 1 to 65 apes per site. Gibbons were the most common taxon in the sample, followed by chimpanzees, gorillas, orangutans and bonobos. The number of solitary apes in the sample was small: 18 apes, or less than 1% of the total. Given their social needs and capabilities, all apes in captivity should be part of groups of compatible individuals.

A small, slowly declining number of apes and other primates are still used in circuses or other unsuitable settings in Europe, although Italy, Norway and Scotland are set to consider or implement bans (Banks, 2016; Born Free Foundation, 2016a, 2016b; Tyson, Draper and Turner, 2016). Other countries have opted for "white lists" of species that are approved for private ownership; these lists do not include apes, meaning that private individuals or companies

FIGURE 8.2

Apes in Captivity in the United States, by Taxon, 2012 and 2016

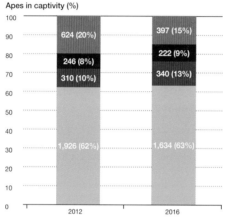

Key: ■ Chimpanzee ■ Gorilla ■ Orangutan ■ Gibbon

Note: Gibbon figures include all gibbons and siamangs; chimpanzee figures include bonobos.

Data sources: Center for Great Apes (n.d.); ChimpCARE (n.d.); Durham (2015, Figure 8.3); Durham and Phillipson (2014, Table 10.6); Gibbon SSP, unpublished data (2016); Gorilla SSP (n.d.); Orangutan SSP (n.d.); Species360 (2016)

TABLE 8.5

Number of Apes in Captivity and Number of Facilities Housing Apes in Japan, October 2016

Taxon	Number of apes	Number of facilities
Bonobos	6	1
Chimpanzees	317	50
Gorillas	20	7
Orangutans	49	21
Gibbons	181	43
Total	**573**	**64***

Note: * Some facilities hold more than one type of ape.

Data source: GAIN (n.d.)

FIGURE 8.3

Apes in Selected European Zoos, by Taxon, 2012 and 2016

Key: ■ Bonobo ■ Chimpanzee ■ Gorilla ■ Orangutan ■ Gibbon

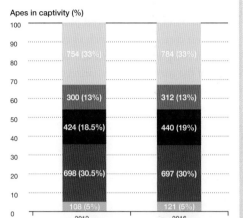

Note: Figures are drawn from aggregate data presented in species-holding reports submitted to the International Species Information System, which was rebranded as Species360 in 2016. Some figures may reflect holdings from prior years.

Data sources: Durham (2015, Figure 8.1); Species360 (2016)

cannot legally own them (Durham and Phillipson, 2014).

Better science has also been a key to positive change. In response to data that revealed a significant proportion of hybrids ("generic" chimpanzees) in its captive breeding program, the European Association of Zoos and Aquaria (EAZA) decided to focus its ongoing efforts on the western and central subspecies (*Pan troglodytes verus* and *Pan t. troglodytes*), while instituting a breeding moratorium for other chimpanzees, including hybrids (Carlsen and de Jongh, 2015; Hvilsom *et al.*, 2013). Despite such progress, a number of challenges remain, including with respect to international cooperation on priorities and good practices for the care and welfare of apes in captivity.

The need for global cooperation is particularly apparent given how regulations and actions in one country or jurisdiction can have unexpected implications in another. A case in point involves the U.S. Fish and Wildlife Service, which, following a regulatory decision by the U.S. National Institutes of Health to retire chimpanzees from medical labs, authorized the transfer of eight generic chimpanzees from the Yerkes National Primate Research Center in the United States to an unaccredited zoo in the United Kingdom, Wingham Wildlife Park. The agency appears to have granted the permit at least in part based on Yerkes' pledge to make a donation to initiate a new project led by a British charity in Uganda, rather than any potential enhancement of the species through the transfer itself, as would be expected under the U.S. Endangered Species Act (Gorman, 2016). A number of organizations—including the United Nations Great Apes Survival Partnership and the Wildlife Conservation Society—had previously declined the donation offer from Yerkes (Bale, 2016).

A range of global stakeholders opposed the transfer during the protracted permitting process (Gorman, 2015b, 2016). The Pan African Sanctuary Alliance cited concerns regarding a precedent that would make fighting the commercial trade in apes even more difficult, especially with respect to illegal markets for infant apes (PASA, 2016b). The EAZA noted challenges related to zoo and sanctuary capacity in Europe, stating: "There are still many chimpanzees in Europe that need outplacement and not enough good places to put them" (Carlsen and de Jongh, 2015). A lawsuit to block the transfer eventually failed and seven chimpanzees (the eighth had died in the interim) were cleared for export to Wingham Wildlife Park in September 2016 (Gorman, 2016). As this case highlights, stakeholders have not yet reached consensus on priorities or on what constitutes good practice in the management of captive apes. Better international cooperation and articulation of scientifically and ethically sound practices would help to close regulatory loopholes, reduce risk and accelerate progress towards global protection.

Captive Apes in Range States and Surrounding Regions

Updated figures for sanctuaries in and near ape range states are presented in Tables 8.6 and 8.7. While figures for chimpanzees remained relatively stable overall, there were increases for both bonobos and gorillas relative to 2011 figures reported in the first volume of *State of the Apes* (Durham and Phillipson, 2014, tables 10.7, 10.8).

Another change is the inclusion of a Liberian facility that was recently reclassified as a sanctuary. From 1976 until 2007, it had served as a research laboratory for the New York Blood Center, carrying out invasive biomedical experiments on chimpanzees. As mentioned above, the Blood Center withdrew funding for the chimpanzee colony in 2015; the decision triggered public outcry for their care and the launch of an intensive fundraising effort (Gorman, 2015a). The fate of the surviving chimpanzees in Liberia has since improved, particularly now that the sanctuary is taking shape and the NGO Liberia Chimpanzee Rescue has been

TABLE 8.6
Number of Apes in African Sanctuaries, by Taxon and Country, 2011 vs. 2015

Country	Number of sanctuaries	Bonobos			Chimpanzees			Gorillas		
		2011	2015	% change	2011	2015	% change	2011	2015	% change
Cameroon	4				244	245	0	33	36	9
DRC*	6	55	72	31	85	104	22	30	18	-40
Gabon	3				20	20	0	9	45	400
Gambia	1				77	106	38			
Guinea	1				38	49	29			
Ivory Coast	1				n/a	1				
Kenya	1				44	39	-11			
Liberia	1				n/a	63				
Nigeria	1				28	30	7			
Republic of Congo	3				156	145	-7	5	28	460
Rwanda*	1				0	0	0	6	0	-100
Sierra Leone	1				101	75	-26			
South Africa	1				33	13	-61			
Uganda	1				45	48	7			
Zambia	1				120	126	5			
Total	27	55	72	31	1,071	1,065	-1	83	127	53

Note: Figures account for total sanctuary population inclusive of births, deaths, transfers and new arrivals. The dark shaded rows are not range states. * Some 2011 figures for DRC and Rwanda include counts from jointly ascribed transboundary operations. For details, see Durham and Phillipson (2014).

Data sources: Durham and Phillipson (2014); PASA (2015); Wanshel (2016)

launched to ensure their well-being (Palm, 2015). Another chimpanzee, the lone survivor of a group that the Blood Center reportedly abandoned on an island off the Ivory Coast in the early 1980s, is now receiving care funded by an organization, which is also attempting to secure international transfer since placement with Chimfunshi sanctuary in Zambia was denied in 2016 (Wanshel, 2016; T. Calvi, personal communication, 2016).

African zoos also hold apes, although far fewer than sanctuaries; 59 apes were reported for zoos on the continent: 33 chimpanzees, 5 gorillas, 20 gibbons and 1 orangutan (Species360, 2016). Sanctuaries and rescue centers thus account for more than 95.5% of all apes reported to be in captivity within Africa.

Range state sanctuaries in Africa have received a slow but steady trickle of new residents through rescue; in some cases, they have transferred or consolidated apes among themselves. In contrast, Asian sanctuaries have continued to experience a staggering demand for care. A recent analysis of data on great apes seized between 2005 and 2016 revealed that 67% of known cases were orangutans (GRASP, 2016).

The continuing challenges that face orangutan rescue centers are illustrated in the first volume of this series, in a case study on the Borneo Orangutan Survival Foundation (BOSF), which at that time had approximately 820 orangutans in its care (Durham and Phillipson, 2014, p. 303). Given that Indonesia's government aims to release all healthy orangutans, BOSF efforts have continued to focus on rehabilitation (Indonesia MoF, 2009). Since 2012, BOSF has reintroduced 234 orangutans—39 of them between January and November 2016; the organization was aiming to release another 250 by the end of 2017 (N. Hermanu, personal communication, 2016). At this writing, 667 orangutans were at BOSF facilities: 471 at Nyaru Menteng and 196 at Samboja Lestari. About 150 of these apes were not in reintroduction training because of their health. Of the remainder, 114 were on pre-release islands and more than 400 had been deemed eligible for release—that is, healthy (N. Hermanu, personal communication, 2016).

In contrast, the GFAS-accredited sanctuary IAR Ketapang saw an increase in its orangutan numbers in 2016. The team released 18 orangutans that year, yet 28 were taken in, resulting in a total of 106 resident orangutans (K. Sánchez, personal communication, 2017). A similar pattern of growth was apparent for the gibbon- and siamang-focused sanctuary Kalaweit, which was featured in the second volume of *State of the Apes* (Durham, 2015). In 2014, Kalaweit reported that it had rescued 16 apes over the prior year, and that the number of residents had thus grown by 6%, to 254 (Durham, 2015, pp. 237–9). By August 2016, the apes in residence had increased to 293—a rise of 15%—not counting apes that had been released since 2014 (Kalaweit France, 2016).

As rescues and successful law enforcement efforts continue, the obligations associated with new arrivals are offsetting reintroduction efforts of Asian sanctuaries such as BOSF, IAR Ketapang and Kalaweit. Reintroduction is fraught with a series of complex challenges, as discussed above. Sanctuaries must juggle priorities such as field staffing, the garnering of representation at international stakeholder meetings and participation in land use planning, all while ensuring the health and welfare of apes in captivity and in their natural habitats. Table 8.7 lists the number of orangutans and gibbons in residence at sanctuaries and rescue centers in Asia in 2016.

In Asia, much like in Europe, a substantial proportion of captive apes resides in zoos. Excluding the data presented for Japan in Table 8.5, zoos that use Species360

TABLE 8.7

Number of Orangutans and Gibbons in Asian Sanctuaries, by Country, 2016

Country	Orangutans	Gibbons
Cambodia		77
Indonesia	1,147	293
Malaysia	98	
Thailand	2	229
Viet Nam		45
Total	**1,247**	**644**

Notes: Figures may include pre-2016 holdings. Median used in instances where a range was reported. Figures account for total sanctuary population inclusive of births, deaths and new arrivals from rescue or transfer.

Data sources: Durham (2015); Highland Farm (n.d.); Kalaweit France (2016); OFI (n.d.); Orangutan Appeal UK (n.d.); Species360 (2016); personal communication: Gibbon Rehabilitation Project (2017); N. Hermanu (2016); M. Kenyon (2016); Orangutan Project (2017); E. Pollard (2016); K. Sánchez (2017)

for voluntary reporting housed 24 gorillas, 344 gibbons, approximately 200 chimpanzees and 130 orangutans (Species360, 2016).

Conclusion

Around the globe, thousands of apes are illegally hunted, traded and exploited for private or commercial ends. We may not know precisely what percentage of these apes are seized or found and then placed in captive care, but there is growing recognition that the sanctuaries that take them in face significant challenges and that these outcomes are insufficiently tracked at both the national and international levels (D'Cruze and Macdonald, 2016).

As states develop stronger legal and regulatory frameworks for ape protection, and as care practitioners continue to enhance their standards and capacity, the opportunities to reduce the harm and improve the quality of life for captive apes are certain to increase. Together with accredited zoos, the sanctuaries that provide care for rescued apes have an important role to play in driving these practices forward, not least by joining forces with strong partners.

If care is to be maintained and improved, ensuring that these facilities have resources and are recognized as essential stakeholders in policymaking and scientific research must be seen as high priorities. Given the sustained—and growing—demand for sanctuary space and services, sanctuaries will require reliable support and partnerships, so that they may focus on providing the same high standard of care for incoming apes as for existing residents.

Acknowledgments

Principal authors of Section I: Julie Sherman[5] and David Greer[6]

Section I contributors: Marc Ancrenaz, Nicholas Bachand, Susan Cheyne, Christelle Colin, Debby Cox, Doug Cress, Kay Farmer, Erik Meijaard, Kari Nienstedt, Tamar Ron, Anne Russon, Albert Schenk, Steve Unwin, Itsaso Vélez del Burgo, Liz Williamson and individual sanctuaries and rescue centers.

Principal author of Section II: Debra Durham[7]

Author's acknowledgments: For the statistics update, the author gratefully acknowledges the following parties for sharing data: Species360, the Ape Taxon Advisory Group and affiliated SSPs, ChimpCARE, PASA and the individual sanctuaries and rescue centers that provided figures and reports.

Reviewers: Meredith Bastian, Kay Farmer and Karmele Llano Sánchez

Endnotes

1. To protect the confidentiality of communication conducted for this research, this review refrains from citing certain sources that would reveal the identity and location of reviewed facilities.

2. For more information, see Ancrenaz *et al.* (2016); Campbell *et al.* (2015); Fruth *et al.* (2016); Humle *et al.* (2016); Maisels *et al.* (2016a); Plumptre, Robbins and Williamson (2016c) and Singleton *et al.* (2016). The Wildlife Conservation and National Parks Act predates South Sudan's independence but is still in force as a 2015 revision has yet to be enacted into law (CANS, 2013; A. Schenk, personal communication, 2017).

3. For details, see Ancrenaz *et al.* (2015b); Brou Yao *et al.* (2005); Campbell *et al.* (2008); Geissmann *et al.* (2013); Hockings and Humle (2009); Imong *et al.* (2014a); Indonesia MoF (2009); Lao MAF (2011); Molur *et al.* (2005); Rawson *et al.* (2011); SWD (2011); Turvey *et al.* (2015); White and Fa (2014); Wich *et al.* (2012b); Williamson *et al.* (2014).

4. For more information, see Campbell *et al.* (2008); Dunn *et al.* (2014); Geissmann *et al.* (2013); Gumal and Braken Tisen (2015); Indonesia MoF (2009); Lao MAF (2011); Lu and Tianxiao (2012); Maldonado and Fourrier (2015); Molur *et al.* (2005); Morgan *et al.* (2011); Plumptre *et al.* (2010); Rawson *et al.* (2011); SWD (2011); Turvey *et al.* (2015).

5. Wildlife Impact – https://wildlifeimpact.org/

6. WWF – http://wwf.panda.org/what_we_do/endangered_species/great_apes/apes_programme/

7. Save the Chimps – http://www.savethechimps.org/

Annex I

Power Grid Electrocution: A Risk to Primates in Rural, Suburban and Urban Environments

Medium-sized mammals rarely endure in anthropogenic environments. Primates are an exception, as they occur in a number of towns and cities in Africa, Asia, and Central and South America. In areas with a matrix of human development, indigenous trees and green spaces, primates can successfully persist as either urban exploiters or urban adaptors, often transforming such locations into important conservation centers. These settings are not free from danger, however. Among the threats to primates is power infrastructure. While overhead distribution lines may serve as aerial pathways for individual primates and groups, particularly in areas that are fraught with terrestrial hazards, they also pose a substantial risk of electrocution.

Numerous anecdotes and reports attribute primate injuries and deaths to electrocutions (Ampuero and Sá Lilian, 2012; Chetry *et al.*, 2010; Rodrigues and Martinez, 2014; M. Ancrenaz, personal communication, 2017; S. Cheyne, personal communication, 2017). Given the difficulty of recording electrocutions, few studies have been able to quantify the impact of electrocutions on a population. Recently, however, researchers have begun to analyze electrocution data, identifying patterns based on primate species, size, age, sex and locomotion of victims, as well as seasonal variations in electrocutions (Katsis, 2017; Kumar and Kumar, 2015; Ram, Sharma and Rajpurohit, 2015; Slade, 2016).

The findings show that electrocutions occur in at least 28 species across eight primate families ranging from marmosets to orangutans (Slade and Cunneyworth, 2017, table 1). Electrocutions are not limited to arboreal species; they also affect those classified as terrestrial in their natural habitat, albeit at a lower rate. Mortality rates are generally high, as would be expected. Surviving individuals often have catastrophic injuries with poor to very poor prognoses, even with veterinary intervention (Kumar and Kumar, 2015; Slade, 2016).

A number of mitigation methods can be applied to address the animal welfare issues associated with electrocutions. These vary in duration of effectiveness. As an emergency or short-term measure, pruning trees around transformers and power cables can serve to disassociate vegetation from the power grid (Lokschin *et al.*, 2007). In the medium term, aerial bridges can be installed in strategic locations to provide alternatives to the use of

FIGURE AX.1
Uninsulated vs. Insulated Transformer

Key:
1. Hot clamp is uncovered
2. Primary inputs (bushings) are exposed
3. Secondary output (terminals) and wire connection are exposed
4. Cables are uncovered
5. Lightening arrestor is uncovered
6. Fuse cutout is uncovered

Uninsulated

Insulated
(with protection equipment made of grey silicone rubber)

Source: Courtesy of Refuge for Wildlife (refugeforwildlife.org)

power cables (Jacobs, 2015; Lokschin *et al.*, 2007). Insulating power cables and transformers in known electrocution hotspots is a long-term measure that can nearly eliminate the risk of electrocution (Printes, 1999; Refuge for Wildlife, n.d.; see Figure AX.1).

The financial burden of these mitigations generally lies solely with animal welfare organizations. As these organizations rely mostly on grants and donations, the monitoring and implementation of such measures is potentially unsustainable. Consequently, mitigations can have limited effectiveness, especially with the ongoing expansion of power infrastructure into new areas. Nevertheless, two initiatives have had encouraging results.

The first, the Programa Macacos Urbanos in Porto Alegre, Brazil, lodged a legal complaint against the state power company, citing primate electrocutions as an environmental crime. The judgment was entered in favor of the plaintiff and obliged the company to install approximately US$30,000 worth of insulation materials on the grid (Printes *et al.*, 2010).

The second, Colobus Conservation in Diani, Kenya, worked with Kenya Power on identifying electrocution hotspots. The parastatal voluntarily insulated cables at a cost of US$115,000. The negative impact of the power infrastructure on primates violated Kenya Power's code of ethics, which committed the company to addressing the issue in a substantive manner (J. Guda, personal communication, 2017).

Primate electrocutions have a cost implication for power distribution companies in terms of power infrastructure maintenance and associated customer power outages. Costs to a company for insulating cables can be offset to an extent that is dependent on the number of electrocutions in an area. Increased power reliability leads to improved customer good will, which is a benefit for a company.

Mitigation measures to prevent electrocutions are most effective when they are part of an integrated approach involving the state, power companies, residents, and animal welfare, conservation and research organizations. When these stakeholders work together, significant steps can be made towards creating valuable conservation areas in anthropogenic habitats.

Annex II

The Potential of the Global Forest Watch Platform to Transform the Use of Satellite Imagery to Monitor Tree Cover Loss

The detection of tree cover loss over time using satellite imagery allows for the location, visualization and comparison of forest change before and after infrastructure development. However, using satellite data has historically required substantial expertise and funding to acquire, process and interpret the raw information.

For example, Curran *et al.* (2004) compiled, classified and manually edited images of Gunung Palung National Park in Indonesia at six time periods to document extensive "and accelerating" deforestation over 14 years. Similarly, Laporte *et al.* (2007) tracked the progression of logging roads in the Republic of Congo over more than 25 years by compiling, geometrically correcting and visually enhancing more than 300 Landsat images, and then manually digitizing and cross-checking each of the roads detected in the images. Gaveau *et al.* (2009a) analyzed 98 Landsat images to track deforestation in Sumatra from 1990 to 2000 and showed that protected areas lost some forest but promoted protection overall, both inside and outside their boundaries. Each of these efforts provided valuable evidence of the effects of human activity on forests, but the requisite effort and costs prevented widespread use of satellite data.

Global Forest Watch (GFW), a new forest change analysis platform, has transformed the monitoring process and increased access to the power of satellite imagery. It provides a publicly available tool for monitoring changes in ape habitat in near-real time (GFW, 2014; see Chapter 7). The global reach of this information enables GFW to provide a standardized way of analyzing change in tree cover, enabling comparison across sites.

Launched in 2014, GFW provides spatially explicit, high-resolution tree cover change data, derived from thousands of satellite images that are updated annually for the entire world. GFW offers free access to annually updated tree cover and tree cover change data at a resolution of 30 m × 30 m using Landsat imagery; by late 2017, updates of tree cover change for most ape range states will be available on a weekly basis (Hansen *et al.*, 2013; M. Hansen, personal communication, 2017).[1]

Ape range stakeholders can use the online GFW tools to monitor ape habitat regularly, analyze tree cover loss and gain data for a country or protected area, create custom maps or download data for their target region. Users can upload or draw a specific area, such as a species range or road corridor, and analyze tree cover within it over time. GFW thus allows those involved in ape conservation, including at the grassroots level, to monitor changes and generate critical information at various spatial scales as a means of enhancing their conservation efforts.

Several communities within chimpanzee habitat have started to use the Forest Watcher mobile app to find and validate GFW forest loss data on their village lands and report on human activities responsible for deforestation. The free app requires a smartphone but allows non-technical users to download data, go offline and collect location data on forest change, and upload upon return to Internet access. The next version of the Forest Watcher app will have a weekly forest loss alert capability, which will help community participants to monitor, verify and manage road impacts in near-real time.

Endnote

1. Satellite imagery technology and access are constantly developing; for example, the European Space Agency now provides open access to Sentinel Online. This technical website includes 10-m resolution imagery for most parts of the globe and one of their thematic areas is forest monitoring – https://sentinel.esa.int/web/sentinel/thematic-areas/land-monitoring/forest-monitoring.

Annex III

Data Sets and Detailed Methods

Method

The first three case studies in Chapter 3 examine the effects of road improvement projects on ape habitat in Indonesia and Tanzania using open-access satellite imagery. The analysis quantified the extent of loss each year, which enabled comparison of the area or percent of habitat loss in years before and after the construction or upgrade of a road, as well as a comparison of loss rates at various distances from the road. The chapter explores tree cover change in zones within 5 km and within 10 km of each road; in Tanzania, the buffer zone extends to 30 km, as no other roads existed in the area to confound results. ArcGIS was used to create and display the buffer areas and to overlay them with Global Forest Watch (GFW) tree cover data.

The global forest change 2000–14 data set, freely available at GFW, served as the basis for the analysis (Hansen *et al.*, 2013). Canopy cover in the year 2000 served as the baseline forest coverage, and cover in each of the subsequent 14 years provided annual change data. The annually updated 30 m × 30 m Landsat tree cover data allowed changes in available habitat to be quantified over time. Each road project began after 2000, the initial year of the data set, and concluded before 2014, the most recent year for which the global forest change data are available.

Definition of "Forest"

Each case study used a "canopy density" value—the percent tree cover in each pixel of the satellite image data—that reflected the general forest type in the area and the species' ecological requirements and tolerance of canopy openness (GFW, 2014; IUCN, 2016a; see annexes VIII and IX).

The GFW platform allows users to select canopy density values. Ape habitat for the Sumatra case studies included all pixels with 75% or more tree cover, which reflects the dense, interconnected canopy required for movement by resident gibbons. Chimpanzees in Tanzania have evolved in drier forests than those supporting gibbons and orangutans in Sumatra, and chimpanzees are believed to tolerate a more open canopy (Kano, 1972). Habitat in western Tanzania was therefore defined as pixels with 30% or more tree cover, to include dry forests and savannah–woodland habitats of chimpanzees. Tanzanian roads were digitized on screen in ArcGIS Desktop using historical DigitalGlobe satellite images; the Landsat satellite images were accessed through Google Earth and Earth Engine.

The global forest change data set measures "tree cover," which in some areas may overestimate forest cover by including mature tree plantations as well as natural forest (Tropek *et al.*, 2014). Therefore, this analysis incorporates boundaries of agricultural plantations in Sumatra, which were located and delineated using visual interpretations of satellite imagery, primarily Landsat, supplemented by high-resolution imagery from Google Maps, Bing Maps or DigitalGlobe, if available (Transparent World, 2015). Detectable areas of agricultural plantations were considered tree cover loss in the first year of the study (2001). Dates of plantation establishment were unknown, however, and some mature plantations may have been included in the 2000 forest cover values. In these cases, the inclusion of some mature plantation land may have overestimated initial natural forest cover and loss in the 2000–14 period (Tropek et al., 2014).[2]

In contrast, the data set may underestimate forest cover in dry forest, such as savannah–woodland chimpanzee habitats (Achard et al., 2014). Piel et al. (2015a) compare changes in chimpanzee density in Tanzania to local forest loss assessed using GFW data. They find lower mean chimpanzee densities associated with increasing habitat loss, suggesting that chimpanzee distribution and abundance may decrease with forest loss and that the GFW platform could be useful for assessing their status in chimpanzee habitats.

Data Sets and Tools

Hutan, Alam dan Lingkungan Aceh (HAkA) provided the north Sumatran road data. Tanzanian roads were digitized in ArcGIS Desktop using DigitalGlobe historical satellite images at a spatial resolution of 60 cm; 30 m Landsat satellite images were accessed through the Google Earth and Earth Engine platforms. High-resolution UrtheCast satellite imagery, available from GFW, and high-resolution DigitalGlobe satellite image base maps helped assess GFW's capacity to verify forest loss associated with roads, identify causes of deforestation and visualize their impact in this study's areas of interest. The GFW tree cover change data were presented in maps produced using ArcGIS.

Input from ape and regional experts and a review of related scientific literature (Barber et al., 2014; Clements et al., 2014; Laurance, Goosem and Laurance, 2009) assisted in estimating the expected effects on ape populations, based on recorded distances from roads traveled by hunters, farmers, loggers and others who threaten forests. This straightforward method can help expose damage, predict potential further loss and target mitigation efforts to reduce adverse effects on surrounding ape habitat. Regularly updated imagery allows for the detection of forest loss, although detection of forest degradation, including hunting and other below-canopy effects, remains a frontier in the application of remote sensing.

Endnote

2. Where accuracy is essential, when monitoring canopy cover using satellite imagery, ground truthing is important. This is particularly relevant in areas that include plantations, as shaded coffee and cocoa plantations resemble good forest habitat from above. Shaded coffee and cocoa plantations are used by many species of animals, but this habitat is of limited or no value to arboreal primates, including gibbons and orangutans (M. Coroi, personal communication, 2017).

Annex IV

Global Land Analysis and Discovery Alerts for Early Detection of Forest Loss and Focusing of On-The-Ground Response

Global Forest Watch (GFW) currently provides annual updates of forest cover and forest change at a resolution of 30 m. More importantly for detecting loss of ape habitat, managers will soon be able to use the power of Global Land Analysis & Discovery (GLAD) alerts, products of an "as-it-happens" deforestation alarm system, which are triggered when a threshold portion of a 30 m × 30 m pixel changes from forest to non-forest cover (GLAD, n.d.). The platform allows users to receive forest loss alerts in their email inboxes for areas across the tropics at a 30-m resolution, updated weekly. By helping to detect habitat loss at the very onset of road building, alerts can facilitate more timely, and therefore more effective and efficient, interventions (Hansen et al., 2016).

Subscribers—such as NGO staff, concession holders, park directors or other officials—can receive these near-real-time alerts of detections of large-scale forest loss for whatever areas they select—be it a country, reserve, conservation landscape, road buffer or other stretch or expanse. Internet access is needed to receive alerts. GLAD started with the Congo Basin, Indonesia and Malaysia, and is now available to help managers easily and consistently monitor change across most tropical forest (M. Hansen, personal communication, 2017).

In the near future, assessments of infrastructure effects on forest habitat could include evaluations of patterns of GLAD alerts as possible indicators of the intensity and direction of imminent forest loss. Analyses could also compare areas experiencing GLAD alerts to factors associated with forest loss, such as slope or distance to clearings, roads and towns. The rapid alert system can help guide associated development and enforcement to ensure no additional illegal development happens along roads where restrictions or planning regulations have been established.

Annex V

Conservation Action Planning Review Process Results for Chimpanzees in Tanzania

Tanzania's National Chimpanzee Management Plan workshop has identified roads as a "high" threat to both core chimpanzee habitats and chimpanzee corridors in the country. In some locations that are critical for chimpanzees, the threat could be "very high" (TAWIRI, in preparation). Roads already cut through most chimpanzee corridor areas in Tanzania. Roads alone are not a major threat to chimpanzee corridors since migrating chimps can cross isolated roads, but they are much more of a threat if the forest on either side of the road is not maintained. The roads of greatest concern are those that fragment core chimpanzee habitat areas, such as the road currently being planned and constructed along the eastern border of Mahale Mountains National Park.

A key driver of deforestation across the region is the arrival of settlers via roads and footpaths, as they clear riverine forests for farming. These forests are sought after because they grow on the region's more fertile soils, which are also suited for farming. Riverine forest represents a small but critical habitat for chimpanzees in western Tanzania—roughly 2% of the total chimpanzee range. No chimpanzees in the region live in miombo woodlands without some access to riverine forest patches (Pusey *et al.*, 2007).

Findings from the 2016 conservation action planning review process reveal that efforts have successfully protected chimpanzee habitats in dry forests and miombo woodlands (see Figure AX.2a). However, interventions so far have been less successful at protecting the evergreen forest habitats that are critical for chimpanzee survival in the region (TAWIRI, in preparation; see Figure AX.2b). Wildlife managers in the region will need to continue to monitor and develop conservation strategies that prevent loss along and near these riverine forests.

In developing the plan, the authors applied the Open Standards methodology to assess and rank the threat roads pose to chimpanzee core habitat conservation targets and corridor conservation targets. Specifically, they looked at:

- **Scope**: The proportion of the core habitat and corridors that can reasonably be expected to be damaged within ten years if current circumstances and trends persist.
- **Severity**: Within the scope, the expected level of damage to the core habitat and corridors if current circumstances and trends persist.
- **Irreversibility**: The degree to which projected damage may be reversed and the damaged core habitat and corridors restored (CMP, 2013).

The immediate footprint of a road is considered a threat of very high severity to chimpanzee core habitat since it directly eliminates woodland and forest habitat. Very high severity ranking implies that the threat is likely to destroy or eliminate chimpanzee core habitat in the near future. The area around a road is considered a high- or medium-severity threat, depending on the degree to which other human activities can be controlled in the area. Roads by themselves generally have a medium irreversibility rating since the road footprint itself could be restored within 50 years.

FIGURE AX.2
Status of Core Chimpanzee Habitat Areas and Corridors, Tanzania, 2000 vs. 2014

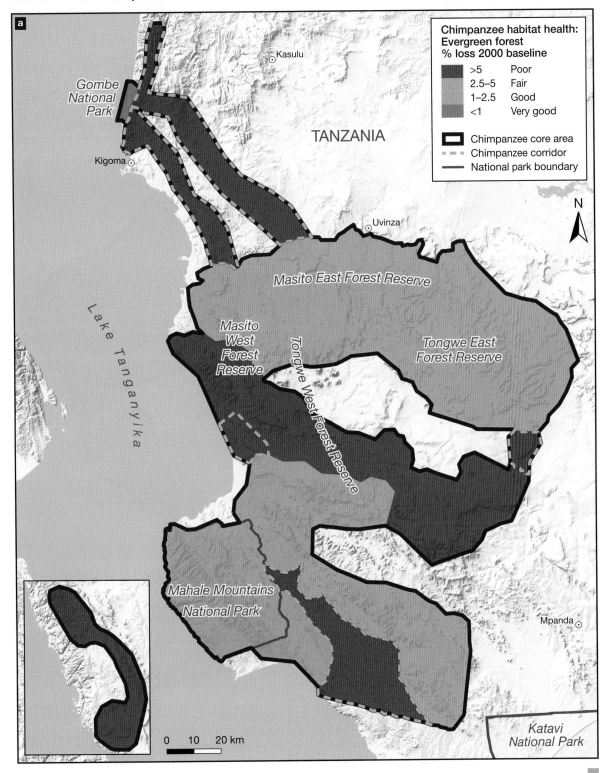

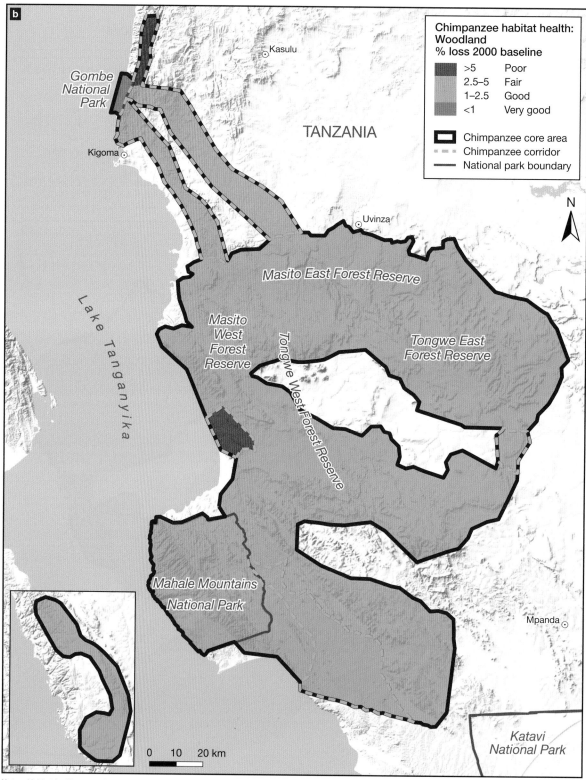

Notes: Habitat health was estimated using Global Forest Watch data to show the percentage of loss of (a) forest and woodland and (b) evergreen forest.

Data sources: GFW (2014); Hansen *et al*. (2013); TAWIRI (in preparation); courtesy of JGI.

Annex VI

The World Bank's Safeguard Policies and the Pro-Routes Project

The World Bank's safeguard policy framework is designed to avoid negative environmental impacts, or otherwise to minimize, reduce, mitigate or compensate for them by integrating environmental considerations into project planning. The policies also require best practice in regard to public participation in decision-making (World Bank, n.d.-b). At the time of writing, the Bank's Environmental and Social Framework included:

- operational policies (OP): concise statements of policy objectives and operational principles, including the roles and obligations of the borrower and the World Bank; and
- Bank procedures (BP): mandatory OP-related procedures to be followed by the borrower and the World Bank.[3]

The World Bank screens each proposed project to determine what type of environmental assessment is appropriate. As part of this process, the Bank classifies each project based on its potential environmental impacts and related factors. In accordance with OP/BP 4.01 on environmental assessments, the World Bank classified the Pro-Routes plan as a Category A project, which the Bank defines as "likely to have significant adverse environmental impacts that are sensitive, diverse, or unprecedented." For proposed projects in this category, borrowers are responsible for preparing an EIA or comparable report (World Bank, n.d.-a, 2013a, p. 2).

In line with OP/BP 4.01, borrowers for Category A projects that are "highly risky or contentious or that involve serious and multidimensional environmental concerns" are encouraged to "engage an advisory panel of independent, internationally recognized environmental specialists to advise on all aspects of the project relevant to the [environmental assessment]" (World Bank, 2013a, p. 1). Accordingly, the Pro-Routes project planners established an environmental and social advisory panel.

With reference to the environmental assessment, under OP/BP 4.01, the borrowers for the Pro-Routes project are also required to (World Bank, n.d.-b):

- inform decision-makers of the nature of environmental and social risks;
- ensure that projects proposed for Bank financing are environmentally and socially sound and sustainable (so as to promote positive impacts and avoid or mitigate negative impacts); and
- increase transparency and provide mechanisms for the participation of stakeholders in the decision-making process for the project.

Under OP/BP 4.04, policies relating to natural habitats further require borrowers to:

- protect, maintain and restore natural habitats and their biodiversity;
- ensure sustainability of the services and products that natural habitats provide to human society;
- involve local communities in planning and implementation; and
- take a precautionary approach to natural resource management.

Under OP/BP 4.36, concerning forests, borrowers must also:

- contribute to sustainable development and meet the demand for forest products and services through sustainable forest management;
- protect and maintain the rights of communities to use their traditional forest areas in a sustainable manner;
- protect global environmental services and values of forests;
- avoid encroachment on significant areas of forest; and
- ensure forest restoration projects maintain or enhance biodiversity and ecosystem function.

Endnote

3. The World Bank approved a new Environmental and Social Framework in August 2016. It will launch the framework in 2018 and will apply it to all new investment projects; the Bank's current safeguards are expected to run parallel to the new framework for about seven years, in the context of projects approved before the launch of the new framework (World Bank, n.d.-d; see Box 5.1).

Annex VII

The Decommissioning of Dams

Hydropower infrastructure, such as dams, provides energy to many communities, cities and countries around the world. The ecological, economic and social impacts of dams are often inadequately considered, but they have been well documented. Factors that often garner greater attention from decision-makers are the economic and safety concerns associated with building and maintaining dam structures. As is the case with any type of hard infrastructure, dams must be maintained to certain standards in order to ensure the safety of downstream communities, including animals living in riparian corridors (Brown *et al.*, 2009; WCD, 2000).

Hundreds of thousands of people around the globe have perished following the failure of dams (Si, 1998); such disasters can result from any number of design and deterioration issues (ASDSO, n.d.). In the United States alone, hundreds of fatalities have resulted from people climbing on, paddling over, fishing near or otherwise interacting with dam structures and ignoring risks associated with potential hydraulic undertows at the base of dam structures (Tschantz, 2014).

Dam owners who decide to remove a structure typically cite socioeconomic concerns as the motivating factors behind their decision (Engberg, 2002). For example, when a hydropower dam no longer produces enough power to justify its existence on economic grounds, it is decommissioned.

In the United States, the Federal Power Act and the Magnuson Stevens Fishery Conservation and Management Act require hydropower regulators to consult with agencies charged with protecting natural resources before issuing hydropower licenses. These requirements exist to protect access to important habitat for migratory fish species (McDavitt, 2016). If installing a required fish passage system is cost-prohibitive compared with the income from power production, a hydropower producer may abandon a project and the dam might eventually be removed. A case in point involves the removal of two dams on the Elwha River in Washington State. Both federal environmental agencies and native tribes had insisted that the dams incorporate adequate fish passage for salmon, which would have been difficult to ensure and cost-prohibitive (Gowan, Stephenson and Shabman, 2006). After years of deliberation, the Elwha and Glines Canyon dams were removed in 2011 and 2014, respectively.

A non-hydropower dam may become obsolete when associated mills shut down, an impoundment fills with sediment, its structure breaks down, its design proves to be ineffective for its intended purpose or if it no longer has an identifiable purpose. If a dam has become obsolete, the owner may be particularly unwilling or unable to bear maintenance costs and shoulder liabilities associated with keeping it intact, and the dam may be removed (Engberg, 2002).

Given that the cost of dam removal can vary widely, and that the economic valuation of the ecosystem services provided by a free-flowing river can be difficult to quantify, conducting a true cost–benefit analysis of dam removal can be challenging (Whitelaw and MacMullen, 2002). As of 2015, more than 1,200 of about 80,000 dams in the United States had been removed, but fewer than 10% of those removals were scientifically studied—and most of the studies that were conducted did not examine broader ecosystem responses (Bellmore *et al.*, 2016).

Research that did consider ecosystem impacts has shown that the potential benefits of dam removal include the following:

- reconnection of river habitat for fish and other aquatic species;
- reestablishment of more natural flow patterns of water, sediment and nutrients through the ecosystem;

- reduction of thermal impacts of impoundments;
- improvement of aquatic organism access to upstream refugia from increasing water temperatures;
- elimination of liability and maintenance costs;
- decreases in flooding upstream; and
- increases in connectivity of the river for recreational purposes (Lejon, Malm-Renöfält and Nilsson, 2009; Magilligan *et al.*, 2016; Wildman, 2013).

Ultimately, the removal of a dam means bringing a river closer to a more natural, functional state. When a dam is constructed, the flow of animals, nutrients, sediment and other natural processes is stopped or limited indefinitely (O'Connor, Duda and Grant, 2015). Downstream communities are the most impacted by dams; too often, they are disadvantaged or disenfranchised, with no capacity to defend themselves against political pressures to build new infrastructure (WCD, 2000). By taking the services provided by a functioning river into account, along with the likely impacts to downstream communities, decision-makers can help to avert or at least minimize negative impacts on local populations and biodiversity, be it with reference to the construction of a dam or its removal.

Annex VIII

Data Sets Used

The assessment of the status of ape habitat at two spatial scales (see Figure 7.1) included the analysis of several global data sets:

- **The Global Forest Change 2000–14 data set.** Provided by the University of Maryland in association with Google, Inc., and the foundation of GFW 2.3, these data on annual forest cover and tree cover loss are presented at 30-m resolution. Development of this data set included verification using very high-resolution spatial data, such as Quickbird imagery and existing percent tree cover data sets derived from Landsat data (Hansen *et al.*, 2013). Data are available online at: http://earthenginepartners.appspot.com/science-2013-global-forest.
- **IUCN Red List data on geographic ranges of each recognized subspecies.** The range delineations vary in terms of precision. Ranges of 22 subspecies—including most gibbons in mainland Asia and gorillas—reflect the recent extent of their distribution, given the historical loss of forest cover. In contrast, ranges of the other 16 subspecies have less refined boundaries and reflect the apes' historical distributions, which include areas of urban development that are no longer home to apes (IUCN, 2016c).
- **The World Database on Protected Areas (WDPA).** A joint effort between IUCN and the UN Environment Programme (UNEP), the WDPA is managed by UNEP's World Conservation Monitoring Centre (IUCN and UNEP-WCMC, 2016). The 2016 PA boundaries may include areas that were not protected throughout the study period. The analysis included all reserves and PAs within IUCN protected area categories I–VI, except those with a status of "not reported" or "proposed" and those designated as UNESCO biosphere reserves (IUCN, 2016c, n.d.-b; UNESCO, n.d.-b). The analysis retained all PAs and other reserves with no IUCN classification, including thousands of forest reserves and a large number of community reserves. The WDPA does not yet contain all of the world's community reserves, but it is the most comprehensive database currently available at the global scale.
- **Unpublished data on industrial plantations.** Mapped by the World Resources Institute and Transparent World, these data were used to account for the tree cover created by mature oil palm, timber, pulpwood and rubber plantations, particularly in Southeast Asia (GFW, 2014; see Annex XI).

Errors in any of these data sets could influence results, but the global scale of this analysis precludes inclusion of finer-scale data for each ape range country. Annex XII suggests additional data layers that could refine this analysis.

Annex IX

Forest Habitat Definitions for 38 Ape Subspecies

Geographically unique ape subspecies have adapted to particular environmental conditions, including canopy openness. This analysis uses different values of tree cover density to define habitat and to estimate forest change in the ranges of each subspecies. For each subspecies, it defines a threshold canopy density (the percent tree cover per pixel) below which the subspecies may not be viable.

The thresholds reflect each subspecies' ecology, based on IUCN habitat data and scientific literature (IUCN, 2016c), tree cover within their range and in protected areas (PAs) with known occupancy; regional expert opinion on chimpanzees, gibbons, gorillas and orangutans was also taken into consideration, as was the ability of certain subspecies to survive in a variety of forest types, including habitat degraded by humans (see Annex X).

- **Chimpanzees.** Among the apes, chimpanzees tend to have the most flexible ranging behavior (Maldonado et al., 2012; K. Abernethy, personal communication, 2016). Western chimpanzees occupy areas ranging from closed-canopy moist forest to wooded savannah, indicating a habitat threshold that included pixels with 15% or more tree cover. A lower canopy density was thus selected for all chimpanzee subspecies (see Annex X).
- **Gibbons.** Tropical and subtropical moist broadleaf forests comprise most ape habitat. For gibbons, who require canopy connectivity and a diversity of fruit trees (W. Brockelman, personal communication, 2016), recommended canopy cover is at least 75% (Gaveau et al., 2014; D. Gaveau, personal communication, 2016).
- **Gorillas.** Most gorillas occupy densely forested areas, although habitat for Grauer's and mountain gorillas contains substantial areas of bamboo (K. Abernethy, personal communication, 2016).
- **Orangutans.** A wide range of canopy density was suggested for orangutan habitats. At the high end, based on natural forest cover in Borneo, the proposed density was at least 75% (D. Gaveau, personal communication, 2016). Yet, based on the ability of orangutans to use partially disturbed habitat so long as they are not hunted, the density could be as low as 30% (E. Meijaard, personal communication, 2016).

Where the ranges of subspecies overlap, forest was defined in the overlapping regions using the requirements of the more exacting subspecies—that is, the one with the higher canopy density percentage.

The availability of IUCN ape range data and the Google Earth Engine scripts on GFW will enable other users to run such analyses at a higher or lower percent tree cover per pixel and thus tailor the parameters to specific environments (see Annex VIII).

In this analysis, a single taxon-specific threshold value of canopy density was used to exclude areas where forest structure or composition may not be suitable for a given ape subspecies. This approach may not adequately reflect the ecological variability found among populations of some subspecies, particularly chimpanzees. Eastern and western chimpanzees occupy regions dominated by either denser forest or savannah woodlands and forest mosaic.

In the range of western chimpanzees, most deforestation occurs in zones of higher canopy density, and the low canopy density percentage assigned to this subspecies to reflect its ecological flexibility probably underestimates forest loss in the wetter southern half of its range (L. Pintea, unpublished data, 2016). Use of a denser canopy threshold (for example, 30%, instead of 15%) for western chimpanzees would have decreased detected habitat loss by 2.5% over the period under review, mainly because the baseline forest in 2000 (564,000 km² or 56.4 million ha at 15% canopy cover) would have covered a much smaller area (355,000 km² or 35.5 million ha at 30%).

Analysis of forest change for these and other ape taxa will therefore benefit from the addition of environmental layers, such as potential or suitable habitat, ecoregions and elevation (see Annex XII).

Annex X

Canopy Density Percentages Used for 38 Ape Subspecies

Common name	Scientific name	Threshold percentage	Source
Bonobo	Pan paniscus	50	occupancy
Central chimpanzee	Pan troglodytes troglodytes	30	occupancy, expert
Eastern chimpanzee	Pan t. schweinfurthii	30	occupancy, expert
Nigeria–Cameroon chimpanzee	Pan t. ellioti	30	occupancy
Western chimpanzee	Pan t. verus	15	occupancy
Cross River gorilla	Gorilla gorilla diehli	50	occupancy
Grauer's gorilla	Gorilla beringei graueri	50	occupancy
Mountain gorilla	Gorilla b. beringei	50	occupancy
Western lowland gorilla	Gorilla g. gorilla	75	occupancy, expert
Northeast Bornean orangutan	Pongo pygmaeus morio	50	expert, range
Northwest Bornean orangutan	Pongo p. pygmaeus	50	expert, range
Southwest Bornean orangutan	Pongo p. wurmbii	50	expert, range
Sumatran orangutan	Pongo abelii	50	expert, range
Eastern hoolock	Hoolock leuconedys	75	ecology
Western hoolock	Hoolock hoolock	75	ecology
Abbott's gray gibbon	Hylobates abbotti	75	ecology
Agile gibbon	Hylobates agilis	75	ecology
Bornean gray gibbon	Hylobates funereus	75	ecology
Bornean white-bearded gibbon	Hylobates albibarbis	75	ecology
Carpenter's lar gibbon	Hylobates lar carpenteri	75	ecology
Central lar gibbon	Hylobates l. entelloides	75	ecology
Kloss's gibbon	Hylobates klossii	75	ecology
Malaysian lar gibbon	Hylobates l. lar	75	ecology
Moloch gibbon	Hylobates moloch	75	ecology
Müller's gibbon	Hylobates muelleri	75	ecology
Pileated gibbon	Hylobates pileatus	75	ecology
Sumatran lar gibbon	Hylobates l. vestitus	75	ecology
Yunnan lar gibbon	Hylobates l. yunnanensis	75	ecology
Cao Vit gibbon	Nomascus nasutus	75	ecology
Central Yunnan black-crested gibbon	Nomascus concolor jingdongensis	75	ecology
Hainan gibbon	Nomascus hainanus	75	ecology
Laotian black-crested gibbon	Nomascus c. lu	75	ecology
Northern white-cheeked crested gibbon	Nomascus leucogenys	75	ecology
Southern white-cheeked crested gibbon	Nomascus siki	75	ecology
Southern yellow-cheeked crested gibbon	Nomascus gabriellae	75	ecology
Tonkin black-crested gibbon	Nomascus c. concolor	75	ecology
West Yunnan black-crested gibbon	Nomascus c. furvogaster	75	ecology
Siamang	Symphalangus syndactylus	75	ecology

Notes: Percentages reflect overall vegetation cover and tolerance of canopy openness. IUCN geographic ranges of most gibbon species are highly fragmented; much of their former range was already converted to non-forest land uses by 2000.

Annex XI

Value and Limitations of the GFW Tool for Monitoring Forest Cover Change

Value

Launched in 2014, Global Forest Watch provides access to spatially explicit, high-resolution forest change data, which is derived from thousands of satellite images and updated annually for the entire world (GFW, 2014; Hansen *et al.*, 2013). GFW's online forest monitoring and alert system combines cutting-edge algorithms, satellite technology and cloud computing to identify where trees are growing and disappearing in near-real time.

In a few minutes, a user can obtain up-to-date information about the status of forest landscapes worldwide or for a particular region, such as a country, species range or PA. The user can also modify the percent tree cover per pixel (the canopy density) GFW uses to determine forest cover, so as to adjust any analysis to reflect a more open or closed canopy. Users can generate regular, accurate summaries of forest extent and change, create custom maps, analyze forest trends, subscribe to near-real-time alerts of forest loss or download data for their local area, country or region.

GFW provides free access to the annually updated global forest change data at relatively high resolution, along with tools with which to analyze tree cover loss and gain data. The analysis presented in Chapter 7 shows the data's broad application with respect to species using both closed-canopy and more open-canopy forest. GFW forest change information is highly scalable, from a forest corridor to all ape species' ranges. Applications include not only monitoring of PAs and range polygons, but also identification and monitoring of forest corridors and other areas of concern. As discussed in the chapter, GLAD alerts will identify areas of likely forest loss on a weekly basis, facilitating each of these activities and helping resource managers to monitor forest cover consistently.

Global Forest Watch Fires is a related platform that provides near-real-time information on forest fires in Southeast Asia. Fires have led to sweeping forest loss in Indonesia, especially in peat forests that are critical orangutan habitat. The tool's daily, spatially explicit updates on fire occurrence empower people to enhance their monitoring of and responses to fires before they burn out of control and to hold accountable those who may have burned forests illegally (GFW, n.d.-a).

Limitations

Despite its advantages, the use of GFW alone has certain limitations. For example, annual updates at 30-m resolution or bi-weekly updates at 500-m resolution will not necessarily provide the precision needed to determine effects on a given primate population, the cause of clearance or associated impacts, such as poaching and non-timber forest product collection. Reliance on remotely sensed information further constrains the GFW platform's ability to help explain drivers of forest change.

The Global Forest Change 2000–14 data set created by Hansen *et al.* (2013) may underestimate forest

---

BOX AX.1

Detecting Plantations at a Broad Scale

A criticism of the Global Forest Change 2000–14 data set is that it does not always distinguish between natural forest and industrial plantations, such as oil palm. To address this issue and identify industrial plantations in the areas under review, the analysis presented in Chapter 7 incorporates data from an industrial plantation mapping project undertaken by the World Resources Institute and Transparent World (Transparent World, 2015). Overall forest loss figures in this analysis include these plantations; to avoid double counting an area once its forest was converted to plantation, any tree cover loss within plantations was ignored.

Areas under plantation are counted as "lost" as of 2001, so that cumulative loss includes all plantation areas, regardless of whether an area was developed. The map that shows forest cover in 2000 includes any vegetation in these properties that was tall and dense enough to be considered tree cover by GFW. An unknown proportion of forest within these areas had already been converted to tree plantations by 2000; without knowledge of the dates of initial conversion of each area, it is impossible to discern whether the loss from 2000 to 2001 was all, mostly, or only partially loss of natural forest.

Although the plantations were recently digitized from high-resolution satellite imagery, the data lack information on the year in which each plantation was established. Consequently, the annual forest loss data in Figure 7.5 ignore the presence of plantations and thus show less loss than actually occurred in creating plantations in all cases.

This limitation of the global forest change data may affect results in areas of high plantation density. Agricultural plantations (of oil palm, rubber and timber) overlapped with portions of the ranges of 15 ape subspecies (13 of which are in Indonesia and Malaysia) and correspond with more than 50% of the forest loss in 12 of these ranges.

cover in dry forest habitats, such as those used by chimpanzees in Mali and Senegal (Achard *et al.*, 2014). Setting canopy density threshold values of 30% or 15% for regions with sparser tree cover helped control for this limitation, while acknowledging that most deforestation occurs in zones of higher canopy density (K. Abernethy, personal communication, 2016; L. Pintea, unpublished data, 2016). This looser definition of forest may lead to underestimating of forest loss in more densely forested portions of the ranges of *Pan* spp.

Conversely, the data created by Hansen *et al.* (2013) measure "tree cover," which in some areas may include mature tree plantations as well as natural forest (Tropek *et al.*, 2014). In addition to excluding known plantations (see Box AX.1), setting a high canopy density threshold (50% or 75%) consistent with tropical moist forest canopy helps to screen out young oil palm plantations, which have gaps in canopy cover due to the low height and small crowns of palm saplings. For the ranges of some ape species, however, plantation data were unavailable; for others, dates of plantation establishment were unknown (Transparent World, 2015). Consequently, mature plantations may have been included in the 2000 forest cover values in some areas, causing both initial forest cover in 2000 and loss in the 2000–14 period to be overestimated. Despite its limitations in distinguishing vegetation types at the local scale, the global forest change data set developed by Hansen *et al.* (2013) may provide valuable forest cover information for areas where local data may be lacking (Burivalova *et al.*, 2015).

Annex XII

Future Refinements to the Ape Habitat Assessment

Aggregating and summarizing the data for 38 ape species and subspecies of 33 countries with varying canopy cover requirements and numerous vegetation types for 2000–14 inherently involved accepting error. Given the following unknowns and shortcomings, additional data or analysis would improve this assessment:

- Forest use patterns by various subspecies within their range polygons are not fully known.
- While IUCN geographic ranges represent the best available data on subspecies at the global scale, ape populations are not evenly distributed within them; range maps, therefore, are prone to commission (false presence) errors (Rondinini *et al.*, 2006). Moreover, updates to range map polygon boundaries rely on scarce species presence data and are not consistent across subspecies (Wich *et al.*, 2016).
- The selection of a single canopy density threshold value for areas of range overlap of subspecies with differing forest cover requirements created some discrepancy in the aggregations of forest presence and loss for countries and across all ape ranges.
- PAs were created in different years, so certain forested areas may not have been fully protected until later in the period under review.
- Plantations were created in different years, so while they were considered a loss from 2001, they may have been created before 2000 and would thus reduce the initial forest cover extent.

Assessing the status and trends in forest habitat is a first step in estimating the status of ape populations. Future assessments of ape habitat will benefit from including additional data sets as they become available and accessible. These may include:

- species-specific habitat suitability maps that incorporate on-the-ground verification (Hickey *et al.*, 2013; Jantz *et al.*, 2016; Torres *et al.*, 2010; Wich *et al.*, 2012b);
- higher-resolution satellite imagery of important primate areas, via platforms such as Planet or DigitalGlobe, which increasingly provide remotely sensed data that can help conservationists determine drivers of deforestation;
- the normalized difference vegetation index (NDVI) and other satellite-derived vegetation cover data, which could help quantify forest degradation and may be particularly important for gibbons, as they require intact canopy;
- elevation data (Tracewski *et al.*, 2016);
- information on forest structure, including cover, height, stand age and intactness;
- land use data, including legal and illegal agriculture and settlements in addition to forestry;

- zoning data from both official (government) sources and unofficial sources—such as Global Witness, Greenpeace and MightyEarth—to help assess emerging or future drivers of forest loss, such as rubber, oil palm or logging concessions that may not yet be active but that have been allocated;
- ground-truthed information on land cover, species presence and human activity, including roads, to help determine drivers of forest loss; and
- important ape landscape boundaries (Max Planck Institute, n.d.-b), which were not available for Asia and were not used in the analysis in this chapter, as these would allow future habitat assessments to be conducted within their boundaries.

For this global-scale analysis, no attempt was made to determine suitable environmental conditions (SEC), which suggest the occurrence of an ape subspecies within its range (Junker *et al.*, 2012). An Africa-wide great ape suitability analysis, published in 2012 and presented in the first *State of the Apes* volume, acknowledges various limitations of a modeling approach at that scale (Funwi-Gabga *et al.*, 2014; Junker *et al.*, 2012). These included presence locations geographically biased to PAs, outdated vegetation and road data, and a lack of true absence data, each of which may have skewed determination of suitable habitat.

Habitat suitability models use a variety of factors, including forest canopy cover, to help predict and map potential habitats, but integrating these data has limited previous modeling efforts to small regions or coarse spatial and temporal resolutions. Jantz *et al.* (2016) combined global forest change data with other Landsat Enhanced Thematic Mapper Plus (ETM+) satellite imagery to model and map changes in habitat suitability between 2001 and 2014 over the entire chimpanzee range at 30-m resolution (see Annex X). However, the animal presence data on which suitability maps are based do not exist across the ranges of most ape subspecies, and current global satellite data do not allow for refinement of these results to account for suitability.

Reliable habitat suitability maps will allow future assessments to exclude areas of unsuitable land cover—such as illegal oil palm and other plantations, pressure from human activity and natural barriers—and reduce the amount of commission errors (Beresford *et al.*, 2011). With broad applicability to a range of ape taxa, this type of assessment will allow for more effective and efficient interventions.

Annex XIII

Applying Habitat Assessment to Conservation Action Plans for Apes

Transparency to reveal the relative condition of forest habitat at high spatial and temporal resolution will be increasingly central—not only to halting forest loss directly, but also to planning effective conservation strategies for apes and other forest-dependent species.

Conservation action plans (CAPs) have been developed for at least 30 ape taxa. Some CAPs have identified conservation units that, if successfully implemented, would protect majorities of the subspecies' range (Plumptre *et al.*, 2010). However, ape-range authorities and conservationists often lack the means to monitor forest cover status of these critical areas (Kühl, 2008).

CAPs that use the Open Standards process developed for chimpanzees and gorillas in Tanzania and eastern DRC have used GFW data to assess viability of ape conservation targets, prioritize threats and measure conservation success (TANAPA *et al.*, 2015). For example, the process in Tanzania considered chimpanzee habitats to be in very good or good condition if the areas lost less than 1% and 2.5%, respectively, of their forest with tree cover density greater than 30% (TAWIRI, in preparation). For the period 2000–14, forest loss between 2.5% and 5% or greater than 5%, respectively, would lead chimpanzee habitat viability to be categorized as fair or poor.

Tanzanian authorities will apply these standardized criteria in continuous monitoring of the viability of chimpanzee habitats in the country, as new forest loss data are added to the GFW platform. Together with the chimpanzee conservation community, and with an eye to supporting the conservation action planning process, GFW is developing its new Map Builder platform to allow for comparisons of thresholds of tree cover loss across customizable areas (GFW, n.d.-b).

ACRONYMS AND ABBREVIATIONS

AIIB	Asian Infrastructure Investment Bank
A.P.E.S.	Ape Populations, Environments and Surveys
AWA	Animal Welfare Act
AWF	African Wildlife Foundation
AZA	Association of Zoos and Aquariums
B–K	Blangkejeren–Kutacane
BAP	biodiversity action plan
BEGES	Bureau d'Études Spécialisé en Gestion Environnementale et Sociale, DRC
BOSF	Borneo Orangutan Survival Foundation
BP	bank procedures
BUPAC	Bili–Uélé Protected Area Complex
CAP	conservation action plan
CAR	Central African Republic
CBG	Compagnie des Bauxites de Guinée
CI	Cellule Infrastructures
CIA	cumulative impact assessment
CITES	Convention on International Trade in Endangered Species of Wild Fauna and Flora
cm	centimeter
DDNP	Deng Deng National Park
DFID	United Kingdom Department for International Development
DNA	deoxyribonucleic acid
DOPA	Digital Observatory for Protected Areas
DRC	Democratic Republic of Congo
DSEZ	Dawei Special Economic Zone
EAZA	European Association of Zoos and Aquaria
EIA	environmental impact assessment
ESAP	environmental and social advisory panel
ESIA	environmental and social impact assessment
FCFA	Franc Communauté Financière Africaine (Central African franc)
FFI	Fauna and Flora International
FOIA	Freedom of Information Act
FPIC	free, prior and informed consent
FPP	Forest Peoples Programme
FSC	Forestry Stewardship Council
GAC	Guinea Alumina Corporation
GEI	Global Environmental Institute
GFAS	Global Federation of Animal Sanctuaries
GFW	Global Forest Watch

GLAD	Global Land Analysis and Discovery
GLNP	Gunung Leuser National Park
GP	good practice
GRanD Database	Global Reservoir and Dam Database
gROADS	Global Roads Open Access Data Set
GW	gigawatt
I–R–K	Ilagala–Rukoma–Kashagulu
IAR	International Animal Rescue
ICCN	Institut Congolais pour la Conservation de la Nature
IFC	International Finance Corporation
IGCP	International Gorilla Conservation Programme
IHA	International Hydropower Association
IIED	International Institute for Environment and Development
IUCN	International Union for Conservation of Nature
ha	hectare
HKU	University of Hong Kong
HSUS	Humane Society of the United States
km	kilometer
km²	square kilometer
kV	kilovolt
ILO	International Labour Organization
JRC	European Commission Joint Research Centre
KBNP	Kahuzi-Biega National Park
kg	kilogram
KNU	Karen National Union
Lao PDR	Lao People's Democratic Republic
LAPSSET	Lamu Port, South Sudan, Ethiopia Transport (corridor)
LPHP	Lom Pangar Hydropower Project
LRA	Lord's Resistance Army
m	meter
MAAP	Monitoring of the Andean Amazon Project
MDB	multilateral development bank
MMNP	Mahale Mountains National Park
MW	megawatt
MYR	Malaysian ringgit
NCR	native customary rights
NGO	non-governmental organization
OP	operational policy
OSM	OpenStreetMap
OVAG	Orangutan Veterinary Advisory Group
PA	protected area
PAD	project appraisal document
PADDD	protected area downgrading, downsizing and degazettement

PASA	Pan African Sanctuary Alliance
Pro-Routes	High-Priority Roads Reopening and Maintenance
RAEL	Renewable and Appropriate Energy Laboratory
REDD	Reducing Emissions from Deforestation and Forest Degradation
RRI	Rights and Resources Initiative
RSPO	Roundtable on Sustainable Palm Oil
Save Rivers	Save Sarawak Rivers Network
SCORE	Sarawak Corridor of Renewable Energy
SEB	Sarawak Energy Berhad
sp.	species (singular)
spp.	species (plural)
SSP	Species Survival Plan
TANROADS	Tanzania National Roads Agency
TBC	The Biodiversity Consultancy
TCL	tiger conservation landscape
TH–L	Tamiang Hulu–Lokop
THV	terrestrial herbaceous vegetation
TNC	The Nature Conservancy
TRIDOM	Tri-National Dja–Odzala–Minkébé
UGM	Universitas Gadjah Mada (university)
UNEP	United Nations Environment Programme
UNEP-WCMC	United Nations Environment Programme World Conservation Monitoring Centre
UNESCO	United Nations Education, Science and Culture Organization
WARN	Wild Animal Rescue Network
WB	World Bank
WCS	Wildlife Conservation Society
WISER	World Indigenous Summit on Environment and Rivers
WWF	World Wide Fund for Nature/World Wildlife Fund

GLOSSARY

Algorithm: A set of instructions or rules for performing tasks such as calculations, data processing and automated reasoning.

Alternative energy: Usable power derived from sources other than fossil fuels, often with a focus on avoiding pollution and greenhouse gas emissions. See also: **clean energy** and **renewable energy**.

Anthropocene: A buzzword used to refer to the current geological epoch in view of humankind's profound impact on Earth. The term was popularized by atmospheric chemist Paul Crutzen in 2000 and recommended for adoption by a dedicated working group of the International Union of Geological Sciences in 2016. Scholars disagree about the start date of the Anthropocene, with suggestions ranging from 8,000 years ago to about 1950.

ArcGIS Desktop: A mapping and spatial data analysis application produced by Esri.

ArcGIS Online: An Internet-based mapping platform developed by Esri that enables users to access, create and share maps, scenes, apps, layers, analytics and spatial data. Available at: www.esri.com/software/arcgis/arcgisonline.

Artificial Intelligence for Ecosystem Services (ARIES): A suite of computer models that support science-based decision-making to promote environmental sustainability. Available at: aries.integratedmodelling.org.

Association of Southeast Asian Nations (ASEAN) Heritage Parks: Sites throughout the ASEAN region—Brunei, Cambodia, Indonesia, Laos PDR, Malaysia, Myanmar, the Philippines, Singapore, Thailand and Viet Nam—that are designated conservation areas in recognition of their rich biodiversity; four of the 37 sites are also UNESCO World Heritage sites.

Base-load demand: The power required to run facilities, electronics and appliances that are always on, such as hospitals and refrigerators, in contrast to peak-load demand, which is the power needed to run appliances and machines that can be turned on and off, such as computers and televisions.

Behavioral enrichment: Conditions or stimuli introduced to establish and support species-specific natural behaviors and reduce aberrant ones, with the goal of improving the psychological and physiological well-being of captive animals.

Bimaturism: Development characterized by differing stages or timings within a species or within a sex; among orangutans, mature males are flanged or unflanged (see **flanged**).

Bing Maps: An online mapping service that is part of Microsoft's Bing suite of search engines and that offers worldwide orthographic aerial and satellite imagery. Available at: www.bing.com/maps.

Biodiversity: The variety of plant and animal life on earth or in a particular habitat.

Biodiversity hotspot: A significant reservoir of biodiversity that is threatened with destruction.

Biota: Plant and animal life of a specific region.

Blood diamond: A diamond mined in a war zone and sold to finance an insurgency, an invading army's war efforts or a warlord's activity.

Blood gold: Gold mined by enslaved or otherwise victimized populations, including brutalized Congolese and the impoverished illegal miners of South Africa.

Boom-and-bust cycle: Alternating periods of economic growth and contraction. An increase in business activity, for example in connection with exploitation of a valuable natural resource, may be followed by sharp price declines for the resource or its overexploitation, a spike in unemployment and a drop in returns to investors.

Brachiation: Arboreal locomotion that relies exclusively on the arms to propel the body forward. Related term: brachiate.

Catchment: A rural or urban area where the natural landscape collects water from rain, or other precipitation. This gradually drains into a common outlet, such as a river, a bay or another body of water. Also referred to as a drainage area, river basin or watershed.

Circular economy: An economic model that aims to transform waste into resources and to bridge gaps between production and consumption.

Clean energy: Usable power generated with little or no pollution or greenhouse gas emissions, as derived from renewable sources such as sunlight, wind, biomass and waves, in contrast to "dirty" sources of energy, such as fossil fuels (coal, natural gas and oil). Not all geothermal and hydroelectric power is clean energy. See also: **alternative energy** and **renewable energy**.

Cleaner production: Processes and services that are characterized by the continual application of strategies that increase efficiency and reduce risks to the environment.

Cloud computing (or cloud technology): The use of a network of remote servers hosted on the Internet to store, manage and process data.

Conspecific: A member of the same species.

Core area: The most heavily used portion of the home range of a group or individual.

Corridor: See **Wildlife Corridor**.

Cost–benefit analysis: A process by which the benefits of a given situation or action are summed and the corresponding costs are subtracted; opportunity costs may also be factored in.

Cost engineering: The practice of managing project costs by using approaches such as estimating, cost control, cost forecasting, investment appraisal and risk analysis.

Critical habitat: An area of high biodiversity value. The International Finance Corporation defines it as habitat of significant importance to critically endangered, endangered, endemic, or restricted-range species; habitat that supports globally significant concentrations of migratory or congregatory species; a highly threatened or unique ecosystem; or an area associated with critical evolutionary processes (IFC, 2012a, p. 4).

Critically endangered: Facing an extremely high risk of extinction in the wild.

Cumulative impact: The incremental effects of one project, combined with the past, present and foreseeable future effects arising from other developments (such as infrastructure, extractive or agricultural activities) in selected areas.

Cumulative impact assessment (CIA): An evaluation that factors in the combined effects of past, present and foreseeable human activities, over time and on the environment, economy and society in a particular place.

Customary title to land: Recognition of a community's rights to access, use and control an area of land, usually based on long-established, traditional patterns or norms. Customary forest and community territory fall within this category. Customary or international law may be the source of such titles, particularly if relevant rights are not enshrined in a country's statutory legislation.

Deciduous: Pertaining to trees that lose their leaves for part of the year.

Decommissioning: In relation to dams, full decommissioning is the full removal of a dam; partial decommissioning is the partial removal of a dam.

Delegated management contractor: A public or private entity that is hired to implement a specified aspect of a development project on behalf of a state or other stakeholder.

Deterrent: A punishment or other measure established to discourage future attempts at breaking the law.

Developer: In the context of infrastructure, a firm that undertakes any of a variety of tasks related to developing a project, such as planning, finance, engineering, construction, hiring and management of assessors, compliance with regulations and coordination with partners.

Development corridor: An area characterized by major integrated infrastructure, such as paved roads, railroads, power lines and gas lines that run in parallel and are designed to open up regions for increased economic activity and land use, such as in Africa and other parts of the developing world.

Dichromatic: Exhibiting two color variations independent of sex and age.

DigitalGlobe: A commercial vendor of high-resolution satellite imagery and geospatial content. Available at: www.digitalglobe.com.

Digital Observatory for Protected Areas (DOPA): A global database of protected areas and their characteristics, operated by the European Commission's Joint Research Centre. Available at: dopa.jrc.ec.europa.eu.

Dimorphic: Having two distinct forms.

Dipterocarp: A tall hardwood tree of the family Dipterocarpaceae that grows primarily in Asian rainforests and that is the source of valuable timber, aromatic oils and resins.

Dispatchable renewables: Renewable electricity whose output can be adjusted to meet demand.

Dispersing sex: Either male or female apes who, upon reaching sexual maturity, depart from their birth area to establish their own range.

Diurnal: Daily, or active during the day.

Downstream: Towards the mouth of a river.

Ecosystem representativeness: The degree to which an ecosystem unit is representative of a biological or physical class to which it belongs, typically in accordance with biological and physical criteria. A patch of old-growth rainforest in a protected area may have a high or low representative value with respect to the vegetation type in the surrounding landscape.

Endangered: Facing a very high risk of extinction in the wild.

Endemic: Native to or only found in a certain place; indigenous.

Endemism: The state of being unique to a particular geographic area.

Environmental (and social) impact assessment (EIA or ESIA): An analytical tool used to identify and assess potential environmental (and social) impacts of a project, development or policy. The parameters for drawing up an EIA (or ESIA) are typically laid out in an environmental (and social) management framework. ESIAs are sometimes referred to as social and environmental impact assessments (SEIAs).

Environmental (and social) management framework: A plan that specifies what procedures to use in the preparation and approval of a site-specific environmental (and social) impact assessment or environmental (and social) management plan for a development project.

Environmental (and social) management plan: Guidance that identifies a set of mitigation, management, monitoring and institutional measures and explains how to apply them during the implementation and operation phases of a development project. Drawn up in accordance with an environmental (and social) management framework, the plan is designed to eliminate, offset and reduce adverse environmental (and social) impacts identified in an environmental (and social) impact assessment.

Environmental protection function: The ability of a forest or other ecosystem to contribute to the conservation of a landscape, habitat, soil or hydrogeological area, or to the preservation of human settlements or other assets, by preventing or reducing the impact of destructive natural events.

Externality: A positive or negative consequence of an economic activity as experienced by a party that is not directly related to the production or consumption.

Fission–fusion: Pertaining to communities whose size and composition are dynamic due to the coming together (fusion) and moving away (fission) of individuals.

Flanged: Pertaining to one of two morphs of adult male orangutan, the other being "unflanged"; characterized by large cheek pads, greater size, a long coat of dark hair on the back and a throat sac used for "long calls."

Floodplain: Relatively flat land that stretches out from either side of a river and that may flood during heavy rain or snowmelt. Since floodplain soil integrates materials deposited by a river, it is typically rich in nutrients and ideal for cultivation.

Folivore: Any chiefly leaf-eating animal. Related terms: folivorous, folivory.

Forest-smart approach: A strategy that aims to maximize the benefits from development investments while minimizing the negative impacts to forests and forest biodiversity.

Free, prior and informed consent (FPIC): The principle that a community has the right to give or withhold its consent to proposed projects that may affect the lands they customarily own, occupy or otherwise use. It is a normative obligation whereby a state must seek the voluntary consent of indigenous peoples (including indigenous forest-dependent communities) as a precondition to allowing or engaging in activities that could significantly affect the communities' substantive rights, such as the right to property. While there is no internationally agreed definition of FPIC or any single mechanism for its implementation, international human rights instruments and other treaty obligations grant potentially impacted peoples the right to give or withhold their consent to proposed actions.

Frugivore: Any chiefly fruit-eating animal. Related terms: frugivorous, frugivory.

Functional connectivity: The degree to which the land that divides and separates natural habitats facilitates or impedes the habitats' ability to allow movements of animals and to perform ecosystem functions. See also: **structural connectivity**.

G20: An international forum for the leaders, finance ministers and central bank governors of 20 major economies. Its members are the eight leading industrialized nations—in descending order, the United States, Japan, Germany, the United Kingdom, France, Italy, Canada and Russia; 11 emerging-market and smaller industrialized countries, namely Argentina, Australia, Brazil, China, India, Indonesia, Mexico, Saudi Arabia, South Africa, South Korea and Turkey; and the European Union.

Gallery forest: A narrow stretch of forest along the banks of a body of water, such as a river or wetland, that projects into non-forested landscapes. See also: **riparian forest**.

Geobrowser: A geographical web browser designed to access satellite and aerial imagery, ocean bathymetry and other geographic data over the Internet to represent Earth as a three-dimensional globe.

Geographic information system (GIS): A tool that allows users to capture, store, manipulate, analyze, manage and present spatial and geographic data.

Georeferencing: The process of aligning geographic data to a known coordinate system so it can be viewed and analyzed with other geographic data.

Gigawatt: A unit of power equal to one billion watts.

GLAD (Global Land Analysis & Discovery): A laboratory in the Department of Geographical Sciences at the University of Maryland that investigates methods, causes and impacts of global land surface change. GLAD's primary data source is Earth-observation imagery; its key focus area is land cover extent. Available at: glad.geog.umd.edu.

Global Accessibility Map: A mapping tool designed to estimate the travel time from any point on Earth to the nearest city exceeding 50,000 people. Developed by the European Commission's Joint Research Centre and first published by the World Bank in 2008, it can serve as a proxy for rural populations' access to **services and resources** in urban areas. Available at: forobs.jrc.ec.europa.eu/products/gam. See also: **Roadless Forest**.

Global Forest Watch (GFW): An open-access initiative of the World Resources Institute that provides a range of remote-sensing and other databases designed to monitor forests globally. Available at: www.globalforestwatch.org.

Global Positioning System (GPS): A US-owned tool that allows users access to positioning, navigation and timing services. The US Air Force maintains and develops the utility.

Global Roadfree Areas Map: Launched in 2012 under the aegis of the **RoadFree** initiative, this collaboration between Google, the Society for Conservation Biology and the European Parliament assesses the status, quality and extent of all protected areas. Available at: roadfree.org.

Global Roadmap: An initiative run by an alliance of environmental scientists, geographers, planners and agricultural specialists whose aim is to enhance planning for roads in ways that reduce the environmental impacts of roads, limit construction of new roads and road improvements to where they will have the greatest social and economic benefits, assist environmental managers to better plan and prioritize roads, and educate the general public about the environmental risks of poorly planned roads and transportation projects. Available at: www.global-roadmap.org.

Global Roads Open Access Data Set (gROADS): A freely available data set on roads. Horizontal-accuracy limitations (±2 km) restrict its use to general comparisons. Available at: sedac.ciesin.columbia.edu/data/set/groads-global-roads-open-access-v1.

GLOBIO tool: Designed to facilitate environmental assessments and provide policy support, this global biodiversity model serves to evaluate past, present and future impacts of human activity on biodiversity. Available at: www.globio.info.

Google Earth: A geobrowser released by Google in 2005. Available at: www.google.com/earth/index.html.

Google Earth Engine: A cloud computing platform that processes satellite imagery and other Earth-observation data and analyzes geospatial information. It provides access to a large catalog of satellite imagery and the computational power needed to analyze those images. Available at: earthengine.google.com.

Google Maps: Launched in 2005, this online mapping service offers satellite imagery, street maps, panoramic views of streets, information on traffic conditions and route planning. Data for rural areas are patchier than for urban centers. Available at: maps.google.com.

Green procurement: The acquisition of products and services that cause minimal adverse environmental impacts. The approach incorporates human health and environmental concerns into the search for high-quality products and services at competitive prices.

Ground truth: Empirical evidence collected on location, as opposed to information inferred from other sources, such as satellite imagery.

Habitat: The natural and required environment of an animal, plant or other organism

Herbivore: Any plant-eating animal. Related terms: herbivorous, herbivory.

Home range: An area that is used by an individual or group on a regular basis and, in territorial species, is defended from others. Not to be confused with **ape range**, which is the extent of occurrence (EOO) of each species, as explained in the Notes to Readers, p. ix.

Hybrid: The offspring of two different species or varieties of plant or animal; something that is formed by combining different elements.

Hydrological connectivity: The degree to which water, and the organisms, matter and energy within it, is able to freely move with natural timing through the hydrologic cycle, including along and between longitudinal (river length), lateral (floodplain) and vertical (groundwater) dimensions.

Impounding reservoir: An artificial lake formed by constructing a dam across a natural watercourse so that water builds up behind it.

Impoundment: The accumulation of water in a reservoir or other storage area.

Indigenous: Originating from or occurring naturally in a certain place.

Induced access: Project-related encroachment into a landscape.

Infanticide: The act of killing an infant.

Integrated Biodiversity Assessment Tool (IBAT) for Business: A database for accessing biodiversity information, including Key Biodiversity Areas and legally protected areas. Via an interactive mapping tool, decision-makers can identify biodiversity risks and opportunities within a project's boundary. Available at: www.ibatforbusiness.org.

Integrated Valuation of Ecosystem Services and Tradeoffs (InVEST): A suite of open source software models used to map and value goods and services from nature. Available at: www.naturalcapitalproject.org/invest.

Interbirth interval: The biologically determined period of time between consecutive births.

Karst: A landscape formed through the dissolution of soluble rocks, such as limestone, dolomite and gypsum, and characterized by underground drainage systems with sinkholes, dolines and caves.

Keystone species: A species that plays a crucial role in the way an ecosystem functions, and whose presence and role has a disproportionately large effect on other organisms within the ecosystem.

Landsat imagery: Medium-resolution (30 m × 30 m) satellite images acquired by any of the six satellites in the **Landsat program**. Landsat images can be viewed and downloaded for free from the United States Geological Survey Earth Explorer website. Available at: earthexplorer.usgs.gov.

Landsat program: The longest-running endeavor to capture satellite imagery of Earth. Since the program's launch in 1972, its satellites have acquired millions of images. See also: **Landsat imagery** and **Landsat Thematic Mapper**.

Landsat Thematic Mapper (TM): A Earth-observing sensor placed aboard a satellite in the **Landsat program**. A TM features seven bands of image data (in visible and infrared wavelengths), most of which have a resolution of 30 m. An Enhanced Thematic Mapper Plus (ETM+) sensor, which includes an eighth (panchromatic) band with a spatial resolution of 15 m, was onboard Landsat 7 when it successfully launched in 1999. See also: **Landsat imagery**.

Landscape metrics: Measurements of one or more sections of an area of land, such as patches of forest or mosaics, used to quantify composition and spatial configuration, including forest size and fragmentation.

Logging Roads: By combining **OpenStreetMap** and satellite imagery, this initiative maps and dates logging roads, particularly in the Congo Basin. Available at: loggingroads.org.

Mast fruiting: The simultaneous production of fruit by a large numbers of trees every 2–10 years, without any seasonal change in temperature or rainfall.

Megawatt: A unit of power equal to one million watts.

Metapopulation: A group of spatially separated populations of the same species that interact at some level.

Miombo: An oak-like tree (genus *Brachystegia*); a type of savannah woodland found across eastern and southern Africa dominated by these trees.

Mitigation: The act of making a condition or consequence less severe.

Mitigation hierarchy: A tool used to limit the negative impacts of development projects on biodiversity.

Monodominant forest: A forest in which more than 60% of the tree canopy consists of a single species of tree.

Monogamy: The practice of having a single mate over a period of time.

Morph: A distinct form of an organism or species.

My DigitalGlobe: A web-based application through which users can view, analyze and download DigitalGlobe's high-resolution satellite images. Available at: services.digitalglobe.com/myDigitalGlobe.

National strategic areas: In line with Indonesia's 2007 law on spatial planning, domestic conservation areas recognized for their rich biodiversity. In an effort to protect the ecosystems and curb rent-seeking among officials, the law stipulates that individuals who engage in or facilitate illegal activities in such areas may be charged with criminal offences.

Natural resource management: The application of scientific and technical principles to control environmental assets such as land, water, soil, plants and animals so as to meet ecological, economic, social and policy objectives.

Net gain: In an ecological context, a positive outcome for biodiversity following a development project and the application of targeted conservation measures.

No deforestation: A corporate policy aimed at protecting forest and peatland while minimizing the impact of operations on biodiversity and local communities. Implementation of the policy requires assessments to be conducted for high conservation value forest areas prior to the clearing of any land. Related term: zero deforestation.

No net loss: In an ecological context, an outcome that avoids an overall loss of biodiversity and ecosystem services following a development project and targeted conservation activities. This term is often used in association with the **mitigation hierarchy**.

Offset: Compensation for loss of biodiversity due to a development project.

Open Standards for the Practice of Conservation: An adaptive planning framework utilized by governments and non-governmental organizations around the world to conserve flora and fauna collaboratively and systematically. Available at: cmp-openstandards.org.

OpenStreetMap (OSM): Launched in 2004, this free and editable map of the world is continually updated by registered members. OSM data feeds into the **Roadless Forest** and **Logging Roads** mapping initiatives, among other programs focused on environmental crises. Available at: www.openstreetmap.org.

OpenStreetMap (OSM) Analytics: Released in 2016, this platform enables the tracking of mapping activity for roads and buildings at the global level. Available at: osm-analytics.org.

Optimism bias: A systematic tendency to underestimate the probability of negative events.

Outstanding Universal Value: A UNESCO designation used to recognize heritage of exceptional cultural or natural significance and signal that its permanent protection is of the highest importance to the international community.

Pathogen: A virus, bacteria or other microorganism that can cause disease.

Pathogenic: Capable of causing disease.

Pelage: Fur; coat.

Pith: The spongy tissue in the stems and branches of many plants.

Pixel: The smallest unit of information in an image; the fundamental unit of data collection in remote sensing.

Planet: A commercial vendor of high-resolution satellite imagery and geospatial content. Available at: www.planet.com.

Pollution haven: A jurisdiction that attracts polluting industries due to limited environmental restrictions, as posited by the pollution haven hypothesis (or pollution haven effect).

Polyandrous: Pertaining to a mating system that involves one female and two or more males.

Polygynandrous: Pertaining to an exclusive mating system that involves two or more males and two or more females. The numbers of males and females are not necessarily equal.

Polygynous: Pertaining to a mating system that involves one male and two or more females.

Preputial: Relating to the foreskin or clitoral hood.

Project appraisal document (PAD): A comprehensive, continually updated record of a development project, covering aspects such as the development problem to be addressed, the technical approach to be followed, the expected results, the financial plan and budget, the overall project implementation and procurement plan, and the monitoring and evaluation plan.

Protected area downgrading, downsizing and degazettement (PADDD): Legal reductions in the size or protection status of parks and other protected areas, typically to gain access to natural resources inside those parks or to permit infrastructure projects to cut through them.

Protection forest: An area of forest that is characterized by average slope gradients of at least 40° and on which commercial logging is illegal; and/or is managed primarily for its beneficial effects on water and soil movement; and/or managed for its ability to protect people or assets against the impacts of destabilizing natural events or adverse climates.

Radar: A system that detects the presence, direction, distance and speed of aircraft, ships and other objects by sending out pulses of high-frequency electromagnetic waves that are reflected off the object back to the source.

Ramsar wetlands: Water-saturated land areas designated under the Convention on Wetlands, known as the Ramsar Convention, an intergovernmental treaty that provides the framework for national action and international cooperation for the conservation and wise use of wetlands and their resources.

Range: In the context of 'ape ranges', the extent of occurrence (EOO) of each species. An EOO includes all known populations of a species contained within the shortest possible continuous imaginary boundary. It is important to note that some areas within these boundaries are unsuitable and unoccupied.

Reducing Emissions from Deforestation and Forest Degradation plus (REDD+): A United Nations initiative that goes beyond curbing the release of greenhouse gases to include the role of conservation, sustainable management of forests and enhancement of forest carbon stocks.

Regression line: An approach for modeling the relationship between two variables.

Reintroduction: The managed release of an organism into its natural habitat after life in captivity.

Remote sensing: The science of identifying, measuring and observing objects or areas from a distance, typically from aircraft or satellites.

Renewable energy: Usable power derived from natural sources whose supply is not depleted when used, such as sunlight, geothermal heat, tides and wind, in contrast to finite resources such as oil and coal. See also: **alternative energy** and **clean energy**.

Resettlement action plan: A detailed, legally binding strategy that developers must fulfill when relocating and compensating people affected by an infrastructure project.

Residual impact: In the context of the mitigation hierarchy, a negative effect that remains after the implementation of mitigation measures.

Riparian (or riverine) forest: A forest that grows alongside a body of water, such as rivers, streams and lakes. See also: **gallery forest**.

River basin development: Use, control or diversion of elements of a network of watercourses to promote economic growth, often with significant impacts on quantity, quality, sediment load, timing and predictability of the water regime, such as through hydropower development.

River reach: A segment of a river that can be distinguished from other segments by its width, habitat composition, vegetation coverage, the presence of dams or other structures, and other characteristics; distinct reaches exhibit differing natural resource problems and are evaluated separately.

River system: The natural structure within which a river flows, typically including a watershed.

RoadFree: An initiative designed to highlight the importance of roadless wilderness areas for biodiversity conservation and reductions in atmospheric carbon emissions. RoadFree helped to give rise to the **Global Roadfree Areas Map**. Available at: roadfree.org.

Roadless Forest: Designed to evaluate the benefits of road-free forests, this European Union initiative is strongly linked to EU policies on reducing illegal logging and carbon emissions resulting from forest disruption. To inform decision-making, it makes use of the **Global Accessibility Map** to identify which areas are most likely to benefit from infrastructure development and to highlight associated risks to protected areas. Available at: roadlessforest.eu.

Roadside zoo: An unaccredited zoo or roadside attraction engaged in commercial exhibition and other activities with animals, potentially including apes.

Run-of-river scheme: A hydroelectric power scheme that operates without water storage, using the flow of a river channel.

Sanctuary: A non-profit facility dedicated to providing care for orphaned, confiscated or injured wildlife.

Sentinels: A family of satellites developed for the operational needs of Copernicus, an Earth-observation program headed by the European Commission in partnership with the European Space Agency. The Sentinels provide observations such as radar images, high-resolution optical images, and data for the monitoring of atmospheric composition and global sea-surface height.

Silverback: An adult male gorilla that has reached maturity and developed silver hairs on the saddle of his back.

Smart green infrastructure: Facilities that avoid critical habitats, minimize and mitigate adverse impacts on communities and biodiversity, and compensate for any inadvertent or unavoidable damage.

Spatial resolution: The level of detail in a digital (usually satellite) image; often expressed in meters, measuring the edge length of a pixel, the smallest unit of the image. The smaller the pixel size, or the greater the number of pixels in an image, the higher the spatial resolution. Satellite images tend to be grouped into three resolution categories: low resolution (>30 m), medium (2–30 m) and high (<2 m).

Standing sale: The selling of timber as it stands in the forest, in advance of harvesting and generally by weight.

Stranded asset: An economic resource that has become obsolete or nonperforming before the end of its useful life and that is recorded as a loss.

Structural connectivity: The makeup of a landscape based on ecological attributes of the area (habitat type and composition) and its connectivity (vs. fragmentation) across a landscape, excluding behavioral patterns of organisms in the area. See also: **functional connectivity**.

Subadult: A stage of development where an individual has not yet acquired all adult characteristics.

Sustainable Development Goals (SDGs): Seventeen global aims established by the United Nations to end poverty, protect the planet and result in peace and prosperity for all. The SDGs were adopted by the 193 countries of the UN General Assembly in 2015, with specific targets to be achieved by 2030.

Swamp forest (or freshwater swamp forest): Natural forest that stands in waterlogged soil and has more than 30% canopy cover.

Sympatric: Pertaining to species or populations that occupy the same geographic ranges.

Taxon: Any unit used in the science of biological classification or taxonomy (plural: taxa).

Terawatt: A unit of power equal to one trillion watts, or one million megawatts.

Terra nullius: In international law, land that does not officially belong to anyone or any state, and that can be acquired through occupation.

Terrestrial herbaceous vegetation: Species of herbs that are staple food items for apes, such as Marantaceae and Zingiberaceae.

Terrestriality: Adaptation to living on the ground.

Toolkit for Ecosystem Service Site-based Assessment (TESSA): Guidance on low-cost methods for evaluating the benefits people receive from nature at a particular site to generate information that can be used to influence decision-making. Available at: tessa.tools.

Translocation: In conservation, the process of moving an organism from one area to another, in captive or wild settings.

Tropical Rainforest Heritage of Sumatra: A 25,000-km² (2.5 million-ha) conservation site that comprises three of Indonesia's national parks: Bukit Barisan Selatan, Gunung Leuser and Kerinci Seblat. It is home to many endangered species, including the endemic Sumatran orangutan (*Pongo abelii*).

UNESCO Man and Biosphere Reserve: Any of the 669 sites around the word that are internationally recognized for the simultaneous conservation and sustainable use of their ecosystems. Each reserve has three interrelated zones: a strictly protected core zone; a buffer zone that is used in ways that facilitate scientific research, monitoring, training and education; and a transition area that fosters sustainable human and economic development.

UNESCO World Heritage site: An area of internationally recognized cultural and natural significance, including geological and physiographical formations and precisely delineated areas that constitute the habitat of threatened animal and plant species that are of outstanding value to science or conservation.

Upstream: Towards the source of a river.

Upstream planning: Advance target setting and coordination of feasibility studies, design, implementation and operation of an investment project, usually involving collaboration among public authorities and other stakeholders, often with technical assistance.

UrtheCast: An Earth-imaging system company that specializes in geospatial analysis. Its high-resolution satellite imagery is made available on **Global Forest Watch**. Available at: www.urthecast.com.

Watershed: A tract of land drained by a river and its tributaries.

Wean: To accustom a young animal to nourishment other than the mother's milk.

Wetlands: Areas where water is sometimes or always above, at or near the surface of the soil.

Wildlife corridor: Habitat that joins two or more larger areas of similar habitat and thus allows wildlife movement, supports the viability of populations and maintains ecological processes. Corridors can occur naturally, such as riparian forests, or be created through habitat management practices.

Zero-sum game: A situation in which each participant's gain or loss of utility is exactly balanced by the losses or gains of the utility of the other participants. Suggesting that the earth's carrying capacity is a zero-sum game implies that any land, food or resources consumed or degraded by humans must ultimately incur a comparable cost to other species or ecosystems.

REFERENCES

Abel, D. (2017). Animal advocates say removal of database hurts efforts to prevent abuse. *Boston Globe*, July 9, 2017. Available at: http://www.bostonglobe.com/metro/2017/07/19/animal-advocates-say-removal-database-hurts-efforts-prevent-abuse.

Abell, R., Thieme, M.L., Revenga, C., *et al.* (2008). Freshwater ecoregions of the world: a new map of biogeographic units for freshwater biodiversity conservation. *BioScience*, **58**, 403–14. DOI: 10.1641/B580507.

Abernethy, K., Maisels, F. and White, L.J.T. (2016). Environmental issues in central Africa. *Annual Review of Environment and Resources*, **41**, 1–33. DOI: 10.1146/annurev-environ-110615-085415.

Abood, S.A., Lee, J.S.H., Burivalova, Z., Garcia-Ulloa, J. and Koh, L.P. (2015). Relative contributions of the logging, fiber, oil palm, and mining industries to forest loss in Indonesia. *Conservation Letters*, **8**, 58–67. DOI: 10.1111/conl.12103.

Abram, N.K., Meijaard, E., Wells, J.A., *et al.* (2015). Mapping perceptions of species' threats and population trends to inform conservation efforts: the Bornean orangutan case study. *Diversity and Distributions*, **21**, 487–99. DOI: 10.1111/ddi.12286.

Abutu, A. and Charles, E. (2016). C/River communities reject superhighway. *Daily Trust*, March 9, 2016. Available at: https://www.dailytrust.com.ng/news/environment/c-river-communities-reject-superhighway/137088.html.

Achard, F., Beuchle, R., Mayaux, P., *et al.* (2014). Determination of tropical deforestation rates and related carbon losses from 1990 to 2010. *Global Change Biology*, **20**, 2540–54. DOI: 10.1111/gcb.12605.

ADB (2008). *Preparing the Cumulative Impact Assessment for the Nam Ngum 3 Hydropower Project*. Manila, Philippines: Asian Development Bank (ADB). Available at: https://www.adb.org/projects/documents/preparing-cumulative-impact-assessment-nam-ngum-3-hydropower-project-financed-j-0.

ADB (2011). *Asia Solar Energy Initiative: A Primer*. Manila, Philippines: Asian Development Bank (ADB).

ADB (2012). *Environment Safeguards: A Good Practice Sourcebook*. Manila, Philippines: Asian Development Bank (ADB). Available at: http://www.adb.org/sites/default/files/institutional-document/33739/files/environment-safeguards-good-practices-sourcebook-draft.pdf.

ADB (2017). *Meeting Asia's Infrastructure Needs*. Manila, Philippines: Asian Development Bank (ADB). Available at: https://www.adb.org/sites/default/files/publication/227496/special-report-infrastructure.pdf.

Adeney, J.M., Christensen, N. and Pimm, S.L. (2009). Reserves protect against deforestation fires in the Amazon. *PLoS One*, **4**, e5014.

ADF (2011). *Lom-Pangar Hydroelectric Project. Republic of Cameroon. Project Appraisal Report. November 2011*. Tunis, Tunisia: African Development Fund (ADF), African Development Bank Group. Available at: http://www.afdb.org/fileadmin/uploads/afdb/Documents/Project-and-Operations/Cameroon%20-%20AR%20-%20Lom-Pangar%20Hydroelectric%20Project%20(Final).pdf.

AfDB (2011a). *Africa in 50 Years' Time: The Road Towards Inclusive Growth*. Tunis, Tunisia: African Development Bank (AfDB). Available at: https://www.afdb.org/fileadmin/uploads/afdb/Documents/Publications/Africa%20in%202050%20Years%20Time.pdf.

AfDB (2011b). *Republic of Cameroon. Lom-Pangar Hydroelectric Project. Summary of the Environmental and Social Impact Assessment (ESIA)*. Abidjan, Ivory Coast: African Development Bank (AfDB), Energy, Environment and Climate Change Department. Available at: https://www.afdb.org/fileadmin/uploads/afdb/Documents/Environmental-and-Social-Assessments/2011%20Lom-Pangar%20R%C3%A9sum%C3%A9%20Environnemental%20et%20Social_EN.pdf.

AfDB (2013). *Safeguards and Sustainability Series: African Development Bank Group's Integrated Safeguards System. Policy Statement and Operational Safeguards*. Tunis, Tunisia: African Development Bank (AfDB). Available at: https://www.afdb.org/fileadmin/uploads/afdb/Documents/Policy-Documents/December_2013_-_AfDB%E2%80%99S_Integrated_Safeguards_System__-_Policy_Statement_and_Operational_Safeguards.pdf.

AfDB (2015). *Multinational Cameroon-Congo-Ketta Djoum Road Project and Facilitation of Transportation on the Yaoundé-Brazzaville Corridor_Phase 2 – Summary ESIA – 06 2015*. Tunis, Tunisia: African Development Bank (AfDB). Available at: https://www.afdb.org/en/documents/document/multinational-cameroon-congo-ketta-djoum-road-project-and-facilitation-of-transportation-on-the-yaounde-brazzaville-corridor_phase-2-summary-esia-06–2015–54091/.

AfDB, OECD and UNDP (2015). *African Economic Outlook 2015: Regional Development and Spatial Inclusion*. African Development Bank (AfDB), Organisation for Economic Co-operation and Development (OECD) and United Nations Development Programme (UNDP). Available at: www.africaneconomicoutlook.org/sites/default/files/content-pdf/AEO2015_EN.pdf.

Africa-EU Energy Partnership (2013). *Country Power Market Brief: Cameroon*. Brussels, Belgium: Alliance for Rural Electrification. Available at: https://www.ruralelec.org/publications/country-power-market-brief-cameroon.

AgDevCo (2013). *Developing Sustainable Agriculture in Africa*. London, UK: African Agriculture Development Company (AgDevCo). Available at: http://www.agdevco.com/uploads/reports/AgDevCo%20Brochure_June%202013%20low%20res(1).pdf.

AgDevCo (2017). *About Us*. London, UK: African Agricultural Development Company (AgDevCo). Available at: http://www.agdevco.com/about-us.html. Accessed July, 2017.

Agence Ecofin (2012). *Cameroun: 2000 Emplois Camerounais pour le Barrage de Lom Pangar*. Agence Ecofin. Available at: https://www.agenceecofin.com/hydroelectricite/0701–2818-cameroun-2000-emplois-camerounais-pour-le-barrage-de-lom-pangar.

AGRECO (2007). *Etude d'Impact Social et Environnemental de la Réhabilitation de Routes en RDC, Projet Pro-Routes, Cadre Stratégique, Rapport Final*. Brussels, Belgium: Ministère des Travaux Publics et Infrastructures & Union Européenne.

AIIB (2016). *Environmental and Social Framework*. Beijing, China: Asian Infrastructure Investment Bank (AIIB). Available at: https://www.aiib.org/en/policies-strategies/_download/environment-framework/20160226043633542.pdf.

Akpan, A. (2016a). Communities, NGOs tackle Cross River government over superhighway project. *The Guardian*, August 26, 2016. Available at: https://guardian.ng/features/communities-ngos-tackle-cross-river-government-over-superhighway-project.

Akpan, A. (2016b). Government, groups fault Cross River's EIA on highway project. *The Guardian*, June 6, 2016. Available at: https://guardian.ng/property/government-groups-fault-cross-rivers-eia-on-highway-project.

Akpan, A. (2016c). Pressure mounts against Cross River super highway project. *The Guardian*, September 25, 2016. Available at: https://guardian.ng/news/pressure-mounts-against-cross-river-super-highway-project.

Akpan, A. (2017). Groups fault FG's EIA to Cross River on super highway project. *The Guardian*, July 17, 2017. Available at: https://guardian.ng/property/groups-fault-fgs-eia-to-cross-river-on-super-highway-project.

Alcamo, J. (2008). *Environmental Futures: The Practice of Environmental Scenario Analysis*. Boston, MA: Elsevier.

Alden Wily, L. (2011a). 'The law is to blame': the vulnerable status of common property rights in sub-Saharan Africa. *Development and Change*, **42**, 733–57. DOI: 10.1111/j.1467–7660.2011.01712.x.

Alden Wily, L. (2011b). *Whose Land Is It? The Status of Customary Land Tenure in Cameroon*. Centre for Environment and Development (CED), FERN and Rainforest Foundation UK.

Alden Wily, L. (2016). *Traditional forest communities as owner-conservators: is this a viable way forward?* Unpublished paper.

Alexander, N. (2014). *The Emerging Multi-Polar World Order*. Washington DC: Heinrich Böll Foundation North America.

Alexandratos, N. and Bruinsma, J. (2012). *World Food and Agriculture Towards 2030/2050: The 2012 Revision*. ESA Working Paper No. 12–03. Rome, Italy: Food and Agriculture Organization of the United Nations (FAO).

Alshuwaikhat, H.M. (2005). Strategic environmental assessment can help solve environmental impact assessment failures in developing countries. *Environmental Impact Assessment Review*, **25**, 307–17. DOI: https://doi.org/10.1016/j.eiar.2004.09.003.

Ampuero, F. and Sá Lilian, R.M. (2012). Electrocution lesions in wild brown howler monkeys (*Alouatta guariba clamitans*) from São Paulo city: importance for conservation of wild populations. *ESVP/ECVP Proceedings*, **146**, 88.

Ancrenaz, M., Ambu, L., Sunjoto, I., *et al.* (2010). Recent surveys in the forests of Ulu Segama Malua, Sabah, Malaysia, show that orang-utans (*P. p. morio*) can be maintained in slightly logged forests. *PLoS One*, **5**, e11510. DOI: 10.1371/journal.pone.0011510.

Ancrenaz, M., Calaque, R. and Lackman-Ancrenaz, I. (2004). Orangutan nesting behavior in disturbed forest of Sabah, Malaysia: implications for nest census. *International Journal of Primatology*, **25**, 983–1000.

Ancrenaz, M., Cheyne, S., Humle, T. and Robbins, M.M. (2015a). Impacts of industrial agriculture on ape ecology. In *State of the Apes: Industrial Agriculture and Ape Conservation*, ed. Arcus Foundation. Cambridge, UK: Cambridge University Press, pp. 165–92. Available at: https://www.stateoftheapes.com/volume-2-industrial-agriculture/.

Ancrenaz, M., Dabek, L. and O'Neil, S. (2007). The costs of exclusion: recognizing a role for local communities in biodiversity conservation. *PLoS Biology*, **5**, e289. DOI: 10.1371/journal.pbio.0050289.

Ancrenaz, M., Gumal, M., Marshall, A.J., *et al.* (2016). Pongo pygmaeus. *The IUCN Red List of Threatened Species 2016*: e.T17975A17966347. Gland, Switzerland: International Union for Conservation of Nature (IUCN). Available at: http://dx.doi.org/10.2305/IUCN.UK.2016-1.RLTS.T17975A17966347.en. Accessed August 14, 2017.

Ancrenaz, M., Oram, F., Ambu, L., *et al.* (2015b). Of pongo, palms and perceptions a multidisciplinary assessment of Bornean orang-utans *Pongo pygmaeus* in an oil palm context. *Oryx*, **49**, 465–72. DOI: 10.1017/S0030605313001270.

Ancrenaz, M., Sollmann, R., Meijaard, E., *et al.* (2014). Coming down from the trees: is terrestrial activity in Bornean orangutans natural or disturbance driven? *Scientific Reports*, **4**, 4024. DOI: 10.1038/srep04024. Available at: https://www.nature.com/articles/srep04024#supplementary-information.

Anderson, D.P., Nordheim, E.V. and Boesch, C. (2006a). Environmental factors influencing the seasonality of estrus in chimpanzees. *Primates*, **47**, 43–50.

Anderson, J., Benjamin, C., Campbell, B. and Tiveau, D. (2006b). Forests, poverty and equity in Africa: new perspectives on policy and practice. *International Forestry Review*, **8**, 44–53.

André, C., Kamate, C., Mbonzo, P., Morel, D. and Hare, B. (2008). The Conservation Value of Lola ya Bonobo Sanctuary. In *Bonobos Revisited: Ecology, Behavior, Genetics, and Conservation*, ed. I. Takesi and J. Thompson. New York, NY: Springer, pp. 303–22.

Andrews, A. (1990). Fragmentation of habitat by roads and utility corridors: a review. *Australian Zoologist*, **26**, 130–41. DOI: 10.7882/az.1990.005.

Angelsen, A., Jagger, P., Babigumira, R., *et al.* (2014). Environmental income and rural livelihoods: a global-comparative analysis. *World Development*, **64**, S12-S28. DOI: https://doi.org/10.1016/j.worlddev.2014.03.006.

Angelsen, A. and Kaimowitz, D. (1999). Rethinking the causes of deforestation: lessons from economic models. *The World Bank Research Observer*, **14**, 73–98. DOI: 10.1093/wbro/14.1.73.

Ansar, A., Flyvbjerg, B., Budzier, A. and Lunn, D. (2014). Should we build more large dams? The actual costs of hydropower megaproject development. *Energy Policy*, **69**, 43–56. DOI: https://doi.org/10.1016/j.enpol.2013.10.069.

Antara News (2015). Indonesian govt. to focus on geothermal energy development: President Jokowi. *Antara News*, August 22, 2015. Available at: http://www.antaranews.com/en/news/100118/indonesian-govt-to-focus-on-geothermal-energy-development-president-jokowi. Accessed January 22, 2017.

Anthony, N.M., Johnson-Bawe, M., Jeffery, K., *et al.* (2007). The role of Pleistocene refugia and rivers in shaping gorilla genetic diversity in central Africa. *Proceedings of the National Academy of Sciences*, **104**, 20432–6. DOI: 10.1073/pnas.0704816105.

Arcus Foundation (2015). *State of the Apes: Industrial Agriculture and Ape Conservation*. Cambridge, UK: Cambridge University Press. Available at: https://www.stateoftheapes.com/volume-2-industrial-agriculture/.

Arunmart, P. (1996). *Feasibility Study on Burma Link*. Bangkok, Thailand: Bangkok Post.

ASI (2015). *Integrated Resource Corridors Initiative, Scoping and Business Plan*. Adam Smith International (ASI), World Wide Fund for Nature (WWF), Department for International Development (DFID). Available at: http://www.adamsmithinternational.com/documents/resource-uploads/IRCI_Scoping_Report_Business_Plan.pdf.

ASM-PACE and Phillipson, A. (2014). Artisanal and small-scale mining and apes. In *State of the Apes: Extractive Industries and Ape Conservation*, ed. Arcus Foundation. Cambridge, UK: Cambridge University Press, pp. 162–95. Available at: https://www.stateoftheapes.com/themes/artisanal-and-small-scale-mining/.

Asner, G.P., Llactayo, W., Tupayachi, R. and Luna, E.R. (2013). Elevated rates of gold mining in the Amazon revealed through high-resolution monitoring. *Proceedings of the National Academy of Sciences*, **110**, 18454–9. DOI: 10.1073/pnas.1318271110.

Association of State Dam Safety Officials (2016). *Dam Failures and Incidents.* Lexington, KY: Association of State Dam Safety Officials. Available at: http://damsafety.org/what-are-causes-dam-failures. Accessed October 6, 2016.

AU (2009). *Africa Mining Vision.* African Union (AU). Available at: http://www.africaminingvision.org/amv_resources/AMV/Africa_Mining_Vision_English.pdf.

AU (2015). *Agenda 2063: The Africa We Want.* African Union (AU). Available at: https://au.int/sites/default/files/pages/3657-file-agenda2063_popular_version_en.pdf.

Auty, R. (2002). *Sustaining Development in Mineral Economies: The Resource Curse Thesis.* Oxford, UK: Routledge.

AWF (2015). *AWF Kickstarts Efforts in Bili Uele Protected Area.* Nairobi, Kenya: African Wildlife Foundation (AWF). Available at: https://www.awf.org/blog/awf-kickstarts-efforts-bili-uele-protected-area.

AWF (2016). *Bili Uele Landscape Strategy: 2016–2021.* Nairobi, Kenya: African Wildlife Foundation (AWF).

Ayres, J.M. and Clutton-Brock, T.H. (1992). River boundaries and species range size in Amazonian primates. *The American Naturalist*, **140**, 531–7. DOI: 10.1086/285427.

Baabud, S.F., Griffiths, M., Afifuddin and Safriansyah, R. (2016). *Total Economic Value (TEV) of Aceh's Forests.* European Union Delegation for Indonesia and Brunei Darussalam. Vienna, Austria: CEU Consulting GmbH.

Babbitt, B. (2002). What goes up, may come down. *BioScience*, **52**, 656–8. DOI: 10.1641/0006–3568(2002)052[0656:WGUMCD]2.0.CO;2.

Baeckler Davis, S. (2016). *A Journey Begins.* Blue Ridge, GA: Project Chimps. Available at: http://projectchimps.org/a-journey-begins/. Accessed October 2, 2016.

Bale, R. (2016). *Controversial Plan Would Send Lab Chimps to Unaccredited Zoo.* Washington DC: National Geographic. Available at: http://news.nationalgeographic.com/2016/06/yerkes-research-chimpanzees-controversy/. Accessed October 1, 2016.

Banks, M. (2016). *New Bill from Scottish Government set to Ban Wild Animals in Circuses.* London, UK: EU Today. Available at: http://eutoday.net/news/circus-2983. Accessed October 1, 2016.

Barbash, F. (2014). Six killed in Thailand when elephant collides with traffic. *Washington Post*, March 12, 2014. Available at: https://www.washingtonpost.com/news/morning-mix/wp/2014/03/12/six-killed-in-thailand-when-van-crashes-into-wild-elephant/.

Barber, C.P., Cochrane, M.A., Souza, C.M. and Laurance, W.F. (2014). Roads, deforestation, and the mitigating effect of protected areas in the Amazon. *Biological Conservation*, **177**, 203–9. DOI: https://doi.org/10.1016/j.biocon.2014.07.004.

Barr, R., Burgués Arrea, I., Asuma, S., Behm Masozera, A. and Gray, M. (2015). *Pave the Impenetrable? An Economic Analysis of Potential Ikumba-Ruhija Road Alternatives In and Around Uganda's Bwindi Impenetrable National Park.* Conservation Strategy Fund Technical Series No. 35. Sebastopol, CA: Conservation Strategy Fund. Available at: http://www.conservation-strategy.org/en/publication/pave-impenetrable-economic-analysis-potential-ikumba-ruhija-road-alternatives-and-arou-0.

Barra, A.F., Burnouf, M.M.J., Damania, R. and Russ, J.D. (2016). *Economic Boom or Ecologic Doom? Using Spatial Analysis to Reconcile Road Development with Forest Conservation.* Washington DC: World Bank. Available at: http://documents.worldbank.org/curated/en/952691468195575937/Economic-boom-or-ecologic-doom-using-spatial-analysis-to-reconcile-road-development-with-forest-conservation.

Barreto Jr, P., Brandão, A., Baima, S. and Souza Jr, C. (2014). O risco de desmatamento associado a doze hidrelétricas na Amazônia. In *Tapajós: Hidrelétricas, Infraestrutura e Caos*, ed. W. C. de Sousa. São José, Brazil: Instituto Aeronáutica, pp. 149–75.

Bartlett, T.Q. (2001). Extra-group copulations by sub-adult gibbons: implications for understanding gibbon social organisation. *American Journal of Physical Anthropology*, **114** (supple 32), 36.

Bartlett, T.Q. (2007). The Hylobatidae: small apes of Asia. In *Primates in Perspective*, ed. C. Campbell, A. Fuentes, K. C. Mackinnon, M. Panger and S. K. Bearder. New York, NY: Oxford University Press, pp. 274–89.

Bashaw, M.J., Gullott, R.L. and Gill, E.C. (2010). What defines successful integration into a social group for hand-reared chimpanzee infants? *Primates*, **51**, 139–47. DOI: 10.1007/s10329–009–0176–8.

Baskaran, N. and Boominathan, D. (2010). Road kill of animals by highway traffic in the tropical forests of Mudumalai Tiger Reserve, southern India. *Journal of Threatened Taxa*, **2**, 753–9.

Bassey, E., Nkonyu, L. and Dunn, A. (2010). *A Reconnaissance Survey of the Bushmeat Trade in Eight Border Communities of South-East Nigeria, September–October 2009*. Nigeria: Wildlife Conservation Society.

BBOP (2009–2012). *Standard & Guidelines*. Washington DC: Business and Biodiversity Offsets Programme (BBOP). Available at: bbop.forest-trends.org/pages/guidelines.

BBOP (2012). *Resource Paper: No Net Loss and Loss–Gain Calculations in Biodiversity Offsets*. Washington DC: Business and Biodiversity Offsets Programme (BBOP). Available at: http://www.forest-trends.org/documents/files/doc_3103.pdf.

Beastall, C.A. and Bouhuys, J. (2016). *Apes in Demand: For Zoo and Wildlife Attractions in Peninsular Malaysia and Thailand*. Selangor, Malaysia: TRAFFIC. Available at: http://www.trafficj.org/publication/16_Apes_in_Demand.pdf.

Beck, B., Walkup, K., Rodrigues, M., *et al.* (2007). *Best Practice Guidelines for the Re-introduction of Great Apes*. Gland, Switzerland: International Union for Conservation of Nature Species Survival Commission (IUCN SSC), Primate Specialist Group.

Bellmore, J.R., Duda, J.J., Craig, L.S., *et al.* (2017). Status and trends of dam removal research in the United States. *Wiley Interdisciplinary Reviews: Water*, **4**, e1164-n/a. DOI: 10.1002/wat2.1164.

Benítez-López, A., Alkemade, R. and Verweij, P.A. (2010). The impacts of roads and other infrastructure on mammal and bird populations: a meta-analysis. *Biological Conservation*, **143**, 1307–16. DOI: https://doi.org/10.1016/j.biocon.2010.02.009.

Bennett, E.L. (2011). Another inconvenient truth: the failure of enforcement systems to save charismatic species. *Oryx*, **45**, 476–9. DOI: 10.1017/S003060531000178X.

Bennett, E.L. (2015). Legal ivory trade in a corrupt world and its impact on African elephant populations. *Conservation Biology*, **29**, 54–60. DOI: 10.1111/cobi.12377.

Bennett, G., Gallant, M. and ten Kate, K. (2017). *State of Biodiversity Mitigation 2017. Markets and Compensation for Global Infrastructure Development*. Washington DC: Forest Trends' Ecosystem Marketplace. Available at: http://forest-trends.org/releases/p/sobm2017.

Beresford, A.E., Buchanan, G.M., Donald, P.F., *et al.* (2011). Poor overlap between the distribution of protected areas and globally threatened birds in Africa. *Animal Conservation*, **14**, 99–107. DOI: 10.1111/j.1469-1795.2010.00398.x.

Berg, A., Portillo, R., Yang, S.S. and Zanna, L. (2012). *Public Investment in Resource-Abundant Developing Countries*. Washington DC: International Monetary Fund (IMF). Available at: https://www.imf.org/external/pubs/ft/wp/2012/wp12274.pdf.

Berg, C.N., Deichmann, U., Liu, D. and Selod, H. (2015). *Transport Policies and Development. Policy Research Working Paper 7366*. Washington DC: World Bank. Available at: http://documents.worldbank.org/curated/en/893851468188672137/pdf/WPS7366.pdf.

Bergl, R.A. (2006). *Conservation Biology of the Cross River Gorilla (Gorilla gorilla diehli)*. New York, NY: City University of New York.

Bergl, R.A., Oates, J.F. and Fotso, R. (2007). Distribution and protected area coverage of endemic taxa in west Africa's Biafran forests and highlands. *Biological Conservation*, **134**, 195–208. DOI: https://doi.org/10.1016/j.biocon.2006.08.013.

Bhagabati, N.K., Ricketts, T., Sulistyawan, T.B.S., *et al.* (2014). Ecosystem services reinforce Sumatran tiger conservation in land use plans. *Biological Conservation*, **169**, 147–56. DOI: https://doi.org/10.1016/j.biocon.2013.11.010.

BIC (2016). *World Bank's Updated Safeguards a Missed Opportunity to Raise the Bar for Development Policy*. Banking Information Center (BIC). Available at: http://www.bankinformationcenter.org/world-banks-updated-safeguards-a-missed-opportunity-to-raise-the-bar-for-development-policy/.

Blake, S., Deem, S.L., Strindberg, S., *et al.* (2008). Roadless wilderness area determines forest elephant movements in the Congo Basin. *PLoS One*, **3**, e3546. DOI: 10.1371/journal.pone.0003546.

Blake, S., Strindberg, S., Boudjan, P., *et al.* (2007). Forest elephant crisis in the Congo Basin. *PLoS Biology*, **5**, e111. DOI: 10.1371/journal.pbio.0050111.

Blaser, M., Feit, H.A. and McRae, G. (2004). *In the Way of Development: Indigenous Peoples, Life Projects, and Globalization.* London, UK: Zed Books.

Bleisch, B. and Geissmann, T. (2008). Nomascus nasutus. *The IUCN Red List of Threatened Species 2008: e.T41642A10526189.* Gland, Switzerland: International Union for Conservation of Nature (IUCN). Available at: http://dx.doi.org/10.2305/IUCN.UK.2008.RLTS.T41642A10526189.en. Accessed December 11, 2016.

Bleisch, B., Geissmann, T., Timmins, R.J. and Xuelong, J. (2008). Nomascus concolor. *The IUCN Red List of Threatened Species 2008: e.T39775A10265349.* Gland, Switzerland: International Union for Conservation of Nature (IUCN). Available at: http://dx.doi.org/10.2305/IUCN.UK.2008.RLTS.T39775A10265349.en.

Blomley, T. (2013). *Lessons Learned from Community Forestry in Africa and Their Relevance for REDD+.* Washington DC: USAID-supported Forest Carbon, Markets and Communities (FCMC) Program.

Boakes, E.H., Mace, G.M., McGowan, P.J.K. and Fuller, R.A. (2010). Extreme contagion in global habitat clearance. *Proceedings of the Royal Society B: Biological Sciences,* **277**, 1081–5. DOI: 10.1098/rspb.2009.1771.

Bonn, A., Allott, T., Evans, M., Joosten, H. and Stoneman, R. (2016). *Peatland Restoration and Ecosystem Services: Science, Policy and Practice.* Cambridge, UK: Cambridge University Press.

Born Free Foundation (2016a). *Italian Government Proposes an End to Animal Circuses.* Horsham, UK: Born Free Foundation. Available at: http://www.bornfree.org.uk/index.php?id=34&tx_ttnews%5Btt_news%5D=2123&cHash=764e4075f905659779faae1cbfa53735. Accessed October 1, 2016.

Born Free Foundation (2016b). *Wild Animal Circuses Look Set to be Outlawed in Norway from Early 2017.* Horsham, UK: Born Free Foundation. Available at: http://www.bornfree.org.uk/campaigns/zoo-check/zoo-news/article/?no_cache=1&tx_ttnews%5Btt_news%5D=2285.

Bortolamiol, S., Cohen, M., Jiguet, F., *et al.* (2016). Chimpanzee non-avoidance of hyper-proximity to humans. *The Journal of Wildlife Management,* **80**, 924–34. DOI: 10.1002/jwmg.1072.

Boutot, L., Lino, M., Ntep, J. and Essam, S. (2005). *Etude Environnementale du Barrage de Lom Pangar: Rapport Après Consultation. Thème no. 12: Zones d'Emprunt, Accès, Cité et Zone de Chantier.* Yaoundé, Cameroon: Ministry of Water Resources and Energy. Available at: https://www.iucn.org/sites/dev/files/import/downloads/t12_chantier_carrieres_lom_pangar.pdf.

Bowland, C. and Otto, L. (2012). *Implementing Development Corridors: Lessons from the Maputo Corridor. South African Foreign Policy and African Drivers Programme.* Johannesburg, South Africa: South African Institute of International Affairs. Available at: http://www.saiia.org.za/policy-briefings/implementing-development-corridors-lessons-from-the-maputo-corridor.

Bradley, B.J., Doran-Sheehy, D.M., Lukas, D., Boesch, C. and Vigilant, L. (2004). Dispersed male networks in western gorillas. *Current Biology,* **14**, 510–3. DOI: https://doi.org/10.1016/j.cub.2004.02.062.

BRM (2017). Lom Pangar dam officially delivered and handed over to Cameroonian authorities. *Business in Cameroon,* July 7, 2017. Available at: www.businessincameroon.com/electricity/0707-7249-lom-pangar-dam-officially-delivered-and-handed-over-to-cameroonian-authorities.

Brncic, T., Amarasekaran, B. and McKenna, A. (2010). *Final Report of the Sierra Leone National Chimpanzee Census Project.* Freetown, Sierra Leone: Tacugama Chimpanzee Sanctuary.

Brncic, T., Amarasekaran, B., McKenna, A., Mundry, R. and Kühl, H.S. (2015). Large mammal diversity and their conservation in the human-dominated land-use mosaic of Sierra Leone. *Biodiversity and Conservation,* **24**, 2417–38. DOI: 10.1007/s10531-015-0931-7.

Brockelman, W. and Geissmann, T. (2008). Hylobates lar. *The IUCN Red List of Threatened Species 2008: e.T10548A3199623.* Gland, Switzerland: International Union for Conservation of Nature (IUCN). Available at: http://dx.doi.org/10.2305/IUCN.UK.2008.RLTS.T10548A3199623.en. Accessed November 15, 2015.

Brodie, J.F., Giordano, A.J., Dickson, B., *et al.* (2015). Evaluating multispecies landscape connectivity in a threatened tropical mammal community. *Conservation Biology,* **29**, 122–32. DOI: 10.1111/cobi.12337.

Brou Yao, T., Oszwald, J., Bigot, S. and Servat, E. (2005). Risques de déforestation dans le domaine permanent de l'état en Côte d'ivoire: quel avenir pour ces derniers massifs forestiers? *Télédétection,* **5**, 3.

Brown, C., King, S., Ling, M., *et al.* (2016). *Natural Capital Assessments at the National and Sub-National Level.* Cambridge, UK: UNEP World Conservation Monitoring Centre.

Brown, P.H., Tullos, D., Tilt, B., Magee, D. and Wolf, A.T. (2009). Modeling the costs and benefits of dam construction from a multidisciplinary perspective. *Journal of Environmental Management*, **90**, S303-S11. DOI: https://doi.org/10.1016/j.jenvman.2008.07.025.

Brulliard, K. (2016). Chimp is freed from 'solitary confinement,' meets Jane Goodall, retires in Florida. *Washington Post*, April 29, 2016. Available at: https://www.washingtonpost.com/news/animalia/wp/2016/04/29/chimp-is-freed-from-solitary-confinement-meets-jane-goodall-retires-in-florida/. Accessed November 26, 2016.

Brulliard, K. (2017a). People who care about animal welfare are demanding information from USDA. *Washington Post*, August 10, 2017. Available at: https://www.washingtonpost.com/news/animalia/wp/2017/08/10/people-who-care-about-animal-welfare-are-demanding-information-from-usda/. Accessed August 12, 2017.

Brulliard, K. (2017b). USDA abruptly purges animal welfare information from its website. *Washington Post*, February 3, 2017. Available at: https://www.washingtonpost.com/news/animalia/wp/2017/02/03/the-usda-abruptly-removes-animal-welfare-information-from-its-website/. Accessed February 8, 2017.

Brulliard, K. (2017c). USDA removed animal welfare reports from its site. A showhorse lawsuit may be why. *Washington Post*, February 9, 2017. Available at: https://www.washingtonpost.com/news/animalia/wp/2017/02/09/usda-animal-welfare-records-purge-may-have-been-triggered-by-horse-industry-lawsuit.

Brunner, J., Talbott, K. and Elkin, C. (1998). *Logging Burma's Frontier Forests: Resources and the Regime*. Washington DC: World Resources Institute.

Bruno Manser Fonds (2012a). *Sold Down River*. Basel, Switzerland: Bruno Manser Fonds. Available at: http://www.bmf.ch/upload/berichte/sold_down_the_river_bmf_dams_report.pdf. Accessed November 4, 2016.

Bruno Manser Fonds (2012b). *The Taib Timber Mafia*. Basel, Switzerland: Bruno Manser Fonds. Available at: http://bmf.ch/upload/berichte/bmf_taib_family_report_2012_09_20_2.pdf. Accessed November 15, 2016.

Bryson-Morrison, N., Tzanopoulos, J., Matsuzawa, T. and Humle, T. (2017). Chimpanzee (*Pan troglodytes verus*) activity and patterns of habitat use in the anthropogenic landscape of Bossou, Guinea, West Africa. *International Journal of Primatology*, **38**, 282–302. DOI: 10.1007/s10764-016-9947-4.

Burgess, N.D., Balmford, A., Cordeiro, N.J., *et al.* (2007). Correlations among species distributions, human density and human infrastructure across the high biodiversity tropical mountains of Africa. *Biological Conservation*, **134**, 164–77. DOI: https://doi.org/10.1016/j.biocon.2006.08.024.

Burivalova, Z., Bauert, M.R., Hassold, S., Fatroandrianjafinonjasolomiovazo, N.T. and Koh, L.P. (2015). Relevance of global forest change data set to local conservation: case study of forest degradation in Masoala National Park, Madagascar. *Biotropica*, **47**, 267–74. DOI: 10.1111/btp.12194.

BurmaNet News (2000). *KNU: Hundreds of Civilians Forced to Construct Bongti-Tavoy Highway*. KNU Mergui-Tavoy District Information Department. *BurmaNet News*, June 3, 2000.

Byiers, B. and Vanheukelom, J. (2014). *What Drives Regional Economic Integration? Lessons from the Maputo Development Corridor and the North-South Corridor*. Maastricht, the Netherlands: European Centre for Development Policy Management (ECDPM). Available at: http://ecdpm.org/wp-content/uploads/DP-157-Regional-Economic-Integration-Maputo-Development-Corridor-2014.pdf.

Bynens, E., Ellenberg, L., Oldorff, D., *et al.* (2007). *Projet d'appui à la réhabilitation et l'entretien de la route Bukavu – Walikale. Etude technique détaillé et étudé d'impact socio-économique et environnemental*. GTZ. Unpublished report.

Byron, R.N. and Arnold, J.E.M. (1999). What futures for the people of the tropical forests? *World Development*, **27**, 789–805.

Caillaud, D., Ndagijimana, F., Giarrusso, A.J., Vecellio, V. and Stoinski, T.S. (2014). Mountain gorilla ranging patterns: influence of group size and group dynamics. *American Journal of Primatology*, **76**, 730–46. DOI: 10.1002/ajp.22265.

Caldecott, J.O., Bennett, J.G. and Ruitenbeek, H.J. (1989). *Cross River National Park (Oban Division): Plan for Developing the Park and its Support Zone*. Godalming, UK: World Wide Fund for Nature (WWF).

Cam Iron and Rainbow Environment Consult (2010). *Etude d'Impact Environnemental et Social du Projet de Minerai de Fer de Mbala*m. Yaoundé, Cameroun: Cam Iron and Rainbow Environment Consult.

Cameron, K.N., Reed, P., Morgan, D.B., *et al.* (2016). Spatial and temporal dynamics of a mortality event among central African great apes. *PLoS One*, **11**, e0154505. DOI: 10.1371/journal.pone.0154505.

Campbell, C., Andayani, N., Cheyne, S., *et al.* (2008). *Indonesian Gibbon Conservation and Management Workshop Final Report.* Apple Valley, MN: International Union for Conservation of Nature Species Survival Commission (IUCN SSC), Conservation Breeding Specialist Group. Available at: http://www.gibbons.asia/wp-content/uploads/2015/03/Indonesian-Gibbon-Conservation-Workshop.pdf.

Campbell, C.O., Cheyne, S.M. and Rawson, B.M. (2015). *Best Practice Guidelines for the Rehabilitation and Translocation of Gibbons.* Gland, Switzerland: International Union for Conservation of Nature Species Survival Commission (IUCN SSC), Primate Specialist Group.

Campbell-Smith, G., Campbell-Smith, M., Singleton, I. and Linkie, M. (2011a). Apes in space: saving an imperilled orangutan population in Sumatra. *PLoS One*, **6**, e17210. DOI: 10.1371/journal.pone.0017210.

Campbell-Smith, G., Campbell-Smith, M., Singleton, I. and Linkie, M. (2011b). Raiders of the lost bark: orangutan foraging strategies in a degraded landscape. *PLoS One*, **6**, e20962. DOI: 10.1371/journal.pone.0020962.

Cannon, J.C. (2017a). Cross River superhighway changes course in Nigeria. *Mongabay*, April, 2017. Available at: https://news.mongabay.com/2017/04/cross-river-superhighway-changes-course-in-nigeria.

Cannon, J.C. (2017b). Not out of the woods: concerns remain with Nigerian superhighway. *Mongabay*, May, 2017. Available at: https://news.mongabay.com/2017/05/not-out-of-the-woods-concerns-remain-with-nigerian-superhighway.

Cannon, J.C. (2017c). Scrapping Nigerian superhighway buffer isn't enough, say conservation groups. *Mongabay*, February, 2017. Available at: https://news.mongabay.com/2017/02/scrapping-nigerian-superhighway-buffer-isnt-enough-say-conservation-groups.

CANS (2003). *Wild Life Conservation and National Parks Act, 2003.* Civil Authority of New Sudan (CANS).

Carleton-Hug, A. and Hug, J.W. (2010). Challenges and opportunities for evaluating environmental education programs. *Evaluation and Program Planning*, **33**, 159–64. DOI: https://doi.org/10.1016/j.evalprogplan.2009.07.005.

Carlsen, F. and de Jongh, T. (2015). *EAZA EEP Opposes Transfer of Chimpanzees from YNPRC to Wingham Wildlife Park.* European Association of Zoos and Aquaria (EAZA), European Endangered Species Programme EEP. Available at: https://assets.documentcloud.org/documents/2843283/EEP-Letter-Regarding-Wingham.pdf.

Carne, C., Semple, S., Morrogh-Bernard, H., Zuberbühler, K. and Lehmann, J. (2014). The risk of disease to great apes: simulating disease spread in orang-utan (*Pongo pygmaeus wurmbii*) and chimpanzee (*Pan troglodytes schweinfurthii*) association networks. *PLoS One*, **9**, e95039. DOI: 10.1371/journal.pone.0095039.

CARP (n.d.). *Central African Regional Program for the Environment GIS and Remote Sensing Datasets.* Gombe, DRC: Central African Regional Program for the Environment (CARPE)/US Agency for International Development (USAID). Available at: http://carpe.umd.edu/. Accessed January, 2018.

Cartwright, B. (2010). *Pan African Sanctuary Alliance Education Resource Manual.* Portland, OR: Pan African Sanctuary Alliance (PASA).

CEE Bankwatch Network (2015). *New Beijing-Backed Asian Infrastructure Investment Bank Struggles to Convince on Environment and Sustainability Issues.* CEE Bankwatch Network. Available at: http://bankwatch.org/bwmail/63/new-beijing-backed-asian-infrastructure-investment-bank-struggles-convince-environment.

Center for Great Apes (n.d.). *Meet the Orangutans.* Wauchula, FL: Center for Great Apes. Available at: http://www.centerforgreatapes.org/meet-apes/orangutans/. Accessed October 3, 2016.

Chaléard, J.-L., Chanson-Jabeur, C. and Béranger, C. (2006). *Le Chemin de Fer en Afrique.* Paris, France: KARTHALA Editions.

Chan, B., Fellowes, J., Geissmann, T. and Zhang, J. (2005). *Status Survey and Conservation Action Plan for the Hainan Gibbon–VERSION I (Last Updated November 2005).* Kadoorie Farm & Botanic Garden Technical Report No. 3. Hong Kong: Kadoorie Farm & Botanic Garden (KFBG).

Chan, M. (2017). The government purged animal welfare data. So this guy is publishing it. *Time*, February 17, 2017. Available at: http://time.com/4673506/russ-kick-usda-animal-welfare-data.

Chan, S.W.D. (2016). *Asymmetric bargaining between Myanmar and China in the Myitsone Dam controversy: social opposition akin to David's stone against Goliath.* Presented at: International Studies Association Conference, 2016, Hong Kong. Available at: http://web.isanet.org/Web/Conferences/AP%20Hong%20Kong%202016/Archive/71f82563-316f-415d-85ce-6458e57111a6.pdf.

Channa, P. and Gray, T. (2009). *The Status and Habitat of Yellow-Cheeked Crested Gibbon* Nomascus gabriellae *in Phnom Prich Wildlife Sanctuary, Mondulkiri*. Phnom Penh, Cambodia: World Wide Fund for Nature (WWF) Greater Mekong-Cambodia Country Programme.

Chapman, C.A., Lawes, M.J. and Eeley, H.A.C. (2006). What hope for African primate diversity? *African Journal of Ecology*, **44**, 116–33. DOI: 10.1111/j.1365-2028.2006.00636.x.

Chase, J., Benoy, G., Hann, S. and Culp, J. (2016). Small differences in riparian vegetation significantly reduce land use impacts on stream flow and water quality in small agricultural watersheds. *Journal of Soil and Water Conservation*, **71**, 194–205.

Chetry, D., Chetry, R., Ghosh, K. and Singh, A.K. (2010). Status and distribution of the eastern hoolock gibbon (*Hoolock leuconedys*) in Mehao Wildlife Sanctuary, Arunachal Pradesh, India. *Primate Conservation*, **25**, 87–94. DOI: 10.1896/052.025.0113.

Cheyne, S.M. (2008). Feeding ecology, food choice and diet characteristics of gibbons in a disturbed peat-swamp forest, Indonesia. In *XXII Congress of the International Primatological Society, Edinburgh, UK*, ed. P. C. Lee, P. Honess, H. Buchanan-Smith, A. MaClarnon and W. I. Sellers. Bristol, UK: Top Copy, pp. 3–8.

Cheyne, S.M. (2010). Behavioural ecology of gibbons (*Hylobates albibarbis*) in a degraded peat-swamp forest. In *Indonesian Primates*, ed. S. Gursky and J. Supriatna. New York, NY: Springer, pp. 121–56. DOI: 10.1007/978-1-4419-1560-3_8. Available at: https://doi.org/10.1007/978-1-4419-1560-3_8.

Cheyne, S.M., Gilhooly, L.J., Hamard, M.C., et al. (2016). Population mapping of gibbons in Kalimantan, Indonesia: correlates of gibbon density and vegetation across the species range. *Endangered Species Research*, **30**, 133–43.

Cheyne, S.M., Höing, A., Rinear, J. and Sheeran, L.K. (2012). Sleeping site selection by agile gibbons: the influence of tree stability, fruit availability and predation risk. *Folia Primatologica*, **83**, 299–311.

Chhatre, A. and Agrawal, A. (2009). Trade-offs and synergies between carbon storage and livelihood benefits from forest commons. *Proceedings of the National Academy of Sciences*, **106**, 17667–70. DOI: 10.1073/pnas.0905308106.

ChimpCARE (n.d.). *Where are Our Amazing Chimpanzees in the United States*. Lincoln Park Zoo, Chicago, IL: ChimpCARE. Available at: http://www.chimpcare.org/map. Accessed October 2, 2016.

Choudhury, A. (2013). Description of a new subspecies of hoolock gibbon *Hoolock hoolock* from northeast India. *Newsletter and Journal of the Rhino Foundation for Nature in Northeast India*, **9**, 49–59.

Cibot, M., Bortolamiol, S., Seguya, A. and Krief, S. (2015). Chimpanzees facing a dangerous situation: a high-traffic asphalted road in the Sebitoli area of Kibale National Park, Uganda. *American Journal of Primatology*, **77**, 890–900. DOI: 10.1002/ajp.22417.

CIESIN and ITOS (2013). *Global Roads Open Access Data Set, Version 1 (gROADSv1)*. Palisades, NY: NASA Socioeconomic Data and Applications Center (SEDAC), Center for International Earth Science Information Network (CIESIN; Columbia University), Information Technology Outreach Services (ITOS, University of Georgia). Available at: http://dx.doi.org/10.7927/H4VD6WCT. Accessed April, 2018.

CIFOR (2015). *Sumatran Road Plan Could Spell a Dark New Chapter for Storied Ecosystem: Study*. Bogor, Indonesia: Center for International Forestry Research (CIFOR). Available at: http://blog.cifor.org/27018/leuser-ecosystem-aceh-spatial-plan-ladia-galaska-road?fnl=en.

CITES (2010a). *Disposal of Confiscated Live Specimens of Species Included in the Appendices, Resolution Conf. 10.7 (Rev. CoP15)*. Geneva, Switzerland: Convention on International Trade in Endangered Species of Wild Fauna and Flora (CITES). Available at: https://cites.org/eng/res/10/10-07R15.php.

CITES (2010b). *National Laws for Implementation of the Convention, Resolution Conf. 8.4 (Rev. CoP15)*. Geneva, Switzerland: Convention on International Trade in Endangered Species of Wild Fauna and Flora (CITES). Available at: https://cites.org/eng/res/08/08-04R15.php.

CITES (2012). *Interpretation and Implementation of the Convention: Compliance and Enforcement Matters, SC62 Doc. 29*. Geneva, Switzerland: Convention on International Trade in Endangered Species of Wild Fauna and Flora (CITES). Available at: https://cites.org/eng/com/sc/62/E62-29.pdf.

CITES (2014). *Great Apes Exported from Guinea to China from 2009 to 2011. Statement by Secretariat. January, 2014*. Geneva, Switzerland: Convention on International Trade in Endangered Species of Wild Fauna and Flora (CITES). Available at: https://cites.org/sites/default/files/common/docs/CITES-Guinea-China-great-apes.pdf. Accessed November 15, 2016.

CITES (2016a). *Draft Resolution on Review of Trade in Animal Specimens Reported as Produced in Captivity, CoP 17 Doc. 32, CoP17 Com. II.18*. Geneva, Switzerland: Convention on International Trade in Endangered Species of Wild Fauna and Flora (CITES). Available at: https://cites.org/sites/default/files/eng/cop/17/Com_II/E-CoP17-Com-II-18.pdf.

CITES (2016b). *Status of Legislative Progress for Implementing CITES (updated on 1 September 2016) (English only), CoP17 Doc. 22, Annex 3 (Rev. 1)*. Geneva, Switzerland: Convention on International Trade in Endangered Species of Wild Fauna and Flora (CITES). Available at: https://cites.org/sites/default/files/eng/cop/17/WorkingDocs/E-CoP17-22-A3-R1.pdf.

CITES (2017). *Appendices I, II and III. Valid from April 4, 2017*. Geneva, Switzerland: Convention on International Trade in Endangered Species of Wild Fauna and Flora (CITES). Available at: https://cites.org/eng/app/appendices.php. Accessed April 5, 2017.

CITES (n.d.-a). *CITES Appendices*. Geneva, Switzerland: Convention on International Trade in Endangered Species of Wild Fauna and Flora (CITES). Available at: https://www.cites.org/eng/app/index.php. Accessed May, 2017.

CITES (n.d.-b). *CITES Trade Database*. Geneva, Switzerland: Convention on International Trade in Endangered Species of Wild Fauna and Flora (CITES). Available at: https://trade.cites.org. Accessed November 17, 2016.

Clements, G.R. (2013). *The environmental and social impacts of roads in Southeast Asia*. Doctoral thesis. Cairns, Australia: James Cook University.

Clements, G.R., Lynam, A.J., Gaveau, D., et al. (2014). Where and how are roads endangering mammals in southeast Asia's forests? *PLoS One*, **9**, e115376. DOI: 10.1371/journal.pone.0115376.

CMP (2013). *Open Standards for the Practice of Conservation. Version 3.0*. Washington DC: Conservation Measures Partnership (CMP).

Cochard, R. (2017). Degradation and biodiversity losses in Aceh Province, Sumatra. In *Redefining Diversity and Dynamics of Natural Resources Management in Asia. Volume 1. Sustainable Natural Resources Management in Dynamic Asia*, ed. G. Shivakoti, U. Pradhan and H. Helmi. Amsterdam, the Netherlands: Elsevier, pp. 231–71.

Coffin, A.W. (2007). From roadkill to road ecology: a review of the ecological effects of roads. *Journal of Transport Geography*, **15**, 396–406. DOI: https://doi.org/10.1016/j.jtrangeo.2006.11.006.

Colchester, M., Wee, A.P., Wong, M.C. and Jalong, T. (2007). *Land is Life: Land Rights and Oil Palm Development in Sarawak*. Forest Peoples Programme and Sawit Watch. Available at: http://www.forestpeoples.org/sites/fpp/files/publication/2010/08/sarawaklandislifenov07eng.pdf. Accessed November 14, 2016.

Collier, P., Kirchberger, M. and Söderbom, M. (2015). *The Cost of Road Infrastructure in Low and Middle Income Countries*. Washington DC: World Bank. Available at: http://documents.worldbank.org/curated/en/124841468185354669/pdf/WPS7408.pdf.

Commitante, R., Unwin, S., Jaya, R., et al. (2015). *Orangutan Veterinary Advisory Group Workshop Report*. Los Angeles, CA: Orangutan Conservancy.

Connette, G., Oswald, P., Songer, M. and Leimgruber, P. (2016). Mapping distinct forest types improves overall forest identification based on multi-spectral landsat imagery for Myanmar's Tanintharyi region. *Remote Sensing*, **8**, 882.

Corridor Partnership (n.d.). *Southern Cameroun*. Nairobi, Kenya: The Corridor Partnership. Available at: http://thecorridorspartnership.com/pages/pilot-corridors/southern-cameroun-corridor/.

Cotula, L. (2016). *Foreign Investment, Law and Sustainable Development: A Handbook on Agriculture and Extractive Industries*. IIED Natural Resource Issues No. 31. London, UK: International Institute for Environment and Development (IIED).

Cotula, L., Jokubauskaite, G., Sutz., P. and Singleton, I. (2015). Legal frameworks at the interface between industrial agriculture and ape conservation. In *State of the Apes: Industrial Agriculture and Ape Conservation*, ed. Arcus Foundation. Cambridge, UK: Cambridge University Press, pp. 105–33. Available at: https://www.stateoftheapes.com/volume-2-industrial-agriculture/.

Council of the European Union (1999). 1999/22/EC Keeping of wild animals in zoos. *Official Journal L 094*, **09/04/1999**, 0024–6. Available at: http://europa.eu/legislation_summaries/environment/nature_and_biodiversity/l28069_en.htm.

Cowlishaw, G. and Dunbar, R.I. (2000). *Primate Conservation Biology*. Chicago, IL: University of Chicago Press.

Cox, D.A., Poaty, P., Tolliday, C.M., *et al.* (2014). *Impact of public awareness campaign for reduction in the illegal trade of great apes in Congo Republic*. Presented at: International Primatological Society, XXV Congress, August 11–16, 2014, Hanoi, Vietnam. Madison: International Primatological Society.

Cuny, P. (2011). *Etat de lieux de la foresterie communautaire et communale au Cameroun*. The Netherlands: Tropenbos. Available at: http://www.tropenbos.org/publications/current+status+of+community+forestry+in+cameroon.

Curran, L.M., Trigg, S.N., McDonald, A.K., *et al.* (2004). Lowland forest loss in protected areas of Indonesian Borneo. *Science*, **303**, 1000–3. DOI: 10.1126/science.1091714.

Currey, K. (2013). *Social and environmental safeguard policies at the World Bank: historical lessons for a changing context*. Draft internal briefing note. New York, NY: Ford Foundation.

Dai, Y. (2013). Outlook for energy supply and demand in China. In *Green Low-Carbon Development in China*, ed. J. Xue, Z. Zhao, Y. Dai and B. Wang. Cham, Switzerland: Springer, pp. 81–102. DOI: 10.1007/978-3-319-01153-0.

Daily, G. and Ellison, K. (2012). *The New Economy of Nature: The Quest to Make Conservation Profitable*. Washington DC: Island Press.

Daily, G.C., Zhiyun, O., Hua, Z., *et al.* (2013). Securing natural capital and human well-being: innovation and impact in China. *Acta Ecologica Sinica*, **33**, 677–85.

Daly, N. and Bale, R. (2017). *We asked the Government Why Animal Welfare Records Disappeared. They sent 1,700 Blacked-Out Pages*. Washington DC: National Geographic. Available at: http://news.nationalgeographic.com/2017/05/usda-animal-welfare-records-foia-black-out-first-release/. Accessed May 1, 2017.

Damania, R., Barra, A.F., Burnouf, M. and Russ, J.D. (2016). *Transport, Economic Growth, and Deforestation in the Democratic Republic of Congo: A Spatial Analysis*. Washington DC: World Bank. Available at: https://openknowledge.worldbank.org/handle/10986/24044. Accessed May, 2017.

Damarad, T. and Bekker, G.J. (2003). *COST 341: Habitat Fragmentation due to Transportation Infrastructure: Findings of the COST Action 341*. Luxembourg, Luxembourg: Office for Official Publications of the European Communities.

Dames and Moore (1997). Annexe: etude sur les ressources biologiques – Cameroon. In *Projet d'Exportation Tchadien*. Available at: http://web.worldbank.org/archive/website01210/WEB/IMAGES/MULTI0-7.PDF.

Das, J., Biswas, J., Bhattacherjee, P.C. and Rao, S.S. (2009). Canopy bridges: an effective conservation tactic for supporting gibbon populations in forest fragments. In *The Gibbons: New Perspectives on Small Ape Socioecology and Population Biology*, ed. D. Whittaker and S. Lappan. New York, NY: Springer, pp. 467–75. DOI: 10.1007/978-0-387-88604-6_22. Available at: https://doi.org/10.1007/978-0-387-88604-6_22.

Davis, J.T., Mengersen, K., Abram, N.K., *et al.* (2013). It's not just conflict that motivates killing of orangutans. *PLoS One*, **8**, e75373. DOI: 10.1371/journal.pone.0075373.

D'Cruze, N. and Macdonald, D.W. (2016). A review of global trends in CITES live wildlife confiscations. *Nature Conservation*, **15**. DOI: 10.3897/natureconservation.15.10005.

DDA (2014). *Voices from the Ground: Concerns Over the Dawei Special Economic Zone and Related Projects*. Dawei Township, Myanmar: Dawei Development Association (DDA). Available at: http://www.burmalibrary.org/docs19/Voices_from_the_ground-en-red.pdf.

DDA, TYG and TripNet (2015). *We Used to Fear Bullets Now We Fear Bulldozers: Dirty Coal Mining by Military Cronies and Thai Companies*. Burma Partnership: Dawei Development Association (DDA), Tarkapaw Youth Group (TYG) and Tenasserim River and Indigenous People Networks (TripNet). Available at: http://www.burmapartnership.org/2015/10/we-used-to-fear-bullets-now-we-fear-bulldozers-dirty-coal-mining-by-military-cronies-thai-companies-ban-chaung-dawei-district-myanmar/.

De Koninck, R., Bernard, S. and Girard, M. (2012). Aceh's forests as an asset for reconstruction. In *From the Ground Up: Perspectives on Post-Tsunami and Post-Conflict Aceh*, ed. P. Daly, R. M. Feener and A. J. Reid. Singapore: Institute of Southeast Asian Studies, pp. 156–75.

de Wasseige C., Flynn, J., Louppe, D., Hiol Hiol, F. and Mayaux, P. (2013). *The Forests of the Congo Basin: State of the Forest 2013*. Neufchateau, Belgium: Weyrich. Available at: http://www.observatoire-comifac.net/docs/edf2013/EN/EDF2013_EN.pdf.

Dean, J.M., Lovely, M.E. and Wang, H. (2009). Are foreign investors attracted to weak environmental regulations? Evaluating the evidence from China. *Journal of Development Economics*, **90**, 1–13. DOI: https://doi.org/10.1016/j.jdeveco.2008.11.007.

Debonnet, G. and Vié, J.C. (2010). *Rapport de Mission de Suivi au Parc National de Kahuzi-Biega (RDC)*. United Nations Educational, Scientific and Cultural Organization (UNESCO), International Union for Conservation of Nature (IUCN).

Delgado, R.A. (2010). Communication, culture and conservation in orangutans. In *Indonesian Primates*, ed. S. Gursky and J. Supriatna. New York, NY: Springer, pp. 23–40. DOI: 10.1007/978-1-4419-1560-3_3. Available at: https://doi.org/10.1007/978-1-4419-1560-3_3.

Delgado, R.A. and Van Schaik, C.P. (2000). The behavioral ecology and conservation of the orangutan (*Pongo pygmaeus*): a tale of two islands. *Evolutionary Anthropology: Issues, News, and Reviews*, **9**, 201–18. DOI: 10.1002/1520-6505(2000)9:5<201::AID-EVAN2>3.0.CO;2-Y.

DFID (2010). *Annual Review: PRO ROUTES AUGUST 2010*. Department for International Development (DFID). Available at: iati.dfid.gov.uk/iati_documents/4109130.xls.

Dierkers, G. and Mattingly, J. (2009). *How States and Territories Fund Transportation: An Overview of Traditional and Nontraditional Strategies*. Washington DC: National Governors Association (NGA) Center for Best Practices. Environment, Energy & Natural Resources Division. Available at: https://www.nga.org/files/live/sites/NGA/files/pdf/0907TRANSPORTATIONSTRATEGIES.PDF.

DigitalGlobe (n.d.). *DigitalGlobe.* Westminster, CO: DigitalGlobe. Available at: https://www.digitalglobe.com. Accessed December, 2016–April, 2017.

Dinsi, S.C. and Eyebe, S.A. (2016). *Great Ape Conservation in Cameroon: Mapping Institution and Policies. Poverty and Conservation Learning Group (PCLG) Research Report*. London, UK: International Institute for Environment and Development (IIED). Available at: http://pubs.iied.org/pdfs/G04017.pdf.

Dkamela, G.P. (2011). *The Context of REDD+ in Cameroon: Drivers, Agents and Institutions*. Bogor, Indonesia: Center for International Forestry Research (CIFOR).

Doran-Sheehy, D., Mongo, P., Lodwick, J. and Conklin-Brittain, N.L. (2009). Male and female western gorilla diet: preferred foods, use of fallback resources, and implications for ape versus old world monkey foraging strategies. *American Journal of Physical Anthropology*, **140**, 727–38. DOI: 10.1002/ajpa.21118.

Doumenge, C. and Heymer, A. (1992). *Evaluation de l'impact environnemental de la route Kisangani – Bukavu/ Goma (Zaire)*. GTZ/International Union for Conservation of Nature (IUCN). Unpublished report.

DSU (2016). *Better Decision-Making about Large Dams with a View to Sustainable Development. 22nd December 2016. Reference 7199*. Utrecht, the Netherlands: Dutch Sustainability Unit (DSU), Netherlands Commission for Environmental Assessment. Available at: http://api.commissiemer.nl/docs/os/i71/i7199/7199_advice_on_better_decision-making_about_large_dams.pdf.

Dubois, G., Bastin, L., Martinez Lopez J., *et al.* (2015). *The Digital Observatory for Protected Areas (DOPA) Explorer 1.0*. Luxembourg, Luxembourg: Publications Office of the European Union. Available at: http://dopa.jrc.ec.europa.eu/sites/default/files/test/2015%20Online%20LB-NA-27162-EN-N2%20.pdf.

Dudley, N., Stolton, S. and Elliott, W. (2013). Wildlife crime poses unique challenges to protected areas. *Parks*, **19**, 7–12.

Dulac, J. (2013). *Global Land Transport Infrastructure Requirements to 2050*. Paris, France: International Energy Agency.

Dunn, A. (2016). On a road to nowhere? *Gorilla Journal*, **53**. Available at: http://www.berggorilla.org/en/journal/issues/journal-no-51/article-view/?tx_ttnews%5Btt_news%5D=859&cHash=d71bc775cf973adfcd1141544ced400a.

Dunn, A., Bergl, R., Byler, D., *et al.* (2014). *Revised Regional Action Plan for the Conservation of the Cross River Gorilla (Gorilla gorilla diehli) 2014–2019*. New York, NY: International Union for Conservation of Nature Species Survival Commission (IUCN SSC), Primate Specialist Group and Wildlife Conservation Society. Available at: http://static1.1.sqspcdn.com/static/f/1200343/24396223/1392763857780/CRG_action_plan_2014.pdf?token=qRnBxb%2Fev%2BkjCcK7a1OdITpCpq0%3D.

Dunn, A. and Imong, I. (2017). A brief update on the proposed superhighway in Cross River State. *Gorilla Journal*, **54**. Available at: www.berggorilla.org/en/journal/issues/journal-no-54/article-view/?tx_ttnews[tt_news]=917&cHash=1831ff7c0faadd33c6f7a922d808b072.

Durán, A.P., Rauch, J. and Gaston, K.J. (2013). Global spatial coincidence between protected areas and metal mining activities. *Biological Conservation*, **160**, 272–8. DOI: https://doi.org/10.1016/j.biocon.2013.02.003.

Durham, D. (2015). The status of captive apes. In *State of the Apes: Industrial Agriculture and Ape Conservation*, ed. Arcus Foundation. Cambridge, UK: Cambridge University Press, pp. 228–59. Available at: http://www.stateoftheapes.com/themes/the-status-of-captive-apes/.

Durham, D. and Phillipson, A. (2014). Status of captive apes across Africa and Asia: the impact of extractive industry. In *State of the Apes: Extractive Industries and Ape Conservation*, ed. Arcus Foundation. Cambridge, UK: Cambridge University Press, pp. 279–305. Available at: http://www.stateoftheapes.com/volume-1-extractive-industries/.

Ebrahim-zadeh, C. (2003). Back to basics: when countries get too much of a good thing. Christine Ebrahim-zadeh explains Dutch disease. *Finance and Development-English Edition*, **40**, 50–1. Available at: http://www.webcitation.org/5YeSchvbI.

ECD (2016). *EIA Procedure*. Nay Pyi Taw, Myanmar: Myanmar Environmental Conservation Department (ECD). Available at: http://www.ecd.gov.mm/?q=policy.

EDC (2011a). *Lom Pangar Hydroelectric Project Environmental and Social Assessment, Executive Summary*. Yaoundé, Cameroon: Electricity Development Corporation (EDC).

EDC (2011b). *Projet Hydroélectrique de Lom Pangar; Evaluation Environnementale et Sociale (EES). Volume 1. Evaluation des Impacts Environnementaux et Sociaux (EIES)*. Yaoundé, Cameroon: Electricity Development Corporation (EDC).

EDC (2011c). *Réformulation de l'étude d'impacts et du Plan de Gestion Environnementale et Sociale du Barrage de Lom Pangar: Mise en Oeuvre de la Compensation Biodiversité: Parc National de Deng-Deng*. Yaoundé, Cameroon: Electricity Development Corporation (EDC). Available at: http://www.edc-cameroon.org/IMG/pdf/sde/ANNEXE%204%20PNDD%20projet%20110111.pdf.

EDC (n.d.-a). *La Pêche s'Organise Autour de Lom Pangar*. Yaoundé, Cameroon: Electricity Development Corporation (EDC). Available at: http://www.edc-cameroon.org/francais/societe/nos-activites/article/la-peche-s-organise-autour-de-lom. Accessed August, 2017.

EDC (n.d.-b). *Mise en Eau Partielle de Lom Pangar: Pari Tenu!* Yaoundé, Cameroon: Electricity Development Corporation (EDC). Available at: http://www.edc-cameroon.org/francais/projets/lom-pangar/actualite/article/mise-en-eau-partielle-de-lom. Accessed August, 2017.

Eddy, T. (2015). Embodying ecological policy in defending the Leuser ecosystem area for sustaining collective life. *International Journal of Humanities and Social Science*, **5**, 252–64.

Edelman, M. and Haugerud, A., ed. (2005). *The Anthropology of Development and Globalization*. Oxford, UK: Blackwell Publishing Inc.

EDG (2007). *Etude Détaillée de l'Impact Socio-Environnemental de la Route Allant de Kisangani et Bunduki*. Kinshasa, DRC: Ministère des Travaux Publics et Infrastructures & DFID, Environment and Development Group (EDG).

Edwards, D.P., Sloan, S., Weng, L., *et al.* (2014). Mining and the African environment. *Conservation Letters*, **7**, 302–11. DOI: 10.1111/conl.12076.

Ehrlich, P.R., Ehrlich, A.H. and Daily, G.C. (1997). *The Stork and the Plow: The Equity Answer to the Human Dilemma*. New Haven, CT: Yale University Press.

EIB (2013). *Environmental and Social Handbook. Volume I: EIB Environmental and Social Standards*. Luxembourg: European Investment Bank (EIB). Available at: http://www.eib.org/attachments/strategies/environmental_and_social_practices_handbook_en.pdf.

Elder, A.A. (2009). Hylobatid diets revisited: the importance of body mass, fruit availability, and interspecific competition. In *The Gibbons: New Perspectives on Small Ape Socioecology and Population Biology*, ed. D. Whittaker and S. Lappan. New York, NY: Springer, pp. 133–59. DOI: 10.1007/978-0-387-88604-6_8. Available at: https://doi.org/10.1007/978-0-387-88604-6_8.

Elkan, P., Elkan, S., Moukassa, A., *et al.* (2006). Managing threats from bushmeat hunting in a timber concession in the Republic of Congo. In *Emerging Threats to Tropical Forests*, ed. W. F. Laurance and C. A. Peres. Chicago, IL: University of Chicago Press, pp. 393–415.

Ellioti.org (n.d.). *P.T. Ellioti by the Numbers*. Ellioti.org. Available at: http://www.ellioti.org/p.t.ellioti-by-the-numbers.html. Accessed August, 2017.

Emery Thompson, M. and Wrangham, R.W. (2008). Diet and reproductive function in wild female chimpanzees (*Pan troglodytes schweinfurthii*) at Kibale National Park, Uganda. *American Journal of Physical Anthropology*, **135**, 171–81. DOI: 10.1002/ajpa.20718.

Emery Thompson, M. and Wrangham, R.W. (2013). *Pan troglodytes* robust chimpanzee. In *Mammals of Africa. Volume II: Primates*, ed. T. M. Butynski, J. Kingdon and J. Kalina. London, UK: Bloomsbury Publishing, pp. 55–64.

Emery Thompson, M., Zhou, A. and Knott, C.D. (2012). Low testosterone correlates with delayed development in male orangutans. *PLoS One*, **7**, e47282. DOI: 10.1371/journal.pone.0047282.

Encyclopaedia Britannica (1998). *Baram River*. Chicago, IL: Encyclopaedia Britannica Inc. Available at: https://www.britannica.com/place/Baram-River. Accessed September 14, 2017.

Engberg, C.C. (2002). The dam owner's guide to retirement planning: assessing owner liability for downstream sediment flow from obsolete dams. *Stanford Environmental Law Journal*, **21**, 177.

Engel, R. and Petropoulos, A. (2016). Saving Cobra: the rescue of an orphaned chimpanzee. *NBC*, May 9, 2016. Available at: http://www.nbcnews.com/dateline/saving-cobra-rescue-orphaned-chimpanzee-n570416. Accessed September, 2016.

Environmental Justice Atlas (n.d.). *Exploitation of Forests, Cameroon*. Environmental Justice Atlas. Available at: https://ejatlas.org/print/exploitation-of-forests-cameroon.

ERI (2009). *Total Impact: The Human Rights, Environmental, and Financial Impacts of Total and Chevron's Yadana Gas Project in Military-Ruled Burma (Myanmar)*. Bangkok, Thailand: Earth Rights International (ERI).

ERM (2016). *Calabar-Ikom-Katsina Ala superhighway ESIA gap analysis: gap analysis report.* Report to World Wide Fund for Nature (WWF) UK May, 2016.

ESI Africa (2016). The dawn of wind energy in Africa. *ESI Africa*. Available at: https://www.esi-africa.com/magazine_articles/dawn-wind-energy-africa/.

Espinosa, S., Branch, L.C. and Cueva, R. (2014). Road development and the geography of hunting by an Amazonian indigenous group: consequences for wildlife conservation. *PLoS One*, **9**, e114916. DOI: 10.1371/journal.pone.0114916.

Esri (2016). *ArcGIS Desktop: Release 10.4.1.* Redlands, CA: Environmental Systems Research Institute.

ETP (n.d.). *Malaysia Economic Transformation Programme.* Kuala Lumpur, Malaysia: Economic Transformation Programme (ETP). Available at: http://etp.pemandu.gov.my/About_ETP-@-Overview_of_ETP.aspx.

European Commission (2015). *EU Zoos Directive Good Practices Document.* Luxembourg, Luxembourg: Publications Office of the European Union.

European Commission (n.d.). *Digital Observatory for Protected Areas (DOPA)*. Ispra, Italy: European Commission. Available at: dopa.jrc.ec.europa.eu/en.

Fa, J.E., Seymour, S., Dupain, J., *et al.* (2006). Getting to grips with the magnitude of exploitation: bushmeat in the Cross–Sanaga rivers region, Nigeria and Cameroon. *Biological Conservation*, **129**, 497–510. DOI: https://doi.org/10.1016/j.biocon.2005.11.031.

Fan, P., Fei, H., Xiang, Z., *et al.* (2010). Social structure and group dynamics of the Cao Vit gibbon (*Nomascus nasutus*) in Bangliang, Jingxi, China. *Folia Primatologica*, **81**, 245–53.

Fan, P.-F., He, K., Chen, X., *et al.* (2017). Description of a new species of hoolock gibbon (Primates: Hylobatidae) based on integrative taxonomy. *American Journal of Primatology*, **79**. https://doi.org/10.1002/ajp.22631.

Fan, P.-F. and Jiang, X.-L. (2008). Effects of food and topography on ranging behavior of black crested gibbon (*Nomascus concolor jingdongensis*) in Wuliang Mountain, Yunnan, China. *American Journal of Primatology*, **70**, 871–8. DOI: 10.1002/ajp.20577.

Fan, P.-F. and Jiang, X.-L. (2010). Maintenance of multifemale social organization in a group of *Nomascus concolor* at Wuliang Mountain, Yunnan, China. *International Journal of Primatology*, **31**, 1–13. DOI: 10.1007/s10764-009-9375-9.

Fan, P.-F., Jiang, X.-L. and Tian, C.-C. (2009). The critically endangered black crested gibbon *Nomascus concolor* on Wuliang Mountain, Yunnan, China: the role of forest types in the species conservation. *Oryx*, **43**, 203–8. DOI: 10.1017/S0030605308001907.

FAO (2015). *Global Forest Resources Assessment: How Are the World's Forests Changing?* Rome, Italy: Food and Agriculture Organization of the United Nations (FAO). Available at: http://www.fao.org/3/a-i4793e.pdf.

FAO, Action Against Hunger, Action Aid, International Federation of Red Cross and Red Crescent Societies and World Vision International (2016). *Free Prior and Informed Consent: An Indigenous Peoples' Right and a Good Practice for Local Communities. Manual for Project Practitioners.* Rome, Italy: Food and Agriculture Organization of the United Nations (FAO). Available at: www.fao.org/3/a-i6190e.pdf.

Farmer, K.H. (2002). Pan-African Sanctuary Alliance: status and range of activities for great ape conservation. *American Journal of Primatology*, **58**, 117–32. DOI: 10.1002/ajp.10054.

Farmer, K.H. (2012). *Building Sustainable Sanctuaries.* Cambridge, UK: Arcus Foundation. Available at: http://www.sanctuaryfederation.org/gfas/wp-content/uploads/2013/09/Arcus_Building_Sustainable_Sanctuaries.pdf.

Farmer, K.H., Jamart, A. and Goossens, B. (2010). The re-introduction of chimpanzees *Pan troglodytes troglodytes* to the Conkouati-Douli National Park, Republic of Congo. In *Global Re-Introduction Perspectives: Additional Case-Studies from Around the Globe*, ed. P. S. Soorae. Abu Dhabi, UAE: International Union for Conservation of Nature Species Survival Commission (IUCN SSC), Re-introduction Specialist Group, pp. 231–7.

Farmer, K.H., Unwin, S., Cress, D., *et al.* (2009). *Pan African Sanctuary Alliance (PASA) Operations Manual.* Portland, OR: Pan African Sanctuary Alliance (PASA).

Faust, L.J., Cress, D., Farmer, K.H., Ross, S.R. and Beck, B.B. (2011). Predicting capacity demand on sanctuaries for African chimpanzees (*Pan troglodytes*). *International Journal of Primatology*, **32**, 849–64. DOI: 10.1007/s10764-011-9505-z.

Fawcett, K. (2000). *Female relationships and food availability in a forest community of chimpanzees.* PhD thesis. Edinburgh, UK: University of Edinburgh.

Fearnside, P.F. (2006). Containing destruction from Brazil's Amazon highways: now is the time to give weight to the environment in decision-making. *Environmental Conservation*, **33**, 181–3. DOI: 10.1017/S0376892906003109.

Fearnside, P.M. (2016a). Greenhouse gas emissions from Brazil's Amazonian hydroelectric dams. *Environmental Research Letters*, **11**, 011002.

Fearnside, P.M. (2016b). Tropical dams: to build or not to build? *Science*, **351**, 456–7.

Fearnside, P.M. and de Alencastro Graça, P.M.L. (2006). BR-319: Brazil's Manaus-Porto Velho Highway and the potential impact of linking the arc of deforestation to central Amazonia. *Environmental Management*, **38**, 705–16. DOI: 10.1007/s00267-005-0295-y.

Fears, D. (2016). NIH vowed to move its research chimps from labs, but only 7 got safe haven in 2015. *Washington Post*, February 26, 2016. Available at: https://www.washingtonpost.com/news/animalia/wp/2016/02/26/despite-nihs-promise-only-7-research-chimps-got-safe-haven-in-2015/.

Federal Republic of Nigeria (1992). *Environmental Impact Assessment Decree, No 86 of 1992, Laws of the Federation of Nigeria.* Federal Republic of Nigeria. Available at: http://www.nigeria-law.org/Environmental%20Impact%20Assessment%20Decree%20No.%2086%201992.htm.

Ferraro, P.J. and Pattanayak, S.K. (2006). Money for nothing? A call for empirical evaluation of biodiversity conservation investments. *PLoS Biology*, **4**, e105. DOI: 10.1371/journal.pbio.0040105.

Ferrie, G.M., Farmer, K.H., Kuhar, C.W., *et al.* (2014). The social, economic, and environmental contributions of Pan African Sanctuary Alliance primate sanctuaries in Africa. *Biodiversity and Conservation*, **23**, 187–201. DOI: 10.1007/s10531-013-0592-3.

Fishpool, L.D.C. and Evans, M.I. (2001). *Important Bird Areas in Africa and Associated Islands: Priority Sites for Conservation.* Cambridge, UK: Birdlife International.

FLEGT (2016). *Forest Law Enforcement, Governance and Trade (FLEGT).* Luxembourg, Luxembourg: Publications Office of the European Union. Available at: http://ec.europa.eu/environment/forests/illegal_logging.htm.

Flyvbjerg, B. (2009). Survival of the unfittest: why the worst infrastructure gets built—and what we can do about it. *Oxford Review of Economic Policy*, **25**, 344–67. DOI: 10.1093/oxrep/grp024.

Foley, J.A., DeFries, R., Asner, G.P., *et al.* (2005). Global consequences of land use. *Science*, **309**, 570–4. DOI: 10.1126/science.1111772.

Forman, R.T.T. and Alexander, L.E. (1998). Roads and their major ecological effects. *Annual Review of Ecology and Systematics*, **29**, 207–31. DOI: 10.1146/annurev.ecolsys.29.1.207.

Forrest, J.L., Mascia, M.B., Pailler, S., *et al.* (2015). Tropical deforestation and carbon emissions from protected area downgrading, downsizing, and degazettement (PADDD). *Conservation Letters*, **8**, 153–61. DOI: 10.1111/conl.12144.

Foster, V. and Briceño-Garmendia, C.M. (2010). *Africa Infrastructure: A Time for Transformation.* Washington DC: World Bank. Available at: http://documents.worldbank.org/curated/en/246961468003355256/Africas-infrastructure-a-time-for-transformation.

FPP, IIFB and CBD (2016). *Local Biodiversity Outlooks. Indigenous Peoples' and Local Communities' Contributions to the Implementation of the Strategic Plan for Biodiversity 2011–2020. A Complement to the Fourth Edition of the Global Biodiversity Outlook.* Moreton-in-Marsh, UK: Forest Peoples Programme (FPP), The International Indigenous Forum on Biodiversity (IIFB) and The Secretariat of the Convention on Biological Diversity (CBD). Available at: http://localbiodiversityoutlooks.net.

FPP, Pusaka and Pokker SHK (2014). *Securing Forests, Securing Rights: Report of the International Workshop on Deforestation and the Rights of Forest Peoples.* Forest Peoples' Programme (FPP), Pusaka and Pokker SHK. Available at: https://www.forestpeoples.org/sites/default/files/private/publication/2014/09/prreport.pdf.

Franco, J. (2014). *Reclaiming Free Prior and Informed Consent (FPIC) in the Context of Global Land Grabs.* Amsterdam, the Netherlands: Transnational Institute. Available at: https://www.tni.org/files/download/reclaiming_fpic_0.pdf.

Frankfurt School–UNEP Centre/BNEF (2017). *Global Trends in Renewable Energy Investment 2017.* Frankfurt, Germany: Frankfurt School of Finance and Management. Available at: http://fs-unep-centre.org/sites/default/files/publications/globaltrendsinrenewableenergyinvestment2017.pdf.

Freudenthal, E., Nnah, S. and J., K. (2011). *REDD and Rights In Cameroon: A Review of the Treatment of Indigenous Peoples and Local Communities in Policies and Projects.* Moreton-in-Marsh, UK: Forest Peoples Programme and Centre for Environment and Development. Available at: http://www.forestpeoples.org/topics/forest-carbon-partnership-facility-fcpf/publication/2011/redd-and-rights-cameroon-review-trea.

Fruth, B., Hickey, J.R., André, C., *et al.* (2016). Pan paniscus. *The IUCN Red List of Threatened Species 2016: e.T15932A17964305.* Gland, Switzerland: International Union for Conservation of Nature (IUCN). Available at: http://www.iucnredlist.org/details/15932/0.

Fruth, B., Williamson, E.A. and Richardson, M.C. (2013). Bonobo *Pan paniscus.* In *Handbook of the Mammals of the World. Volume 3: Primates*, ed. R. A. Mittermeier, A. B. Rylands and D. E. Wilson. Barcelona, Spain: Lynx Edicions, pp. 853–4.

FSC (2015). *FSC Principles and Criteria for Forest Stewardship.* Bonn, Germany: Forest Stewardship Council (FSC). Available at: http://www.fsc-uk.org/en-uk/business-area/fsc-certificate-types/forest-management-fm-certification/what-standard-is-used.

Fund for Animals (n.d.). *About The Cleveland Amory Black Beauty Ranch.* Available at: http://www.fundforanimals.org/blackbeauty/about/.

Fünfstück, T., Arandjelovic, M., Morgan, D.B., *et al.* (2014). The genetic population structure of wild western lowland gorillas (*Gorilla gorilla gorilla*) living in continuous rain forest. *American Journal of Primatology*, **76**, 868–78. DOI: 10.1002/ajp.22274.

Funwi-Gabga, N., Kuehl, H., Maisels, F., *et al.* (2014). The status of apes across Africa and Asia. In *State of the Apes: Extractive Industries and Ape Conservation*, ed. Arcus Foundation. Cambridge, UK: Cambridge University Press, pp. 253–77.

GAIN (n.d.). *Great Ape Information Network.* Kyoto, Japan: Great Ape Information Network (GAIN). Available at: http://www.shigen.nig.ac.jp/gain/index.jsp.

Galeano, E. (2009). *Open Veins of Latin America: Five Centuries of the Pillage of a Continent.* London, UK: Serpent's Tail.

Galinato, G.I. and Galinato, S.P. (2013). The short-run and long-run effects of corruption control and political stability on forest cover. *Ecological Economics*, **89**, 153–61. DOI: https://doi.org/10.1016/j.ecolecon.2013.02.014.

Ganas, J., Robbins, M.M., Nkurunungi, J.B., Kaplin, B.A. and McNeilage, A. (2004). Dietary variability of mountain gorillas in Bwindi Impenetrable National Park, Uganda. *International Journal of Primatology*, **25**, 1043–72. DOI: 10.1023/b:ijop.0000043351.20129.44.

Garcia, L.C., Ribeiro, D.B., de Oliveira Roque, F., Ochoa-Quintero, J.M. and Laurance, W.F. (2017). Brazil's worst mining disaster: corporations must be compelled to pay the actual environmental costs. *Ecological Applications*, **27**, 5–9. DOI: 10.1002/eap.1461.

Gartland, A. (2017). NGOs urge UNESCO to intervene to save rainforest heritage site. *Changing Times*, March 24, 2017. Available at: https://changingtimes.media/2017/03/24/ngos-urge-unesco-to-intervene-to-save-sumatras-leuser-ecosystem/. Accessed February 2, 2018.

Gascon, C., Malcolm, J.R., Patton, J.L., *et al.* (2000). Riverine barriers and the geographic distribution of Amazonian species. *Proceedings of the National Academy of Sciences*, **97**, 13672–7. DOI: 10.1073/pnas.230136397.

Gauvey Herbert, D. (2017). Kony 2017: from guerrilla marketing to guerrilla warfare. *Foreign Policy*, May, 2017. Available at: http://foreignpolicy.com/2017/03/02/kony-2017-from-guerilla-marketing-to-guerrilla-warfare-invisible-children-africa/.

Gaveau, D.L.A., Epting, J., Lyne, O., *et al.* (2009a). Evaluating whether protected areas reduce tropical deforestation in Sumatra. *Journal of Biogeography*, **36**, 2165–75. DOI: 10.1111/j.1365-2699.2009.02147.x.

Gaveau, D.L.A., Kshatriya, M., Sheil, D., *et al.* (2013). Reconciling forest conservation and logging in Indonesian Borneo. *PLoS One*, **8**, e69887. DOI: 10.1371/journal.pone.0069887.

Gaveau, D.L.A., Sheil, D., Husnayaen, *et al.* (2016). Rapid conversions and avoided deforestation: examining four decades of industrial plantation expansion in Borneo. *Scientific Reports*, **6**, 32017. DOI: 10.1038/srep32017. Available at: https://www.nature.com/articles/srep32017#supplementary-information.

Gaveau, D.L.A., Sloan, S., Molidena, E., *et al.* (2014). Four decades of forest persistence, clearance and logging on Borneo. *PLoS One*, **9**, e101654. DOI: 10.1371/journal.pone.0101654.

Gaveau, D.L.A., Wandono, H. and Setiabudi, F. (2007). Three decades of deforestation in southwest Sumatra: have protected areas halted forest loss and logging, and promoted re-growth? *Biological Conservation*, **134**, 495–504. DOI: https://doi.org/10.1016/j.biocon.2006.08.035.

Gaveau, D.L.A., Wich, S., Epting, J., *et al.* (2009b). The future of forests and orangutans (*Pongo abelii*) in Sumatra: predicting impacts of oil palm plantations, road construction, and mechanisms for reducing carbon emissions from deforestation. *Environmental Research Letters*, **4**, 034013.

GEF (2013). *Project Identification Form: Strengthening Forest and Ecosystem Connectivity in RIMBA Landscape of Central Sumatra through Investing in Natural Capital, Biodiversity Conservation, and Land-based Emission Reductions*. Washington DC: Global Environment Facility (GEF).

GEI (2013). *Environmental and Social Challenges of China's Going Global.* Beijing, PRC: Global Environmental Institute (GEI). Available at: www.geichina.org/_upload/file/book/2013Goingout_EN.pdf.

GEI (2015). *Understanding China's Overseas Investments Governance and Analysis of Environmental and Social Policies.* Beijing, PRC: Global Environmental Institute (GEI). Available at: http://www.geichina.org/_upload/file/book/goingoutreport/Understanding_China's_Overseas_Investments_Governance_and_Anyalsis%20of_E&S_Policies.pdf.

GEI (2016). *Cambodian FDI Policy and Management System: Analysis of Chinese Investments in Cambodia*. Beijing, PRC: Global Environmental Institute (GEI). Available at: http://www.geichina.org/_upload/file/report/China_Going_Global_Cambodia.pdf.

Geissmann, T. (1991). Reassessment of age of sexual maturity in gibbons (*Hylobates* spp.). *American Journal of Primatology*, **23**, 11–22. DOI: 10.1002/ajp.1350230103.

Geissmann, T. (2007). Status reassessment of the gibbons: results of the Asian primate red list workshop 2006. *Gibbon Journal*, **3**, 5–15.

Geissmann, T. and Bleisch, W. (2008). Nomascus hainanus. *The IUCN Red List of Threatened Species 2008: e.T41643A10526461*. Gland, Switzerland: International Union for Conservation of Nature (IUCN). Available at: http://dx.doi.org/10.2305/IUCN.UK.2008.RLTS.T41643A10526461.en.

Geissmann, T., Grindley, M., Ngwe, L., *et al.* (2013). *The Conservation Status of Hoolock Gibbons in Myanmar*. Zürich, Switzerland: Gibbon Conservation Alliance.

Geissmann, T., Manh Ha, N., Rawson, B., *et al.* (2008). Nomascus gabriellae. *The IUCN Red List of Threatened Species 2008: e.T39776A10265736*. Gland, Switzerland: International Union for Conservation of Nature (IUCN). Available at: http://www.iucnredlist.org/details/39776/0.

Geissmann, T. and Nijman, V. (2008a). Hylobates muelleri *ssp.* funereus. *The IUCN Red List of Threatened Species 2008: e.T39890A10271063*. Gland, Switzerland: International Union for Conservation of Nature (IUCN). Available at: http://dx.doi.org/10.2305/IUCN.UK.2008.RLTS.T39890A10271063.en.

Geissmann, T. and Nijman, V. (2008b). Hylobates muelleri *ssp.* muelleri. *The IUCN Red List of Threatened Species 2008: e.T39888A10270564*. Gland, Switzerland: International Union for Conservation of Nature (IUCN). Available at: http://dx.doi.org/10.2305/IUCN.UK.2008.RLTS.T39888A10270564.en.

Geist, H.J. and Lambin, E.F. (2002). Proximate causes and underlying driving forces of tropical deforestation. *BioScience*, **52**, 143–50. DOI: 10.1641/0006-3568(2002)052[0143:PCAUDF]2.0.CO;2.

Geldmann, J., Barnes, M., Coad, L., *et al.* (2013). Effectiveness of terrestrial protected areas in reducing habitat loss and population declines. *Biological Conservation*, **161**, 230–8. DOI: https://doi.org/10.1016/j.biocon.2013.02.018.

GFAS (2013a). *Standards for Great Ape Sanctuaries*. Phoenix, AZ: Global Federation of Animal Sanctuaries (GFAS). Available at: http://www.sanctuaryfederation.org/gfas/wp-content/uploads/2016/07/GreatApe Standards_Dec2015.pdf. Accessed September, 2016.

GFAS (2013b). *Standards for Old World Primates*. Phoenix, AZ: Global Federation of Animal Sanctuaries (GFAS). Available at: http://www.sanctuaryfederation.org/gfas/wp-content/uploads/2016/07/OldWorldMonkey Standards_Dec2015.pdf. Accessed September, 2016.

GFAS (n.d.-a). *Accreditation Frequently Asked Questions*. Phoenix, AZ: Global Federation of Animal Sanctuaries (GFAS). Available at: http://www.sanctuaryfederation.org/gfas/for-sanctuaries/faq/#visit.

GFAS (n.d.-b). *GFAS Accredited Sanctuaries and GFAS Verified Sanctuaries*. Phoenix, AZ: Global Federation of Animal Sanctuaries (GFAS). Available at: http://www.sanctuaryfederation.org/gfas/about-gfas/gfas-sanctuaries/.

GFAS (n.d.-c). *How to Apply*. Phoenix, AZ: Global Federation of Animal Sanctuaries (GFAS). Available at: http://www.sanctuaryfederation.org/gfas/for-sanctuaries/how-to-apply/.

GFW (2014). *Tree Cover Loss (Hansen/UMD/Google/USGS/NASA). Version 1.2*. Washington DC: Global Forest Watch (GFW) World Resources Institute. Available at: data.globalforestwatch.org/datasets/63f9425c45404c36a23495 ed7bef1314?uiTab=metadata.

GFW (n.d.-a). *GFW Fires*. Washington DC: Global Forest Watch (GFW) World Resources Institute. Available at: http://fires.globalforestwatch.org/about/. Accessed March, 2017.

GFW (n.d.-b). *GFW Map Builder*. Washington DC: Global Forest Watch (GFW) World Resources Institute. Available at: http://developers.globalforestwatch.org/map-builder/. Accessed March, 2017.

GFW (n.d.-c). *GFW Open Data Portal*. Washington DC: Global Forest Watch (GFW) World Resources Institute. Available at: http://data.globalforestwatch.org/datasets/16dae98167264b8abfbd13e23802e4f3_0.

Gilardi, K.V., Gillespie, T.R., Leendertz, F.H., *et al.* (2015). *Best Practice Guidelines for Health Monitoring and Disease Control in Great Ape Populations*. Gland, Switzerland: International Union for Conservation of Nature Species Survival Commission (IUCN SSC), Primate Specialist Group.

Gillanders, R. (2014). Corruption and infrastructure at the country and regional level. *The Journal of Development Studies*, **50**, 803–19. DOI: 10.1080/00220388.2013.858126.

Gillespie, T.R. and Chapman, C.A. (2008). Forest fragmentation, the decline of an endangered primate, and changes in host–parasite interactions relative to an unfragmented forest. *American Journal of Primatology*, **70**, 222–30. DOI: 10.1002/ajp.20475.

GLAD (n.d.). *Global Forest Watch*. College Park, MD: Global Land Analysis & Discovery (GLAD), University of Maryland. Available at: http://glad.umd.edu/projects/global-forest-watch. Accessed December, 2016.

Global Commission on the Economy and Climate (2016). *The Sustainable Infrastructure Imperative: Financing for Better Growth and Development. The 2016 New Climate Economy Report*. Washington DC: New Climate Economy. Available at: http://newclimateeconomy.report/2016/wp-content/uploads/sites/4/2014/08/NCE_ 2016Report.pdf.

Global Road Map (n.d.). *Key Facts about Roads*. Cairns, Australia: James Cook University. Available at: http://www.global-roadmap.org/about/. Accessed January, 2017.

Global Wind Report (2015). *Global Wind Report 2015: Annual Market Update*. Brussels, Belgium: Global Wind Energy Council.

Global Witness (2012). *In the Future, There Will be No Forests Left*. London, UK: Global Witness. Available at: https://www.globalwitness.org/sites/default/files/library/HSBC-logging-briefing-FINAL-WEB_0.pdf. Accessed November 10, 2016.

Golder Associates (2015). *SETRAG - Programme de Maintenance Des Voies et Des Installations Connexes: Plan d'Action Sur La Biodiversité*. Libreville, Gabon: SETRAG.

Google Earth (n.d.). *Google Earth*. Google Earth. Available at: https://www.google.com/earth/index.html. Accessed January, 2017.

Google Earth Engine Team (n.d.). *Google Earth Engine: A Planetary-Scale Geospatial Analysis Platform*. Google Earth Engine. Available at: https://earthengine.google.com. Accessed 2017.

Goossens, B., Kapar, M.D., Kahar, S. and Ancrenaz, M. (2011). First sighting of Bornean orang-utan twins in the wild. *Asian Primates Journal*, **2**, 10–2.

Goossens, B., Setchell, J.M., Tchidongo, E., *et al.* (2005). Survival, interactions with conspecifics and reproduction in 37 chimpanzees released into the wild. *Biological Conservation*, **123**, 461–75. DOI: https://doi.org/10.1016/j.biocon.2005.01.008.

Gorilla SSP (n.d.). *Gorilla Species Survival Plan*. The Gorilla Species Survival Plan (SSP). Available at: http://www.gorillassp.org/aboutSSP.html. Accessed October 10, 2014.

Gorman, J. (2015a). Chimpanzees in Liberia, used in New York blood center research, face uncertain future. *The New York Times*, May 28, 2015. Available at: http://www.nytimes.com/2015/05/29/science/chimpanzees-liberia-new-york-blood-center.html.

Gorman, J. (2015b). Plan to export chimps tests law to protect species. *The New York Times*, November 14, 2015. Available at: http://www.nytimes.com/2015/11/15/science/plan-to-export-chimps-tests-law-to-protect-species.html.

Gorman, J. (2016). 2nd lawsuit filed in US to block chimps' move to England. *The New York Times*, April 26, 2016. Available at: http://www.nytimes.com/2016/04/27/science/2nd-lawsuit-filed-in-us-to-block-chimps-move-to-england.html.

Goufan, J.-M. and Adeline, T. (2005). *Etude Environnementale du Barrage de Lom Pangar. Etude de l'Urbanisation (thème 10) – Volet 'Afflux de population' – Rapport Après Consultation*. Yaoundé, Cameroon: Ministry of Water Resources and Energy. Available at: https://www.iucn.org/sites/dev/files/import/downloads/t10_urbanisation_vol1_v4_lom_pangar.pdf.

Gowan, C., Stephenson, K. and Shabman, L. (2006). The role of ecosystem valuation in environmental decision making: hydropower relicensing and dam removal on the Elwha River. *Ecological Economics*, **56**, 508–23. DOI: https://doi.org/10.1016/j.ecolecon.2005.03.018.

GRASP (2016). *Apes Seizure Database Reveals True Extent of Illegal Trade*. Great Apes Survival Project (GRASP). Available at: http://www.un-grasp.org/apes-seizure-database-reveals-true-extent-of-illegal-trade/. Accessed October, 2016.

Gray, M., Roy, J., Vigilant, L., *et al.* (2013). Genetic census reveals increased but uneven growth of a critically endangered mountain gorilla population. *Biological Conservation*, **158**, 230–8. DOI: https://doi.org/10.1016/j.biocon.2012.09.018.

Green, J.M.H., Cranston, G.R., Sutherland, W.J., *et al.* (2017). Research priorities for managing the impacts and dependencies of business upon food, energy, water and the environment. *Sustainability Science*, **12**, 319–31. DOI: 10.1007/s11625-016-0402-4.

Griffiths, M. and Van Schaik, C.P. (1993). The impact of human traffic on the abundance and activity periods of Sumatran rain forest wildlife. *Conservation Biology*, **7**, 623–6.

Gron, K. (2010). *Lar gibbon Hylobates lar*. University of Wisconsin-Madison, Madison, WI: Primate Info Net. Available at: http://pin.primate.wisc.edu/factsheets/entry/lar_gibbon. Accessed July 12, 2017.

Grueter, C.C., Ndamiyabo, F., Plumptre, A.J., *et al.* (2013). Long-term temporal and spatial dynamics of food availability for endangered mountain gorillas in Volcanoes National Park, Rwanda. *American Journal of Primatology*, **75**, 267–80. DOI: 10.1002/ajp.22102.

Grumbine, R.E., Dore, J. and Xu, J. (2012). Mekong hydropower: drivers of change and governance challenges. *Frontiers in Ecology and the Environment*, **10**, 91–8. DOI: 10.1890/110146.

Gubelman, E. (1995). *Proposal for the contraction of the Kacwamuhoro-Kiyebe-Katoma-Ruhija road around Bwindi Impenetrable National Park: findings of a feasibility study and preliminary project proposal for CARE*. Unpublished report.

Guerry, A.D., Polasky, S., Lubchenco, J., *et al.* (2015). Natural capital and ecosystem services informing decisions: from promise to practice. *Proceedings of the National Academy of Sciences*, **112**, 7348–55. DOI: 10.1073/pnas.1503751112.

Gullett, W. (1998). Environmental impact assessment and the precautionary principle: legislating caution in environmental protection. *Australian Journal of Environmental Management*, **5**, 146–58. DOI: 10.1080/14486563.1998.10648411.

Gumal, M. and Braken Tisen, O. (2015). *Orangutan Strategic Action Plan: Trans-Boundary Biodiversity Conservation Area.* Wildlife Conservation Society, Malaysia Program and Sarawak Forestry Corporation, Sarawak Forest Department.

Guy, A.J., Curnoe, D. and Banks, P.B. (2014). Welfare based primate rehabilitation as a potential conservation strategy: does it measure up? *Primates,* **55,** 139–47. DOI: 10.1007/s10329-013-0386-y.

GVC, BIC and IRN (2006). *In Whose Interest? The Lom Pangar Dam and Energy Sector Development in Cameroon.* June, 2006. Global Village Cameroon (GVT), Bank Information Center (BIC) and International Rivers Network (IRN). Available at: https://www.internationalrivers.org/sites/default/files/attached-files/whoseinterest.pdf.

HAkA, KPHA, OIC, *et al.* (2016). *The Importance of the Kappi Area in the Gunung Leuser National Park and Further Support for its Current Core Area Status.* Medan, Indonesia: ALERT, Forest, Nature, and Environment of Aceh (HAkA), Koalisi Peduli Hutan Aceh (KPHA), Orangutan Information Centre (OIC), PanEco Foundation, Sumatran Orangutan Society (SOS) and Yayasan Ekosistem Lestari (YEL) Available at: https://static1.squarespace.com/static/51b078a6e4b0e8d244dd9620/t/586645cc414fb5c4d35e8c87/1483097562011/Report+to+Minister.pdf.

Halleson, D. (2016). *The central African Iion ore corridor: transforming the Congo Basin.* Presented at: Integrated Resources Corridor Partnership Workshop, June 21–22, Dar es Salaam, Tanzania. Cameroon: World Wide Fund for Nature (WWF).

Hamard, M., Cheyne, S.M. and Nijman, V. (2010). Vegetation correlates of gibbon density in the peat-swamp forest of the Sabangau catchment, Central Kalimantan, Indonesia. *American Journal of Primatology,* **72,** 607–16. DOI: 10.1002/ajp.20815.

Hanafiah, J. (2016). Aceh governor eyes geothermal project in Indonesia's leuser ecosystem. *Mongabay,* August, 2016. Available at: https://news.mongabay.com/2016/08/aceh-governor-eyes-geothermal-project-in-indonesias-leuser-ecosystem/. Accessed April 5, 2017.

Hansen, A.J. and DeFries, R. (2007). Ecological mechanisms linking protected areas to surrounding lands. *Ecological Applications,* **17,** 974–88. DOI: 10.1890/05-1098.

Hansen, M.C., Alexander, K., Alexandra, T., *et al.* (2016). Humid tropical forest disturbance alerts using Landsat data. *Environmental Research Letters,* **11,** 034008.

Hansen, M.C., Potapov, P.V., Moore, R., *et al.* (2013). High-resolution global maps of 21st-century forest cover change. *Science,* **342,** 850–3. DOI: 10.1126/science.1244693. Available at: http://science.sciencemag.org/content/sci/342/6160/850.full.pdf. Data available on-line from: http://earthenginepartners.appspot.com/science-2013-global-forest. Accessed through Global Forest Watch September 2016-February 2017. www.globalforestwatch.org.

Harcourt, A.H. and Greenberg, J. (2001). Do gorilla females join males to avoid infanticide? A quantitative model. *Animal Behaviour,* **62,** 905–15. DOI: https://doi.org/10.1006/anbe.2001.1835.

Harcourt, A.H. and Wood, M.A. (2012). Rivers as barriers to primate distributions in Africa. *International Journal of Primatology,* **33,** 168–83. DOI: 10.1007/s10764-011-9558-z.

Hart, J. (2014). *Summary results of elephant surveys in North Central DR Congo 2007–2013.* Submission to African Elephant Data Base.

Harvey, P. and Knox, H. (2015). *Roads: An Anthropology of Infrastructure and Expertise.* Ithaca, NY: Cornell University Press.

Haskoning (Nederland B.V. Environment) (2011). *Specific Environmental Impact Assessment (SEIA) for the Interaction between the Chad-Cameroon Pipeline Project and the Lom Pangar Dam Project.* Yaoundé, Cameroon: Cameroon Oil Transportation Company.

Head, J.S., Boesch, C., Makaga, L. and Robbins, M.M. (2011). Sympatric chimpanzees (*Pan troglodytes troglodytes*) and gorillas (*Gorilla gorilla gorilla*) in Loango National Park, Gabon: dietary composition, seasonality, and intersite comparisons. *International Journal of Primatology,* **32,** 755–75. DOI: 10.1007/s10764-011-9499-6.

Head, J.S., Boesch, C., Robbins, M.M., *et al.* (2013). Effective sociodemographic population assessment of elusive species in ecology and conservation management. *Ecology and Evolution,* **3,** 2903–16. DOI: 10.1002/ece3.670.

Helsingen, H., Sai Nay Won Myint, Bhagabati, N., *et al.* (2015). *A Better Road to Dawei: Protecting Wildlife, Sustaining Nature, Benefiting People.* Yangon, Myanmar: World Wide Fund for Nature (WWF).

Hernandez-Aguilar, R.A. (2009). Chimpanzee nest distribution and site reuse in a dry habitat: implications for early hominin ranging. *Journal of Human Evolution*, **57**, 350–64. DOI: https://doi.org/10.1016/j.jhevol.2009.03.007.

Hettige, H. (2006). *When Do Rural Roads Benefit the Poor and How? An In-Depth Analysis.* Manila, Philippines: Asian Development Bank (ADB).

Hickey, J.R., Nackoney, J., Nibbelink, N.P., *et al.* (2013). Human proximity and habitat fragmentation are key drivers of the rangewide bonobo distribution. *Biodiversity and Conservation*, **22**, 3085–104. DOI: 10.1007/s10531-013-0572-7.

Hicks, T.C., Darby, L., Hart, J., *et al.* (2010). Trade in orphans and bushmeat threatens one of the Democratic Republic of the Congo's most important populations of eastern chimpanzees (*Pan troglodytes schweinfurthii*). *African Primates (Print)*, **7**, 1–18.

Hicks, T.C. and van Boxel, J.H. (2010). *The Study Region and a Brief History of the Bili Project.* Amsterdam, the Netherlands: Institute for Biodiversity and Ecosystem Dynamics (IBED).

Highland Farm (n.d.). *Meet our Gibbons.* Tambol Chongkab, Thailand: Gibbon at Highland Farm. Available at: http://www.gibbonathighlandfarm.org/content-Meetourgibbons-5-7334-2.html. Accessed February 10, 2017.

Hobbs, J. and Kumah, F. (2015). The extractives slowdown in Africa: a window of opportunity? *Africa Policy Review*. Available at: http://africapolicyreview.com/the-extractives-slowdown-in-africa-a-window-of-opportunity/.

Hobbs, J.B. and Butkovic, L. (2016). *Proceedings of the Integrated Resources Corridor Workshop 2016.* Nairobi, Kenya: World Wide Fund for Nature (WWF) Regional Office for Africa.

Hockings, K.J. (2011). Behavioral flexibility and division of roles in chimpanzee road-crossing. In *The Chimpanzees of Bossou and Nimba*, ed. T. Matsuzawa, T. Humle and Y. Sugiyama. Tokyo, Japan: Springer, pp. 221–9.

Hockings, K.J., Anderson, J.R. and Matsuzawa, T. (2006). Road crossing in chimpanzees: a risky business. *Current Biology*, **16**, R668–R70. DOI: 10.1016/j.cub.2006.08.019.

Hockings, K.J., Anderson, J.R. and Matsuzawa, T. (2009). Use of wild and cultivated foods by chimpanzees at Bossou, Republic of Guinea: feeding dynamics in a human-influenced environment. *American Journal of Primatology*, **71**, 636–46. DOI: 10.1002/ajp.20698.

Hockings, K.J. and Humle, T. (2009). *Best Practice Guidelines for the Prevention and Mitigation of Conflict Between Humans and Great Apes.* Gland, Switzerland: International Union for Conservation of Nature Species Survival Commission (IUCN SSC), Primate Specialist Group.

Hockings, K.J., McLennan, M.R., Carvalho, S., *et al.* (2015). Apes in the Anthropocene: flexibility and survival. *Trends in Ecology & Evolution*, **30**, 215–22. DOI: https://doi.org/10.1016/j.tree.2015.02.002.

Hockings, K.J. and Sousa, C. (2013). Human–chimpanzee sympatry and interactions in Cantanhez National Park, Guinea-Bissau: current research and future directions. *Primate Conservation*, **26**, 57–65. DOI: 10.1896/052.026.0104.

Hohmann, G., Fowler, A., Sommer, V. and Ortmann, S. (2006). Frugivory and gregariousness of Salonga bonobos and Gashaka chimpanzees: the influence of abundance and nutritional quality of fruit. In *Feeding Ecology in Apes and Other Primates*, ed. G. Hohmann, M. M. Robbins and C. Boesch. Cambridge, UK: Cambridge University Press, pp. 123–59.

Hohmann, G., Gerloff, U., Tautz, D. and Fruth, B. (1999). Social bonds and genetic ties: kinship, association and affiliation in a community of bonobos (*Pan paniscus*). *Behaviour*, **136**, 1219–35. DOI: https://doi.org/10.1163/156853999501739.

Honjiang, C. (2016). *The Belt and Road Initiative and Its Impact on Asia-Europe Connectivity.* Beijing, PRC: China Institute of International Studies. Available at: http://www.ciis.org.cn/english/2016-07/21/content_8911184.htm.

Horta, K. (2012). Public-private-partnership and institutional capture: the state, international institutions and indigenous peoples in Chad and Cameroon. In *The Politics of Resource Extraction: Indigenous Peoples, Multinational Corporations, and the State*, ed. S. Sawyer and E. T. Gomez. UN Research Institute for Social Development. Available at: https://urgewald.org/sites/default/files/the_politics_of_resource_extraction.pdf.

Houghton, R.A., House, J.I., Pongratz, J., *et al.* (2012). Carbon emissions from land use and land-cover change. *Biogeosciences*, **9**, 5125–42. DOI: https://doi.org/10.5194/bg-9-5125-2012.

Human Rights Watch (2016). *World Report 2017: Malaysia, Events of 2016.* New York, NY: Human Rights Watch. Available at: https://www.hrw.org/world-report/2017/country-chapters/malaysia. Accessed April 6, 2017.

Humle, T. (2011). Location and ecology. In *Chimpanzees of Bossou and Nimba*, ed. T. Matsuzawa, T. Humle and Y. Sugiyama. Tokyo, Japan: Springer-Verlag, pp. 371–80.

Humle, T. (2015). *The Dimensions of Ape–Human Interactions in Industrial Agricultural Landscapes. Background Paper for State of the Apes: Industrial Agriculture and Ape Conservation. Arcus Foundation.* Cambridge, UK: Cambridge University Press. Available at: http://www.stateoftheapes.com/wp-content/uploads/2016/03/Ape–Human-Interactions-in-Industrial-Agricultural-Landscapes.pdf.

Humle, T. and Farmer, K. (2015). Primate rehabilitation in Africa: myths and realities. *African Conservation Telegraph*, **9**, 2.

Humle, T. and Hill, C. (2016). People–primate interactions: implications for primate conservation. In *Introduction to Primate Conservation*, ed. S. A. Wich and A. J. Marshall. Oxford, UK: Oxford University Press, pp. 219–40.

Humle, T., Boesch, C., Campbell, G., *et al.* (2016a). Pan troglodytes *ssp. verus (errata version published in 2016). The IUCN Red List of Threatened Species 2016: e.T15935A102327574.* Gland, Switzerland: International Union for Conservation of Nature (IUCN). Available at: http://dx.doi.org/10.2305/IUCN.UK.2016–2.RLTS.T15935A17989872.en.

Humle, T., Colin, C., Laurans, M. and Raballand, E. (2011). Group release of sanctuary chimpanzees (*Pan troglodytes*) in the Haut Niger National Park, Guinea, west Africa: ranging patterns and lessons so far. *International Journal of Primatology*, **32**, 456–73. DOI: 10.1007/s10764–010–9482–7.

Humle, T., Maisels, F., Oates, J.F., Plumptre, A. and Williamson, E.A. (2016b). Pan troglodytes *(errata version published in 2016). The IUCN Red List of Threatened Species 2016: e.T15933A102326672.* Gland, Switzerland: International Union for Conservation of Nature (IUCN). Available at: http://www.iucnredlist.org/details/15933/0. Accessed November 15, 2016.

Humphrey, C., Griffith-Jones, S., Xu, J., Carey, R. and Prizzon, A. (2015). *Multilateral Development Banks in the 21st Century. Three Perspectives on China and the Asian Infrastructure Investment Bank.* London, UK: Overseas Development Institute. Available at: https://www.odi.org/publications/10159-multilateral-development-banks-21st-century-three-perspectives-china-asian-infrastructure-investment-bank.

Hunsberger, C., Corbera, E., Borras Jr, S.M., *et al.* (2015). *Land-Based Climate Change Mitigation, Land Grabbing and Conflict: Understanding Intersections and Linkages, Exploring Actions for Change.* MOSAIC Research Project: MOSAIC Working Paper Series No 1. Available at: https://www.iss.nl/sites/corporate/files/CMCP_72-Hunsberger_et_al.pdf.

Hvilsom, C., Frandsen, P., Børsting, C., *et al.* (2013). Understanding geographic origins and history of admixture among chimpanzees in European zoos, with implications for future breeding programmes. *Heredity*, **110**, 586. DOI: 10.1038/hdy.2013.9. Available at: https://www.nature.com/articles/hdy20139#supplementary-information.

IBAT (n.d.). *Global Biodiversity Decision Support Platform.* Integrated Biodiversity Assessment Tool (IBAT) for Business. Available at: https://www.ibatforbusiness.org.

Ibisch, P.L., Hoffmann, M.T., Kreft, S., *et al.* (2016). A global map of roadless areas and their conservation status. *Science*, **354**, 1423–7. DOI: 10.1126/science.aaf7166.

ICA (2014). *Infrastructure Financing Trends in Africa.* Infrastructure Consortium for Africa (ICA), African Development Bank. Available at: http://www.icafrica.org/fileadmin/documents/Annual_Reports/INFRASTRUCTURE_FINANCING_TRENDS_IN_AFRICA_%E2%80%93_2014.pdf.

ICCN (2009). *Plan General de Gestion du Parc National de Kahuzi-Biega.* Gombe–Kinshasa, DRC: Institut Congolais pour la Conservation de la Nature (ICCN). Available at: http://www.kahuzi-biega.org/publications/publications/.

ICCN (2015). *Rapport Annuel des Activités du Parc National de Kahuzi-Biega.* Gombe–Kinshasa, DRC: Institut Congolais pour la Conservation de la Nature (ICCN).

ICCN (2016). *Rapport à l'UNESCO sur l'Etat de Conservation des Sites de patrimoine mondial en RDC.* Gombe–Kinshasa, DRC: Institut Congolais pour la Conservation de la Nature (ICCN). Available at: http://whc.unesco.org/en/list/137/documents/.

ICOLD (n.d.). *Definition of a Large Dam.* Paris, France: International Commission on Large Dams (ICOLD). Available at: www.icold-cigb.org/GB/dams/definition_of_a_large_dam.asp.

IDB (n.d.). *Project Snapshot: Hydroelectric Project takes Unprecedented Measures to Protect Habitat, Reventazón Hydroelectric Project, Costa Rica.* Washington DC: Inter-American Development Bank (IDB). Available at: http://www.iadb.org/en/topics/sustainability/project-snapshot-hydroelectric-project-takes-unprecedented-measures-to-protect-habitat,7998.html. Accessed June 28, 2017.

IEA (2012). *Technology Roadmap: Hydropower.* Paris, France: International Energy Agency (IEA). Available at: http://www.iea.org/publications/freepublications/publication/2012_Hydropower_Roadmap.pdf.

IEA (2016). *World Energy Outlook 2016.* Paris, France: International Energy Agency (IEA). Available at: https://www.docdroid.net/IOBt86G/world-energy-outlook-2016.pdf#page=4.

IFC (2012a). *Guidance Note 6: Biodiversity Conservation and Sustainable Management of Living Natural Resources.* Washington DC: International Finance Corporation (IFC), World Bank Group. Available at: https://www.ifc.org/wps/wcm/connect/a359a380498007e9a1b7f3336b93d75f/Updated_GN6-2012.pdf?MOD=AJPERES.

IFC (2012b). *Performance Standard 1: Assessment and Management of Environmental and Social Risks and Impacts.* Washington DC: International Finance Corporation (IFC), World Bank Group.

IFC (2012c). *Performance Standard 6: Biodiversity Conservation and Sustainable Management of Living Natural Resources.* Washington DC: International Finance Corporation (IFC), World Bank Group. Available at: http://www.ifc.org/wps/wcm/connect/bff0a28049a790d6b835faa8c6a8312a/PS6_English_2012.pdf?MOD=AJPERES.

IFC (2013). *Cumulative Impact Assessment and Management: Guidance for the Private Sector in Emerging Markets.* Washington DC: International Finance Corporation (IFC), World Bank Group.

IFC (n.d.). *Chad–Cameroon Pipeline Project Documentation.* Washington DC: International Finance Corporation (IFC), World Bank Group. Available at: http://www.ifc.org/wps/wcm/connect/region__ext_content/regions/sub-saharan+africa/investments/chadcameroon. Accessed October 6, 2016.

IGF (2013). *A Mining Policy Framework. Mining and Sustainable Development, Managing One to Advance Another.* Ottawa, Canada: Intergovernmental Forum on mining, minerals, metals and sustainable development (IGF). Available at: http://www.globaldialogue.info/MPFOct2013.pdf.

IHA (2010). *Hydropower Sustainability Assessment Protocol.* London, UK: International Hydropower Association (IHA). Available at: http://www.hydrosustainability.org/Protocol/Protocol.aspx.

Ihua-Maduenyi, M. (2016). FG stops work on Cross River superhighway. *Punch*, March 14, 2016. Available at: http://punchng.com/fg-stops-work-on-cross-river-superhighway/.

Ihua-Maduenyi, M. (2017). W'Bank manager lauds rerouting of C'River superhighway. *Punch*, April 3, 2017. Available at: http://punchng.com/wbank-manager-lauds-rerouting-of-criver-superhighway/.

ILO (n.d.). *C169: Indigenous and Tribal Peoples Convention, 1989 (No. 169).* Geneva, Switzerland: International Labour Organization (ILO). Available at: http://www.ilo.org/dyn/normlex/en/f?p=NORMLEXPUB:12100:0::NO::P12100_ILO_CODE:C169.

Imong, I., Kuhl, H.S., Robbins, M.M. and Mundry, R. (2016a). Evaluating the potential effectiveness of alternative management scenarios in ape habitat. *Environmental Conservation*, **43**, 1–11.

Imong, I., Robbins, M.M., Mundry, R., Bergl, R. and Kühl, H.S. (2014b). Distinguishing ecological constraints from human activity in species range fragmentation: the case of Cross River gorillas. *Animal Conservation*, **17**, 323–31. DOI: 10.1111/acv.12100.

Imong, I., Robbins, M.M., Mundry, R., Bergl, R. and Kühl, H.S. (2014). Informing conservation management about structural versus functional connectivity: a case-study of Cross River gorillas. *American Journal of Primatology*, **76**, 978–88. DOI: 10.1002/ajp.22287.

Indonesia CMEA (2011). *Masterplan for Acceleration and Expansion of Indonesia Economic Development 2011–2025.* Jakarta, Indonesia: Coordinating Ministry for Economic Affairs (CMEA), Republic of Indonesia Ministry of National Development Planning/National Development Planning Agency.

Indonesia Investments (2011). *Master Plan for Acceleration and Expansion of Indonesia's Development (MP3EI).* Jakarta, Indonesia: Department of Planning (BAPPENAS). Available at: https://www.indonesia-investments.com/projects/government-development-plans/masterplan-for-acceleration-and-expansion-of-indonesias-economic-development-mp3ei/item306.

Indonesia MoF (2009). *Orangutan Indonesia Conservation Strategies and Action Plan 2007–2017.* Jakarta, Indonesia: Ministry of Foreign Affairs of the Republic of Indonesia (MoF). Directorate General of Forest Protection and Nature Conservation, Ministry of Forestry of the Republic of Indonesia. Available at: http://static1.1.sqspcdn.com/static/f/1200343/20457144/1348934583800/Indonesian_Orangutan_NAP_2007–2017.pdf?token=zWCXNjRaYJv5Lx7C30sEQgryqWI%3D.

Ingle, N. (2016). Op-Ed. Will the ax fall on Nigeria's national parks? *NY Times*, November 3, 2016. Available at: http://www.nytimes.com/2016/11/04/opinion/will-the-ax-fall-on-nigerias-national-parks.html?_r=0.

Integrated Environments (2010). *Environmental Management Plan: Trung Son Hydropower Project (TSHPP)*. Prepared for the Trung Son Project Management Board. Calgary, Alberta: Integrated Environments (2006) Ltd. Available at: http://siteresources.worldbank.org/INTVIETNAM/Resources/A7.pdf.

International Rivers (n.d.-a). *Bakun Dam*. Berkeley, CA: International Rivers. Available at: https://www.internationalrivers.org/campaigns/bakun-dam. Accessed September 12, 2017.

International Rivers (n.d.-b). *Economic Impact of Dams*. Oakland, CA: International Rivers. Available at: https://www.internationalrivers.org/economic-impacts-of-dams. Accessed May, 2017.

International Rivers (n.d.-c). *Grand Inga Hydroelectric Project: An Overview.* Berkeley, CA: International Rivers. Available at: https://www.internationalrivers.org/resources/grand-inga-hydroelectric-project-an-overview-3356. Accessed August, 2017.

International Rivers (n.d.-d). *Murum Dam*. Berkeley, CA: International Rivers. Available at: https://www.internationalrivers.org/campaigns/murum-dam. Accessed September 12, 2017.

IRENA (2015). *Africa 2030: Roadmap for a Renewable Energy Future*. Abu Dhabi, UAE: International Renewable Energy Agency (IRENA).

ITALTHAI (n.d.). *Overview*. Bangkok, Thailand: ITALTHAI. Available at: www.italthaigroup.com/en/overview.

ITD (2011). *Dawei project overview and work in progress*. Internal Company Report, July. Bangkok, Thailand: Italian-Thai Development Co. (ITD).

ITD (2012). *Dawei deep seaport and industrial estate development project*. PowerPoint slides. Bangkok, Thailand: Italian-Thai Development Co (ITD).

IUCN (2013). *The IUCN Red List of Threatened Species. Version 2013.1*. Gland, Switzerland: International Union for Conservation of Nature (IUCN). Available at: http://www.iucnredlist.org.

IUCN (2014a). *Biodiversity Offsets Technical Study Paper*. Gland, Switzerland: International Union for Conservation of Nature (IUCN).

IUCN (2014b). *Plan d'Action Régional Pour la Conservation des Gorilles de Plaine de l'Ouest et des Chimpanzés d'Afrique Centrale 2015–2025*. Gland, Switzerland: Groupe de Spécialistes des Primates de la CSE/UICN.

IUCN (2014c). *Regional Action Plan for the Conservation of Western Lowland Gorillas and Central Chimpanzees 2015–2025*. Gland, Switzerland: International Union for Conservation of Nature Species Survival Commission (IUCN SSC), Primate Specialist Group. Available at: https://d2ouvy59p0dg6k.cloudfront.net/downloads/wea_apes_plan_2014_7mb.pdf.

IUCN (2014d). *The IUCN Red List of Threatened Species. Version 2014.3*. Gland, Switzerland: International Union for Conservation of Nature (IUCN). Available at: http://www.iucnredlist.org.

IUCN (2016a). *IUCN was at the 3rd Asia Ministerial Conference on Tiger Conservation*. Gland, Switzerland: International Union for Conservation of Nature (IUCN). Available at: https://www.iucn.org/news/species/201607/iucn-was-3rd-asia-ministerial-conference-tiger-conservation.

IUCN (2016b). *The IUCN Red List of Threatened Species. Version 2016.2*. Gland, Switzerland: International Union for Conservation of Nature (IUCN). Available at: http://www.iucnredlist.org. Accessed September, 2016.

IUCN (2016c). *The IUCN Red List of Threatened Species. Version 2016.3*. Gland, Switzerland: International Union for Conservation of Nature (IUCN). Available at: http://www.iucnredlist.org. Accessed December 7, 2016.

IUCN (2016d). *Understanding Human Dependence on Forests: An Overview of IUCN's Efforts and Findings, and their Implications*. Gland, Switzerland: International Union for Conservation of Nature (IUCN). Available at: https://www.iucn.org/news/forests/201611/understanding-human-dependence-forests-overview-iucn%E2%80%99s-efforts-and-findings-and-their-implications.

IUCN (2017). *The IUCN Red List of Threatened Species. Version 2017.1*. Gland, Switzerland: International Union for Conservation of Nature (IUCN). Available at: http://www.iucnredlist.org.

IUCN (n.d.-a). *Categories and Criteria*. Gland, Switzerland: International Union for Conservation of Nature (IUCN). Available at: www.iucnredlist.org/technical-documents/categories-and-criteria. Accessed March, 2017.

IUCN (n.d.-b). *Protected Areas Categories*. Gland, Switzerland: International Union for Conservation of Nature (IUCN). Available at: https://www.iucn.org/theme/protected-areas/about/protected-areas-categories.

IUCN SSC (2013). *Guidelines for Reintroductions and Other Conservation Translocations. Version 1.0*. Gland, Switzerland: International Union for Conservation of Nature Species Survival Commission (IUCN SSC). Available at: http://www.issg.org/pdf/publications/RSG_ISSG-Reintroduction-Guidelines-2013.pdf.

IUCN SSC Primate Specialist Group (2006). *Primates of Peru: Taxonomy and Conservation Status*. Gland, Switzerland: International Union for Conservation of Nature Species Survival Commission (IUCN SSC). Available at: http://www.primate-sg.org/primates_of_peru/.

IUCN and UNEP-WCMC (2016). *The World Database on Protected Areas (WDPA)*. Cambridge, UK: United Nations Environment Programme (UNEP), World Conservation Monitoring Centre (WCMC). Available at: http://www.protectedplanet.net.

Jacobs, A. (2015). *Factors affecting the prevalence of road and canopy bridge crossings by primates in Diani Beach, Kenya*. Masters thesis. Canterbury, UK: University of Kent.

Jaeger, J.A.G., Bowman, J., Brennan, J., *et al.* (2005). Predicting when animal populations are at risk from roads: an interactive model of road avoidance behavior. *Ecological Modelling*, **185**, 329–48. DOI: https://doi.org/10.1016/j.ecolmodel.2004.12.015.

Jaeger, J.A., Fahrig, L. and Ewald, K.C. (2006). *Does the configuration of road networks influence the degree to which roads affect wildlife populations?* In *Proceedings of the 2005 International Conference on Ecology and Transportation*, ed. C. L. Irwin, P. Garrett and K. P. McDermott. Center for Transportation and the Environment, North Carolina State University, pp. 151-63. Available at: http://www.icoet.net/ICOET_2005/05proceedings_directory.asp.

Jakob-Hoff R.M., MacDiarmid S.C. and C., L. (2014). *Manual of Procedures for Wildlife Disease Risk Analysis*. Paris, France: World Organization for Animal Health in association with the International Union for Conservation of Nature Species Survival Commission (IUCN SSC).

Jalil, M.F., Cable, J., Sinyor, J., *et al.* (2008). Riverine effects on mitochondrial structure of Bornean orang-utans (*Pongo pygmaeus*) at two spatial scales. *Molecular Ecology*, **17**, 2898–909. DOI: 10.1111/j.1365–294X.2008.03793.x.

Jantz, S., Pintea, L., Nackoney, J. and Hansen, M. (2016). Landsat ETM+ and SRTM data provide near real-time monitoring of chimpanzee (*Pan troglodytes*) habitats in Africa. *Remote Sensing*, **8**, 427.

Jong, H.N. (2016). Govt to revise wildlife law as protected animals face extinction. *Jakarta Post*, March 1, 2016. Available at: http://www.thejakartapost.com/news/2016/03/01/govt-revise-wildlife-law-protected-animals-face-extinction.html. Accessed February 23, 2017.

Joshi, A.R., Dinerstein, E., Wikramanayake, E., *et al.* (2016). Tracking changes and preventing loss in critical tiger habitat. *Science Advances*, **2**. DOI: 10.1126/sciadv.1501675.

Junker, J., Blake, S., Boesch, C., *et al.* (2012). Recent decline in suitable environmental conditions for African great apes. *Diversity and Distributions*, **18**, 1077–91. DOI: 10.1111/ddi.12005.

Kabukuru, W. (2016). World Bank's monkey business. *New African*, March, 2016.

Kahler, M., Henning, C.R., Bown, C.P., *et al.* (2016). *Global Order and the New Regionalism*. Discussion Paper Series on Global and Regional Governance. (September). New York, NY: Council on Foreign Relations. Available at: https://www.cfr.org/report/global-order-and-new-regionalism.

Kalaweit France (2016). *Compte rendu de l'Assemblée Générale du Samedi 17 Septembre 2016*. Paris, France: Kalaweit. Available at: http://kalaweit.org/gestion/Modules/tiny_mce/plugins/filemanager/files/CR_AG_2015.pdf.

Kallang, P. (2016). *Murum Another Dam Which Does Not Make Economic Sense*. Kuala Lumpur, Malaysia: Malaysiakini. Available at: http://www.malaysiakini.com/letters/342275. Accessed November 15, 2016.

Kalpers, J., Gray, M., Asuma, S., *et al.* (2011). Buffer zone and human–wildlife conflict management. In *20 Years of IGCP: Lessons Learned in Mountain Gorilla Conservation*, ed. M. Gray and E. Rutagarama. Kigali, Rwanda: International Gorilla Conservation Programme, pp. 105–37.

Kampala, L.E. (2012). Request for expression of interest for the development of 1900 km of roads supporting primary growth sectors through contractor facilitated financing mechanism, procurement reference number: UNRA/Works/2011–2012/00002/02/01–05. Uganda National Road Authority (UNRA). *Daily Monitor*.

Kano, T. (1972). Distribution and adaptation of the chimpanzee on the eastern shore of Lake Tanganyika. In *African Studies VII*. Kyoto, Japan: Kyoto University, pp. 37–129.

Katsis, L. (2017). *Spatial patterns of primate electrocutions in Diani, Kenya*. Masters thesis. Bristol, UK: University of Bristol.

Kelly, A.S., Connette, G., Helsingen, H. and Soe, P. (2016). *Wildlife Crossing: Locating Species' Movement Corridors in Tanintharyi*. Yangon, Myanmar: World Wide Fund for Nature (WWF).

Kenrick, J. (2006). Equalising processes, processes of discrimination and the Forest People of Central Africa. In *Property and Equality. Volume 2: Encapsulation, Commercialization, Discrimination*, ed. T. Widlock and W. Tadesse. Oxford, UK: Berghahn, pp. 104–28.

Kenrick, J. and Lewis, J. (2004). Indigenous peoples' rights and the politics of the term 'indigenous'. *Anthropology Today*, **20**, 4–9. DOI: 10.1111/j.0268-540X.2004.00256.x.

KFBG (n.d.). *Wild Animal Rescue Centre (WARC)*. Hong Kong: Kadoorie Farm & Botanic Garden (KFBG). Available at: http://www.kfbg.org/eng/warc.aspx. Accessed September, 2016.

Kidd, C. and Kenrick, J. (2009). The Forest People of Africa: land rights in context. In *Land Rights and the Forest Peoples of Africa: Historical, Legal and Anthropological Perspectives*, ed. V. Couillard, J. Gilbert, J. Kenrick and C. Kidd. Moreton-in-Marsh, UK: Forest Peoples Programme, pp. 4–27.

Kidd, C. and Kenrick, J. (2011). Mapping everyday practices as rights of resistance: indigenous peoples in central Africa. In *The Politics of Indigeneity: Dialogues and Reflections on Indigenous Activism*, ed. S. Venkateswar and E. Hughes. London, UK: Zed Books.

Kigula, J. (2015). *Participatory forest management: Tanzania's experience*. Presented at: Colloquium between Forest Dwellers and the Kenya Forestry Service, March 5, 2015, Eldoret.

Kikawasi, G.J. (2012). Causes and effects of delays and disruptions in construction projects in Tanzania, 2012. *Australasian Journal of Construction Economics and Building, Conference Series*, **1**, 52–9.

Killeen, T.J. (2007). *A Perfect Storm in the Amazon Wilderness: Development and Conservation in the Context of the Initiative for the Integration of the Regional Infrastructure of South America (IIRSA)*. Washington DC: Conservation International.

Kindlmann, P. and Burel, F. (2008). Connectivity measures: a review. *Landscape Ecology*, **23**, 879–90. DOI: 10.1007/s10980-008-9245-4.

King, T., Chamberlan, C. and Courage, A. (2012). Assessing initial reintroduction success in long-lived primates by quantifying survival, reproduction and dispersal parameters: western lowland gorillas (*Gorilla gorilla gorilla*) in Congo and Gabon. *International Journal of Primatology*, **33**, 134–49. DOI: 10.1007/s10764-011-9563-2.

Kis-Katos, K. and Suharnoko Sjahrir, B. (2014). *The Impact of Fiscal and Political Decentralization on Local Public Investments in Indonesia. Discussion Paper 7884*. Bonn, Germany: Institute for the Study of Labor (IZA).

Kitzes, J. and Shirley, R. (2016). Estimating biodiversity impacts without field surveys: a case study in northern Borneo. *Ambio*, **45**, 110–9. DOI: 10.1007/s13280-015-0683-3.

Kleinschroth, F., Gourlet-Fleury, S., Sist, P., Mortier, F. and Healey, J.R. (2015). Legacy of logging roads in the Congo Basin: how persistent are the scars in forest cover? *Ecosphere*, **6**, 1–17. DOI: 10.1890/ES14-00488.1.

Kleinschroth, F., Healey, J.R., Gourlet-Fleury, S., Mortier, F. and Stoica, R.S. (2017). Effects of logging on roadless space in intact forest landscapes of the Congo Basin. *Conservation Biology*, **31**, 469–80. DOI: 10.1111/cobi.12815.

Knott, C.D. (1998). Changes in orangutan caloric intake, energy balance, and ketones in response to fluctuating fruit availability. *International Journal of Primatology*, **19**, 1061–79. DOI: 10.1023/a:1020330404983.

Knott, C.D. (2005). Energetic responses to food availability in the great apes: implications for hominin evolution. In *Seasonality in Primates Studies of Living and Extinct Human and Non-Human Primates*, ed. D. K. Brockman and C. P. Van Schaik. New York, NY: Cambridge University Press, pp. 351–78.

Knowledge@Wharton (2017). *Where Will China's 'One Belt, One Road' Initiative Lead?* Available at: http://knowledge.wharton.upenn.edu/article/can-chinas-one-belt-one-road-initiative-match-the-hype/#. March 22, 2017.

KNU (2012). *The KNU Press Release on 1st Meeting between KNU Delegation and Union-Level Peace Delegation*. Karen National Union (KNU). Available at: http://www.knuhq.org/the-knu-press-release-on-1st-meeting-between-knu-delegation-and-union-level-peace-delegation.

Köndgen, S., Kühl, H., N'Goran, P.K., *et al.* (2008). Pandemic human viruses cause decline of endangered great apes. *Current Biology*, **18**, 260–4. DOI: https://doi.org/10.1016/j.cub.2008.01.012.

Kondolf, G.M., Rubin, Z.K. and Minear, J.T. (2014). Dams on the Mekong: cumulative sediment starvation. *Water Resources Research*, **50**, 5158–69. DOI: 10.1002/2013WR014651.

Kormos, R., Boesch, C., Bakarr, M.I. and Butynski, T.M. (2003). *West African Chimpanzees: Status, Survey and Conservation Action Plan*. Gland, Switzerland: International Union for Conservation of Nature (IUCN) World Conservation Union.

Kormos, R., Kormos, C.F., Humle, T., *et al.* (2014). Great apes and biodiversity offset projects in Africa: the case for national offset strategies. *PLoS One*, **9**, e111671. DOI: 10.1371/journal.pone.0111671.

KPMG (2014). *Cameroon Country Mining Guide*. Switzerland: KPMG Global Mining Institute.

Krief, S., Cibot, M., Bortolamiol, S., *et al.* (2014). Wild chimpanzees on the edge: nocturnal activities in croplands. *PLoS One*, **9**, e109925. DOI: 10.1371/journal.pone.0109925.

Krief, S., Jamart, A., Mahé, S., *et al.* (2008). Clinical and pathologic manifestation of oesophagostomosis in African great apes: does self-medication in wild apes influence disease progression? *Journal of Medical Primatology*, **37**, 188–95. DOI: 10.1111/j.1600-0684.2008.00285.x.

Kühl, H. (2008). *Best Practice Guidelines for the Surveys and Monitoring of Great Ape Populations. No. 36*. Gland, Switzerland: International Union for Conservation of Nature (IUCN).

Kumar, A., Tormod, S., Ahenkorah, A., *et al.* (2011). Hydropower. In *IPCC Special Report on Renewable Energy Sources and Climate Change Mitigation*, ed. O. Edenhofer, R. Pichs-Madruga, Y. Sokona, *et al.* Cambridge, UK and New York, NY: Cambridge University Press, pp. 437–96.

Kumar, P. (2011). *The Economics of Ecosystems and Biodiversity: Environmental and Economic Foundations*. Oxford, UK: Routledge.

Kumar, V. and Kumar, V. (2015). Seasonal electrocution fatalities in free-range rhesus macaques (*Macaca mulatta*) of Shivalik hills area in northern India. *Journal of Medical Primatology*, **44**, 137–42. DOI: 10.1111/jmp.12168.

Kummer, D.M. and Turner, B.L. (1994). The human causes of deforestation in southeast Asia. *BioScience*, **44**, 323–8. DOI: 10.2307/1312382.

Lambi, C.M., Kimengsi, J.N., Kometa, C.G. and Tata, E.S. (2012). The management and challenges of protected areas and the sustenance of local livelihoods in Cameroon. *Environment and Natural Resources Research*, **2**, 10.

Lanjouw, A. (2014). Mining/oil extraction and ape populations and habitats. In *State of the Apes: Extractive Industries and Ape Conservation*, ed. Arcus Foundation. Cambridge, UK: Cambridge University Press, pp. 127–62. Available at: http://www.stateoftheapes.com/themes/industrial-mining-oil-and-gas/.

Lao MAF (2011). *Gibbon Conservation Action Plan for Lao PDR*. Vientiane, Lao PDR: Ministry of Agriculture and Forestry of the Lao People's Democratic Republic (MAF) Division of Forest Resource Conservation, Department of Forestry.

Laporte, N.T., Stabach, J.A., Grosch, R., Lin, T.S. and Goetz, S.J. (2007). Expansion of industrial logging in central Africa. *Science*, **316**, 1451. DOI: 10.1126/science.1141057.

Lappan, S. (2008). Male care of infants in a siamang (*Symphalangus syndactylus*) population including socially monogamous and polyandrous groups. *Behavioral Ecology and Sociobiology*, **62**, 1307–17. DOI: 10.1007/s00265-008-0559-7.

LAPSSET (2017). *Coordinated Approach Key to the Success of the LAPSSET Corridor*. Nairobi, Kenya: Lamu Port, South Sudan, Ethiopia Transport Corridor (LAPSSET). Available at: http://www.lapsset.go.ke/coordinated-approach-key-to-the-success-of-the-lapsset-corridor/.

Lasch, C., Pintea, L., Traylor-Holzer, K. and Kamenya, S. (2011). *Tanzania Chimpanzee Conservation Action Planning Workshop Report*. Dar es Salaam, Tanzania: The Jane Goodall Institute.

Laurance, S.G.W., Stouffer, P.C. and Laurance, W.F. (2004). Effects of road clearings on movement patterns of understory rainforest birds in central Amazonia. *Conservation Biology*, **18**, 1099–109. DOI: 10.1111/j.1523-1739.2004.00268.x.

Laurance, W.F. (2004). The perils of payoff: corruption as a threat to global biodiversity. *Trends in Ecology & Evolution*, **19**, 399–401. DOI: https://doi.org/10.1016/j.tree.2004.06.001.

Laurance, W.F. (2005). When bigger is better: the need for Amazonian mega-reserves. *Trends in Ecology & Evolution*, **20**, 645–8. DOI: https://doi.org/10.1016/j.tree.2005.10.009.

Laurance, W.F. (2007). Road to ruin. *Himalayan Journal of Sciences*, **4**, 9.

Laurance, W.F. (2008). Expect the unexpected. *New Scientist*, April 12, 2008.

Laurance, W.F. (2014). *How Global Forest-Destroyers are Turning Over a New Leaf.* London, UK: The Conversation. Available at: https://theconversation.com/how-global-forest-destroyers-are-turning-over-a-new-leaf-22943. February 12, 2014.

Laurance, W.F. (2016a). *Conservationists Aghast at Scheme to Degrade 'Heart of Biodiversity'*. Alliance of Leading Environmental Researchers and Thinkers (ALERT). Available at: http://alert-conservation.org/issues-research-highlights/2016/12/30/conservationists-aghast-at-scheme-to-degrade-heart-of-biodiversity. Accessed January 5, 2017.

Laurance, W.F. (2016b). Lessons from research for sustainable development and conservation in Borneo. *Forests*, **7**, 314.

Laurance, W.F. (2016c). *Perils for the Last Place Where Tigers, Orangutans, Elephants and Rhinos Survive*. Alliance of Leading Environmental Researchers and Thinkers (ALERT). Available at: http://alert-conservation.org/issues-research-highlights/2016/9/28/dangers-loom-for-the-last-place-where-tigers-orangutans-elephants-and-rhinos-survive-in-the-wild. Accessed September 15, 2016.

Laurance, W.F., Achard, F., Peedell, S. and Schmitt, S. (2016). Big data, big opportunities. *Frontiers in Ecology and the Environment*, **14**, 347.

Laurance, W.F. and Balmford, A. (2013). A global map for road building. *Nature*, **495**, 308. DOI: 10.1038/495308a.

Laurance, W.F., Campbell, M.J., Alamgir, M. and Mahmoud, M.I. (2017a). Road expansion and the fate of Africa's tropical forests. *Frontiers in Ecology and Evolution*, **5**. DOI: 10.3389/fevo.2017.00075.

Laurance, W.F., Clements, G.R., Sloan, S., *et al.* (2014a). A global strategy for road building. *Nature*, **513**, 229. DOI: 10.1038/nature13717. Available at: https://www.nature.com/articles/nature13717#supplementary-information.

Laurance, W.F., Cochrane, M.A., Bergen, S., *et al.* (2001). The future of the Brazilian Amazon. *Science*, **291**, 438–9. DOI: 10.1126/science.291.5503.438.

Laurance, W.F., Croes, B.M., Guissouegou, N., *et al.* (2008). Impacts of roads, hunting, and habitat alteration on nocturnal mammals in African rainforests. *Conservation Biology*, **22**, 721–32. DOI: 10.1111/j.1523–1739.2008.00917.x.

Laurance, W.F., Croes, B.M., Tchignoumba, L., *et al.* (2006). Impacts of roads and hunting on central African rainforest mammals. *Conservation Biology*, **20**, 1251–61. DOI: 10.1111/j.1523–1739.2006.00420.x.

Laurance, W.F. and Edwards, D.P. (2014). Saving logged tropical forests. *Frontiers in Ecology and the Environment*, **12**, 147.

Laurance, W.F., Goosem, M. and Laurance, S.G.W. (2009). Impacts of roads and linear clearings on tropical forests. *Trends in Ecology & Evolution*, **24**, 659–69. DOI: https://doi.org/10.1016/j.tree.2009.06.009.

Laurance, W.F., Mahmoud, M.I. and Kleinschroth, F. (2017b). Infrastructure expansion and the fate of Central African forests. In *Central African Forests Forever*, ed. M. Brouwer. Berlin, Germany: Central African Forests Commission (COMIFAC) and German Development Bank (KfW), pp. 88–95.

Laurance, W.F., Peletier-Jellema, A., Geenen, B., *et al.* (2015a). Reducing the global environmental impacts of rapid infrastructure expansion. *Current Biology*, **25**, R259-R62. DOI: https://doi.org/10.1016/j.cub.2015.02.050.

Laurance, W.F. and Peres, C.A. (2006). *Emerging Threats to Tropical Forests.* Chicago, IL: University of Chicago Press.

Laurance, W.F., Sayer, J. and Cassman, K.G. (2014b). Agricultural expansion and its impacts on tropical nature. *Trends in Ecology & Evolution*, **29**, 107–16. DOI: https://doi.org/10.1016/j.tree.2013.12.001.

Laurance, W.F., Sloan, S., Weng, L. and Sayer, J.A. (2015b). Estimating the environmental costs of Africa's massive 'development corridors'. *Current Biology*, **25**, 3202–8. DOI: https://doi.org/10.1016/j.cub.2015.10.046.

Laurance, W.F., Useche, D.C., Rendeiro, J., *et al.* (2012). Averting biodiversity collapse in tropical forest protected areas. *Nature*, **489**, 290. DOI: 10.1038/nature11318. Available at: https://www.nature.com/articles/nature11318#supplementary-information.

Lawson, K. and Vines, A. (2014). *Global Impacts of the Illegal Wildlife Trade: The Costs of Crime, Insecurity and Institutional Erosion.* London, UK: Chatham House (The Royal Institute of International Affairs).

Lawson, S. (2014). *Consumer Goods and Deforestation: An Analysis of the Extent and Nature of Illegality in Forest Conversion for Agriculture and Timber Plantations [Forest Trends Report Series].* Forest Trends and DFID. Available at: http://www.forest-trends.org/documents/files/doc_4718.pdf. Accessed November 15, 2016.

Le Saout, S., Hoffmann, M., Shi, Y., *et al.* (2013). Protected areas and effective biodiversity conservation. *Science*, **342**, 803–5. DOI: 10.1126/science.1239268.

Ledec, G.C. and Johnson, S.D.R. (2016). *Biodiversity Offsets: A User Guide.* Washington DC: World Bank. Available at: http://documents.worldbank.org/curated/en/344901481176051661/Biodiversity-offsets-a-user-guide.

Lee, T., Jalong, T. and Wong, M.C. (2014). *No Consent to Proceed: Indigenous Peoples' Rights Violations at the Proposed Baram Dam in Sarawak.* Fact finding mission report. Sarawak, Malaysia: Save Sarawak Rivers Network. Available at: http://www.forestpeoples.org/sites/fpp/files/publication/2014/08/noconsenttoproceedbaramreport2014-1.pdf. Accessed November 15, 2016.

Leendertz, F.H., Lankester, F., Guislain, P., *et al.* (2006). Anthrax in western and central African great apes. *American Journal of Primatology*, **68**, 928–33. DOI: 10.1002/ajp.20298.

Lehner, B., Liermann, C.R., Revenga, C., *et al.* (2011). High-resolution mapping of the world's reservoirs and dams for sustainable river-flow management. *Frontiers in Ecology and the Environment*, **9**, 494–502. DOI: 10.1890/100125.

Leighton, D.S.R. (1987). Gibbons: territoriality and monogamy. In *Primate Societies*, ed. B. B. Smuts, D. L. Cheyney, R. M. Seyfarth, R. W. Wrangham and T. T. Struhsaker. Chicago IL: University of Chicago Press.

Leighty, K.A., Valuska, A.J., Grand, A.P., *et al.* (2015). Impact of visual context on public perceptions of non-human primate performers. *PLoS One*, **10**, e0118487. DOI: 10.1371/journal.pone.0118487.

Lejon, A., Malm Renöfält, B. and Nilsson, C. (2009). Conflicts associated with dam removal in Sweden. *Ecology and Society*, **14**, 4–22.

Leroy, E.M., Rouquet, P., Formenty, P., *et al.* (2004). Multiple Ebola virus transmission events and rapid decline of central African wildlife. *Science*, **303**, 387–90. DOI: 10.1126/science.1092528.

Liden, R. and Lyon, K. (2014). *The Hydropower Sustainability Assessment Protocol for Use by World Bank Clients: Lessons Learned and Recommendations.* Water Papers 89147. World Bank, Water Partnership Program. Available at: http://documents.worldbank.org/curated/en/870411468336660190/pdf/891470REVISED00Box0385238B00PUBLIC0.pdf,.

Lima, I.B.T., Ramos, F.M., Bambace, L.A.W. and Rosa, R.R. (2008). Methane emissions from large dams as renewable energy resources: a developing nation perspective. *Mitigation and Adaptation Strategies for Global Change*, **13**, 193–206. DOI: 10.1007/s11027-007-9086-5.

Liu, D.S., Iverson, L.R. and Brown, S. (1993). Rates and patterns of deforestation in the Philippines: application of geographic information system analysis. *Forest Ecology and Management*, **57**, 1–16. DOI: https://doi.org/10.1016/0378-1127(93)90158-J.

Liu, L. (2003). *Study on Infrastructure and Its Contributions to Economic Growth.* Report 76. Nanchang, PRC: Jiangxi University of Finance and Economics.

Live Science (2011). *Gorilla Stronghold Found, Apes Still In Danger.* Available at: http://www.livescience.com/13436-cameroon-gorilla-count.html. Accessed March 28, 2011.

Loken, B., Boer, C. and Kasyanto, N. (2015). Opportunistic behaviour or desperate measure? Logging impacts may only partially explain terrestriality in the Bornean orang-utan *Pongo pygmaeus morio*. *Oryx*, **49**, 461–4. DOI: 10.1017/S0030605314000969.

Loken, B., Spehar, S. and Rayadin, Y. (2013). Terrestriality in the Bornean orangutan (*Pongo pygmaeus morio*) and implications for their ecology and conservation. *American Journal of Primatology*, **75**, 1129–38. DOI: 10.1002/ajp.22174.

Lokschin, L.X., Rodrigo, C.P., Hallal Cabral, J.N. and Buss, G. (2007). Power lines and howler monkey conservation in Porto Alegre, Rio Grande do Sul, Brazil. *Neotropical Primates*, **14**, 76–80. DOI: 10.1896/044.014.0206.

LRA Crisis Tracker (2016). *The State of the LRA in 2016.* Gland, Switzerland: Invisible Children and The Resolve LRA Crisis Initiative. Available at: https://reports.lracrisistracker.com/pdf/2016-The-State-of-the-LRA.pdf. March, 2016.

Lu, Y. and Tianxiao, Z. (2012). *The Conservation Action Plan of Western Black Crested Gibbon in Yunnan Province (2012–2015).* Western Black Crested Gibbon Conservation Network.

MAAP (2016). *MAAP #40: Early Warning Deforestation Alerts in the Peruvian Amazon.* Monitoring of the Andean Amazon Project (MAAP). Available at: http://maaproject.org/2016/gladalerts/.

MAAP (n.d.). *Methodology.* Monitoring of the Andean Amazon Project (MAAP). Available at: http://maaproject.org/methodology. Accessed February–March, 2017.

MacArthur, R.H. and Wilson, E.O. (1967). *The Theory of Island Biogeography.* Princeton, NJ: Princeton University Press.

Macfie, E.J. and Williamson, E.A. (2010). *Best Practice Guidelines for Great Ape Tourism.* Gland, Switzerland: International Union for Conservation of Nature Species Survival Commission (IUCN SSC), Primate Specialist Group.

MacKay, F. (2017). *Indigenous Peoples' Rights and Conservation: Recent Developments in Human Rights Jurisprudence.* Moreton-in-Marsh, UK: Forest Peoples Programme. Available at: http://www.forestpeoples.org/en/rights-based-conservation/news-article/2017/indigenous-peoples-rights-and-conservation-recent.

Mackinnon, J. (1974). The behaviour and ecology of wild orang-utans (*Pongo pygmaeus*). *Animal Behaviour,* **22**, 3–74. DOI: https://doi.org/10.1016/S0003-3472(74)80054-0.

Magilligan, F., Graber, B., Nislow, K., *et al.* (2016). River restoration by dam removal: enhancing connectivity at watershed scales. *Elementa: Science of the Anthropocene,* **4**, 1–14.

Maiorano, L., Falcucci, A. and Boitani, L. (2008). Size-dependent resistance of protected areas to land-use change. *Proceedings of the Royal Society B: Biological Sciences,* **275**, 1297–304. DOI: 10.1098/rspb.2007.1756.

Maisels, F., Bergl, R.A. and Williamson, E.A. (2016a). Gorilla gorilla *(errata version published in 2016). The IUCN Red List of Threatened Species 2016: e.T9404A102330408.* Gland, Switzerland: International Union for Conservation of Nature (IUCN). Available at: http://www.iucnredlist.org/details/9404/0.

Maisels, F., Strindberg, S., Blake, S., *et al.* (2013). Devastating decline of forest elephants in central Africa. *PLoS One,* **8**, e59469. DOI: 10.1371/journal.pone.0059469.

Maisels, F., Strindberg, S., Breuer, T., *et al.* (2016b). Gorilla gorilla *ssp.* gorilla. *The IUCN Red List of Threatened Species 2016: e.T9406A102328866.* Gland, Switzerland: International Union for Conservation of Nature (IUCN). Available at: http://www.iucnredlist.org/details/9406/0.

Maldonado, O., Aveling, C., Cox, D., *et al.* (2012). *Grauer's Gorillas and Chimpanzees in Eastern Democratic Republic of Congo (Kahuzi-Biega, Maiko, Tayna and Itombwe Landscape): Conservation Action Plan 2012-2022.* Gland, Switzerland: International Union for Conservation of Nature (IUCN).

Maldonado, O. and Fourrier, M. (2015). *Conservation Action Plan for Great Apes in Eastern Democratic Republic of the Congo — Revised version - March-July 2015.* Jane Goodall Institute, Ministry of Environment, Nature Conservation & Tourism and the ICCN.

Mandle, L., Bryant, B.P., Ruckelshaus, M., *et al.* (2016a). Entry points to considering ecosystem services within infrastructure planning: How to integrate conservation with development in order to aid them both. *Conservation Letters,* **9**, 221–7. DOI: 10.1111/conl.12201.

Mandle, L., Wolny, S., Hamel, P., *et al.* (2016b). *Natural Connections: How Natural Capital Supports Myanmar's People and Economy.* Washington DC: World Wide Fund for Nature (WWF).

Mansoer, W.R. and Idral, A. (2015). *Geothermal resources development in Indonesia: a history.* Presented at: World Geothermal Congress 2015, April 19–25, Melbourne, Australia.

March, J.G., Benstead, J.P., Pringle, C.M. and Scatena, F.N. (2003). Damming tropical island streams: problems, solutions, and alternatives. *BioScience,* **53**, 1069–78. DOI: 10.1641/0006-3568(2003)053[1069:DTISPS]2.0.CO;2.

Maron, M., Hobbs, R.J., Moilanen, A., *et al.* (2012). Faustian bargains? Restoration realities in the context of biodiversity offset policies. *Biological Conservation,* **155**, 141–8. DOI: https://doi.org/10.1016/j.biocon.2012.06.003.

Marshall, A.J., Ancrenaz, M., Brearley, F.Q., *et al.* (2009). The effects of forest phenology and floristics on populations of Bornean and Sumatran orangutans. In *Orangutans: Geographic Variation in Behavioral Ecology and Conservation,* ed. S. A. Wich, S. Utami Atmoko, T. Mitra Setia and C. P. Van Schaik. Oxford, UK: Oxford University Press, pp. 97–117.

Marshall, A.J. and Leighton, M. (2006). How does food availability limit the population density of white-bearded gibbons? In *Feeding Ecology of the Apes,* ed. G. Hohmann, M. Robbins and C. Boesch. Cambridge, UK: Cambridge University Press, pp. 313–35.

Marshall, A.J., Nardiyono, Engström, L.M., *et al.* (2006). The blowgun is mightier than the chainsaw in determining population density of Bornean orangutans (*Pongo pygmaeus morio*) in the forests of East Kalimantan. *Biological Conservation,* **129**, 566–78. DOI: https://doi.org/10.1016/j.biocon.2005.11.025.

Martinez-Alier, J. (2002). *The Environmentalism of the Poor.* Cheltenham, UK: Edward Elgar.

Martini, M. (2013). Wildlife crime and corruption: in which way does corruption exacerbate the problem of poaching and illegal wildlife trade in southern Africa and how can anti-corruption contribute to the fight against it? *U4 Expert Answer,* **367**, February 15, 2013. U4 Anti-Corruption Resource Centre. Available at: https://www.transparency.org/files/content/corruptionqas/367_Wildlife_Crimes_and_Corruption.pdf.

Mascia, M.B. and Pailler, S. (2011). Protected area downgrading, downsizing, and degazettement (PADDD) and its conservation implications. *Conservation Letters,* **4**, 9–20. DOI: 10.1111/j.1755-263X.2010.00147.x.

Masi, S., Cipolletta, C. and Robbins, M.M. (2009). Western lowland gorillas (*Gorilla gorilla gorilla*) change their activity patterns in response to frugivory. *American Journal of Primatology,* **71**, 91–100. DOI: 10.1002/ajp.20629.

Masi, S., Mundry, R., Ortmann, S., *et al.* (2015). The influence of seasonal frugivory on nutrient and energy intake in wild western gorillas. *PLoS One,* **10**, e0129254. DOI: 10.1371/journal.pone.0129254.

Mason, M. (1999). *Environmental Democracy: A Contextual Approach.* London, UK: Earthscan Publications.

Matsuzawa T, H.T., Sugiyama Y. (2011). *The Chimpanzees of Bossou and Nimba.* Tokyo, Japan: Springer.

Max Planck Institute (n.d.-a). *A.P.E.S. Database.* Available at: http://apesportal.eva.mpg.de/database/archiveTable. Accessed October, 2017.

Max Planck Institute (n.d.-b). *A.P.E.S. Portal Dashboard.* Munich, Germany: Max Planck Institute. Available at: http://mapper.eva.mpg.de/status/tools/dashboard.

Mbodiam, B. (2010). *Mines: Une Nouvelle Société à l'Assaut de l'Or à Bétaré Oya.* Cameroonvoice.com. Available at: http://www.cameroonvoice.com/news/article-news-1954.html.

Mbodiam, B.R. (2016). Cameroon: onslought of more than 6000 fishermen on fish-filled waters of Lom Pangar Dam. *Business in Cameroon,* June 21, 2016. Available at: www.businessincameroon.com/fish/2106-6317-cameroon-onslought-of-more-than-6000-fishermen-on-fish-filled-waters-of-lom-pangar-dam.

McCarthy, J.F. (2000). *'Wild Logging': The Rise and Fall of Logging Networks and Biodiversity Conservation Projects on Sumatra's Rainforest Frontier.* Bogor, Indonesia: Center for International Forestry Research (CIFOR).

McCarthy, J.F. (2002). Power and interest on Sumatra's rainforest frontier: clientelist coalitions, illegal logging and conservation in the Alas Valley. *Journal of Southeast Asian Studies,* **33**, 77–106. DOI: 10.1017/S0022463402000048.

McCarthy, N. (2017). Solar employs more people in US electricity generation than oil, coal and gas combined. *Forbes Magazine Online,* January 25, 2017. Available at: https://www.forbes.com/sites/niallmccarthy/2017/01/25/u-s-solar-energy-employs-more-people-than-oil-coal-and-gas-combined-infographic/#47ac39f28000.

McConkey, K.R. (2000). Primary seed shadow generated by gibbons in the rain forests of Barito Ulu, central Borneo. *American Journal of Primatology,* **52**, 13–29. DOI: 10.1002/1098-2345(200009)52:1<13::AID-AJP2>3.0.CO;2-Y.

McConkey, K.R. (2005). The influence of gibbon primary seed shadows on post-dispersal seed fate in a lowland dipterocarp forest in central Borneo. *Journal of Tropical Ecology,* **21**, 255–62.

McConkey, K.R. and Chivers, D.J. (2007). Influence of gibbon ranging patterns on seed dispersal distance and deposition site in a Bornean forest. *Journal of Tropical Ecology,* **23**, 269–75. DOI: 10.1017/S0266467407003999.

McDavitt, B. (2016). *Ladders and Licenses: Fish Passages Play Role in Relicensing Hydroelectric Facilities.* Gloucester, MA: NOAA Fisheries. Available at: https://www.greateratlantic.fisheries.noaa.gov/stories/2015/october/27_ladders_and_licenses__fish_passages_play_role_in_relicensing_hydroelectric_facilities.html?utm_source=Hydropower+Habitat+Story&utm_campaign=Hydropower&utm_medium=email. Accessed October 7, 2016.

McGrew, W.C., Baldwin, P.J. and Tutin, C.E.G. (1981). Chimpanzees in a hot, dry and open habitat: Mt Assirik, Senegal, west Africa. *Journal of Human Evolution,* **10**, 227–44. DOI: https://doi.org/10.1016/S0047-2484(81)80061-9.

McKenzie, E., Rosenthal, A., *et al.* (2012). *Developing Scenarios to Assess Ecosystem Service Tradeoffs: Guidance and Case Studies for InVEST Users.* Washington DC: World Wide Fund for Nature (WWF).

McLennan, M.R. (2008). Beleaguered chimpanzees in the agricultural district of Hoima, western Uganda. *Primate Conservation,* **23**, 45–54.

McLennan, M.R. and Asiimwe, C. (2016). Cars kill chimpanzees: case report of a wild chimpanzee killed on a road at Bulindi, Uganda. *Primates,* **57**, 377–88. DOI: 10.1007/s10329-016-0528-0.

McLennan, M.R. and Ganzhorn, J.U. (2017). Nutritional characteristics of wild and cultivated foods for chimpanzees (*Pan troglodytes*) in agricultural landscapes. *International Journal of Primatology*, **38**, 122–50. DOI: 10.1007/s10764-016-9940-y.

McLennan, M.R. and Hill, C.M. (2012). Troublesome neighbours: changing attitudes towards chimpanzees (*Pan troglodytes*) in a human-dominated landscape in Uganda. *Journal for Nature Conservation*, **20**, 219–27. DOI: https://doi.org/10.1016/j.jnc.2012.03.002.

McLennan, M.R. and Hockings, K.J. (2016). The aggressive apes? Causes and contexts of great ape attacks on local persons. In *Problematic Wildlife: A Cross-Disciplinary Approach*, ed. F. M. Angelici. Cham, Switzerland: Springer, pp. 373–94. DOI: 10.1007/978-3-319-22246-2_18. Available at: https://doi.org/10.1007/978-3-319-22246-2_18.

McRae, B.H., Dickson, B.G., Keitt, T.H. and Shah, V.B. (2008). Using circuit theory to model connectivity in ecology, evolution, and conservation. *Ecology*, **89**, 2712–24. DOI: 10.1890/07-1861.1.

MCRB (2016). *Environmental Impact Assessment Procedures*. Yangon, Myanmar: Myanmar Centre for Responsible Business (MCRB). Available at: http://www.myanmar-responsiblebusiness.org/resources/environmental-impact-assessment-procedures.html.

McSweeney, K., Nielsen, E.A., Taylor, M.J., *et al.* (2014). Drug policy as conservation policy: narco-deforestation. *Science*, **343**, 489–90. DOI: 10.1126/science.1244082.

MEA (2005). *Ecosystems and Human Well-Being: Synthesis*. Washington DC: Island Press. Millennium Ecosystem Assessment (MEA).

Meehan, D. (2013). *The Mbalam–Nabeba Iron Ore Project: Developing Central Africa's Iron Ore Region*. Presentation by Sundance Resources Ltd. Presented at Cameroon Mining Forum (CIMEC) 2013. Available at: https://cameroonminingopportunities.files.wordpress.com/2013/12/presentation-acc80-cimec-d-meehan-presentation-cameroon-mining-forum-2013.pdf.

Megevand, C. (2013). *Deforestation Trends in the Congo Basin: Reconciling Economic Growth and Forest Protection*. Washington DC: World Bank.

Meijaard, E., Abram, N.K., Wells, J.A., *et al.* (2013). People's perceptions about the importance of forests on Borneo. *PLoS One*, **8**, e73008. DOI: 10.1371/journal.pone.0073008.

Meijaard, E., Albar, G., Nardiyono, *et al.* (2010a). Unexpected ecological resilience in Bornean orangutans and implications for pulp and paper plantation management. *PLoS One*, **5**, e12813. DOI: 10.1371/journal.pone.0012813.

Meijaard, E., Buchori, D., Hadiprakarsa, Y., *et al.* (2011). Quantifying killing of orangutans and human-orangutan conflict in Kalimantan, Indonesia. *PLoS One*, **6**, e27491. DOI: 10.1371/journal.pone.0027491.

Meijaard, E., Welsh, A., Ancrenaz, M., *et al.* (2010b). Declining orangutan encounter rates from Wallace to the present suggest the species was once more abundant. *PLoS One*, **5**, e12042. DOI: 10.1371/journal.pone.0012042.

Meijaard, E. and Wich, S. (2014). *Extractive Industries and Orangutans. Occasional Paper for State of the Apes, Volume 1*. Cambridge, UK: Arcus Foundation. Available at: https://www.stateoftheapes.com/wp-content/uploads/2014/07/Extractive-Industries-and-Orangutans1.pdf.

METI (2015). *Infrastructure System Export Promotion Survey, 2014 Fiscal Year: Infrastructure and Mining* [in Japanese]. Tokyo, Japan: Ministry of Economy, Trade and Industry (METI).

Milman, O. (2016). Mass chimpanzee transfer begins in effort to protect endangered species. *The Guardian*, September 9, 2016. Available at: https://www.theguardian.com/world/2016/sep/09/georgia-chimpanzees-sanctuary-project-chimps-transfer. Accessed October 2, 2016.

MINFOF (2015). *Annual Report of the Deng Deng National Park*. Yaoundé, Cameroon: Ministère des Forêts et de la Faune (MINFOF).

Mining Review Africa (2016). *Sundance Resources Secures New Funding for Mbalam Nabeba Development*. Rondebosch, South Africa: Mining Review Africa. Available at: https://www.miningreview.com/news/sundance-resources-funding-for-mbalam-nabeba-development/. Accessed November 25, 2016.

Mitani, J.C. (2009). Male chimpanzees form enduring and equitable social bonds. *Animal Behaviour*, **77**, 633–40. DOI: https://doi.org/10.1016/j.anbehav.2008.11.021.

Mitani, J.C., Watts, D.P. and Amsler, S.J. (2010). Lethal intergroup aggression leads to territorial expansion in wild chimpanzees. *Current Biology*, **20**, R507-8. DOI: https://doi.org/10.1016/j.cub.2010.04.021.

Mitchard, E. (2012). Are Cameroon's forests doomed? *Deforestationwatch*, October 23, 2012. Available at: http://deforestationwatch.wordpress.com/2012/10/23/are-cameroons-forests-doomed.

Mitchell, M.W., Locatelli, S., Sesink Clee, P.R., Thomassen, H.A. and Gonder, M.K. (2015). Environmental variation and rivers govern the structure of chimpanzee genetic diversity in a biodiversity hotspot. *BMC Evolutionary Biology*, **15**, 1. DOI: 10.1186/s12862-014-0274-0.

Mittermeier, R.A., Rylands, A.B. and Wilson, D.E., ed. (2013). *Handbook of the Mammals of the World. Volume 3: Primates*. Barcelona, Spain: Lynx Edicions.

MLUD (2016). *Public Notice: Notice of Revocation of Rights of Occupancy for Public Purpose Land Use Act 1978. January 22*. Calabar, Nigeria: Ministry of Lands and Urban Development (MLUD), Government of Cross River State of Nigeria.

MME (n.d.). *Mission and Values*. Littoral, Cameroun: Mississauga Mining & Exploration (MME). Available at: http://mississaugamining.com/mission-and-values/. Accessed December, 2017.

MNRT (2012). *Participatory Forest Management in Tanzania, Facts and Figures*. Dar es Salaam, Tanzania: Ministry of Natural Resources and Tourism of Tanzania (MNRT).

MoC (2007). *Notice on the Release of 'A Guide for Chinese Enterprises on Sustainable Silviculture Overseas' by Ministry of Commerce and State Forestry Administration*. Beijing, PRC: Ministry of Commerce (MoC) of the People's Republic of China. Available at: http://www.mofcom.gov.cn/aarticle/b/g/200712/20071205265858.html.

MoC (2014). *Statistical Bulletin of China's Outward Foreign Direct Investment*. Beijing, PRC: Ministry of Commerce (MoC) of the People's Republic of China, China Statistics Press.

MoC (2016a). *Officials Talk about China's Overseas Investment and Cooperation in 2015*. Beijing, PRC: Ministry of Commerce (MoC) of the People's Republic of China. Available at: http://www.mofcom.gov.cn/article/ae/ai/201601/20160101235603.shtml.

MoC (2016b). *Statistical Bulletin of China's Outward Foreign Direct Investment*. Beijing, PRC: Ministry of Commerce (MoC) of the People's Republic of China, China Statistics Press.

MoC (2016c). *Statistical Bulletin of China's Outward Foreign Direct Investment*. [in Chinese]. Beijing, PRC: Ministry of Commerce (MoC) of the People's Republic of China, China Statistics Press.

Modus Aceh (2016). *Kenapa harus ngotot proyek PT Hitay Panas Energy di Lapangan Kafi*. Banda Aceh, Indonesia: Modus Aceh. Available at: http://www.modusaceh.co/news/kenapa-harus-ngotot-proyek-pt-hitay-panas-energy-di-lapangan-kafi/index.html.

Moehrenschlager, A., Shier, D.M., Moorhouse, T.P. and Stanley Price, M.R. (2013). Righting past wrongs and ensuring the future: challenges and opportunities for effective reintroductions amidst a biodiversity crisis. In *Key Topics in Conservation Biology 2*, ed. D. W. MacDonald and K. Willis. John Wiley & Sons, pp. 405–29. DOI: 10.1002/9781118520178.ch22. Available at: http://dx.doi.org/10.1002/9781118520178.ch22.

Molina, S., Cerdas Vegas, G., Jarrín Hidalgo, S., Torres, V. and Rivasplata Cabrera, F. (2015). *De IIRSA a COSIPLAN, Cambios y Continuidades. Boletín No. 2*. La Paz, Bolivia: Centro de Estudios para el Desarrollo Laboral y Agrario (CEDLA).

Molur, S., Walker, S., Islam, A., *et al.*, ed. (2005). *Conservation of Western Hoolock Gibbon (*Hoolock hoolock hoolock*) in India and Bangladesh*. Coimbatore, India: Zoo Outreach Organisation/CBSG-South Asia.

Mongabay (2016a). Controversial dam officially cancelled in Borneo after Indigenous protests. *Mongabay*, March, 2016. Available at: https://news.mongabay.com/2016/03/controversial-dam-officially-canceled-in-borneo-after-indigenous-protests/. Accessed September 14, 2017.

Mongabay (2016b). Palm oil giant defends its deforestation in Gabon, points to country's 'right to develop'. *Mongabay*, December, 2016. Available at: https://news.mongabay.com/2016/12/palm-oil-giant-defends-its-deforestation-in-gabon-points-to-countrys-right-to-develop/.

MONUSCO (2015). *North Kivu*. Kinshasa, RDC: Mission de l'Organisation des Nations Unies pour la Stabilisation en République démocratique du Congo (MONUSCO). Available at: https://monusco.unmissions.org/sites/default/files/north_kivu.factsheet.eng_.pdf.

Moore, P., Prompinchompoo, C. and Beastall, C.A. (2016). *CITES Implementation in Thailand: A Review of the Legal Regime Governing the Trade in Great Apes and Gibbons and Other CITES-Listed Species*. Selangor, Malaysia: TRAFFIC.

Moorthy, E. (1997). With the Karen on the Thai border. *Wall Street Journal*.

Morgan, B., Adeleke, A., Bassey, T., et al. (2011). *Regional Action Plan for the Conservation of the Nigeria-Cameroon Chimpanzee (*Pan troglodytes ellioti*).* New York, NY: International Union for Conservation of Nature Species Survival Commission (IUCN SSC), Primate Specialist Group and Zoological Society of San Diego, CA. Available at: http://static1.1.sqspcdn.com/static/f/1200343/20456353/1348922758247/NCCAP.pdf?token=whcKSia0%2F%2BabiK74%2BLfCDsBQPMc%3D.

Morgan, D. and Sanz, C. (2006). Chimpanzee feeding ecology and comparisons with sympatric gorillas in the Goualougo Triangle, Republic of Congo. In *Primates: Feeding Ecology in Apes and Other Primates: Ecological, Physiological, and Behavioural Aspects*, ed. G. Hohmann, M. M. Robbins and C. Boesch. Cambridge Studies in Biological and Evolutionary Anthropology, **48** Cambridge, UK: Cambridge University Press, pp. 97–122.

Morgan, D. and Sanz, C. (2007). *Best Practice Guidelines for Reducing the Impact of Commercial Logging on Great Apes in Western Equatorial Africa.* Gland, Switzerland: International Union for Conservation of Nature Species Survival Commission (IUCN SSC), Primate Specialist Group.

Morgan, D., Sanz, C., Onononga, J.R. and Strindberg, S. (2006). Ape abundance and habitat use in the Goualougo Triangle, Republic of Congo. *International Journal of Primatology*, **27**, 147–79. DOI: 10.1007/s10764-005-9013-0.

Morrogh-Bernard, H., Husson, S., Page, S.E. and Rieley, J.O. (2003). Population status of the Bornean orang-utan (*Pongo pygmaeus*) in the Sebangau peat swamp forest, Central Kalimantan, Indonesia. *Biological Conservation*, **110**, 141–52. DOI: https://doi.org/10.1016/S0006-3207(02)00186-6.

Mosse, D. (2005). *Cultivating Development: An Ethnography of Aid Policy and Practice.* London, UK: Pluto Press.

Moutinho Sá, R.M., Ferreira da Silva, M., Sousa, F.M. and Minhós, T. (2012). The trade and ethnobiological use of chimpanzee body parts in Guinea-Bissau. *TRAFFIC Bulletin*, **24**, 31–4.

Moyer, D., Plumptre, A.J., Pintea, L., *et al.* (2006). *Surveys of Chimpanzees and Other Biodiversity in Western Tanzania.* Arlington, VA: United States Fish and Wildlife Service (USFWS).

Muehlenbein, M.P. and Ancrenaz, M. (2009). Minimizing pathogen transmission at primate ecotourism destinations: the need for input from travel medicine. *Journal of Travel Medicine*, **16**, 229–32. DOI: 10.1111/j.1708-8305.2009.00346.x.

Mueller, N.D., Gerber, J.S., Johnston, M., *et al.* (2012). Closing yield gaps through nutrient and water management. *Nature*, **490**, 254. DOI: 10.1038/nature11420. Available at: https://www.nature.com/articles/nature11420#supplementary-information.

Mulavwa, M.N., Yangozene, K., Yamba-Yamba, M., *et al.* (2010). Nest groups of wild bonobos at Wamba: selection of vegetation and tree species and relationships between nest group size and party size. *American Journal of Primatology*, **72**, 575–86. DOI: 10.1002/ajp.20810.

Murai, M., Ruffler, H., Berlemont, A., *et al.* (2013). Priority areas for large mammal conservation in Equatorial Guinea. *PLoS One*, **8**, e75024. DOI: 10.1371/journal.pone.0075024.

Myers, N. (1998). Lifting the veil on perverse subsidies. *Nature*, **392**, 327. DOI: 10.1038/32761.

Myers, N., Mittermeier, R.A., Mittermeier, C.G., da Fonseca, G.A.B. and Kent, J. (2000). Biodiversity hotspots for conservation priorities. *Nature*, **403**, 853. DOI: 10.1038/35002501. Available at: https://www.nature.com/articles/35002501#supplementary-information.

Nater, A., Mattle-Greminger, M.P., Nurcahyo, A., *et al.* (2017). Morphometric, behavioral, and genomic evidence for a new orangutan species. *Current Biology*, **27**, 3487–98. DOI: 10.1016/j.cub.2017.09.047.

Natural Capital Coalition (2016). *Natural Capital Protocol.* National Capital Coalition. Available at: http://naturalcapitalcoalition.org/protocol/.

Natural Capital Coalition (n.d.). *Natural Capital Coalition.* Natural Capital Coalition. Available at: http://naturalcapitalcoalition.org/.

Naughton-Treves, L. (1997). Farming the forest edge: vulnerable places and people around Kibale National Park, Uganda. *Geographical Review*, **87**, 27–46. DOI: 10.1111/j.1931-0846.1997.tb00058.x.

NBS (n.d.). *China's National Statistics.* Beijing, PRC: National Bureau of Statistics of China (NBS). Available at: http://data.stats.gov.cn/index.htm. Accessed July 6, 2017.

NCFA (n.d.). *Natural Capital Finance Alliance.* Natural Capital Finance Alliance (NCFA). Available at: http://www.naturalcapitaldeclaration.org.

Ndobe, S.N. and Klemm, J. (2014). *The Lom Pangar Hydropower Dam Project. Evaluating the Project's Impacts within the Framework of the World Bank Safeguard Policies. Lessons for the World Bank Safeguards Review.* March. Synchronicity Earth.

Ndobe, S.N. and Mantzel, K. (2014). *Deforestation, REDD and Takamanda National Park in Cameroon: A Case Study.* Forest Peoples Programme (FPP) and Umverteilen.

Nellemann, C. and Newton, A. (2002). *The Great Apes, The Road Ahead: A GLOBIO Perspective on the Impacts of Infrastructure Development on the Great Apes.* United Nations Environment Programme (UNEP), GRID-Arendal, World Conservation Monitoring Centre. Available at: http://www.globio.info/downloads/249/Great+Apes+-+The+Road+Ahead.pdf.

Nelson, A. (2008). *Travel Time to Major Cities: A Global Map of Accessibility.* Luxembourg, Luxembourg: Publications Office of the European Union. DOI: 10.2788/95835.

Nelson, A. and Chomitz, K.M. (2011). Effectiveness of strict vs. multiple use protected areas in reducing tropical forest fires: a global analysis using matching methods. *PLoS One*, **6**, e22722. DOI: 10.1371/journal.pone.0022722.

Nelson, J. (2007). *Securing Indigenous Land Rights in the Cameroon Oil Pipeline Zone.* Moreton-in-Marsh, UK: Forest Peoples Programme. Available at: http://www.forestpeoples.org/sites/fpp/files/publication/2010/08/cameroonpipelinejul07lowreseng.pdf.

Nelson, J., Kenrick J. and Jackson, D. (2001). *Report on a Consultation with Bagyeli Pygmy Communities Impacted by the Chad-Cameroon Oil-Pipeline Project.* Moreton-in-Marsh, UK: Forest Peoples Programme. Available at: http://www.forestpeoples.org/sites/fpp/files/publication/2010/07/ccpbagyeliconsultmay01eng.pdf.

NEPAD (n.d.). *New Partnership for Africa's Development.* Midrand, South Africa: New Partnership for Africa's Development (NEPAD). Available at: http://www.nepad.org/. Accessed January 20, 2017.

Ngano, G. (2010). Three nations, one conservation complex. *ITTO Tropical Forest Update*, **20**, 11–3.

Ngoprasert, D., Lynam, A.J. and Gale, G.A. (2017). Effects of temporary closure of a national park on leopard movement and behaviour in tropical Asia. *Mammalian Biology*, **82**, 65–73. DOI: https://doi.org/10.1016/j.mambio.2016.11.004.

Nguiffo, S. (2016). *La cartographie participative et le droit des espaces et des ressources au Cameroun.* Presented at: RRI Land Tenure Facility workshop, February 29, 2016, Yaoundé, Cameroon.

Nguiffo, S. and Djeukam, R. (2008). Using the law as a tool to secure the land rights of indigenous communities in southern Cameroon. In *Legal Empowerment in Practice: Using Legal Tools to Secure Land Rights in Africa*, ed. L. Cotula and P. Mathieu. London, UK: International Institute for Environment and Development (IIED), pp. 29–44. Available at: http://pubs.iied.org/pdfs/12552IIED.pdf.

Nguiffo, S., Kenfack, P.E., Mballa, N. (2009). *Historical and Contemporary Land Laws and their Impact on Indigenous Peoples' Land Rights in Cameroon. Report No. 2. Land Rights and the Forest Peoples of Africa: Historical, Legal and Anthropological Perspectives.* Moreton-in-Marsh, UK: Forest Peoples Programme (FFP).

Nijman, V. (2009). *An Assessment of Trade in Gibbons and Orang-Utans in Sumatra, Indonesia.* Selangor, Malaysia: TRAFFIC Southeast Asia.

Nijman, V. and Geissmann, T. (2008). Symphalangus syndactylus. *The IUCN Red List of Threatened Species 2008*: e.T39779A10266335. Available at: http://dx.doi.org/10.2305/IUCN.UK.2008.RLTS.T39779A10266335.en. Accessed December 11, 2016.

Noam, Z. (2007). Eco-authoritarian conservation and ethnic conflict in Burma. *Policy Matters: Conservation and Human Rights*, **15**. Available at: http://lib.icimod.org/record/13286/files/1734.pdf.

Normand, E. and Boesch, C. (2009). Sophisticated Euclidean maps in forest chimpanzees. *Animal Behaviour*, **77**, 1195–201. DOI: https://doi.org/10.1016/j.anbehav.2009.01.025.

Nuno, A. and St John, F.A.V. (2015). How to ask sensitive questions in conservation: a review of specialized questioning techniques. *Biological Conservation*, **189**, 5–15. DOI: https://doi.org/10.1016/j.biocon.2014.09.047.

Oates, J.F. (1999). *Myth and Reality in the Rain Forest.* Berkeley, CA: University of California Press.

Oates, J.F., Bergl, R.A. and Linder, J.M. (2004). Africa's Gulf of Guinea forests: biodiversity patterns and conservation priorities. *Advances in Applied Biodiversity Science*, **6**, 1–90.

Oates, J.F., Sunderland-Groves, J.L., Bergl, R., *et al.* (2007). *Regional Action Plan for the Conservation of the Cross River Gorilla (Gorilla gorilla diehli).* Gland, Switzerland: International Union for Conservation of Nature Species Survival Commission (IUCN SSC), and Arlington, VA: Primate Specialist Group and Conservation International.

Oberndorf, R.B. (2012). *Legal Review of Recently Enacted Farmland Law and Vacant, Fallow and Virgin Lands Management Law: Improving the Legal and Policy Frameworks Relating to Land Management in Myanmar.* Forest Trends and Food Security Working Group's Land Core Group.

Ocampo-Peñuela, N., Jenkins, C.N., Vijay, V., Li, B.V. and Pimm, S.L. (2016). Incorporating explicit geospatial data shows more species at risk of extinction than the current Red List. *Science Advances*, **2**. DOI: 10.1126/sciadv.1601367.

O'Connor, J.E., Duda, J.J. and Grant, G.E. (2015). 1000 dams down and counting. *Science*, **348**, 496–7. DOI: 10.1126/science.aaa9204.

OFI (n.d.). *Orangutan Care Center and Quarantine.* Orangutan Foundation International (OFI). Available at: https://orangutan.org/occq/. Accessed March, 21 2017.

Ogawa, H., Yoshikawa, M. and Idani, G. (2014). Sleeping site selection by savanna chimpanzees in Ugalla, Tanzania. *Primates*, **55**, 269–82. DOI: 10.1007/s10329-013-0400-4.

Ojeme, V. (2011). *Why Nigeria Is Underdeveloped, by Dowden.* Nigeria: Vanguard. Available at: https://www.vanguardngr.com/2011/09/why-nigeria-is-underdeveloped-by-dowden.

Okeke, F. (2013). *Land cover change analysis in the Afi-Mbe-Okwangwo landscape, Cross River State, Nigeria.* Wildlife Conservation Society, Nigeria Program. Report to CRSFC and UN-REDD.

Olawoyin, O. (2017). Lagos is Nigeria's most indebted state with highest domestic, foreign debts. *Premium Times*, April 28, 2017. Available at: http://www.premiumtimesng.com/regional/ssouth-west/229830-lagos-nigerias-indebted-state-highest-domestic-foreign-debts.html.

Ondoua Ondoua, G., Beodo Moundjim, E., Mambo Marindo, J.C., *et al.* (2017). *An Assessment of Poaching and Wildlife Trafficking in the Garamba-Bili-Chinko Transboundary Landscape.* Cambridge, UK: TRAFFIC. Available at: http://www.indiaenvironmentportal.org.in/files/file/Garamba-Bili-Chinko.pdf.

OpenStreetMap (n.d.). *OpenStreetMap.* OpenStreetMap. Available at: https://www.openstreetmap.org. Accessed July 21, 2017.

Opperman, J., Grill, G. and Hartmann, J. (2015). *The Power of River: Finding Balance Between Energy and Conservation in Hydropower Development.* Washington DC: The Nature Conservancy.

Opperman, J., Hartmann, J. and Raepple, J. (2017). *The Power of Rivers: A Business Case.* Washington DC: The Nature Conservancy.

Orangutan Appeal UK (n.d.). *Sepilok Orangutan Rehabilitation Centre.* Brockenhurst, UK: Orangutan Appeal UK. Available at: https://www.orangutan-appeal.org.uk/about-us/sepilok-orangutan-rehabilitation-centre. Accessed October 10, 2017.

Orangutan SSP (n.d.). *Orangutan SSP Member Zoos.* Orangutan Species Survival Plan (SSP). Available at: http://www.orangutanssp.org/member-zoos.html. Accessed September 30, 2016.

Osterberg, P., Samphanthamit, P., Maprang, O., Punnadee, S. and Brockelman, W.Y. (2014). Population dynamics of a reintroduced population of captive-raised gibbons (*Hylobates lar*) on Phuket, Thailand. *Primate Conservation*, **28**, 179–88. DOI: 10.1896/052.028.0114.

Ouyang, Z., Zheng, H., Xiao, Y., *et al.* (2016). Improvements in ecosystem services from investments in natural capital. *Science*, **352**, 1455–9. DOI: 10.1126/science.aaf2295.

Owono, J.C. (2001). Case study 8 Cameroon – Campo Ma'an the extent of Bagyeli Pygmy involvement in the development and management plan of the Campo Ma'an UTO. In *From Principles to Practice: Indigenous Peoples and Protected Areas in Africa*, ed. J. Nelson and L. Hossack, pp. 243–68. Available at: http://www.forestpeoples.org/sites/fpp/files/publication/2010/08/camerooncampomaaneng.pdf.

Oxfam, ILC and RRI (2016). *Common Ground: Securing Land Rights and Safeguarding the Earth.* Oxfam, International Land Coalition (ILC), and Rights and Resources Initiative (RRI). Available at: https://rightsandresources.org/en/publication/global-call-common-ground/#.WlEexEx2tZU. Accessed March 1, 2016.

Pacca, S. and Horvath, A. (2002). Greenhouse gas emissions from building and operating electric power plants in the Upper Colorado River Basin. *Environmental Science and Technology*, **36**, 3194–200. DOI: 10.1021/es0155884.

Palm, J. (2015). Fresh start for Liberian chimpanzees used for medical tests. *Reuters*. Available at: http://www.reuters.com/article/us-liberia-chimpanzees-idUSKBN0U61KP20151223. Accessed October 4, 2016.

Palombit, R.A. (1992). *Pair bonds and monogamy in wild siamang (*Hylobates syndactylus*) and white-handed gibbons (*Hylobates lar*) in northern Sumatra*. PhD thesis. University of California.

Palombit, R. (1994). Dynamic pair bonds in hylobatids: implications regarding monogamous social systems. *Behaviour*, **128**, 65–101. DOI: https://doi.org/10.1163/156853994X00055.

Palombit, R.A. (1997). Inter- and intraspecific variation in the diets of sympatric siamang (*Hylobates syndactylus*) and lar gibbons (*Hylobates lar*). *Folia Primatologica*, **68**, 321–37.

Panaligan, R. (2005). Another tragedy in Aceh: illegal logging. *Jakarta Post*, Available at: https://www.seapa.org/another-tragedy-in-aceh-illegal-logging/.

PASA (2015). *2015 census for African sanctuaries*. Portland, OR: Pan African Sanctuary Alliance (PASA). Unpublished report.

PASA (2016a). *Operations Manual*, 2nd edn, December 2016. Portland, OR: Pan African Sanctuary Alliance (PASA). Available at: https://www.pasaprimates.org/manuals-reports/.

PASA (2016b). *US Fish & Wildlife Approves Exporting Chimps to UK Zoo in an Unprecedented 'Pay to Play' Scheme*. Portland, OR: Pan African Sanctuary Alliance (PASA). Available at: https://www.pasaprimates.org/advocacy/yerkes-wingham-update/. Accessed October 27, 2016.

Payne, J. (1988). *Orang-utan Conservation in Sabah*. Report 3759. Kuala Lumpur, Malaysia: World Wide Fund for Nature (WWF), Malaysia International.

Percoco, M. (2014). Quality of institutions and private participation in transport infrastructure investment: evidence from developing countries. *Transportation Research Part A: Policy and Practice*, **70**, 50–8. DOI: https://doi.org/10.1016/j.tra.2014.10.004.

Perram, A. (2015). Consulting Study 10A: institutional framework governing the palm oil sector in Cameroon: a report on laws, regulations and practices. In *HCS+ Consulting Study 10: Overview of Existing Regulatory Mechanisms and Relevant Actors*. Available at: http://www.simedarby.com/sustainability/clients/simedarby_sustainability/assets/contentMS/img/template/editor/HCSReports/Consulting%20Report%20.pdf.

Perram, A. (2016). *Behind the Veil: Transparency, Access to Information and Community Rights in Cameroon's Forestry Sector*. Moreton-in-Marsh, UK: Forest Peoples Programme. Available at: http://www.forestpeoples.org/sites/fpp/files/publication/2016/06/behind-veil-artwork-english-web-1.pdf.

PGM Nigeria (2016a). *Draft environmental impact assessment (EIA) report of the proposed Calabar–Ikom–Katsina Ala superhighway project*. Submitted to Federal Ministry of Environment. Abuja, Nigeria: Government of Cross River State, Nigeria.

PGM Nigeria (2016b). *Final environmental impact assessment (EIA) report of the proposed Calabar–Ikom–Katsina Ala superhighway project*. Submitted to Federal Ministry of Environment. Abuja, Nigeria: Government of Cross River State, Nigeria.

PGM Nigeria (2017). *Proposed Calabar-Ikom-Katsina Ala Superhighway project. Environmental impact assessment, May 2017.* Final report submitted to Federal Ministry of Environment, Abuja. PGM Nigeria.

Phalan, B., Onial, M., Balmford, A. and Green, R.E. (2011). Reconciling food production and biodiversity conservation: land sharing and land sparing compared. *Science*, **333**, 1289–91. DOI: 10.1126/science.1208742.

PIB (2016a). *New Delhi Resolution on Tiger Conservation Adopted: Third Asian Ministerial Conference on Tiger Conservation Concludes*. Press Information Bureau (PIB), Ministry of Environment and Forests, Government of India. Available at: http://pib.nic.in/newsite/PrintRelease.aspx?relid=138879. April 14, 2016.

PIB (2016b). *Pledge for Tiger Conservation*. Press Information Bureau (PIB) Government of India. Available at: http://pib.nic.in/newsite/PrintRelease.aspx?relid=138879.

Piel, A.K., Cohen, N., Kamenya, S., *et al.* (2015). Population status of chimpanzees in the Masito-Ugalla ecosystem, Tanzania. *American Journal of Primatology*, **77**, 1027–35. DOI: 10.1002/ajp.22438.

Piel, A.K., Lenoel, A., Johnson, C. and Stewart, F.A. (2015). Deterring poaching in western Tanzania: the presence of wildlife researchers. *Global Ecology and Conservation*, **3**, 188–99. DOI: https://doi.org/10.1016/j.gecco.2014.11.014.

Planet (n.d.). *Planet*. San Francisco, CA: Planet Labs Inc. Available at: https://www.planet.com. Accessed December, 2016–April, 2017.

Planet Survey and CED (2003). *Extractive Industries and Respect for the World Bank Operational Directives vis-à-vis Indigenous Peoples Case Study on the Implementation of the Chad-Cameroon Pipeline.* Planet Survey and Centre for Environment and Development (CED). Available at: http://www.forestpeoples.org/sites/fpp/files/publication/2010/08/eirinternatwshopcamerooncaseeng.pdf.

Plumptre, A.J., Davenport, T.R.B., Behangana, M., et al. (2007). The biodiversity of the Albertine Rift. *Biological Conservation*, **134**, 178–94. DOI: https://doi.org/10.1016/j.biocon.2006.08.021.

Plumptre, A., Hart, J.A., Hicks, T.C., et al. (2016a). Pan troglodytes *ssp.* schweinfurthii *(errata version published in 2016). The IUCN Red List of Threatened Species 2016: e.T15937A102329417.* Gland, Switzerland: International Union for Conservation of Nature (IUCN). Available at: http://dx.doi.org/10.2305/IUCN.UK.2016–2.RLTS.T15937A17990187.en.

Plumptre, A.J. and Johns, A.G. (2001). Changes in primate communities following logging disturbance. In *The Cutting Edge: Conserving Wildlife in Logged Tropical Forests*, ed. R. Fimbel, A. Grajal and J. G. Robinson. New York, NY: Columbia University Press, pp. 71–92.

Plumptre, A.J., Nixon, S., Critchlow, R., et al. (2015). *Status of Grauer's gorilla and chimpanzees in Eastern Democratic Republic of Congo: historical and current distribution and abundance.* Report to Arcus Foundation, USAID and US Fish and Wildlife Service.

Plumptre, A.J., Nixon, S., Kujirakwinja, D.K., et al. (2016b). Catastrophic decline of world's largest primate: 80% loss of Grauer's gorilla (*Gorilla beringei graueri*) population justifies critically endangered status. *PLoS One*, **11**, e0162697. DOI: 10.1371/journal.pone.0162697.

Plumptre, A.J., Reynolds, V. and Bakuneeta, C. (1997). *The effects of selective logging in monodominant tropical forest on biodiversity.* Report to Project R6057, Overseas Development Administration (ODA), London, UK.

Plumptre, A., Robbins, M. and Williamson, E.A. (2016). Gorilla beringei. *The IUCN Red List of Threatened Species 2016: e.T39994A102325702.* Gland, Switzerland: International Union for Conservation of Nature (IUCN). Available at: http://www.iucnredlist.org/details/39994/0. Accessed November 15, 2016.

Plumptre, A.J., Rose, R., Nangendo, G., et al. (2010). *Eastern Chimpanzee (*Pan troglodytes schweinfurthii*): Status Survey and Conservation Action Plan 2010–2020.* Gland, Switzerland: International Union for Conservation of Nature Species Survival Commission (IUCN SSC), Primate Specialist Group.

Poff, N.L., Allan, J.D., Bain, M.B., et al. (1997). The natural flow regime. *BioScience*, **47**, 769–84. DOI: 10.2307/1313099.

Pohlman, C.L., Turton, S.M. and Goosem, M. (2009). Temporal variation in microclimatic edge effects near powerlines, highways and streams in Australian tropical rainforest. *Agricultural and Forest Meteorology*, **149**, 84–95. DOI: https://doi.org/10.1016/j.agrformet.2008.07.003.

Porter-Bolland, L., Ellis, E.A., Guariguata, M.R., et al. (2012). Community managed forests and forest protected areas: an assessment of their conservation effectiveness across the tropics. *Forest Ecology and Management*, **268**, 6–17. DOI: https://doi.org/10.1016/j.foreco.2011.05.034.

Poulsen, J.R., Clark, C.J., Mavah, G. and Elkan, P.W. (2009). Bushmeat supply and consumption in a tropical logging concession in northern Congo. *Conservation Biology*, **23**, 1597–608. DOI: 10.1111/j.1523–1739.2009.01251.x.

Printes, R. (1999). The Lami Biological Reserve, Rio Grande do Sul, Brazil and the danger of power lines to howlers in urban reserves. *Neotropical Primates*, **4**, 135–6.

Printes, R.C., Buss, G., Jardim, M.M. de A., et al. (2010). The Urban Monkeys Program: a survey of *Alouatta clamitans* in the south of Porto Alegre and its influence on land use policy between 1997 and 2007. *Primate Conservation*, **25**, 11–9. DOI: 10.1896/052.025.0103.

PROFOR (2012). *Poverty-Forests Linkages Toolkit: Overview and National Level Engagement.* Washington DC: World Bank. Available at: https://openknowledge.worldbank.org/handle/10986/12618.

Profundo (2016). *Foreign land acquisitions in Cameroon: International linkages and financial flows.* Internal research paper for Forest Peoples Programme: Profundo Research and Advice.

Property Hunter (2016). *The Pan-Borneo Highway: Making a Strong Connection.* Property Hunter. Available at: https://www.propertyhunter.com.my/news/2016/08/2798/sabah/the-pan-borneo-highway-making-a-strong-connection.

Pruetz, J.D. and Bertolani, P. (2009). Chimpanzee (*Pan troglodytes verus*) behavioral responses to stresses associated with living in a savanna-mosaic environment: implications for hominin adaptations to open habitats. *PaleoAnthropology*, **2009**, 252–62.

Prüfer, K., Munch, K., Hellmann, I., *et al.* (2012). The bonobo genome compared with the chimpanzee and human genomes. *Nature*, **486**, 527. DOI: 10.1038/nature11128. Available at: https://www.nature.com/articles/nature11128#supplementary-information.

PT Hitay and UGM (2016). *Kajian Harmonisasi untuk Pengembangan Energi Terbaharukan di Taman Nasional Gunung Leuser*. Yogyakarta, Indonesia: PT Hitay Panas Energy and the Faculty of Forestry, Universitas Gadjah Mada (UGM).

Pusey, A.E., Pintea, L., Wilson, M.L., Kamenya, S. and Goodall, J. (2007). The contribution of long-term research at Gombe National Park to chimpanzee conservation. *Conservation Biology*, **21**, 623–34. DOI: 10.1111/j.1523-1739.2007.00704.x.

Quintero, J.D., Roca, R., Morgan, A.J., Mathur, A. and Xiaoxin, S. (2010). *Smart Green Infrastructure in Tiger Range Countries: A Multi-Level Approach*. Global Tiger Initiative, GTISGI Working Group, Technical Paper. Discussion Papers. Washington DC: World Bank.

Rabanal, L.I., Kuehl, H.S., Mundry, R., Robbins, M.M. and Boesch, C. (2010). Oil prospecting and its impact on large rainforest mammals in Loango National Park, Gabon. *Biological Conservation*, **143**, 1017–24. DOI: https://doi.org/10.1016/j.biocon.2010.01.017.

Radio Free Sarawak (2015). CM dealing with Blockade. *Radio Free Sarawak*, July, 2015. Available at: https://radiofreesarawak.org/2015/07/cm-dealing-with-baram-blockade/. Accessed November 4, 2016.

Radio Okapi (2013). *Province Orientale: La Route Kisangani-Buta-Dulia Rouverte*. Gombe, DRC: Radio Okapi. Available at: https://www.radiookapi.net/regions/province-orientale/2013/12/22/province-orientale-la-route-kisangani-buta-dulia-rouverte. Accessed December 22, 2013.

Rainer, H. (2014). Avoiding the chainsaws: industrial timber extraction and apes. In *State of the Apes: Extractive Industries and Ape Conservation*, ed. Arcus Foundation. Cambridge, UK: Cambridge University Press, pp. 101–26.

Rainer, H. and Lanjouw, A. (2015). Encroaching on ape habitat: deforestation and industrial agriculture in Cameroon, Liberia and on Borneo. In *State of the Apes: Industrial Agriculture and Ape Conservation*, ed. Arcus Foundation. Cambridge, UK: Cambridge University Press, pp. 41–69.

Rainforest Action Network (2014). *Last Place on Earth: Leuser Ecosystem*. San Francisco, CA: Rainforest Action Network. Available at: https://www.ran.org/lastplaceonearth.

Ram, C., Sharma, G. and Rajpurohit, L.S. (2015). Mortality and threats to hanuman langurs (*Semnopithecus entellus entellus*) in and around Jodhpur (Rajasthan). *The Indian Forester*, **10**, 1042–5.

Rambu Energy (2016). *President Widodo Inaugurates Lahendong Geothermal Power Plant Unit 5 and 6*. Jakarta, Indonesia: Rambu Energy. Available at: https://www.rambuenergy.com/2016/12/president-widodo-inaugurates-lahendong-geothermal-power-plant-unit-5-and-6/. Accessed November 27, 2017.

Rawson, B.M., Insua-Cao, P., Manh Ha, N., *et al.* (2011). *The Conservation Status of Gibbons in Vietnam*. Hanoi, Vietnam: Fauna and Flora International/Conservation International.

Ray, S. (2015). *Infrastructure Finance and Financial Sector Development*. ADBI Working Paper 522. Tokyo, Japan: Asian Development Bank Institute. Available at: https://www.adb.org/sites/default/files/publication/159842/adbi-wp522.pdf.

REDD+ (n.d.). *Reducing Emissions from Deforestation and Forest Degradation in Developing Nations*. New York, NY: United Nations (UN) Framework Convention on Climate Change. Available at: http://redd.unfccc.int/.

Reed, S.E. and Merenlender, A.M. (2008). Quiet, nonconsumptive recreation reduces protected area effectiveness. *Conservation Letters*, **1**, 146–54. DOI: 10.1111/j.1755-263X.2008.00019.x.

Refisch, J. and Koné, I. (2005). Impact of commercial hunting on monkey populations in the Tai region, Cote d'Ivoire. *Biotropica*, **37**, 136–44. DOI: 10.1111/j.1744-7429.2005.03174.x.

Refuge for Wildlife (n.d.). *Stop the Shocks*. Guanacaste Province, Costa Rica: Refuge for Wildlife. Available at: http://refugeforwildlife.com/stop-the-shocks/.

Reichard, U. (1995). Extra-pair copulations in a monogamous gibbon (*Hylobates lar*). *Ethology*, **100**, 99–112. DOI: 10.1111/j.1439-0310.1995.tb00319.x.

Reinartz, G., Ingmanson, E.J. and Vervaecke, H. (2013). *Pan paniscus gracile* chimpanzee (bonobo, pygmy chimpanzee). In *Mammals of Africa. Volume II: Primates*, ed. T. Butynski, J. Kingdon and J. Kalina. London, UK: Bloomsbury Publishing, pp. 64–9.

Republic of Cameroon (2001). *Loi No. 001 du 16 Avril 2001 Portant Code Minier.* Yaoundé, Cameroon: Republic of Cameroon.

Republic of Cameroon (2009a). *Cameroun Vision 2035: Document de Travail.* Yaoundé, Cameroon: Republic of Cameroon, Ministry of Economy, Planning and Regional Development. Available at: http://extwprlegs1.fao.org/docs/pdf/cmr145894.pdf.

Republic of Cameroon (2009b). *Growth and Employment Strategy Paper: Reference Framework for Government Action over the Period 2010–2020. August 2009.* Yaoundé, Cameroon: Republic of Cameroon. Available at: http://www.imf.org/external/pubs/ft/scr/2010/cr10257.pdf.

Republic of Cameroon (2012). *National Biodiversity Strategy and Action Plan Version II (NBSAB II): MINEPDED.* Yaoundé, Cameroon: Ministry of Environment, Protection of Nature and Sustainable Development. Available at: https://www.cbd.int/doc/world/cm/cm-nbsap-v2-en.pdf. December 2012.

Republic of Guinea (n.d.). *Mining-Infrastructure Synergies, Growth Corridors.* Conakry, Guinea: Ministry of Mines and Geology. Available at: http://mines.gov.gn/en/priorities/infrastructure/.

Republik Indonesia (2014). *Undang-Undang Republik Indonesia Nomor 21 Tahun 2014 Tentang Panas Bumi/Law of the Republic of Indonesia 21 Year 2014 About Geothermal.* Republik Indonesia. Available at: http://www.indolaw.org/UU/Law%20No.%2021%20of%202014%20on%20Geothermal.pdf.

Reynolds, V. (2005). *The Chimpanzees of the Budongo Forest: Ecology, Behaviour, and Conservation.* Oxford, UK: Oxford University Press.

Rhodes, J.R., Lunney, D., Callaghan, J. and McAlpine, C.A. (2014). A few large roads or many small ones? How to accommodate growth in vehicle numbers to minimise impacts on wildlife. *PLoS One,* **9**, e91093. DOI: 10.1371/journal.pone.0091093.

Richter, B.D., Postel, S., Revenga, C., *et al.* (2010). Lost in development's shadow: the downstream human consequences of dams. *Water Alternatives,* **3**, 14.

Riedler, B., Millesi, E. and Pratje, P.H. (2010). Adaptation to forest life during the reintroduction process of immature *Pongo abelii. International Journal of Primatology,* **31**, 647–63. DOI: 10.1007/s10764-010-9418-2.

Riesco, I.L. (2005). *After the Tsunami: The EC, the Environment and Rebuilding Indonesia.* European Community Forest Platform-FERN. Available at: http://www.fern.org/sites/fern.org/files/media/documents/document_865_934.pdf.

Rio Tinto Simfer S.A. (2012a). *Social and Environmental Impact Assessment, Simandou Project Mine Component.* Conakry, Republic of Guinea: Rio Tinto Simfer SA.

Rio Tinto Simfer S.A. (2012b). *Social and Environmental Impact Assessment, Simandou Project Rail Component.* Conakry, Republic of Guinea: Rio Tinto Simfer SA.

Ripple, W.J., Abernethy, K., Betts, M.G., *et al.* (2016). Bushmeat hunting and extinction risk to the world's mammals. *Royal Society Open Science,* **3**. DOI: 10.1098/rsos.160498.

Robbins, A.M., Stoinski, T., Fawcett, K. and Robbins, M.M. (2011). Lifetime reproductive success of female mountain gorillas. *American Journal of Physical Anthropology,* **146**, 582–93. DOI: 10.1002/ajpa.21605.

Robbins, M.M. (2011). Gorillas: diversity in ecology and behavior. In *Primates in Perspective,* 2nd edn, ed. C. J. Campbell, A. Fuentes, K. C. MacKinnon, S. Bearder and R. M. Stumpf. Oxford, UK: Oxford University Press, pp. 326–39.

Robbins, M.M. and Sawyer, S. (2007). Intergroup encounters in mountain gorillas of Bwindi Impenetrable National Park, Uganda. *Behaviour,* **144**, 1497–519. DOI: https://doi.org/10.1163/156853907782512146.

Roberts, M., Patel, J. and Minella, G. (2015). *Why Invest in Infrastructure?* Research Report. New York, NY: Deutsche Asset and Wealth Management.

Robertson, Y. (2002). *Briefing Document on Road Network through the Leuser Ecosystem.* Cambridge University: International Union for Conservation of Nature Species Survival Commission (IUCN SSC).

Robinson, J.G., Redford, K.H. and Bennett, E.L. (1999). Wildlife harvest in logged tropical forests. *Science,* **284**, 595–6. DOI: 10.1126/science.284.5414.595.

Robson, S.L. and Wood, B. (2008). Hominin life history: reconstruction and evolution. *Journal of Anatomy,* **212**, 394–425. DOI: 10.1111/j.1469-7580.2008.00867.x.

Rodrigues, N.N. and Martinez, R.A. (2014). Wildlife in our backyard: interactions between Wied's marmoset *Callithrix kuhlii* (Primates: Callithrichidae) and residents of Ilhéus, Bahia, Brazil. *Wildlife Biology*, **20**, 91–6. DOI: 10.2981/wlb.13057.

Rogala, J., Hebblewhite, M., Whittington, J., *et al.* (2011). Human activity differentially redistributes large mammals in the Canadian Rockies National Parks. *Ecology and Society*, **16**, 16.

Rogers, M.E., Abernethy, K., Bermejo, M., *et al.* (2004). Western gorilla diet: a synthesis from six sites. *American Journal of Primatology*, **64**, 173–92. DOI: 10.1002/ajp.20071.

Rondinini, C., Wilson, K.A., Boitani, L., Grantham, H. and Possingham, H.P. (2006). Tradeoffs of different types of species occurrence data for use in systematic conservation planning. *Ecology Letters*, **9**, 1136–45. DOI: 10.1111/j.1461-0248.2006.00970.x.

Ross, S.R., Lukas, K.E., Lonsdorf, E.V., *et al.* (2008). Inappropriate use and portrayal of chimpanzees. *Science*, **319**, 1487. DOI: 10.1126/science.1154490.

Roy, J., Vigilant, L., Gray, M., *et al.* (2014). Challenges in the use of genetic mark-recapture to estimate the population size of Bwindi mountain gorillas (*Gorilla beringei beringei*). *Biological Conservation*, **180**, 249–61. DOI: https://doi.org/10.1016/j.biocon.2014.10.011.

RRI (2016). *Closing the Gap: Strategies and Scale Needed to Secure Rights and Save Forests*. Rights and Resources Initiative (RRI). Available at: http://rightsandresources.org/en/publication/closing-the-gap/#.WQD45FPyuuU.

RRI (2017). *From Risk and Conflict to Peace and Prosperity: The Urgency of Securing Community Land Rights in a Turbulent World*. Rights and Resources Initiative (RRI). Available at: http://rightsandresources.org/en/publication/risk-conflict-to-peace-prosperity/#.WQD4G1PyuuU.

Ruckelshaus, M., McKenzie, E., Tallis, H., *et al.* (2015). Notes from the field: lessons learned from using ecosystem service approaches to inform real-world decisions. *Ecological Economics*, **115**, 11–21. DOI: https://doi.org/10.1016/j.ecolecon.2013.07.009.

Rudel, T.K., Defries, R., Asner, G.P. and Laurance, W.F. (2009). Changing drivers of tropical deforestation create new challenges and opportunities for conservation. *Conservation Biology*, **23**, 1396–405. DOI: 10.1111/j.1523-1739.2009.01332.x.

Russon, A. (2009). Orangutan rehabilitation and reintroduction: successes, failures and role in conservation. In *Orangutans: Geographic Variation in Behavioral Ecology and Conservation*, ed. S. A. Wich, S. S. Utami Atmoko, T. Mitra Setia and C. P. Van Schaik. Oxford, UK: Oxford University Press, pp. 327–50.

Russon, A.E., Smith, J.J. and Adams, L. (2016). Managing human-orangutan relationships in rehabilitation. In *Ethnoprimatology: Primate Conservation in the 21st Century*, ed. M. Waller: Springer, pp. 233–58.

Russon, A.E., Wich, S.A., Ancrenaz, M., *et al.* (2009). Geographic variation in orangutan diets. In *Orangutans: Geographic Variation in Behavioral Ecology and Conservation*, ed. S. A. Wich, S. Utami Atmoko, T. Mitra Setia and C. P. Van Schaik. Oxford, UK: Oxford University Press, pp. 135–56.

Sanusi, L. (2012). *Nigeria's economic development aspirations and the leadership question: is there a nexus?* Speech by Mr Sanusi Lamido Sanusi, Governor of the Central Bank of Nigeria. Presented at: 2nd General Dr Yakubu Gowon Distinguished Annual Lecture, October 19, 2012, Lagos, Nigeria. Available at: http://www.bis.org/review/r121105f.pdf.

Sarawak Report (2014). Bakun turbines running at just 50% capacity. *Sarawak Report*, January, 2014. Available at: http://www.sarawakreport.org/2014/01/bakun-turbines-running-at-just-50-capacity-exclusive/. Accessed April, 6 2017.

Sarawak Report (2016). Plantaion boss wanted over Bill Kayong murder. *Sarawak Report*, July 16, 2016. Available at: http://www.sarawakreport.org/2016/07/plantation-boss-wanted-over-bill-kayong-murder-world-exclusive/.

Satriastanti, F.E. (2016). Indonesian environment ministry shoots down geothermal plan in Mount Leuser national park. *Mongabay*, September, 2016. Available at: https://news.mongabay.com/2016/09/indonesian-environment-ministry-shoots-down-geothermal-plan-in-mount-leuser-national-park/. Accessed September 12, 2016.

Schaumburg, F., Mugisha, L., Peck, B., *et al.* (2012). Drug-resistant human *Staphylococcus aureus* in sanctuary apes pose a threat to endangered wild ape populations. *American Journal of Primatology*, **74**, 1071–5. DOI: 10.1002/ajp.22067.

Scudder, T. (2005). *The Future of Large Dams: Dealing with Social, Environmental, Institutional and Political Costs*. London, UK: Earthscan Publications.

SEDIA (2008). *Sabah Development Corridor Blueprint*. Kota Kinabalu, Malaysia: Sabah Economic Development and Investment Authority (SEDIA). Available at: http://www.sedia.com.my/SDC_Blueprint.html.

Seiler, N., Boesch, C., Mundry, R., Stephens, C. and Robbins, M.M. (2017). Space partitioning in wild, non-territorial mountain gorillas: the impact of food and neighbours. *Royal Society Open Science*, **4**. DOI: 10.1098/rsos.170720.

Seiler, N. and Robbins, M.M. (2016). Factors influencing ranging on community land and crop raiding by mountain gorillas. *Animal Conservation*, **19**, 176–88. DOI: 10.1111/acv.12232.

Seto, K.C., Güneralp, B. and Hutyra, L.R. (2012). Global forecasts of urban expansion to 2030 and direct impacts on biodiversity and carbon pools. *Proceedings of the National Academy of Sciences*, **109**, 16083–8. DOI: 10.1073/pnas.1211658109.

Seymour, F., La Vina, T. and Hite, K. (2014). *Evidence Linking Community-Level Tenure and Forest Condition: An Annotated Bibliography*. San Francisco, CA: Climate and Land Use Alliance.

Sharp, J. (2016). *Mobile Zoo Closes Following USDA Judge's Order*. Available at: http://www.al.com/news/mobile/index.ssf/2016/11/mobile_zoo_closes_following_us.html. Accessed November 17, 2016.

Shearman, P., Bryan, J. and Laurance, W.F. (2012). Are we approaching 'peak timber' in the tropics? *Biological Conservation*, **151**, 17–21. DOI: https://doi.org/10.1016/j.biocon.2011.10.036.

Shepherd, C. and Nijman, V. (2008). *The Wild Cat Trade in Myanmar*. Selangor, Malaysia: TRAFFIC Southeast Asia. Available at: http://www.traffic.org/publications/the-wild-cat-trade-in-myanmar.html.

Sherman, J., Brent, L. and Farmer, K. (2016). *A picture is worth a thousand words: an analysis of animal images posted on the internet by African ape sanctuaries*. Poster presentation. Presented at: International Primatological Society, 26th Congress, August 23, 2016, Chicago, IL. International Primatological Society.

Shipping Position Online (2016). *The Deep Seaport Craze in Nigeria*. Available at: http://shippingposition.com.ng/editorial/the-deep-seaport-craze-in-nigeria. Accessed November 14, 2016.

Shirley, R. and Kammen, D. (2015). Energy planning and development in Malaysian Borneo: assessing the benefits of distributed technologies versus large scale energy mega-projects. *Energy Strategy Reviews*, **8**, 15–29. DOI: https://doi.org/10.1016/j.esr.2015.07.001.

Shirley, R., Kammen, D. and Wynn, G. (2014). *Kampung capacity: analyzing local energy solutions in the Baram River basin, east Malaysia*. Unpublished paper.

Si, Y. (1998). The world's most catastrophic dam failures: the August 1975 collapse of the Banqiao and Shimantan dams. In *Dai Qing, The River Dragon Has Come!*, ed. M. E. Sharpe. New York, NY: San José State University, pp. 25–38. Available at: http://www.sjsu.edu/faculty/watkins/aug1975.htm.

Silveira, L., Sollmann, R., Jácomo, A.T.A., Diniz Filho, J.A.F. and Tôrres, N.M. (2014). The potential for large-scale wildlife corridors between protected areas in Brazil using the jaguar as a model species. *Landscape Ecology*, **29**, 1213–23. DOI: 10.1007/s10980-014-0057-4.

Simpson, A. (2014). *Energy, Governance and Security in Thailand and Myanmar (Burma): A Critical Approach to Environmental Politics in the South*. Farnham, UK: Ashgate Publishing Ltd.

Simpson, A. (2015). Starting from year zero: environmental governance in Myanmar. In *Environmental Challenges and Governance: Diverse Perspectives from Asia*, ed. S. Mukherjee and D. Chakraborty. London, UK: Routledge, pp. 152–65. Available at: https://www.researchgate.net/publication/275973353_Starting_from_year_zero_Environmental_governance_in_Myanmar.

Singleton, I., Knott, C.D., Morrogh-Bernard, H.C., Wich, S.A. and Van Schaik, C.P. (2009). Ranging behavior of orangutan females and social organization. In *Orangutans: Geographic Variation in Behavioral Ecology and Conservation*, ed. S. A. Wich, S. Utami Atmoko, T. Mitra Setia and C. P. Van Schaik. Oxford, UK: Oxford University Press, pp. 205–13.

Singleton, I., Wich, S., Husson, S., *et al.* (2004). *Orangutan Population and Habitat Viability Assessment: Final Report*. Apple Valley, MN: International Union for Conservation of Nature Species Survival Commission (IUCN SSC), Conservation Breeding Specialist Group.

Singleton, I., Wich, S.A., Nowak, M. and Usher, G. (2016). Pongo abelii. *The IUCN Red List of Threatened Species 2016: e.T39780A102329901*. Gland, Switzerland: International Union for Conservation of Nature (IUCN). Available at: http://www.iucnredlist.org/details/121097935/0. Accessed August 14, 2017.

Skinner, J. and Haas, L.J. (2014). *Watered Down? A Review of Social and Environmental Safeguards for Large Dam Projects*. Natural Resource Issues No. 28. London, UK: International Institute for Environment and Development (IIED).

Slade, A. (2016). *Survivorship, demographics and seasonal trends among electrocuted primate species in Diani, Kenya*. Masters thesis. Bristol, UK: University of Bristol.

Slade, A. and Cunneyworth, P. (2017). *Electrocution trends in six sympatric primates in the suburban environment of Diani, Kenya*. Unpublished report. Diani, Kenya: Colobus Conservation.

Sloan, S., Bertzky, B. and Laurance, W.F. (2017). African development corridors intersect key protected areas. *African Journal of Ecology*, **55**, 731–7. DOI: 10.1111/aje.12377.

Smith, J., Obidzinski, K., Subarudi, S. and Suramenggala, I. (2003). Illegal logging, collusive corruption and fragmented governments in Kalimantan, Indonesia. *International Forestry Review*, **5**, 293–302.

Smith, R.J., Biggs, D., St. John, F.A.V., 't Sas-Rolfes, M. and Barrington, R. (2015). Elephant conservation and corruption beyond the ivory trade. *Conservation Biology*, **29**, 953–6. DOI: 10.1111/cobi.12488.

Smith, T. (2013). Cameroon's rich biodiversity is under threat. *UCLA Today*. Available at: http://newsroom.ucla.edu/stories/preserving-camaroon-s-treasures-248074. Accessed August 28, 2013.

Smithsonian Institution (n.d.). *Human Origins*. Washington DC: Smithsonian Institution. Available at: http://humanorigins.si.edu/evidence/genetics.

Sop, T., Cheyne, S.M., Maisels, F.G., Wich, S.A. and Williamson, E.A. (2015). Abundance annex: ape population abundance estimates. In *State of the Apes 2015: Industrial Agriculture and Ape Conservation*, ed. Arcus Foundation. Cambridge, UK: Cambridge University Press, online. Available at: http://www.stateoftheapes.com/volume-2-industrial-agriculture/.

Sovacool, B.K. and Bulan, L.C. (2011). Behind an ambitious megaproject in Asia: the history and implications of the Bakun hydroelectric dam in Borneo. *Energy Policy*, **39**, 4842–59. DOI: https://doi.org/10.1016/j.enpol.2011.06.035.

Species360 (2016). *Species-holding reports (Gorilla, Hylobatidae, Pan, Pongo)*. Unpublished data. Species360.

Spignesi, S.J. (2004). *Catastrophe! The 100 Greatest Disasters of All Time*. New York, NY: Citadel Press.

Spira, C., Kirkby, A., Kujirakwinja, D. and Plumptre, A.J. (2017). The socio-economics of artisanal mining and bushmeat hunting around protected areas: Kahuzi–Biega National Park and Itombwe Nature Reserve, eastern Democratic Republic of Congo. *Oryx*, 1–9. DOI: 10.1017/S003060531600171X.

Spittaels, S. and Hilgert, F. (2010). *Mapping Conflict Motives: Province Orientale (DRC)*. Number 22. Antwerp, Belgium: International Peace Information Service.

Stanley, E.H. and Doyle, M.W. (2003). Trading off: the ecological effects of dam removal. *Frontiers in Ecology and the Environment*, **1**, 15–22. DOI: 10.1890/1540-9295(2003)001[0015:TOTEEO]2.0.CO;2.

Steinmetz, R., Srirattanaporn, S., Mor-Tip, J. and Seuaturien, N. (2014). Can community outreach alleviate poaching pressure and recover wildlife in south-east Asian protected areas? *Journal of Applied Ecology*, **51**, 1469–78. DOI: 10.1111/1365-2664.12239.

Stewart, K.J. (1988). Suckling and lactational anoestrus in wild gorillas (*Gorilla gorilla*). *Journal of Reproduction and Fertility*, **83**, 627–34.

Stokes, E.J., Strindberg, S., Bakabana, P.C., *et al.* (2010). Monitoring great ape and elephant abundance at large spatial scales: measuring effectiveness of a conservation landscape. *PLoS One*, **5**, e10294. DOI: 10.1371/journal.pone.0010294.

Stokstad, E. (2017). New great ape species found, sparking fears for its survival. *Science*. DOI: 10.1126/science.aar3900.

Straumann, L. (2014). *Money Logging: on the Trail of the Asian Timber Mafia*. Basel, Switzerland: Schwabe AG.

Struebig, M.J., Wilting, A., Gaveau, D.L.A., *et al.* (2015). Targeted conservation to safeguard a biodiversity hotspot from climate and land-cover change. *Current Biology*, **25**, 372–8. DOI: https://doi.org/10.1016/j.cub.2014.11.067.

Struhsaker, T.T. (1999). Primate communities in Africa: the consequences of long-term evolution or the artifact of recent hunting. In *Primate Communities*, ed. J. G. Fleagle, C. H. Janson and K. E. Reed. New York, NY: Cambridge University Press, pp. 289–94.

Sulistyawan, B.S., Eichelberger, B.A., Verweij, P., *et al.* (2017). Connecting the fragmented habitat of endangered mammals in the landscape of Riau–Jambi–Sumatera Barat (RIMBA), central Sumatra, Indonesia (connecting the fragmented habitat due to road development). *Global Ecology and Conservation*, **9**, 116–30. DOI: https://doi.org/10.1016/j.gecco.2016.12.003.

Sundance (2016). *Quarterly Activity Report.* Perth, Australia: Sundance Resources Ltd.

Sunderland-Groves, J.L., Slayback, D.A., Bessike Balinga, M.P. and Sunderland, T.C.H. (2011). Impacts of co-management on western chimpanzee (*Pan troglodytes verus*) habitat and conservation in Nialama Classified Forest, Republic of Guinea: a satellite perspective. *Biodiversity and Conservation*, **20**, 2745. DOI: 10.1007/s10531-011-0102-4.

Sunderlin, W.D., Dewi, S. and Puntodewo, A. (2007). *Poverty and Forests: Multi-Country Analysis of Spatial Association and Proposed Policy Solutions.* CIFOR Occasional Paper No. 47. Bogor, Indonesia: Center for International Forestry Research (CIFOR). Available at: http://www.cifor.org/publications/pdf_files/OccPapers/OP-47.pdf.

SWD (2011). *Orangutan Action Plan 2012–2016.* Kota Kinabalu, Malaysia: Sabah Wildlife Department (SWD). Available at: http://static1.1.sqspcdn.com/static/f/1200343/24377541/1392391814550/Sabah_Orangutan_Action_Plan_2012–2016.pdf?token=LgJ5zbZ2eM5CMaiazY5GHhAGbcg%3D.

Tabuchi, H. (2016). How big banks are putting rain forests in peril. *New York Times*, December 3, 2016. Available at: http://www.nytimes.com/2016/12/03/business/energy-environment/how-big-banks-are-putting-rain-forests-in-peril.html?ref=business. Accessed December 4, 2016.

Tagg, N., Willie, J., Duarte, J., Petre, C.A. and Fa, J.E. (2015). Conservation research presence protects: a case study of great ape abundance in the Dja region, Cameroon. *Animal Conservation*, **18**, 489–98. DOI: 10.1111/acv.12212.

TANAPA, TAWIRI, WD-MNRT, *et al.* (2015). *Gombe-Mahale Ecosystem Conservation Action Planning, v 2.0.* Tanzania National Parks (TANAPA), Tanzania Wildlife Research Institute (TAWIRI), Wildlife Division–Ministry of Natural Resources and Tourism (WD–MNRT), *et al.* Available at: http://static1.1.sqspcdn.com/static/f/1200343/26920244/1458214306290/TANAPA_et_al_2015_GME_CAP_2.0.pdf?token=34OmlSmt4LtbEajrZFdmoJMxZto%3D.

Tang, D. and Kelly, A.S. (2016). *Design Manual: Building a Sustainable Road to Dawei: Enhancing Ecosystem Services and Wildlife Connectivity.* Yangon, Myanmar: World Wide Fund for Nature (WWF).

Tasch, B. (2015). The 23 poorest countries in the world. *Business Insider UK*, July, 2015. Available at: http://uk.businessinsider.com/the-23-poorest-countries-in-the-world-2015-7.

Tata, H.L., van Noordwijk, M., Ruysschaert, D., *et al.* (2014). Will funding to Reduce Emissions from Deforestation and (forest) Degradation (REDD+) stop conversion of peat swamps to oil palm in orangutan habitat in Tripa in Aceh, Indonesia? *Mitigation and Adaptation Strategies for Global Change*, **19**, 693–713. DOI: 10.1007/s11027-013-9524-5.

TAWIRI (2017). *Tanzania national chimpanzee management plan.* Unpublished draft. Arusha, Tanzania: Tanzania Wildlife Research Institute (TAWIRI), Ministry of Natural Resources and Tourism.

TBC (2016). *Government Policies on Biodiversity Offsets.* Cambridge, UK: The Biodiversity Consultancy (TBC).

TBC (n.d.). *World Bank ESS6.* Cambridge, UK: The Biodiversity Consultancy (TBC). Available at: http://www.thebiodiversityconsultancy.com/approaches/world-bank-ess6/.

TBC and CSBI (2015). *A Cross-Sector Guide to Implementing the Mitigation Hierarchy.* Cambridge, UK: The Biodiversity Consultancy (TBC) and Cross-Sector Biodiversity Initiative (CSBI).

Teleki, G. (2001). Sanctuaries for ape refugees. In *Great Apes and Humans: The Ethics of Coexistence*, ed. B. Beck, T. Stoinski, M. Hutchins, *et al.* Washington DC: Smithsonian Institution, pp. 133–49.

Tello, I.Z. (2016). En una decisión judicial inédita, la mona cecilia será trasladada de Mendoza a Brasil. *Los Andes*. Available at: http://www.losandes.com.ar/article/tras-una-decision-judicial-inedita-la-mona-cecilia-sera-trasladada-a-brasil?rv=4. Accessed November 6, 2016.

Tempo (2017). *Jokowi Confident in Realization of Renewable Energy Projects.* Jakarta, Indonesia: Tempo. Available at: https://en.tempo.co/read/news/2015/08/22/055694138/Jokowi-Confident-in-Realization-of-Renewable-Energy-Projects. Accessed January 22, 2017.

ten Kate, K. and Crowe, M.L.A. (2014). *Biodiversity Offsets: Policy Options for Governments.* Gland, Switzerland: International Union for Conservation of Nature (IUCN).

Thant, H. (2016). New environmental impact rules released. *The Myanmar Times*.

The Economist (2014). Nigeria: Africa's new number one. *The Economist*, April 12, 2014. Available at: https://www.economist.com/news/finance-and-economics/21600734-revised-figures-show-nigeria-africas-largest-economy-step-change.

The Guardian (n.d.). China GDP: how it has changed since 1980. *The Guardian*. Online data spreadsheet. Available at: https://www.theguardian.com/news/datablog/2012/mar/23/china-gdp-since-1980#data.

Then, S. (2016). New Murum Dam to boost energy output in Sarawak. *The Star Online*. Available at: http://www.thestar.com.my/news/nation/2016/09/27/new-murum-dam-to-boost-energy-output-in-sarawak/. Accessed September 14, 2017.

This Day (2016). Another Chinese firm expresses interest in Bakassi Deep Seaport, superhighway. *This Day*. Available at: https://www.thisdaylive.com/index.php/2016/10/17/another-chinese-firm-expresses-interest-in-bakassi-deep-seaport-superhighway/. Accessed October 17, 2016.

Thompson, M.E. (2013). Reproductive ecology of female chimpanzees. *American Journal of Primatology*, **75**, 222–37. DOI: 10.1002/ajp.22084.

Thouless, C.R., Dublin, H.T., Blanc, J.J., *et al.* (2016). *African Elephant Status Report 2016: An Update from the African Elephant Database*. Occasional Paper Series of the IUCN Species Survival Commission, No. 60. Gland, Switzerland: International Union for Conservation of Nature Species Survival Commission (IUCN SSC), African Elephant Specialist Group. Available at: https://www.iucn.org/ssc-groups/mammals/african-elephant-specialist-group.

Tilman, D., Fargione, J., Wolff, B., *et al.* (2001). Forecasting agriculturally driven environmental change. *Science*, **292**, 281–4. DOI: 10.1126/science.1057544.

Tilt, B., Braun, Y. and He, D. (2009). Social impacts of large dam projects: a comparison of international case studies and implications for best practice. *Journal of Environmental Management*, **90**, S249-S57. DOI: https://doi.org/10.1016/j.jenvman.2008.07.030.

TNC, WWF and UoM (2016). *Improving Hydropower Outcomes through System-Scale Planning: An Example from Myanmar*. United Kingdom, Department for International Development (DFID). The Nature Conservancy (TNC), World Wide Fund for Nature (WWF) and the University of Manchester (UoM).

Torres, J., Brito, J.C., Vasconcelos, M.J., *et al.* (2010). Ensemble models of habitat suitability relate chimpanzee (*Pan troglodytes*) conservation to forest and landscape dynamics in western Africa. *Biological Conservation*, **143**, 416–25. DOI: https://doi.org/10.1016/j.biocon.2009.11.007.

Tracewski, Ł., Butchart, S.H.M., Di Marco, M., *et al.* (2016). Toward quantification of the impact of 21st-century deforestation on the extinction risk of terrestrial vertebrates. *Conservation Biology*, **30**, 1070–9. DOI: 10.1111/cobi.12715.

TRAFFIC (2008). *What's Driving the Wildlife Trade? A Review of Expert Opinion on Economic and Social Drivers of the Wildlife Trade and Trade Control Efforts in Cambodia, Indonesia, Lao PDR and Vietnam. East Asia and Pacific Region Sustainable Development Discussion Papers*. Washington DC: East Asia and Pacific Region Sustainable Development Department, World Bank. Available at: http://www.trafficj.org/publication/08_what%27s_driving_the_wildlife_trade.pdf.

TRAFFIC (2014). *Myanmar: A Gateway For Illegal Trade in Tigers and Other Wild Cats to China*. Cambridge, UK: TRAFFIC.

Tranquilli, S., Abedi-Lartey, M., Abernethy, K., *et al.* (2014). Protected areas in tropical Africa: assessing threats and conservation activities. *PLoS One*, **9**, e114154. DOI: 10.1371/journal.pone.0114154.

Tranquilli, S., Abedi-Lartey, M., Amsini, F., *et al.* (2012). Lack of conservation effort rapidly increases African great ape extinction risk. *Conservation Letters*, **5**, 48–55. DOI: 10.1111/j.1755-263X.2011.00211.x.

Transparent World (2015). *Tree Plantations*. Global Forest Watch. Available at: http://data.globalforestwatch.org/datasets/baae47df61ed4a73a6f54f00cb4207e0_5?uiTab=metadata. Accessed December, 2016.

Trayford, H.R. and Farmer, K.H. (2013). Putting the spotlight on internally displaced animals (IDAs): a survey of primate sanctuaries in Africa, Asia, and the Americas. *American Journal of Primatology*, **75**, 116–34. DOI: 10.1002/ajp.22090.

Tribal Energy and Environmental Information (n.d.). *Geothermal Energy: Construction Impacts.* Office of Indian Energy and Economic Development. Available at: https://teeic.indianaffairs.gov/er/geothermal/impact/construct/index.htm. Accessed March 9, 2017.

Trombulak, S.C. and Frissell, C.A. (2000). Review of ecological effects of roads on terrestrial and aquatic communities. *Conservation Biology*, **14**, 18–30. DOI: 10.1046/j.1523–1739.2000.99084.x.

Tropek, R., Sedláček, O., Beck, J., *et al.* (2014). Comment on 'High-resolution global maps of 21st-century forest cover change'. *Science*, **344**, 981. DOI: 10.1126/science.1248753.

Tschantz, B. (2014). What we know (and don't know) about low-head dams. *Journal of Dam Safety*, **12**, 37–45.

Tsunokawa, K. and Hoban, C. (1997). *Roads and the Environment: A Handbook.* Technical Paper No. 376, Chapter 10. Washington DC: The World Bank.

Tucker, S. (2011). Integration by education: a study of Cameroon's Bakola-Bagyeli. *Journal of Politics and Society*, **21**, 89–116.

Turvey, S.T., Traylor-Holzer, K., Wong, M.H., *et al.* (2015). *International Conservation Planning Workshop for the Hainan Gibbon: Final Report.* London, UK: Zoological Society of London/IUCN SSC Conservation Breeding Specialist Group. Available at: http://www.gibbons.asia/wp-content/uploads/2017/03/Hainan-Gibbon-Action-Plan-2016–2020.pdf.

Tweh, C.G., Lormie, M.M., Kouakou, C.Y., *et al.* (2015). Conservation status of chimpanzees *Pan troglodytes verus* and other large mammals in Liberia: a nationwide survey. *Oryx*, **49**, 710–8. DOI: 10.1017/S0030605313001191.

Tyson, L., Draper, C. and Turner, D. (2016). *The Use of Wild Animals in Performance 2016.* Horsham, UK: Born Free Foundation. Available at: http://www.bornfree.org.uk/fileadmin/user_upload/files/zoo_check/publications/PERFORMING_ANIMALS_REPORT_2016.pdf.

UN Population Division (2015). *World Population Prospects: The 2015 Revision, Key Findings and Advance Tables.* Working Paper No. ESA/P/WP.241. New York, NY: United Nations (UN), Department of Economic and Social Affairs, Population Division.

UN Population Division (2017). *World Population Prospects. Volume 1. 2017 Revision.* New York, NY: United Nations (UN), Department of Economic and Social Affairs, Population Division. Available at: https://esa.un.org/unpd/wpp/Publications/Files/WPP2017_Volume-I_Comprehensive-Tables.pdf.

UN Population Division (n.d.). *World Population Prospects 2017: Fertility Indicators.* New York, NY: United Nations (UN), Department of Economic and Social Affairs, Population Division. Available at: https://esa.un.org/unpd/wpp/Download/Standard/Fertility. Accessed October, 2017.

UNEP/CMS (2009). *Mountain Gorilla* Gorilla beringei beringei *Gorilla Agreement Action Plan. Revised Version of UNEP/CMS/GOR-MOP1/Doc.7d.* Bonn, Germany: Convention on Migratory Species (UNEP/CMS).

UNEP-WCMC and IUCN (n.d.). *Protected Planet: World Database on Protected Areas.* Cambridge, UK: United Nations Environment Programme (UNEP), World Conservation Monitoring Centre (WCMC) and International Union for Conservation of Nature (IUCN). Available at: www.protectedplanet.net. Accessed October, 2016.

UNEP–WCMC and IUCN (2017). *Protected Planet: World Database on Protected Areas.* United Nations Environment Programme (UNEP), World Conservation Monitoring Centre (WCMC) and International Union for Conservation of Nature (IUCN). Available at: www.protectedplanet.net/c/world-database-on-protected-areas. Accessed August, 2017.

UNESCO (n.d.-a). *Decision: 31 COM 7A.5: Kahuzi-Biega National Park (Democratic Republic of the Congo) (N 137).* Paris, France: United Nations Educational, Scientific and Cultural Organization (UNESCO) World Heritage Convention (WHC). Available at: http://whc.unesco.org/en/decisions/1268.

UNESCO (n.d.-b). *Directory of the World Network of Biosphere Reserves (WNBR).* Paris, France: United Nations Educational, Scientific and Cultural Organization (UNESCO). Available at: www.unesco.org/new/en/natural-sciences/environment/ecological-sciences/biosphere-reserves/world-network-wnbr/wnbr/.

UNESCO WHC (2016). *Operational Guidelines for the Implementation of the World Heritage Convention.* Paris, France: United Nations Educational, Scientific and Cultural Organization (UNESCO) World Heritage Convention (WHC). Available at: http://whc.unesco.org/document/156250.

UNESCO WHC (2017). *Tropical Rainforest Heritage of Sumatra.* Paris, France: United Nations Educational, Scientific and Cultural Organization (UNESCO) World Heritage Convention (WHC). Available at: http://whc.unesco.org/en/list/1167. Accessed September 12, 2017.

UNESCO WHC (n.d.). *Bwindi Impenetrable National Park.* Paris, France: United Nations Educational, Scientific and Cultural Organization (UNESCO) World Heritage Convention (WHC). Available at: http://whc.unesco.org/en/list/682.

University of Cambridge (2012). *Capturing the Benefits of Ecosystem Services to Guide Decision-making in the Greater Virungas Landscape of the Albertine Rift Region. Policy Workshop Report.* Cambridge, UK: University of Cambridge.

Unwin, S., Cress, D., Colin, C., Bailey, W. and Boardman, W. (2009). *Primate Veterinary Health Manual*, 2nd edn. Portland, OR: Pan African Sanctuary Alliance (PASA).

Unwin, S., Robinson, I.A.N., Schmidt, V., *et al.* (2012). Does confirmed pathogen transfer between sanctuary workers and great apes mean that reintroduction should not occur? *American Journal of Primatology*, **74**, 1076–83. DOI: 10.1002/ajp.22069.

U-PCLG (2015). *Assessing Options for the Proposed Improvement of the Ikumba to Ruhija Road, U-PCLG Position Paper.* Uganda Poverty and Conservation Learning Group (U-PCLG). Available at: http://igcp.org/wp-content/uploads/U-PCLG-position-on-Ruhija-Road-Mar12_2015.pdf.

USFWS (2015). *US Fish and Wildlife Service Finalizes Rule Listing All Chimpanzees as Endangered Under the Endangered Species Act.* US Fish and Wildlife Service (USFWS). Available at: http://www.fws.gov/news/ShowNews.cfm?ID=E81DA137-BAF2-9619-3492A2972E9854D9. Accessed June 19, 2017.

USGS (n.d.). *Global 30 Arc-Second Elevation (GTOPO30).* Reston, VA: US Geological Survey (USGS). Available at: https://lta.cr.usgs.gov/GTOPO30. Accessed January, 2018.

Uwaegbulam, C. (2016). Stakeholders approve $12m UN-REDD plus strategy for Nigeria. *The Guardian*, September 5, 2016. Available at: https://guardian.ng/property/stakeholders-approve-12m-un-redd-plus-strategy-for-nigeria.

van Beukering, P.J.H., Cesar, H.S.J. and Janssen, M.A. (2003). Economic valuation of the Leuser National Park on Sumatra, Indonesia. *Ecological Economics*, **44**, 43–62. DOI: https://doi.org/10.1016/S0921-8009(02)00224-0.

van den Berg, J. and Biesbrouck, K. (2000). *The Social Dimension of Rainforest Management in Cameroon: Issues for Co-Management. Tropenbos-Cameroon Series 4.* Kribi, Cameroon: The Tropenbos-Cameroon Programme.

Van Der Hoeven, C.A., De Boer, W.F. and Prins, H.H.T. (2010). Roadside conditions as predictor for wildlife crossing probability in a central African rainforest. *African Journal of Ecology*, **48**, 368–77. DOI: 10.1111/j.1365-2028.2009.01122.x.

van der Ree, R., Smith, D.J. and Grilo, C., ed. (2015). *Handbook of Road Ecology.* Oxford, UK: John Wiley & Sons.

Van Gils, H. and Kayijamahe, E. (2010). Sharing natural resources: mountain gorillas and people in the Parc National des Volcans, Rwanda. *African Journal of Ecology*, **48**, 621–7. DOI: 10.1111/j.1365-2028.2009.01154.x.

van Noordwijk, M.A., Sauren, S.E.B., Nuzuar, *et al.* (2009). Development of independence: Sumatran and Bornean orangutans compared. In *Orangutans: Geographic Variation in Behavioral Ecology and Conservation*, ed. S. A. Wich, S. Utami Atmoko, T. Mitra Setia and C. P. Van Schaik. Oxford, UK: Oxford University Press, pp. 189–203.

van Noordwijk, M.A., Willems, E.P., Utami Atmoko, S.S., Kuzawa, C.W. and Van Schaik, C.P. (2013). Multi-year lactation and its consequences in Bornean orangutans (*Pongo pygmaeus wurmbii*). *Behavioral Ecology and Sociobiology*, **67**, 805–14. DOI: 10.1007/s00265-013-1504-y.

Van Schaik, C.P., Monk, K.A. and Robertson, J.M.Y. (2001). Dramatic decline in orang-utan numbers in the Leuser ecosystem, northern Sumatra. *Oryx*, **35**, 14–25. DOI: 10.1046/j.1365-3008.2001.00150.x.

Vancutsem, C. and Achard, F. (2016). *Mapping intact and degraded humid forests over the tropical belt from 32 years of Landsat time series.* Presented at: Living Planet Symposium, 9–13 May, Prague, Czech Republic. Available at: http://lps16.esa.int/files/Contribution2034.pdf.

Vanguard (2015). Imperatives of dredging Calabar Port. *Vanguard News*, August 13, 2015. Available at: https://www.vanguardngr.com/2015/08/imperatives-of-dredging-calabar-port/.

Vanguard (2017). Superhighway: C-River gives FG two weeks ultimatum on EIA. *Vanguard News*, March 11, 2017 Available at: https://www.vanguardngr.com/2017/03/superhighway-c-river-gives-fg-two-weeks-ultimatum-eia/.

Vanthomme, H., Kolowski, J., Korte, L. and Alonso, A. (2013). Distribution of a community of mammals in relation to roads and other human disturbances in Gabon, central Africa. *Conservation Biology*, **27**, 281–91. DOI: 10.1111/cobi.12017.

Vanthomme, H., Kolowski, J., Nzamba, B.S. and Alonso, A. (2015). Hypothesis-driven and field-validated method to prioritize fragmentation mitigation efforts in road projects. *Ecological Applications*, **25**, 2035–46. DOI: 10.1890/14-1924.1.

Varki, A. and Altheide, T.K. (2005). Comparing the human and chimpanzee genomes: searching for needles in a haystack. *Genome Research*, **15**, 1746–58.

Venables, A.J. (2016). Using natural resources for development: why has it proven so difficult? *Journal of Economic Perspectives*, **30**, 161–84.

Venter, O., Sanderson, E.W., Magrach, A., *et al.* (2016). Sixteen years of change in the global terrestrial human footprint and implications for biodiversity conservation. *Nature Communications*, **7**, 12558. DOI: 10.1038/ncomms12558.

Verhegghen, A., Eva, H., Ceccherini, G., *et al.* (2016). The potential of sentinel satellites for burnt area mapping and monitoring in the Congo Basin forests. *Remote Sensing*, **8**, 986.

Virunga National Park (n.d.). *Virunga Alliance*. Kivu Nord, DRC: Virunga National Park. Available at: https://virunga.org/virunga-alliance/. Accessed July 20, 2017.

Wade, R.H. (2011). *Boulevard of Broken Dreams: The Inside Story of the World Bank's Polonoroeste Road Project in Brazil's Amazon*. London, UK: Grantham Research Institute on Climate Change and the Environment. Available at: http://www.lse.ac.uk/GranthamInstitute/wp-content/uploads/2014/02/WP55_world-bank-road-project-brazil.pdf.

Wadman, M. (2017a). Activists battle US government in court over making animal welfare reports public. *Science*, May 24, 2017. Available at: http://www.sciencemag.org/news/2017/05/should-animal-welfare-reports-automatically-be-public-courts-prepare-weigh.

Wadman, M. (2017b). More groups sue to force USDA to restore online animal welfare records. *Science*, February 22, 2017. Available at: http://www.sciencemag.org/news/2017/02/breaking-reversal-usda-reposts-some-animal-welfare-records-it-had-scrubbed-website.

Wallace, A.R. (1849). On the monkeys of the Amazon. *Proceedings of the Zoological Society of London*, **20**, 107–10.

Walsh, P.D., Abernethy, K.A., Bermejo, M., *et al.* (2003). Catastrophic ape decline in western equatorial Africa. *Nature*, **422**, 611. DOI: 10.1038/nature01566. Available at: https://www.nature.com/articles/nature01566#supplementary-information.

Walsh, P.D., Biek, R. and Real, L.A. (2005). Wave-like spread of Ebola Zaire. *PLoS Biology*, **3**, e371. DOI: 10.1371/journal.pbio.0030371.

Walston, J., Robinson, J.G., Bennett, E.L., *et al.* (2010). Bringing the tiger back from the brink: the six percent solution. *PLoS Biology*, **8**, e1000485. DOI: 10.1371/journal.pbio.1000485.

Wanshel, E. (2016). 'World's loneliest chimp', abandoned on a small island, gets cuddly teddy bear after three years alone. *Huffington Post*. Available at: www.huffingtonpost.com/entry/ponso-chimp-abandoned-on-island-gets-teddy-bear_us_56d9f24be4b0ffe6f8e95fc3.

Warigi, G. (2015). Uganda's change of mind on pipeline and the headache it is giving Kenya. *Daily Nation*, Available at: http://allafrica.com/stories/201510250142.html.

Watkins, K., Chansopheaktra, S., Brander, L., *et al.* (2016). *Mapping and Valuing Ecosystem Services in Mondulkiri: Outcomes and Recommendations for Sustainable and Inclusive Land Use Planning in Cambodia*. Phnom Penh, Cambodia: World Wide Fund for Nature (WWF), Cambodia.

Watson, J.E.M., Shanahan, D.F., Di Marco, M., *et al.* (2016). Catastrophic declines in wilderness areas undermine global environment targets. *Current Biology*, **26**, 2929–34. DOI: https://doi.org/10.1016/j.cub.2016.08.049.

Watts, D.P. (1984). Composition and variability of mountain gorilla diets in the central Virungas. *American Journal of Primatology*, **7**, 323–56. DOI: 10.1002/ajp.1350070403.

Watts, D.P. (1989). Infanticide in mountain gorillas: new cases and a reconsideration of the evidence. *Ethology*, **81**, 1–18. DOI: 10.1111/j.1439-0310.1989.tb00754.x.

Watts, D.P., Muller, M., Amsler, S.J., Mbabazi, G. and Mitani, J.C. (2006). Lethal intergroup aggression by chimpanzees in Kibale National Park, Uganda. *American Journal of Primatology*, **68**, 161–80. DOI: 10.1002/ajp.20214.

WBCSD (n.d.). *Natural Capital Protocol Toolkit*. Geneva, Switzerland: World Business Council for Sustainable Development (WBCSD). Available at: www.naturalcapitaltoolkit.org.

WCD (2000). *Dams and Development: A New Framework for Decision-making*. World Commission on Dams (WCD). London, UK: Earthscan Publications. Available at: https://www.internationalrivers.org/sites/default/files/attached-files/world_commission_on_dams_final_report.pdf.

WCS (2011). *Projet pour la Protection des Populations de Gorilles et la Conservation de la Biodiversité dans la Forêt de Deng – Deng. SYNTHESE DES ACTIVITES 2009 – 2010*. Yaoundé, Cameroon: Wildlife Conservation Society (WCS). Available at: https://programs.wcs.org/cameroon/Wild-Places/Deng-Deng-National-Park.aspx.

WCS (2015a). *Biological monitoring, Tanintharyi*. Unpublished survey reports. Yangon, Myanmar: Wildlife Conservation Society (WCS).

WCS (2015b). *Projet pour la Protection des Populations de Gorilles et de Chimpanzés, et Conservation de la Biodiversité dans la Forêt de Deng - Deng Région de l'Est Cameroun. RAPPORT FINAL*. Yaoundé, Cameroon: Wildlife Conservation Society (WCS). Available at: https://programs.wcs.org/cameroon/.

WCS (2015c). *Survey of the Lokofa Block of the Salonga National Park*. Kinshasa, DRC: Wildlife Conservation Society (WCS).

WCS (n.d.). *Superhighway Rerouted and Wildlife Saved!* New York, NY: Wildlife Conservation Society (WCS). Available at: https://secure.wcs.org/campaign/superhighway-thank-you. Accessed August, 2017.

Weinhold, D. and Reis, E. (2008). Transportation costs and the spatial distribution of land use in the Brazilian Amazon. *Global Environmental Change*, **18**, 54–68. DOI: https://doi.org/10.1016/j.gloenvcha.2007.06.004.

Weng, L., Boedhihartono, A.K., Dirks, P.H.G.M., *et al.* (2013). Mineral industries, growth corridors and agricultural development in Africa. *Global Food Security*, **2**, 195–202. DOI: https://doi.org/10.1016/j.gfs.2013.07.003.

Wenstøp, F.E. and Carlsen, A.J. (1988). Ranking hydroelectric power projects with multicriteria decision analysis. *Interfaces*, **18**, 36–48. DOI: 10.1287/inte.18.4.36.

White, A. and Fa, J.E. (2014). The bigger picture: indirect impacts of extractive industries on apes and ape habitat. In *State of the Apes: Extractive Industries and Ape Conservation*, ed. Arcus Foundation. Cambridge, UK: Cambridge University Press, pp. 197–225.

Whitelaw, E. and Macmullan, E. (2002). A framework for estimating the costs and benefits of dam removal. *BioScience*, **52**, 724–30. DOI: 10.1641/0006-3568(2002)052[0724:AFFETC]2.0.CO;2.

Whittaker, D. and Geissmann, T. (2008). Hylobates klossii. *The IUCN Red List of Threatened Species 2008: e.T10547A3199263*. Gland, Switzerland: International Union for Conservation of Nature (IUCN). Available at: http://www.iucnredlist.org/details/10547/0. Accessed November 15, 2016.

Wich, S.A., de Vries, H. and Ancrenaz, M. (2009a). Orangutan life history variation. In *Orangutans: Geographic Variation in Behavioral Ecology and Conservation*, ed. S. A. Wich, S. Utami Atmoko, T. Mitra Setia and C. P. Van Schaik. Oxford, UK: Oxford University Press, pp. 65–75.

Wich, S.A., Fredriksson, G.M., Usher, G., *et al.* (2012a). Hunting of Sumatran orang-utans and its importance in determining distribution and density. *Biological Conservation*, **146**, 163–9. DOI: https://doi.org/10.1016/j.biocon.2011.12.006.

Wich, S.A., Garcia-Ulloa, J., Kühl, Hjalmar S., *et al.* (2014). Will oil palm's homecoming spell doom for Africa's great apes? *Current Biology*, **24**, 1659–63. DOI: https://doi.org/10.1016/j.cub.2014.05.077.

Wich, S.A., Gaveau, D., Abram, N., *et al.* (2012b). Understanding the impacts of land-use policies on a threatened species: is there a future for the Bornean orang-utan? *PLoS One*, **7**, e49142. DOI: 10.1371/journal.pone.0049142.

Wich, S.A., Geurts, M.L., Mitra Setia, T. and Utami Atmoko, S.S. (2006). Influence of fruit availability on Sumatran orangutan sociality and reproduction. In *Feeding Ecology in Apes and Other Primates: Ecological, Physical and Behavioral Aspects*, ed. G. Hohmann, M. M. Robbins and C. Boesch. New York, NY: Cambridge University Press, pp. 337–58.

Wich, S.A., Meijaard, E., Marshall, A.J., *et al.* (2008). Distribution and conservation status of the orang-utan (*Pongo* spp.) on Borneo and Sumatra: how many remain? *Oryx*, **42**, 329–39. DOI: 10.1017/S003060530800197X.

Wich, S., Riswan, J.J., Refish, J. and Nelleman, C. (2011). *Orangutans and the Economics of Sustainable Forest Management in Sumatra*. Norway: UNEP, GRASP, PanEco, YEL, ICRAF, GRID-Arendal, Birkeland Trykkeri AS. Available at: www.grida.no/search?query=Orangutans+and+the+Economics+of+Sustainable+Forest+Management+in+Sumatra.

Wich, S.A., Singleton, I., Nowak, M.G., *et al.* (2016). Land-cover changes predict steep declines for the Sumatran orangutan (*Pongo abelii*). *Science Advances*, **2**. DOI: 10.1126/sciadv.1500789.

Wich, S.A., Utami Atmoko, S., Mitra Setia, M. and Van Schaik, C.P., ed. (2009b). *Orangutans: Geographic Variation in Behavioral Ecology and Conservation*. Oxford, UK: Oxford University Press.

Wich, S.A., Utami Atmoko, S.S., Mitra Setia, M., *et al.* (2004). Life history of wild Sumatran orangutans (*Pongo abelii*). *Journal of Human Evolution*, **47**, 385–98. DOI: https://doi.org/10.1016/j.jhevol.2004.08.006.

Wikramanayake, E., Dinerstein, E., Seidensticker, J., *et al.* (2011). A landscape-based conservation strategy to double the wild tiger population. *Conservation Letters*, **4**, 219–27. DOI: 10.1111/j.1755-263X.2010.00162.x.

Wilcox, B.A. (1978). Supersaturated island faunas: a species-age relationship for lizards on post-Pleistocene land-bridge islands. *Science*, **199**, 996–8. DOI: 10.1126/science.199.4332.996.

Wildlife Impact (2015). *An analysis of the current status, challenges and opportunities in the African ape sanctuary sector: recommendations for prioritizing support and activities*. Internal report for the Arcus Foundation.

Wildlife Impact (2016). *Priority landscapes: conservation planning and captive care capacity*. Internal report for the Arcus Foundation.

Wildman, L. (2013). Dam removal: a history of decision points. In *The Challenges of Dam Removal and River Restoration. Reviews in Engineering Geology, XXI*, ed. J. V. De Graff and J. E. Evans: Geological Society of America, pp. 1–10. Available at: http://geoscienceworld.org/content/the-challenges-of-dam-removal-and-river-restoration.

Wilkie, D., Shaw, E., Rotberg, F., Morelli, G. and Auzel, P. (2000). Roads, development, and conservation in the Congo Basin. *Conservation Biology*, **14**, 1614–22. DOI: 10.1111/j.1523-1739.2000.99102.x.

Willems, W.J.H. and Van Schaik, H.P.J. (2015). *Water and Heritage: Material, Conceptual, and Spiritual Connections*. Leiden, the Netherlands: Sidestone Press. Available at: https://www.sidestone.com/books/water-heritage. Accessed October 7, 2016.

Williams, J.M., Lonsdorf, E.V., Wilson, M.L., *et al.* (2008). Causes of death in the Kasekela chimpanzees of Gombe National Park, Tanzania. *American Journal of Primatology*, **70**, 766–77. DOI: 10.1002/ajp.20573.

Williams, S. (2015). Moving the economy forward. *ABM (African Business Magazine)*, January 14, 2015. Available at: http://africanbusinessmagazine.com/company-profile/african-development-bank/moving-economy-forward/.

Williamson, E.A. (2014). Mountain gorillas: a shifting demographic landscape. In *Primates and Cetaceans: Field Research and Conservation of Complex Mammalian Societies*, ed. J. Yamagiwa and L. Karczmarsk. Tokyo, Japan: Springer.

Williamson, E.A. and Butynski, T.M. (2013a). *Gorilla beringei* eastern gorilla. In *Mammals of Africa. Volume II: Primates*, ed. T. M. Butynski, J. Kingdon and J. Kalina. London, UK: Bloomsbury Publishing, pp. 45–53.

Williamson, E.A. and Butynski, T.M. (2013b). *Gorilla gorilla* western gorilla. In *Mammals of Africa. Volume II: Primates*, ed. T. M. Butynski, J. Kingdon and J. Kalina. London, UK: Bloomsbury Publishing, pp. 39–45.

Williamson, E.A., Maisels, F.G., Groves, C.P., *et al.* (2013). Hominidae. In *Handbook of the Mammals of the World. Volume 3: Primates*, ed. R. A. Mittermeier, A. B. Rylands and D. E. Wilson. Barcelona, Spain: Lynx Edicions, pp. 792–854.

Williamson, E.A., Rawson, B.M., Cheyne, S.M., Meijaard, E. and Wich, S.A. (2014). Ecological impacts of extractive industries on ape populations. In *State of the Apes: Extractive Industries and Ape Conservation*, ed. Arcus Foundation. Cambridge, UK: Cambridge University Press, pp. 65–99.

Wilson, D. and Reeder, D. (2005). *Mammal Species of the World: A Taxonomic and Geographic Reference*. Baltimore, MD: Johns Hopkins University Press.

Wilson, H.B., Meijaard, E., Venter, O., Ancrenaz, M. and Possingham, H.P. (2014a). Conservation strategies for orangutans: reintroduction versus habitat preservation and the benefits of sustainably logged forest. *PLoS One*, **9**, e102174. DOI: 10.1371/journal.pone.0102174.

Wilson, M.L., Boesch, C., Fruth, B., *et al.* (2014b). Lethal aggression in *Pan* is better explained by adaptive strategies than human impacts. *Nature*, **513**, 414. DOI: 10.1038/nature13727.

Wilson, M.L. and Wrangham, R.W. (2003). Intergroup relations in chimpanzees. *Annual Review of Anthropology*, **32**, 363–92. DOI: 10.1146/annurev.anthro.32.061002.120046.

Winders, D. (2017). *USDA Blackout: Scrutinizing the Deletion of Thousands of Animal Welfare Act-Related Records*. American Bar Association Animal Law Committee. Available at: http://apps.americanbar.org/dch/thedl.cfm?filename=/IL201050/relatedresources/Summer2017.pdf.

Winemiller, K.O., McIntyre, P.B., Castello, L., *et al.* (2016). Balancing hydropower and biodiversity in the Amazon, Congo, and Mekong. *Science*, **351**, 128–9. DOI: 10.1126/science.aac7082.

Withanage, H., Masayda, R., Hernandez, R. and Neura, A. (2006). *Development Debacles: A Look into ADB's Involvement in Environmental Degradation, Involuntary Resettlement and Violation of Indigenous People's Rights*. Quezon City, Philippines: NGO Forum on ADB.

Wood, C. (2003). *Environmental Impact Assessment in Developing Countries: An Overview*. Manchester, UK: University of Manchester.

World Bank (2008). *Project Appraisal Document on a Proposed Grant in the Amount of SDR 32 Million (US$50 Million Equivalent) to the Democratic Republic of Congo for a High-Priority Road Reopening and Maintenance Project, Pro-Routes*. Report No. 40028. Africa Transport Sector, Country Department AFCC2, Africa Regional Office. Washington DC: World Bank.

World Bank (2009). *Integrated Safeguards Data Sheet Concept Stage: Lom Pangar Hydropower Project*. Yaoundé, Cameroon: World Bank. Available at: http://documents.worldbank.org/curated/en/346811468232166551/pdf/IntegratedoSaf1oSheet1Concept0Stage.pdf.

World Bank (2012a). *Cameroon: Lom Pangar Hydropower Project (FY12)*. Washington DC: World Bank. Available at: http://documents.worldbank.org/curated/en/540341468017355999/Cameroon-Lom-Pangar-Hydropower-Project-FY12.

World Bank (2012b). *Fact Sheet Lom Pangar Hydropower Project, Cameroon*. Washington DC: World Bank. Available at: http://siteresources.worldbank.org/INTCAMEROON/Resources/LPHP_Fact_Sheet_Mar2012.pdf.

World Bank (2012c). *Project Appraisal Document on a Proposed Credit in the Amount of SDR 85.2 Million (US$132 Million Equivalent) to the Republic of Cameroon for a Lom Pangar Hydropower Project*. Washington DC: World Bank. Available at: http://siteresources.worldbank.org/INTCAMEROON/Resources/LPHP-PAD-Mar2012.pdf.

World Bank (2013a). *Operational Manual OP 4.01: Environmental Assessment. Revised April 2013*. Washington DC: World Bank. Available at: https://policies.worldbank.org/sites/ppf3/PPFDocuments/090224b0822f7384.pdf.

World Bank (2013b). *Operational Manual OP 4.04: Natural Habitats*. Washington DC: World Bank. Available at: https://policies.worldbank.org/sites/ppf3/PPFDocuments/090224b0822f74ac.pdf. Accessed March 8, 2017.

World Bank (2013c). *Operational Manual OP 4.36: Forests*. Washington DC: World Bank. Available at: https://policies.worldbank.org/sites/ppf3/PPFDocuments/090224b0822f8a50.pdf. Accessed March 8, 2018.

World Bank (2016a). *Forest Action Plan FY16-20* Washington DC: World Bank. Available at: documents.worldbank.org/curated/en/240231467291388831/Forest-action-plan-FY16-20.

World Bank (2016b). *Global Tiger Initiative*. Washington DC: World Bank. Available at: http://www.worldbank.org/en/topic/environment/brief/the-global-tiger-initiative.

World Bank (2016c). *New Environmental and Social Framework*. Washington DC: World Bank. Available at: http://www.worldbank.org/en/news/feature/2016/08/05/the-new-environmental-and-social-framework.

World Bank (2016d). *Projet d'Appui à l'Ouverture et l'Entretien des Routes Hautement Prioritaires (Pro Routes)*. RDC, World Bank.

World Bank (2017). *The World Bank Environmental and Social Framework*. Washington DC: World Bank. Available at: http://documents.worldbank.org/curated/en/383011492423734099/pdf/114278-WP-PUBLIC-13-4-2017-11-23-38-EnvironmentalandSocialFrameworkWeb.pdf#page=81&zoom=80.

World Bank (n.d.-a). *Congo DRC: Pro-Routes Project (Additional Financing)*. Washington DC: World Bank. Available at: www.projects.worldbank.org/P120709/congo-drc-pro-routes-project-additional-financing?lang=en&tab=detail.

World Bank (n.d.-b). *Environmental and Social Policies for Projects*. Washington DC: World Bank. Available at: www.worldbank.org/en/programs/environmental-and-social-policies-for-projects. Accessed October, 2017.

World Bank (n.d.-c). *The Chad-Cameroon Petroleum Development and Pipeline Project: Environmental and Social*. Washington DC: World Bank. Available at: http://web.worldbank.org/archive/website01210/WEB/0__CON-5.HTM.

World Bank (n.d.-d). *The Environmental and Social Framework*. Washington DC: World Bank. Available at: http://www.worldbank.org/en/programs/environmental-and-social-policies-for-projects/brief/the-environmental-and-social-framework-esf.

Wrangham, R. (2013). *Presentation on reintroduction of African apes*. Presented at: PASA 2013 Great Ape Reintroduction Workshop, September 14–18, 2013, Chester, UK.

Wrangham, R.W. (1986). Ecology and social relationships in two species of chimpanzee. In *Ecological Aspects of Social Evolution: Birds and Mammals*, ed. D. I. Rubenstein and R. W. Wrangham. Princeton, NJ: Princeton University Press, pp. 352–37.

WRI and MECNT (2010). *Atlas Forestier Interactif de la République Démocratique du Congo: Document de Synthèse*. Washington DC: World Resources Institute (WRI) and Ministry for Environment, Nature Conservation and Tourism (MECNT) of the Republic of Congo. Available at: https://www.wri.org/sites/default/files/pdf/interactive_forest_atlas_drc_fr.pdf.

Wright, E., Grueter, C.C., Seiler, N., *et al.* (2015). Energetic responses to variation in food availability in the two mountain gorilla populations (*Gorilla beringei beringei*). *American Journal of Physical Anthropology*, **158**, 487–500. DOI: 10.1002/ajpa.22808.

WSP (2011). *Water Supply and Sanitation in the Democratic Republic of Congo: Turning Finance into Services for 2015 and Beyond*. Nairobi, Kenya: Water and Sanitation Program (WSP), Africa Region, World Bank.

Wu, S.-S. (2016). Singapore-Kunming rail link: a 'belt and road' case study. *The Diplomat*, June, 2016. Available at: http://thediplomat.com/2016/06/singapore-kunming-rail-link-a-belt-and-road-case-study/.

WWF (2006). *Free-Flowing Rivers: Economic Luxury or Ecological Necessity?* Zeist, the Netherlands: World Wide Fund for Nature (WWF), Netherlands.

WWF (2014). Driving change in Asia: newly open, Myanmar is a treasure trove of natural assets, cultural diversity and enthusiasm for the future. *World Wildlife Magazine*. Available at: https://www.worldwildlife.org/magazine/issues/spring-2014.

WWF (2015a). *A Global Assessment of Extractive Activity within World Heritage Sites*. Gland, Switzerland: World Wide Fund for Nature (WWF)-International.

WWF (2015b). *The Integrated Resources Corridor Initiative: Scoping Study and Business Plan*. Nairobi, Kenya: World Wide Fund for Nature (WWF) Regional Office for Africa. Available at: http://www.adamsmithinternational.com/documents/resource-uploads/IRCI_Scoping_Report_Business_Plan.pdf.

WWF (2016). *Biodiversity survey, Tanintharyi*. Unpublished survey reports. Yangon, Myanmar: World Wide Fund for Nature (WWF).

WWF (n.d.-a). *Borneo Mammals*. World Wide Fund for Nature (WWF). Available at: wwf.panda.org/what_we_do/where_we_work/borneo_forests/about_borneo_forests/borneo_animals/borneo_mammals/. Accessed September 12, 2017.

WWF (n.d.-b). *Borneo Wildlife*. World Wide Fund for Nature (WWF). Available at: http://wwf.panda.org/what_we_do/where_we_work/borneo_forests/about_borneo_forests/borneo_animals/. Accessed September 12, 2017.

WWF and Dalberg (2012). *Fighting Illicit Wildlife Trafficking: A Consultation with Governments*. Gland, Switzerland: World Wide Fund for Nature (WWF) International. Available at: http://www.dalberg.com/documents/WWF_Wildlife_Trafficking.pdf.

WWF and TEREA (2014). *Evaluation Préliminaire des Appuis à l'ICCN en Matière de Gestion Participative des Aires Protégées*. Kinshasa, DRC: Ministère des infrastructures, travaux publics et reconstruction, Cellule Infrastructures, Projet Pro-Routes.

WWF and TRAFFIC (2015). *Strategies for Fighting Corruption in Wildlife Conservation: A Primer*. World Wide Fund for Nature (WWF) and TRAFFIC Wildlife Crime Initiative. Available at: http://d2ouvy59p0dg6k.cloudfront.net/downloads/wci_strategies_for_fighting_corruption_wildlife_conservation.pdf.

Wyatt, T. and Ngoc Cao, A. (2015). *Corruption and Wildlife Trafficking*. U4 Anti-Corruption Resource Centre. Available at: http://issuu.com/cmi-norway/docs/150603144854–73fa106fbd71418f9a0afa481e93443f/1?e=0.

Yamagiwa, J. and Basabose, A.K. (2009). Fallback foods and dietary partitioning among *Pan* and *Gorilla*. *American Journal of Physical Anthropology*, **140**, 739–50. DOI: 10.1002/ajpa.21102.

Yamagiwa, J., Basabose, A.K., Kaleme, K. and Yumoto, T. (2003). Within-group feeding competition and socioecological factors influencing social organisation of gorillas in the Kahuzi-Biega National Park, Democratic Republic of Congo. In *Gorilla Biology*, ed. A. B. Taylor and M. L. Goldsmith. Cambridge, UK: Cambridge University Press, pp. 328–57.

Zarfl, C., Lumsdon, A.E., Berlekamp, J., Tydecks, L. and Tockner, K. (2015). A global boom in hydropower dam construction. *Aquatic Sciences*, **77**, 161–70. DOI: 10.1007/s00027-014-0377-0.

Zhang, W., Hu, Y., Chen, B., *et al.* (2007). Evaluation of habitat fragmentation of giant panda (*Ailuropoda melanoleuca*) on the north slopes of Daxiangling Mountains, Sichuan province, China. *Animal Biology*, **57**, 485–500. DOI: https://doi.org/10.1163/157075607782232107.

Zhou, J., Wei, F., Li, M., Pui Lok, C.B. and Wang, D. (2008). Reproductive characters and mating behaviour of wild *Nomascus hainanus*. *International Journal of Primatology*, **29**, 1037–46. DOI: 10.1007/s10764-008-9272-7.

Index

A

Abbott's gray gibbon (*Hylobates abbotti*) xv, xix, xxii–xxiii, 186, 202, 206, 208, 209, 213, 214, 216–17, 219, *see also* gibbons (Hylobatidae)
Abundance Annex 2, 198, www.stateoftheapes.com
Africa
    deforestation/habitat loss *see* Global Forest Change 2000–14 study
    development corridors 13, 14, 15, 22–26, 108, 110, 111, 134
    distribution of apes xx–xxi
    human population boom 108, 128, 161
    hydropower expansion 170, 171, 172, 175, 176
    monetary cost of infrastructure 81–82
    offset policies 122, 124
    poverty 69–70, 161
    protected areas *see* protected areas in Africa study
    roads 112–15
    sanctuaries *see* sanctuaries
    wildlife trafficking 233, 234, 235
    see also specific countries
African apes *see* bonobo (*Pan paniscus*); chimpanzees (*Pan troglodytes*); gorillas (*Gorilla* spp.)
agile gibbon (*Hylobates agilis*) xix, xxii–xxiii, 202, 206, 208, 209, 211, 213, 214, 216–17, 219, 235, 236, 244, *see also* gibbons (Hylobatidae); *Hylobates* gibbons
agriculture 29, 34, 38, 51, 63, 82, 91, 108, 215, 216, 219, 222, 231–32, *see also* oil palm plantations
Angola 215, 231, 232, 234, 237–38
Ape Populations, Environments and Surveys (A.P.E.S.) Portal 2, 120, 161
Asia
    community tenure of land 76
    deforestation/habitat loss *see* Global Forest Change 2000–14 study
    development corridors 15
    distribution of apes xxii–xxiii
    economic growth, rapid 18
    hydropower expansion 13, 15, 170, 171, 172, 175, 176
    infrastructure planned 13, 81
    offset policies 122, 123, 124
    protected areas, forest loss in 205–6
    sanctuaries *see* sanctuaries
    wildlife trafficking 234
    see also specific countries
Asian apes *see* gibbons (Hylobatidae); orangutans (*Pongo* spp.)
Asian Development Bank (ADB) 24, 30, 81, 191
Asian Infrastructure Investment Bank (AIIB) 4, 15, 18, 20, 21
awareness raising 37, 140, 156, 162, 170, 186–88, 195, 234, 242, 245, 249–50

B

Bagyeli people 63, 65, 66, 67–69, 70, 71, 73
Baka people 63, 66, 68, 69, 70, 71, 72, 73, 75
Bakun Dam, Borneo 182, 183, 184–86, 187
Bangladesh 215, 231, 234
Bantu people 63, 67, 69, 73
Baram Dam, Borneo 182–88
barriers
    artificial 43, 47, 48–49, 50, 52, 57, 58, 173, 232
    natural 16, 17, 110
"Belt and Road" initiative 15, 18, 26
benefits from infrastructure
    environmental 28
    social 2, 24, 27–30, 66, 83, 133, 138
Bili–Mbomu Forest Savanna Mosaic, DRC 158, 159, 163
Bili–Uélé Protected Area Complex (BUPAC), DRC 140, 158–60, 162–63
biodiversity action plans (BAPs) 37, 125, 144–45
biodiversity impact assessments 161–62, 186
Blangkejeren–Kutacane (B–K) road, Sumatra 87–88, 91–94
bonobo (*Pan paniscus*)
    captive 236, 240, 258, 259, 260
    diet xxi, xxvi
    distribution xx–xxi
    distribution factors xviii, 17, 48, 53, 110
    forest cover and loss in ranges 204, 208, 209, 216, 219
    habitat xxi
    home ranges xxvi
    overview xii
    reproduction xxiv–xxv
    socioecology xvii–xxvi
    threats to 22, 51
    travel throughout range xxiv
Bornean gray gibbon (*Hylobates funereus*) xv, xix, xxii–xxiii, 206, 208, 209, 211, 217, 219, 234; *see also* gibbons (Hylobatidae)
Bornean orangutan (*Pongo pygmaeus*)
    captive 235, 236
    diet xxvii
    distribution xviii, xxii–xxiii
    outside protected areas 251
    population xiv
    reproduction xxiv
    threats to 26–27, 182, 206, 208, 209, 213, 217, 219, 232
    see also orangutans (*Pongo* spp.)
Bornean white-bearded gibbon (*Hylobates albibarbis*) xix, xxii–xxiii, 206, 208, 209, 211, 217, 219, 236; *see also* gibbons (Hylobatidae)
Borneo
    community activism 170, 182–88
    corruption 183

deforestation 28, 207, 222
electrocutions 46
infrastructure development 26–27
natural capital and infrastructure planning 130
rainforest and biodiversity 182–83
Sarawak Corridor of Renewable Energy (SCORE) 182–88
translocations 251
see also Indonesia; Malaysia
Borneo Orangutan Survival Foundation (BOSF) 261
Brazil 5, 12, 15, 32, 112, 173–74, 255, 265
Brunei 215
Bukavu–Kisangani Highway, DRC 132–33
Bumbuna dam, Sierra Leone 56
Bureau d'Études Spécialisés en Gestion Environnementale et Sociale (BEGES) 160, 162, 163
Burkina Faso 215
Burundi 215, 232, 234, 237
bushmeat see meat, wild
Business and Biodiversity Offsets Programme (BBOP) 37, 126
Bwindi Impenetrable National Park, Uganda 44–45, 52, 116–17

C

Calabar deep port, Nigeria 138–39, 141
Cam Iron Mine, Cameroon 66, 70, 71, 72
Cambodia
 deforestation/habitat loss 203, 207, 215, 216, 222
 infrastructure development 18
 sanctuaries 236, 262
 wildlife laws 234
Cameroon
 anti-poaching measures 121, 231, 234
 Chad–Cameroon pipeline 63, 65, 66–69, 70, 177, 178, 179
 dams 24, 161, 170, 177–81
 deforestation/habitat loss 63, 208, 215
 drivers of infrastructure expansion 62–65, 66
 economic growth, plans for 177–78
 governance and corruption issues 63–64
 indigenous peoples 63–64, 65, 66, 67–69, 70–76
 infrastructure development 24, 66, 76–78
 Ntam settlement 71–72
 protected areas 24, 65, 68, 69, 71, 72, 142, 161, 178–81
 road and railway impacts 69–73
 sanctuaries 236, 237, 260
Campo Ma'an Reserve, Cameroon 24, 68
Cao Vit gibbon (*Nomascus nasutus*) xvi, xxviii, xix, xxii–xxiii, 205, 207, 208, 210, 213, 217, 219; see also gibbons (Hylobatidae); *Nomascus* gibbons
capacity building 25, 75, 156, 159, 160, 238
captive apes
 breeding programs 234, 243–44, 259
 Europe 258–59
 Japan 257, 258
 key findings 226–27
 overview 255
 pets 231, 235, 242, 243, 245, 257, 258
 range states and surrounding regions 260–62
 regulatory landscape 255, 259
 in sanctuaries see sanctuaries
 types of captivity 225–26, 256
 United States 255–58, 259
Carpenter's lar gibbon (*Hylobates lar carpenteri*) 208, 209, 219; see also gibbons (Hylobatidae); lar gibbon (*Hylobates lar*)
Cellule Infrastructures (CI) 158–59, 160, 162–63
Central Africa 17, 24, 109, 110, 195, 204; see also specific countries
Central African Iron Ore Corridor 24
Central African Republic (CAR) 37, 158, 215, 216, 231, 237
central chimpanzee (*Pan troglodytes troglodytes*) xii, xviii, xx–xxi, 65, 69, 177, 178, 208, 209, 212, 217, 219, 236, 251, 259, see also chimpanzees (*Pan troglodytes*)
central lar gibbon (*Hylobates lar entelloides*) 208, 210, 219; see also gibbons (Hylobatidae); lar gibbon (*Hylobates lar*)
central Yunnan black-crested gibbon (*Nomascus concolor jingdongensis*) 207, 208, 210, 213, 214, 217, 219; see also gibbons (Hylobatidae); *Nomascus* gibbons
certification 59
Chad–Cameroon pipeline 63, 65, 66–69, 70, 177, 178, 179
chimpanzees (*Pan troglodytes*)
 avoiding habitat of 120
 captive 226, 227, 235–36, 237, 244, 255–56, 257, 258, 259, 260–61, 262
 diet xxi, xxvi
 habitat xxi, 274
 habitat health 269, 270
 home ranges xxvi
 infanticide xxiv
 intergroup aggression xxvi, 49, 51
 memorization of locations xxiii
 overview xii–xiii
 reintroduction and translocation 240
 reproduction xxiv–xxv
 road crossing 46, 52
 socioecology xvii–xxvi
 threats to
  disease 52
  habitat loss 25, 203, 204, 211
  human conflicts 232
  human influx and settlement 268
  hunting/poaching 17, 46, 121, 180, 231
  mining 232
  noise pollution 53, 60
  roads 46, 53, 83, 95–99, 268–70
  trafficking 158, 234, 235–36, 237
  water availability, decreased 53
 see also central chimpanzee (*Pan troglodytes troglodytes*); eastern chimpanzee (*Pan troglodytes schweinfurthii*); Nigeria–Cameroon chimpanzee (*Pan troglodytes ellioti*); western chimpanzee (*Pan troglodytes verus*)

China
  "Belt and Road" initiative 15, 18, 26
  "build–operate–transfer projects" 18
  deforestation/habitat loss 207, 215
  economic growth, rapid 18
  illegal animal parts, import of 150
  impact on infrastructure expansion 12, 17, 18–19
  mineral exploitation in Africa 13
  sanctuaries 236
  social and environmental guidelines 18–19, 38
  strategic environmental planning 129–30
climate change 129, 145, 232
communities *see* local communities
Compagnie des Bauxites de Guinée (CBG) 125
compensation measures 27–28, 64, 66, 68–69, 75, 119–20, 144, 145, 147, 190
Congo Basin 13, 22, 24–25, 100, 111, 113, 114, 177; *see also* Central Africa
connectivity
  awareness raising 139–40
  disruption of 43, 51, 52, 111, 134, 137, 173–74
  importance 17, 99, 149
  planning considerations 154–55, 161
  restoration measures 57, 121, 220
conservation action plans (CAPs) 99, 104, 244, 250–51, 278
construction phase 44–46, 162
Convention on International Trade in Endangered Species of Wild Fauna and Flora (CITES) 231, 233, 234
corporate social responsibility (CSR) 146
corruption 3–4, 27, 63, 91, 141, 146, 248
"critical habitat" criterion 20–21
crop raiding 48, 51, 179
Cross River gorilla (*Gorilla gorilla diehli*) xiii, xviii, xx–xxi, 52, 116, 142, 143, 204, 208, 209, 217, 219, 236, 240, *see also* gorillas (*Gorilla* spp.); western gorillas (*Gorilla gorilla*)
Cross River National Park, Nigeria 116, 135, 141–43, 144
Cross River superhighway, Nigeria 138–39, 141–45
cumulative impact assessments (CIAs) 20, 124
cumulative impacts 25, 33, 56, 124–28, 164

D

dams
  Africa 48–49, 56, 108, 127, 161, 170, 171, 177–81, 189
  Asia xxvi, 19, 34–35, 168–69, 171, 174, 182–88, 191, 195
  associated infrastructure 15, 173, 174, 176
  biodiversity offsets 161–62, 174
  case study: Cameroon 177–81
  case study: Malaysian Borneo 182–88
  community resistance 19, 186–88
  decommissioning 272–73
  global scale 171
  Hydropower by Design 189–91
  Hydropower Sustainability Assessment Protocol 189
  impacts on apes
    assessment 174–76
    barriers to dispersal 173–74
    drownings 179
    electrocutions 179–80
    general impacts 50
    habitat loss 13, 56, 112, 173, 195
    human influx and settlement 46–47, 180
    hunting/poaching 179
    mitigation strategies 190
  impacts on environment
    biodiversity loss 186
    deforestation 13, 15, 112, 173, 186
    ecosystem services loss 195
    greenhouse gas (GHG) emissions 173
    hydrological systems 53–54, 168, 172–73
  impacts on people 167–69, 173, 178, 184–85
  low-carbon energy 130–31, 168
  number in ape range states 175, 176
  planning, system-scale approach to 195–96
  projected development 13, 108, 168
  protected areas, threats to 24, 111–12, 178–81
  Virunga, DRC 126–27
Dawei road link, Myanmar 146–57
decommissioning phase 47
deforestation along roads
  addressing effects of road development 90, 94, 99, 102, 103
  case studies
    approach 86
    data sets 266–67
    forest loss alerts and analysis 100–101
    Gunung Leuser National Park, Sumatra 91–94
    Leuser Ecosystem, Sumatra 87–90
    Western Tanzania 95–99
  mitigation hierarchy 102
  monitoring approaches 83–86, 104–5; *see also* Global Forest Watch (GFW) platform
  recommendations
    road planning in local context 103–4
    zoning 86, 102–3
deforestation/habitat loss
  for agriculture 215, 216, 219, 222, 231–32
  for development corridors 111
  Global Forest Change 2000–14 study *see* Global Forest Change 2000–14 study
  impact by infrastructure type 50
  impacts on apes 28, 43, 46–49, 51–52, 85, 138, 214, 219, 221, 231, 232, 268
  for infrastructure development 12, 13, 15, 22, 112, 145, 173, 186; *see also* deforestation along roads
  monitoring 84–86, 100–101, 114, 115, 200–202; *see also* Global Forest Change 2000–14 study
  for oil palm plantations 32–33, 34, 38, 84, 87, 89–90, 100, 183, 184, 208, 216, 222
  in protected areas (PAs) 211–13, 219–20, 221, 223
  for roads *see* deforestation along roads
  see also logging
Democratic Republic of Congo (DRC)
  conservation work 126–27

dams 111–12, 127, 171
deforestation 159, 160, 215, 216
GLAD forest loss alerts 100
human conflicts 126
human population growth 232–33
hunting/poaching 231
mining 158, 180, 232
natural capital and infrastructure planning 130
Pro-Routes project 138, 140, 158–63, 271–72
protected areas 106–7, 126–27, 132–34, 135, 140, 158–60, 162–63
roads 106–7, 132–33, 140, 158–60, 162–63, 271–72
sanctuaries 235, 236, 237, 260
socioeconomic development 126–27
wildlife trafficking 232, 234
Deng Deng National Park, Cameroon 161, 178–81
development banks 4, 20, 54, 73, 104; see also multilateral lenders
development corridors
　Africa 14, 15, 22–26, 108, 110, 111, 134
　Asia 15, 26–27, 182–88
　scale of development 13
Digital Observatory for Protected Areas (DOPA) 118, 161
disease
　reintroduction/translocation risk 240
　roads and rivers as barriers to 17
　transmission to/from humans/domestic animals 44, 46, 48, 50, 51, 52–53, 57, 214, 242
distribution of apes xx–xxi, xxii–xxiii
drivers of infrastructure expansion 12, 18–21, 32, 62–65, 66, 171–72
drownings 46, 179

E

East Africa 15, 22, 111, 204; see also specific countries
eastern chimpanzee (*Pan troglodytes schweinfurthii*) xii, xviii, xx–xxi, 22, 86, 95–99, 116, 132, 158, 208, 209, 216, 218, 219, 236, see also chimpanzees (*Pan troglodytes*)
eastern gorillas (*Gorilla beringei*) xiii, xxiv, xxv, 122; see also Grauer's gorilla (*Gorilla beringei graueri*); mountain gorilla (*Gorilla beringei beringei*)
eastern hoolock (*Hoolock leuconedys*) xiv, xviii, xxii–xxiii, 201, 205, 208, 209, 219; see also hoolocks (*Hoolock* spp.)
Eco Activists for Governance and Law Enforcement (EAGLE) Network, 245–46
ecosystem services 3, 29, 87, 129, 149, 156, 193, 272
education 128, 227, 229, 242, 245, 249–50; see also awareness raising
electrocutions 46, 53, 179, 264–65
environmental and social impact assessments (ESIAs) 20, 32–34, 54, 56, 78, 120, 163, 178; see also environmental impact assessments (EIAs)
environmental impact assessments (EIAs) 19, 29, 32–34, 54, 56–57, 60, 102, 140, 141, 142–43, 144–45, 152–54, 156

environmental impacts of infrastructure development 13, 15–17; see also deforestation/habitat loss
Equatorial Guinea 215, 231, 234, 237
Europe 171, 172, 227, 258–59
European Investment Bank 20, 178, 181
European Union (EU) and Commission 113, 115, 118, 133, 158–59, 193
extractive industries 15, 22–27, 28, 46–47, 53, 60, 63, 64, 65, 66–71, 72, 108, 116, 120, 125, 133, 147, 158, 177, 178, 179, 180, 219–20, 232

F

fires 12, 94, 231–32, 276
food demand, human 29
foraging xxiii, 51
free, prior and informed consent (FPIC) 64, 75, 78, 127, 188
funding for infrastructure 4–5, 18, 20–21, 103, 104, 191; see also development banks; multilateral lenders

G

Gabon
　dams 48–49, 166–67, 189
　deforestation 215, 216
　oil exploration 60
　protected areas 60
　railways 24
　roads 44–45, 80–81
　sanctuaries 236, 237, 260
　TRIDOM landscape 24–25
　wildlife trafficking 234
Gambia 76, 260
Gaoligong hoolock (*Hoolock tianxing*) xiv, xviii
gas extraction see extractive industries
geothermal energy 30, 192–95
Ghana 215, 216, 237
gibbons (Hylobatidae)
　captive 199, 236, 237, 240, 257–58, 259, 261, 262
　dependence on forest xxviii
　diet xxviii
　dispersal xxviii–xxix
　habitat xxvii–xxviii, 274
　home ranges xxviii
　overview xiv–xvi
　reproduction xxviii
　socioecology xxvii–xxix
　threats to
　　dams 15, 170, 175, 176
　　development corridors 26
　　electrocution 53
　　habitat loss 48, 51, 85, 149, 202–3, 207, 213–15, 216, 219, 222
　　human influx and settlement 16
　　hunting/poaching 150, 231
　　roads 87, 150, 232
　　trafficking 152, 234

wildlife bridges 58–59, 121
see also specific genus; specific species
Global Accessibility Map 115
Global Federation of Animal Sanctuaries (GFAS) 229, 230, 241, 242, 243, 256, 261
Global Forest Change 2000–14 study 202–223
    Annex VIII: data sets 273
    Annex IX: forest habitat definitions 274
    Annex X: canopy density percentages 275
    Annex XI: GFW tool's value and limitations 276–77
    Annex XII: habitat assessment, future refinements of 277–78
    Annex XIII: conservation action plans 278
Global Forest Watch Fires 276
Global Forest Watch (GFW) platform 84, 85–86, 89, 90, 92, 93, 95, 97, 98, 100, 101, 105, 115, 202, 207, 216, 222, 265–68, 276–77
Global Roadless Areas Map 113
Google Earth Engine 84, 115
Google Maps 113
gorillas (*Gorilla* spp.)
    captive 228–29, 234, 237, 258, 259, 260, 261, 262
    diet xxiii, xxv
    habitat xxi, xxv, 274
    home ranges xxv
    infanticide xxiv
    intergroup aggression xxv–xxvi
    overview xiii
    reproduction xxiv–xxv
    socioecology xvii–xxvi, 198
    threats to
        development corridors 25
        disease 52–53
        flooding 179
        habitat loss 211
        hunting/poaching 13, 50–51, 179, 180
        mining 180
        noise pollution 53, 60
        roads 52
        trafficking 234, 237
    travel throughout range xxiv, xxv
    see also Cross River gorilla (*Gorilla gorilla diehli*); Grauer's gorilla (*Gorilla beringei graueri*); mountain gorilla (*Gorilla beringei beringei*); western lowland gorilla (*Gorilla gorilla gorilla*)
Grand Inga Hydroelectric Project, DRC 111–12
Grand Poubara Dam, Gabon 48–49
    Grauer's gorilla (*Gorilla beringei graueri*) xviii, xx–xxi, 22, 24, 109, 132–34, 202, 204, 208, 209, 214, 216, 217, 219, 232, 236, *see also* gorillas (*Gorilla* spp.)
great apes *see* bonobo (*Pan paniscus*); chimpanzees (*Pan troglodytes*); gorillas (*Gorilla* spp.); orangutans (*Pongo* spp.)
green energy 12, 30–31
greenhouse gas (GHG) emissions 173
gROADS (Global Roads Open Access Data Set) 112, 118

Guinea
    Chinese infrastructure projects 19
    cumulative impacts case study 125
    dams 161
    deforestation 215
    mining 25–26, 120, 125
    Mount Nimba Biosphere Reserve 116
    rescued apes 235–36
    roads 54–55
    sanctuaries 236, 237, 260
    wildlife trafficking 234, 235–36
Guinea Alumina Corporation (GAC) 125
Guinea-Bissau 215, 234, 237
Gunung Leuser National Park (GLNP), Sumatra 87–94, 192–95, 265

H

habitat xxi, 274
habitat loss *see* deforestation along roads; deforestation/habitat loss; logging
habitat protection 47, 240, 241
Hainan gibbon (*Nomascus hainanus*) xvi, xxviii, xix, xxii–xxiii, 205, 207, 208, 210, 214, 215, 218, 219, *see also* gibbons (Hylobatidae); *Nomascus* gibbons
home ranges xxiii
Hong Kong University (HKU) 140, 147, 152–54, 156–57
hoolocks (*Hoolock* spp.) xiv–xv, xviii, 123, 211; *see also* eastern hoolock (*Hoolock leuconedys*); gibbons (Hylobatidae); western hoolock (*Hoolock hoolock*)
human conflicts/wars 126, 132–33, 146, 158, 232
human influx and settlement
    construction camps 44, 46–47, 66, 179
    river-enabled 16, 17
    road-enabled 50–51, 71–72, 82, 91, 94, 95, 111, 158, 173, 268
human population growth 2, 32, 95, 108, 232–33, 254
human–wildlife conflict 48, 51, 179, 231, 236
Humane Society of the United States (HSUS) 245–46
hunting/poaching
    anti-poaching measures 60, 121, 162, 231, 234, 242, 250
    deforestation-enabled 111, 214, 221
    impact by infrastructure type 50
    law enforcement issues 231, 233
    linked to human conflicts 133
    linked to infrastructure development 44, 133
    in protected areas 215
    road-enabled 12, 13, 15, 50–51, 66, 82, 87, 138, 150, 158, 162, 173
    for wild meat 13, 72, 133, 142, 150, 152, 158, 162, 214, 231
hydrological impacts 50, 53–54, 168, 172–73
hydropower *see* dams
Hydropower by Design 189–91
Hydropower Sustainability Assessment Protocol 189, 191
Hylobates gibbons xv, xix, 123, 211, 213, 235; *see also* gibbons (Hylobatidae); *specific species*

I

Ilagala–Rukoma–Kashagulu (I–R–K) road, Tanzania 95–99
impact assessments 32–37; *see also* environmental and social impact assessments (ESIAs); environmental impact assessments (EIAs); strategic environmental assessments (SEAs)
impacts on apes, infrastructure
    and ape adaptability 50
    barriers to dispersal 43, 47, 48–49, 50, 52, 57, 58, 173, 232
    connectivity loss 43, 51, 52, 111, 134, 137, 173–74
    disease transmission *see* disease
    drownings 46, 179
    electrocutions 46, 53, 179, 264–65
    food availability 46, 48, 49, 51, 232
    general impacts 47–50
    habitat loss *see* deforestation/habitat loss
    human–wildlife conflict 48, 51, 179, 231, 236
    hunting/poaching *see* hunting/poaching
    hydrological impacts 53–54
    by infrastructure phase
        construction 44–46, 53
        decommissioning 47
        utilization/production 46–47
    movement away from human disturbance 44–45
    noise pollution 42, 44–45, 50, 53, 60, 180, 193
    overview 41–42, 43–44
    steps to reduce
        environmental impact assessments (EIAs) 54, 56
        knowledge gaps, reducing 59–60
        mitigation strategies 56–57, 58
    vehicle/equipment collisions 46, 50, 52, 53, 149–50
    wildlife–human impact links 60–62
    wildlife trafficking 150, 231, 234, 237–39, 245–50
impacts on local people, infrastructure 43, 60–64, 66–74, 142, 145, 146, 147, 167–69, 173, 183, 184, 187, 188
India 53, 85, 215, 231, 232–33, 234, 236
Indonesia
    deforestation 28, 84, 199, 202, 203, 215, 216, 220–21, 222, 231–32
    development corridors 26
    geothermal energy projects 192–95
    GLAD forest loss alerts 100
    human population growth 232–33, 254
    hunting/poaching 231
    natural capital and infrastructure planning 130
    offset policies 122, 123
    oil palm plantations 84, 216, 220–21
    rainforests 207
    sanctuaries 235, 236, 261, 262
    wildlife trafficking 234, 235
    see also Borneo; Sumatra
infrastructure, defining 2
Institut Congolais pour la Conservation de la Nature (ICCN) 126–27, 133, 158, 160, 162, 163
Integrated Biodiversity Assessment Tool (IBAT) 161
intergroup aggression xxvi, 49, 51
International Finance Corporation (IFC) 20, 21, 37, 120, 162
International Union for Conservation of Nature (IUCN) xvii, 109, 141, 158, 201, 273, 274
invasive species 44, 57, 121
InVEST (Integrated Valuation of Ecosystem Services and Tradeoffs) Scenario Generator 153
investment in infrastructure
    foreign 4–5, 12, 13, 15, 18–22, 27, 108
    incentives and disincentives 3
    risks 28
    UN Sustainable Development Goals 129
    World Bank safeguard policy framework 162, 271–72
Italian–Thai Development Company (ITALTHAI) 147, 156
Ivory Coast 215, 216, 231, 234, 237, 260

J

Japan 257, 258

K

Kahuzi-Biega National Park (KBNP), DRC 132–34, 135
Kappi geothermal project, Sumatra 192–95
Karen National Union (KNU) 146, 147
Kenya 22, 260, 265
key findings
    captive apes and sanctuaries 226–27
    dams 170
    deforestation along roads 82–83
    Global Forest Change 2000–14 study 202–3, 207
    infrastructure impacts 42–43
    protected areas in Africa 108
    road planning 140
    sustainability 12
Kloss' gibbon (*Hylobates klossii*) xix, xxii–xxiii, 206, 208, 210, 219, 234, 236; *see also* gibbons (Hylobatidae)
knowledge gaps, reducing 59–60
Kribi deepwater port, Cameroon 65, 66, 70

L

Ladia Galaska Road Network, Sumatra 87, 94
Lamu Port, South Sudan, Ethiopia Transport (LAPSSET) corridor 15, 22
land use planning, strategic 34, 38, 56–57, 90, 104, 124, 128, 130, 140
Landsat satellite 85–86, 100, 115, 266, 267
Lao People's Democratic Republic (Lao PDR) 34–35, 168–69, 171, 207, 215, 231, 234, 236
Laotian black-crested gibbon (*Nomascus concolor lu*) 208, 210, 219; *see also* gibbons (Hylobatidae); *Nomascus* gibbons

lar gibbon (*Hylobates lar*) xv, xix, xxii–xxiii, 149, 205, 206, 234, 235, 236, *see also* gibbons (Hylobatidae); *specific subspecies*
law enforcement
    improving 233, 234, 242–43, 245–46, 248–49, 250, 254
    lack of 94, 158, 177, 213–14, 231, 245
    working with sanctuaries 231, 239, 245–46, 249, 250, 254–55
Leuser Ecosystem, Sumatra 16, 87–94, 103, 170, 192–95
Liberia 76, 215, 231, 234, 236, 237, 260–61
light pollution 50, 53, 150
Loango National Park, Gabon 60
local communities
    advocacy 75, 164
    awareness raising *see* awareness raising
    broken promises to 66, 73, 184
    capacity building 25, 75, 156, 160
    community-managed forests 62, 84–85, 188
    community tenure of land 74, 75–76
    compensation *see* compensation measures
    dependence on forest 66, 67–68, 73, 161
    distrust of NGOs/institutions 147, 156
    engagement/participation 25, 37, 67, 74–75, 78, 162, 164, 170
    impact mitigation strategies 74–76
    impacts of infrastructure *see* impacts on local people, infrastructure
    land rights, protection of 75–76
    representative structures 75
    resistance/opposition 4, 75, 182–88
    rights of 62–64, 69, 73, 74–75, 76, 142, 147, 183, 187, 188
logging 56, 63–64, 66, 68, 82, 83–84, 87, 91, 94, 102, 108, 113, 114, 115, 147, 178, 179, 180, 183, 184, 187
Lom Pangar Hydropower Project, Cameroon 161, 170, 177–81

M

Mae Guang Udom Tara dam, Thailand 191
Mahale Mountains National Park (MMNP), Tanzania 95–99, 104
Mahmud, Abdul Taib 183
Maiombe National Park, Angola 237–38
Malaysia
    Central Forest Spine 26
    dams xxvi
    deforestation 84, 202, 203, 215, 216, 220–21, 222, 231
    development corridors 26, 182
    electrocutions 46, 53
    offset policies 122, 123
    sanctuaries 236, 262
    wildlife trafficking 234
    zoos 234, 235
    see also Borneo
Malaysian lar gibbon (*Hylobates lar lar*) 202, 208, 210, 213, 214, 216, 217, 219; *see also* gibbons (Hylobatidae); lar gibbon (*Hylobates lar*)

Mali 215, 234, 237
Maputo Development Corridor, Mozambique 24, 25
Mbalam–Kribi proposed railway 24, 66
Mbe Mountains, Nigeria 141, 142, 143
meat, wild 13, 72, 133, 150, 152, 158, 162, 214, 231
minimization measures 82, 102, 121, 189, 190
mining *see* extractive industries
mitigation hierarchy 36, 82, 102, 119–28, 189–90
mitigation strategies 12, 36–37, 56–57, 58, 78, 140, 156, 164, 264–65; *see also* mitigation hierarchy
modeling 140, 152–57
moloch gibbon (*Hylobates moloch*) xv, xix, xxii–xxiii, 202, 206, 208, 210, 219, 235, 236, *see also* gibbons (Hylobatidae)
mortality, infrastructure-linked 46–47, 48, 49, 51
mosaic landscapes 251–53
Mount Nimba Biosphere Reserve, Guinea 116
mountain gorilla (*Gorilla beringei beringei*) xiii, xviii, xx–xxi, xxiv–xxv, 24, 44–45, 116–17, 126, 204, 208, 209, 212, 214, 217, 219, 232, *see also* gorillas (*Gorilla* spp.)
Mozal aluminum smelter, Mozambique 25
Müller's gibbon (*Hylobates muelleri*) xv, xix, xxii–xxiii, 206, 208, 210, 217, 219, 234, 235, 236, *see also* gibbons (Hylobatidae)
multilateral lenders 4, 12, 18, 20–21, 33, 191
Murum Dam, Borneo 182, 183, 184, 186
Myanmar
    dams 19
    Dawei road link 146–57
    deforestation 149–52, 215
    Environmental Impact Assessments (EIAs) 156
    human conflicts 146, 147
    land policies 147
    logging 146
    natural capital 130
    oil/gas extraction 146
    protected areas 146, 147
    Western Forest Complex 149
    wildlife trafficking 234
Myinmoletkat Nature Reserve, Myanmar 147

N

Nam Ou Cascade Hydropower Project, Lao PDR 34–35, 168–69
natural capital 12, 129–31, 153, 156, 164
natural gas extraction *see* extractive industries
natural resource exploitation *see* extractive industries
Nepal 84–85
nests xxiii, 48, 51, 54, 212
Nigeria
    corruption 141
    Cross River National Park 116, 135, 141–42, 143
    Cross River superhighway 52, 116, 138–39, 141–45
    deforestation 208, 215
    human population 141
    oil reserves 141
    rainforest 141–42

REDD+ project 145
sanctuaries 236, 237, 260
underdevelopment 141
wildlife trafficking 234
Nigeria–Cameroon chimpanzee (*Pan troglodytes ellioti*) xii, xviii, xx–xxi, 142, 208, 209, 217, 219, 236; *see also* chimpanzees (*Pan troglodytes*)
noise pollution 42, 44–45, 50, 53, 60, 180, 193
Nomascus gibbons xv–xvi, xix, 123, 211, 235; *see also* gibbons (Hylobatidae); *specific species*
non-governmental organizations (NGOs) 4, 36, 74, 75, 94, 139, 142–45, 147, 187, 194, 227, 246, 249, 251
northeast Bornean orangutan (*Pongo pygmaeus morio*) 206, 208, 209, 219; *see also* Bornean orangutan (*Pongo pygmaeus*); orangutans (*Pongo* spp.)
northern white-cheeked crested gibbon (*Nomascus leucogenys*) xvi, xix, xxii–xxiii, 205, 208, 210, 219, 236; *see also* gibbons (Hylobatidae); *Nomascus* gibbons
northern yellow-cheeked crested gibbon (*Nomascus annamensis*) xvi, xix, xxii–xxiii, 205, 236; *see also* gibbons (Hylobatidae); *Nomascus* gibbons
northwest Bornean orangutan (*Pongo pygmaeus pygmaeus*) xviii, 206, 208, 209, 213, 219; *see also* Bornean orangutan (*Pongo pygmaeus*); orangutans (*Pongo* spp.)
Ntakata Forest, Tanzania 95, 96, 99
Ntam settlement, Cameroon 71–72

O

Oban Hills, Nigeria 141–42
offsetting 37, 102, 120, 121–24, 156, 174, 189, 190
oil/gas extraction *see* extractive industries
oil palm plantations 32–33, 34, 38, 66, 84, 87, 89–90, 183, 216, 222, 251
OpenStreetMap (OSM) initiative 113, 115
Orangutan Veterinary Advisory Group (OVAG) 229–30
orangutans (*Pongo* spp.)
 aggression between males xxvii
 captive 234, 235, 258, 259, 261, 262
 diet xxi, xxiii, xxvii
 dispersal xxvi–xxvii
 habitat xxi, 274
 home ranges xxvi–xxvii
 overview xiv
 reintroduction and translocation 240
 reproduction xxiv–xxv
 sanctuaries 229–30, 231, 232, 261, 262
 socioecology xviii–xxvii
 threats to
  development corridors 26
  disease 52–53
  electrocution 46, 53
  habitat loss 42, 202–3, 208, 211, 214, 216, 222, 232, 240, 251
  hunting/poaching 231
  noise pollution 53

removal from home territory 251
trafficking 234
travel throughout range xxvii
wildlife bridges 58, 121

P

Palangka Raya Declaration on Deforestation and the Rights of Forest Peoples 62–63
palm plantations *see* oil palm plantations
Pan African Sanctuary Alliance (PASA) 227, 229, 230, 244, 259
"Pan-Borneo Highway" 26
Paris climate accord 129, 130, 145
pathogens *see* disease
Peru 83, 86, 100, 101
pileated gibbon (*Hylobates pileatus*) xix, xxii–xxiii, 205, 207, 208, 210, 214, 217, 219, 235, 236; *see also* gibbons (Hylobatidae)
pipelines
 Chad–Cameroon pipeline 63, 65, 66–69, 70, 177, 178, 179
 impacts on apes 66
 Yadana pipeline 147
planning, infrastructure
 connectivity considerations 154–55, 161
 conservation action planning (CAP) 99
 costs and benefits, optimizing 28–30
 green energy, promoting 30–31
 integrated land use planning 140
 local community participation/engagement 25, 37, 67, 74–75, 78, 162, 164, 170
 Natural Capital Protocol 129–31
 regional planning 12
 road planning *see* road planning
 stakeholder engagement 21, 124, 195
 strategic land use planning 30–31, 34, 38, 56–57, 90, 104, 124, 128, 130, 140
 for sustainability 28–31, 34, 38
 system-scale approach 195–96
 wildlife and ecosystem services, including 149
poaching *see* hunting/poaching
ports 15, 50, 66, 70, 125, 139
poverty 66, 68, 69–70, 78, 126, 158, 161
power lines 40–41, 46, 50, 53, 112, 264–65
Pro-Routes project, DRC 138, 140, 158–63, 271–72
production phase 46–47
protected areas in Africa study 107–35
protected areas (PAs)
 avoidance measures 190
 buffer zones 213
 deforestation 62, 203–6, 208, 211–13, 219–20, 221, 223, 265, 266
 Digital Observatory for Protected Areas (DOPA) 118
 impacts on local people 62–63, 68
 importance 199, 202, 222–23
 infrastructure development within 22, 24–25, 64, 87, 219–20

as offset policy 68, 99, 190
protecting 31, 37
species confined to 142, 143, 202, 218
World Database on Protected Areas (WDPA) 273
see also protected areas in Africa study; specific areas; specific countries; specific protected areas
PT Hitay Panas Energy 193–95

R

Rabi oil field, Gabon 60
railways
    Africa 24, 65, 66, 69–72, 121, 125, 177
    Asia 26
    impacts on apes 50, 51–52, 71
    impacts on local people 66
    inequitable social impacts 69–70
REDD (Reducing Emissions from Deforestation and Forest Degradation) agreement (UN) 62, 142, 145
Renewable and Appropriate Energy Laboratory (RAEL), California 183–84, 186, 188
renewable energy 30–31, 170, 192–95; see also dams; geothermal energy; solar power
Republic of Congo
    awareness raising 245
    chimpanzee releases 240
    deforestation 22, 215, 265
    development corridors 22–23, 24–25, 111
    hunting/poaching 13, 231
    railways 24
    sanctuaries 236, 237, 260
    TRIDOM landscape 24–25
    wildlife trafficking 234
restoration/rehabilitation, habitat 36–37, 47, 59, 102, 121, 129, 189
rivers, lessons from 15–17
road planning
    case studies
        Democratic Republic of Congo 138, 140, 158–63, 271–72
        key findings 140
        Myanmar 146–57
        Nigeria 138–39, 141–45
        overall conclusion 164
        overview 138–40
    external environmental experts, value of 163, 164
    local context application 103–4
    local people, inclusion/engagement of 164
    mitigation strategies
        avoiding critical habitat 31–32, 57, 134, 149, 161
        minimum width, keeping to 57
        wildlife crossings/bridges 57, 58–59
    pre-planning assessments 29, 161, 164
    strategic 29–30, 30–31, 54, 57
    zoning 86, 102
RoadFree initiative 113
Roadless Forest initiative 113
roads
    deforestation see deforestation along roads
    global map for prioritizing building 30–31
    impacts on apes
        behavioral changes 133
        general impacts 12, 50–51, 66, 82, 137–38, 149–50, 152, 158, 161, 170, 173
        habitat loss see deforestation along roads
        human influx and settlement 51, 91, 94, 95, 111
        hunting/poaching, increased 13, 50–51, 87, 232
        trafficking 232
        vehicle collisions 46, 50, 52, 53
    knowledge gaps 59
    length, global 41
    length in ape ranges 176
    links with other infrastructure development 15, 18, 112
    mapping 112–15
    monitoring approaches 83–86, 104–5; see also Global Forest Watch (GFW) platform
    planning see road planning
    rivers, lessons from 15–17
    scale of development 2–3, 11, 13, 81–82
    size and use factors 47, 52
    social benefits, inequitable 69–70
    spreading 31–32
    see also development corridors; specific project; specific road
Rwanda 22, 126, 130, 215, 235, 237, 260

S

sanctuaries
    ape confiscation factors 237–39
    benefits to ape conservation and welfare 241–42
    breeding apes 243–44
    capacity 236, 243–45
    collaborations 230, 251
    corruption issues 248–49
    demand 261
    education and awareness 242, 245, 249
    euthanasia policies 244
    funding 246–48, 252–54
    habitat protection and conservation planning 250–52
    history/scope of range state sanctuaries 227, 229
    intake factors 231–33, 235–36
    intake policies 245–47
    overcrowding 237
    overview 225–26, 254–55
    photos of ape–human contact 243
    regulations and law enforcement 246–50
    reintroduction and translocation 238–41, 251, 261
    roles outside sanctuary 242
    social media use 243, 244
    standards 228–31, 242–43
    state responsibilities 244–45, 246–48

status and outlook 236–39, 262
sustainability 252–54
Sarawak Corridor of Renewable Energy (SCORE), Borneo 182–88
satellite imagery 82, 83, 84–86, 95, 100, 104–5, 112–16; *see also* Global Forest Change 2000–14 study; Global Forest Watch (GFW) platform
Save Rivers, Borneo 186–88
scenario modeling 152–54
Senegal 215, 234, 236, 237
siamang (*Symphlangus syndactylus*) xvi, xix, xxii–xxiii, 86, 87, 202, 206, 208, 210, 211, 213, 216, 217, 219, 234, 235, 236, 258, 261, *see also* gibbons (Hylobatidae)
Sierra Leone 56, 215, 216, 234, 236, 237, 260
Simandou iron ore project, Guinea 25–26, 120
SINOPEC (China Petroleum & Chemical Corporation) 60
social groups, ape xvii, xxviii, 48–49
socioecology,
 gibbons xxii–xxiii, xxvii–xxix
 great apes xvii–xxvii
socioeconomic development 18, 27, 126–27
solar power 30, 171, 178
South America 15, 171, 258; *see also* Brazil; Peru
South Sudan 15, 22, 158, 215, 216, 237
southern white-cheeked crested gibbon (*Nomascus siki*) xvi, xix, xxii–xxiii, 205, 208, 210, 219, 236; *see also* gibbons (Hylobatidae); *Nomascus* gibbons
southern yellow-cheeked crested gibbon (*Nomascus gabriellae*) xvi, xix, xxii–xxiii, 205, 208, 210, 214, 219, 234, 236; *see also* gibbons (Hylobatidae); *Nomascus* gibbons
southwest Bornean orangutan (*Pongo pygmaeus wurmbii*) xviii, 206, 208, 209, 217, 219; *see also* Bornean orangutan (*Pongo pygmaeus*); orangutans (*Pongo* spp.)
stakeholder engagement 21, 124, 195
standards, ecological and social 59
strategic environmental assessments (SEAs) 20, 21, 32–34, 36–37, 129
Sumatra
 deforestation 84, 87–94, 222
 development corridors 26
 geothermal energy plants 30, 170, 192–95
 human settlement, illegal 16–17
 natural capital 130
 oil palm plantations 32–33
 protected areas 91–94
 roads 87–94, 103
 see also Indonesia
Sumatran lar gibbon (*Hylobates lar vestitus*) 86, 87, 90, 208, 210, 217, 219; *see also* gibbons (Hylobatidae); lar gibbon (*Hylobates lar*)
Sumatran orangutan (*Pongo abelii*) xiv, xviii, xxii–xxiii, xxiv, xxvii, 16, 26–27, 86, 87–94, 206, 208, 209, 213, 219, 235, 236, *see also* orangutans (*Pongo* spp.)
sustainability of global infrastructure expansion 12, 28–38 *see also* unsustainability of global infrastructure expansion

T

Tamiang Hulu–Lokop (TH–L) road, Sumatra 87–90
Tana River Primate Reserve 22
Tanintharyi region, Myanmar 146–57
Tanzania
 conservation action plans (CAPs) 278
 deforestation 95–99, 215
 goldfields 22
 human population growth 232–33
 land rights, protection of 76
 National Chimpanzee Management Plan 268–70
 roads 83, 95–99, 103–4, 268–70
 wildlife trafficking 234
Tapanuli orangutan (*Pongo tapanuliensis*) xiv, xviii, 195
temporary infrastructure, dismantling 57, 59
Thailand
 dams 174, 191
 deforestation 215
 forest complexes 149
 roads 146–57
 sanctuaries 236, 262
 vehicle collisions 149–50
 wildlife trafficking 150, 234
 zoos 235
tiger habitat study 84–85
Tonkin black-crested gibbon (*Nomascus concolor concolor*) 208, 210, 217, 219; *see also* gibbons (Hylobatidae); *Nomascus* gibbons
tourism 116, 117, 134, 161, 180
trafficking, wildlife 17, 150, 158, 179, 227, 231, 234, 235, 237–39, 245–50
TRIDOM landscape and development corridor 24–25

U

Uganda
 Bwindi Impenetrable National Park 44–45, 52, 116–17
 dams 171
 deforestation 215
 development corridors 22
 human population growth 232–33
 roads 113, 116–17
 sanctuaries 236, 237, 260
 wildlife trafficking 234
United Nations (UN)
 Declaration on the Rights of Indigenous Peoples 64
 Environmental Program (UNEP) 138
 forest-protection agreements 62
 REDD+ project, Nigeria 142, 145
 Sustainable Development Goals 75, 129
 UNESCO sites 111, 116–17, 126, 132–33, 141, 192, 193
 World Database on Protected Areas (WDPA) 273
United States 226, 227, 245–46, 255–58, 259, 272
unsustainability of global infrastructure expansion 13–28, 32, *see also* sustainability of global infrastructure expansion

V

vehicle/equipment collisions 46, 50, 52, 53, 149–50
vibration pollution 50, 53
Viet Nam 207, 215, 231, 234, 236, 262
Virunga Alliance 127
Virunga National Park, DRC 60, 126–27

W

Walvis Bay Development Corridor, Africa 24
water availability 53–54
West Africa 25–26, 31, 110, 195, 203, 216;
see also specific countries
west Yunnan black-crested gibbon (*Nomascus concolor furvogaster*) 208, 210, 214, 217, 219; see also gibbons (Hylobatidae); *Nomascus* gibbons
western black-crested gibbon (*Nomascus concolor*) xvi, xix, xxii–xxiii, xxviii, 202, 205; see also gibbons (Hylobatidae); *Nomascus* gibbons; specific subspecies
western chimpanzee (*Pan troglodytes verus*) xii, xviii, xx–xxi, 26, 125, 161, 207, 208, 209, 216, 219, 236, 259; see also chimpanzees (*Pan troglodytes*)
western gorillas (*Gorilla gorilla*) xiii, xxv, 17, 122; see also Cross River gorilla (*Gorilla gorilla diehli*); gorillas (*Gorilla* spp.); western lowland gorilla (*Gorilla gorilla gorilla*)
western hoolock (*Hoolock hoolock*) xiv, xviii, xxii–xxiii, 205, 208, 209, 219, 236; see also hoolocks (*Hoolock* spp.)
western lowland gorilla (*Gorilla gorilla gorilla*) xiii, xviii, xx–xxi, 13, 25, 37, 65, 69, 162, 177, 178–80, 181, 204, 208, 209, 212, 214, 217, 219, 235, 236, 251; see also gorillas (*Gorilla* spp.); western gorillas (*Gorilla gorilla*)
Wild Animal Rescue Network (WARN) 229, 230
Wild meat see meat, wild
Wildlife Conservation Society (WCS) 142, 144–45, 147, 178
wildlife crossing sites and bridges 57, 58–59, 121, 149, 154–55
wind power 30
World Bank
　Africa, infrastructure development for 161–62
　Deng Deng National Park, Cameroon 178, 180–81
　environmental safeguard policies 4, 20, 21, 68, 162, 191, 271–72
　Global Accessibility Map 115
　indigenous people, failure towards 67
　Lom Pangar Hydropower Project, Cameroon 178, 180–81
　mitigation hierarchy 82, 120
　Pro-Routes project, DRC 158–60, 271–72
　smart green infrastructure 82
World Indigenous Summit on Environment and Rivers (WISER) 188
World Wide Fund for Nature/World Wildlife Fund (WWF) 140, 141, 144, 147, 152–54, 156–57, 162–63

Y

Yadana pipeline, Myanmar 147
Yunnan lar gibbon (*Hylobates lar yunnanensis*) 207, 208, 210, 219; see also gibbons (Hylobatidae); lar gibbon (*Hylobates lar*)

Z

Zambia 116, 260, 261
zoos 226, 230, 234, 235, 237, 256, 258–59, 261–62